I0049463

Genetics and Genomics of Forest Trees

Genetics and Genomics of Forest Trees

Special Issue Editor

Filippos A. (Phil) Aravanopoulos

MDPI • Basel • Beijing • Wuhan • Barcelona • Belgrade

MDPI

Special Issue Editor
Filippos A. (Phil) Aravanopoulos
Aristotle University of Thessaloniki
Greece

Editorial Office
MDPI
St. Alban-Anlage 66
Basel, Switzerland

This is a reprint of articles from the Special Issue published online in the open access journal *Forests* (ISSN 1999-4907) from 2017 to 2018 (available at: https://www.mdpi.com/journal/forests/special_issues/forests_genetics)

For citation purposes, cite each article independently as indicated on the article page online and as indicated below:

LastName, A.A.; LastName, B.B.; LastName, C.C. Article Title. *Journal Name* **Year**, *Article Number*, Page Range.

ISBN 978-3-03897-298-3 (Pbk)
ISBN 978-3-03897-299-0 (PDF)

Cover images courtesy of Eliades et al.

Articles in this volume are Open Access and distributed under the Creative Commons Attribution (CC BY) license, which allows users to download, copy and build upon published articles even for commercial purposes, as long as the author and publisher are properly credited, which ensures maximum dissemination and a wider impact of our publications. The book taken as a whole is © 2018 MDPI, Basel, Switzerland, distributed under the terms and conditions of the Creative Commons license CC BY-NC-ND (http://creativecommons.org/licenses/by-nc-nd/4.0/).

Contents

About the Special Issue Editor

Filippos A. (Phil) Aravanopoulos is Professor of Forest Genetics and Tree Breeding at the Aristotle University of Thessaloniki. His background couples Forest and Environmental Science (B.Sc.) with Tree Genetics (Ph.D., postdoctoral research), and he has studied in Greece, Canada, and Sweden. His research has been funded by international (European Commission, International Energy Agency) and national organizations of Canada, Greece, and Sweden. His current projects are funded by H2020, FP7, and LIFE+. He is currently the Chairman of the Scientific Council of the Hellenic Agricultural Organization (formerly the National Agricultural Foundation of Greece) overseeing scientific research, research staff recruitment, and promotions for 30 research institutes that focus on agriculture, forestry, fisheries, and veterinary science. He is also the Deputy Coordinator of the Population, Ecological, and Conservation Genetics Unit of the International Union of Forest Research Organizations (IUFRO). Prof. Aravanopoulos has served as Leader in the Genetic Monitoring Working Group of the European Forest Genetic Recourses Network (EUFORGEN), as a member of the FAO Expert Group on the State of the World Forest Genetic Resources, and as the Chair of the Hellenic Scientific Society for Plant Genetics and Breeding. He serves as a reviewer for more than 40 international journals and as an editorial member in four (among them, he acts as Associate Editor in PLoS One). He has given over 30 invited presentation in universities and conferences in Canada, China, Cyprus, Germany, Greece, Serbia, Slovenia, Spain, Sweden, and Turkey. He has published about 200 research papers, including more than 85 in international peer-reviewed journals and 35 in refereed books and international proceedings.

Preface to "Genetics and Genomics of Forest Trees"

This volume provides a first-rate illustration of state-of-the-art research in the advancing area of the genetics and genomics of forest trees, especially as the pace of contemporary environmental change challenges the ability of forest tree populations to adapt. Forest tree genetics and genomics are developing and evolving at an accelerated speed, thanks to recent developments in high-throughput next generation sequencing capabilities and novel biostatistical tools. Population and landscape genetics and genomics have seen the rise of large-scale studies that employ the use of extended or genome-wide sampling. New or newly modified approaches, methodologies, and protocols can be found in this volume. Studies on both neutral and potentially adaptive variation are included at scales ranging from the cell to the landscape, contributing to the unravelling of the genotype-phenotype relationship.

This volume contains an impressive list of 112 authors affiliated with 72 educational and research institutions that have contributed to the 20 papers included. The directions of state-of-the-art forest genetics and genomics research are well reflected in the main topics of this volume. Genetic diversity in nature is dealt with in nine papers. Importantly, an almost equal amount of papers (eight) concerns transcriptomics and includes some of the most recent advances of relevant research in forest trees. The volume is complemented with two papers on quantitative genetics and one paper on tissue culture. Regarding the taxa studied, the volume contains 11 studies on angiosperms and eight on conifers, while a review paper refers to both. Among the genera represented in these studies, *Pinus* dominates the list with seven studies and a total of 12 species studied, followed by *Quercus* (three studies, three species), and then by *Paulownia* (two studies, two species) and *Betula* (two studies, one species). The rest of the genera studied in this volume are: *Cunninghamia, Liquidambar, Passiflora, Picea* and *Xylocarpus*.

The diversity of articles published in this Special Volume underscores the extensive range of contemporary research in this field. Clearly, this volume presents merely a glance, although a very interesting one, of a wide and extensive area of research that strives to promote knowledge in the areas of genetics and genomics of forest trees and to contribute to the management and conservation of forest genetic resources under significant environmental change.

Filippos A. (Phil) Aravanopoulos
Special Issue Editor

forests

MDPI

Review

Do Silviculture and Forest Management Affect the Genetic Diversity and Structure of Long-Impacted Forest Tree Populations?

Filippos A. (Phil) Aravanopoulos

Laboratory of Forest Genetics and Tree Breeding, School of Forestry and Natural Environment, Aristotle, University of Thessaloniki, 54124 Thessaloniki, Greece; aravanop@for.auth.gr

Received: 20 March 2018; Accepted: 26 May 2018; Published: 14 June 2018

Abstract: The consequences of silviculture and management on the genetic variation and structure of long-impacted populations of forest tree are reviewed assessed and discussed, using Mediterranean forests as a working paradigm. The review focuses on silviculture and management systems, regeneration schemes, the consequences of coppicing and coppice conversion to high forest, the effects of fragmentation and exploitation, and the genetic impact of forestry plantations. It emerges that averaging genetic diversity parameters, such as those typically reported in the assessment of forest population genetics, do not generally present significant differences between populations under certain silvicultural systems/forest management methods and "control" populations. Observed differences are usually rather subtler and regard the structure of the genetic variation and the lasting adaptive potential of natural forest tree populations. Therefore, forest management and silvicultural practices have a longer-term impact on the genetic diversity and structure and resilience of long-impacted populations of forest tree; their assessment should be based on parameters that are sensitive to population perturbations and bottlenecks. The nature and extent of genetic effects and impact of silviculture and forest management practices, call for a concerted effort regarding their thorough study using genetic, genomic, as well as monitoring approaches, in order to provide insight and potential solutions for future silviculture and management regimes.

Keywords: genetic diversity; genetic structure; forest management; silviculture; resilience

1. Introduction

Genetic diversity is a crucial biodiversity component that allows species to adapt to local conditions and to evolve in new environments, while securing their long-term adaptive potential, especially under an era of global change. An understanding of the genetic impacts of silvicultural and forest management procedures, is essential for forest genetic resource management and conservation [1,2] and sustainable management in multipurpose forestry. Forest management and silvicultural practices, such as the harvesting system, artificial and natural regeneration, regulation of species mixtures, thinning, and harvesting operations, regeneration planting and management, forest conversion schemes, etc., may impact local environmental conditions and population spatial demographic structure [2,3]. In this respect, they could influence stand population genetics and exercise a potentially strong effect on the major evolutionary forces at microscale of selection, gene flow, mating systems and genetic drift [1,2,4]. As evolutionary forces in natural ecosystems can vary a lot and exert counter-acting effects on local genetic diversity, the impact of silvicultural and forest management practices may better be seen in long-impacted forest ecosystems that have been under constant human exploitation for thousands of years and where silviculture as well as forest management, have been used for a long period. In these conditions, the human management signal is high, and in case it exerts an influence on the genetics of the populations of forest trees, it is more likely to be detected.

Mediterranean forestry fits the above description. Forests have been continuously exploited since the last glaciation period, while Mediterranean forestry today carries the history of ancient and medieval management systems for coppice and high forest management, influenced by the "formal" forest management and silvicultural systems that have been developed originally in central and northern Europe since the 19th century.

Among planet Earth's 34 biodiversity hot spots [5] is the Mediterranean basin. In the Mediterranean area that corresponds to less than 1.5% of the total land mass of the planet, more than 10% of the world's biodiversity in higher plants is present. The Mediterranean higher plant biodiversity is mostly localized within forests which host more than 25,000 species of vascular plants, about 50% of which are endemics [6]. In fact, the total number of both forest tree species and endemic forest species is higher in the Mediterranean that in other Mediterranean type ecosystems, such as in California [7]. Forest tree species richness and endemism are also high (290 indigenous tree species and 201 endemics) and their genetic diversity is extraordinary [6]. Mediterranean conifers, for example, present a higher within population diversity than conifers of any other area [8]. Furthermore, Mediterranean populations of species with a continental distribution frequently are among the most variable in terms of their genetic diversity (e.g., [9]). All of the above have been the outcome of a complex geological, climatic, and anthropogenically influenced evolution of species, ecosystems, and landscapes that in this peculiar geographical setting is manifested even at a small spatial scale. The Mediterranean basin is characterized by an uneven forest distribution. The northern Africa and the Near East parts correspond to 35% of the total forest area, while the European Mediterranean part corresponds to the remaining 65%. Forest cover, as a percentage of the country area, differs among Mediterranean sub regions as well. It ranges from only 1–8% and 5–10% in Southern and Eastern countries, respectively, while in the northern Mediterranean countries it varies between 20% and 30% [10]. Humans populate densely the Mediterranean basin (currently more than 460 million people), while in eastern and southern Mediterranean, human populations are still deeply reliant on terrestrial ecosystem natural resources. Fifteen years ago Bariteau et al. [11], asserted that "the increasing severity of drought, climatic change and a series of anthropogenic influences, contribute to forest decline in the Mediterranean; there is a definite need for the protection of forest genetic resources of low elevation Mediterranean conifers especially in south and east Mediterranean countries".

This notion that stresses the compounding current and anticipated stress in Mediterranean forests brings forward the issue of their resilience. Resilience is defined as the capacity of a system to absorb disturbance and reorganize while undergoing change so as to still retain essentially the same function, structure, identity, and feedbacks [12]. Forest ecosystem resilience can be defined as the capacity of an ecosystem to absorb disturbance and reorganize while undergoing change in order to retain essentially the same function (processes and properties), and ecosystem services. Genetic diversity is regarded as a component that can provide a basis for resilience, while forest management can affect ecosystem resilience, in fact management can be used to ensure ecosystem resilience (resilience-based management [13]).

Modern forest management and silviculture constitute an array of cultivation systems and distinct rules applied in the natural forest. They are widely applied in the Mediterranean region in the middle of the 19th century (in the northern part) and by the conclusion of the 19th century (in the southern and eastern parts). Forest management was principally an adoption of the predominant northern and central European forest management approaches and practices which have been used (with minimal modification) to the most productive temperate forests of the Mediterranean. Their focus is wood production in the realm of "multipurpose forestry", the latter usually confined to the provision of beneficial externalities. Silvicultural management is using an array of guidelines targeted at the development of growing stocks, determining rotation periods and their chronological and spatial distribution, facilitating regeneration (reforestation) and controlling the structural pattern and tree density by thinning [14]. Forest management, as an anthropogenic activity, has been mainly exercised to cover human needs. Upon its application, it has altered progressive forest succession and original forest composition and structure [14].

2. Genetic Impact of Silviculture and Forest Management Practices

2.1. Forest Management and Regeneration

Perhaps the most important issue in forest management and silviculture that may affect the genetic constitution of future forests, concerns the handling of the regeneration. Besides being the most obvious silvicultural practice, it may lead to possibly drastic changes of genetic structure for more than one generation. Natural regeneration may suffer by the potential employment of only a few set-aside trees that have been selected as seeders for the next generation. Such a practice would result in negative genetic consequences, namely the reduction of effective population sizes and the manifestation of genetic drift. Thinning and harvesting operations may also have a stern impact on adaptive and economically significant traits due to positive or negative selection regarding specific phenotypes. Moreover, there is a risk of carryover effects that relate to epigenetic memory regulation induced by temperature during post-meiotic megagametogenesis and seed maturation [15], or by climatic conditions during germination and early growth [16]. This epigenetic impact can have prolonged consequences for instance in (such as bud break, bud set and in the regeneration, and affecting adaptive traits [15,16]. Such phenomena, observed mainly in conifers, are attributed to complex epigenetic inheritance of DNA methylation patterns [17,18]. All of the above may head towards the loss of genetic variation and adaptive potential, particularly if gene flow via pollen is rigorously limited by differential (either considerably low, or high) population densities [4,19]. Such effects have been well documented in forest ecosystems [4,20–22] and are mainly prominent in age-class forestry. They are also apparent in the (common to the Mediterranean) conversion of coppice to seedling forest and regeneration after fire when only few mature trees are left. Continuous-cover forestry where a permanently irregular structure is maintained through the selection and harvesting of individual trees and close-to-nature silviculture where minimum necessary human intervention is used to accelerate the processes that nature would do by itself, rely fully on minimal selective cutting, and hence do not in theory modify genetic diversity, besides any underlying natural selection processes.

The genetic systems of forest trees should be rather resilient to rational harvesting, given forest tree genetic and life history characteristics [23]. However, the practical evaluation of the results of silvicultural procedures on vital genetic processes of the main economically important species, is generally scarce. In the Mediterranean, the consequences of shelterwood and group selection cutting regeneration methods, on the mating systems and pollen movement in *Pinus sylvestris* L. natural populations in Spain, were studied by Robledo-Arnuncio et al. [24]. No significant effects of stand thinning on mating system parameters were found. It has been indicated that the *Pinus sylvestris* pollination system is well adapted to prevent an adverse influence of these regeneration methods. In fact, it appears that density reduction tends to upsurge the effective number of pollen donors and pollen flow, in line with the results of Karlsson and Örlander [25] who showed that *P. sylvestris* seed production increases, after cutting. Robledo-Arnuncio et al. [24] suggest that a raise in pollen donor number and seed production might somewhat counterweigh the decrease in maternal numbers, i.e., compensate for population reduction due to silvicultural thinning. A number of reports imply a minor deficit of low-frequency alleles among residual trees and natural regeneration, following to shelterwood harvesting (e.g., Adams et al. [26]), however, most analyses did not find any notable effects of silvicultural practices on the genetic variation parameters of regenerated stands [24]. Several factors may sustain population genetic structure under forest management and silvicultural measures, such as high gene flow from adjacent non-harvested stands and considerable effective population sizes. The results of Robledo-Arnuncio et al. [24] are in congruence to earlier findings which showed that stand density variation and different cutting methods did not result in substantial mating system parameter modifications, at least regarding wind-pollinated temperate tree species [27–29].

2.2. Post-Fire Natural Regeneration

One particular aspect of regeneration management in fire-prone ecosystems is regeneration after forest fires. Forest management and silvicultural measures aim to promote natural regeneration (usually through protection from grazing and use artificial reforestation only when natural regeneration seems to fail, typically 2–3 years after the fire event). Despite the general notion that forest fires have adverse effects in the forest gene pool, pertinent studies that examine this question are limited. Whenever basic genetic variation parameters that reflect the magnitude of genetic diversity in natural populations were examined, no significant differences between post-fire and control populations have been observed. Nevertheless, when the genetic architecture of post-fire and control populations were studied, it was found that relevant genetic parameters were generally lower in post-fire populations [30].

Forest fire effects on population genetics were studied in Greek populations of *Pinus brutia* and *Pinus halepensis*. The effects of differential forest fire history in the genetic diversity of regeneration after fire of *P. halepensis* were studied in a transect spanning from a general NE to SW direction, that presents significant differences in the frequency and severity of fires ranging from relatively low (NE end) to extremely high (SW end). *P. halepensis* populations presented high genetic variation, absence of inbreeding and Hardy-Weinberg equilibrium in areas with different fire incidence. Gene frequencies in areas with high and relatively low incidence of fires did not show statistically significant differences. High forest fire occurrence and damage (reflected by a forest fire damage index) was not correlated either positively or negatively with genetic diversity parameters [30], in congruence to relevant results by Krauss [31], England et al. [32] and Uchiyama et al. [33]. With regards to *P. brutia*, an assessment of genetic variation (observed heterozygosity and gene diversity) and structure (variation between mature populations that had survived ground fire events ("maternal" stands, or control populations) and populations of the regeneration ("progeny" stands, or post-fire populations), showed a lack of substantial changes in the amounts of genetic variation among the remaining post-fire populations and regeneration [34]. These results are similar to results reported in pertinent literature regarding *Picea* [35]. However, a shift was found in rare allele frequencies (including rare allele loss at a small scale), the occurrence of interspecific hybridization in the post-fire populations and the observation of some genetic bottleneck effects (rare allele frequency change and loss, changes in the frequency of heterozygotes) [34].

Hence, forest fires may not induce genetic erosion in forest tree populations per se, especially when population sizes are large and in reproductive maturity, natural distribution is continuous, and regeneration is abundant. However, there are indications that under particular circumstances, compounded in periods of high forest fire frequency and intensity, forest fires may have adverse consequences for the architecture of forest tree genetic variation and their long-term survival and existence.

2.3. Artificial Regeneration

The need for artificial regeneration (post-fire failure of stand establishment, forest destruction, overgrazing, etc.) is usually considerable, especially in fire-prone areas. The issue of the genetic quality of the plant material used for artificial regeneration is an emerging topic in research and in forest management. Artificial regeneration may suffer by a non-autochthonous origin of the seedlings and a potentially restricted genetic base. Using an allochthonous genetic resource may in theory enrich the local gene pool, but at the same time it may result in outbreeding depression. The use of genetically improved material in reforestations (depending on the protocols used) does not necessarily alleviate the above problems. The major question of concern is the use of certified forest reproductive material (FRM) "source identified, selected, qualified, or tested" [36] (the latter potentially genetically improved) over local randomly chosen seeds and the associated issues of genetic quality, amount of genetic diversity, and adaptive potential of the material chosen. Besides the choice of the material used, the potential adverse genetic implications of nursery practices have long been known [37]. Every step in FRM production, from mother tree phenotypic selection and stand seed collection to artificial planting that includes seed processing and storage, nursery conditions and operations [37], mass production and grading of seed and seedlings, as well as breeding operations and potential FRM transfer, can have an effect on genetic diversity mainly by directional selection [38].

The quality of the genetic material used in artificial regeneration of *Quercus suber* was investigated in Portugal [39], where dwindling reforestation success was attributed to inferior genetic quality of the reproductive material used in reforestation [39,40]. Observable effects of artificial regeneration have also been detected in conifer populations. In a study of *Pinus brutia* in Turkey, Kandedimir et al. [41] reported considerable genetic variation within and among forest stands, but no distinctive genetic variation patterns according to elevation, geography, or breeding zones [42]. They suggest that this finding can be ascribed to intensive forest management and to the substantial use of artificial regeneration [41].

The effects of using genetically improved material in artificial regeneration were investigated by Bouffier et al. [43] and by Icgen et al. [44]. Bouffier et al. [43] studied in successive breeding populations of *Pinus pinaster*, the evolution of genetic variation regarding selected traits in France. By using as a base line the forest where selected trees originated, they have found a notable reduction of quantitative genetic variation in the population of selected plus trees (especially for height and diameter). They further compared the genetic diversity of the latter "plus trees" to the genetic variation of the population that comprised of plus tree progenies (second step of artificial selection). In this case, genetic variation was not significantly altered [43]. Icgen et al. [44] studied the probable influence of forest management and use of artificially selected and genetically improved plant material on established *Pinus brutia* plantations in Turkey. They conveyed that seed source (seed stands, orchards, plantations originating from the same area) genetic relationships differed with respect to seed source locations. In line with Bouffier et al. [43], Icgen et al. [44] reported the presence of genetic changes between initial selection at the seed orchard (plus tree selection phase) level and established plantation level (selected seed/seedling production phase). Generally, it is apparent that the use of genetically improved material for artificial regeneration which has resulted from typical artificial selection and breeding processes may reduce the genetic base of the planting material. It is up to the forest manager to decide the advantages and disadvantages of the use of genetically improved material over a restriction of the genetic base.

3. Management Systems: Seedling vs. Coppice Forests

Coppicing is a traditional vegetative regeneration system which takes advantage of the resprouting ability of many tree species. Coppicing has been prevalent as an ancient forestry practice, in particular in areas where special wood products are needed (wood poles of small dimensions, firewood). Modern forest management and silviculture have been attempting to limit this system and convert coppice forest to seedling high forest. Nevertheless, a considerable amount of coppice forests is still present (for instance more than 8 million ha in the Mediterranean). Stand structure and density differs in the two systems, potentially affecting gene flow and mating patterns, while in rapidly changing environments, local adaptation of lasting coppice forests may be compromised due to the absence of sexual reproduction. These issues reflect a biological basis for investigating genetic differences in differentially managed populations.

The genetic effects of this management method has been studied primarily in Fagaceae species that historically have been managed as coppice forests for hundreds of years, namely chestnut (*Castanea sativa*) and oak (*Quercus* spp.). Various studies have contrasted typical genetic variation parameters (percent polymorphic loci, allelic richness, gene diversity, and observed heterozygosity) between coppice forest and (usually geographically proximal) seedling high-forest. With regards to *C. sativa* there is a general agreement that the genetic diversity parameters between seedling high forest and coppice forest do not differ significantly [45–48]; a result that has been established by using different genetic markers. Nevertheless, a more in-depth analysis of genetic data identified some differences between these management systems. Aravanopoulos and Drouzas [47] have studied high forest-coppice pairs, each of them situated in the same geographical area. Geographic proximity is essential in such comparisons in order to exclude possible provenance effects. The chestnut high forest populations studied were true old-growth seedling populations and the coppice populations have been managed as such for centuries. It was found that the distribution of genetic diversity differed between the two management systems. Coppice populations appear to retain a higher percentage of within population genetic diversity compared to seedling high-forest populations. Some young seedling recruits are also found within coppice forests, but gene flow is clearly higher in the latter. As original stools were old (aged at 250 years on the average) and records suggest that coppice management has been employed since antiquity, these results may indicate a potential for higher within population differentiation and slower evolutionary response for coppice populations in a contemporary time of rapid environmental change [47]. Mattioni et al. [48] have identified weak (nonetheless significant) differences in two-locus allelic correlations between naturalized stands ("natural" stands which originated from abandoned anthropogenically managed populations, most likely old orchards) and coppice forests in a linkage disequilibrium analysis. They suggested that long-standing management methods could affect the population genetic composition, although their results may have possibly been inflated by one-locus disequilibria that could not be assessed by the dominant inter simple sequence repeat (ISSR) markers that they used. In fact, the clonal (to a varying extent) nature of the coppice forest (several ramets originating from the same genet) may increase the global allelic correlation among loci of the forest and decrease effective population size [48,49]. Genetic differences were also found between standards and coppice shoots in a *Quercus cerris* coppice forest under conversion [50]. Therefore, all studies indicated some differences in the genetic structure between natural (seedling) high forest and coppice populations, despite the absence of differences in averaging genetic diversity statistics.

Mating systems were studied in a pair of an old-growth natural seedling *C. sativa* population and one coppice population located in the same geographical area in Greece by Papadima et al. [51], as mating systems parameters are important when planning coppice conversion to high forest. Parental stands were evaluated based on an analysis of 16 codominant loci in 27 trees per population. Twenty seeds per tree from eight of the above trees were also genotyped for the same loci in order to derive mating system parameters. Inbreeding was very low in both population types with an upper bound somewhat higher in the coppice population and was solely attributed to consanguineous matting and bi-parental inbreeding. Pollen and ovule allele frequencies did not show marked differences. Overall, results indicated the absence of strong differences in the mating system parameters of the two population types [51]. A quantitative genetics comparison between an old-growth seedling high forest chestnut population and a coppice population located in the same area was attempted by Alizoti et al. [52]. The genetic structure (variance between and within these populations) of seven seedling quantitative traits was investigated. Heterogeneity was higher within than between the old-growth and coppice populations growing in the same environment. High genetic correlations among all quantitative traits were found in the coppice population in contrast to the seedling one, where a high genetic correlation was observed for seedling height and leaf length only [52]. These results may imply a higher genetic uniformity for the coppice population for quantitative adaptive traits.

The genetic diversity and clonality levels in the oak *Quercus pyrenaica* were studied in Iberian high forest and coppice populations by Valbuena-Carabaña et al. [53]. Results showed that the considerable genetic diversity levels detected were comparable under these different silvicultural systems [53]. Results are in concordance to similar investigations presented above regarding *Castanea sativa* (e.g., [47–49]). Nevertheless, in a preliminary comparative study on the genetic diversity of standards compared to coppice shoots in *Quercus cerris* under conversion, [50] found a reduced genetic variation in the standards. The latter reports differences in averaging genetic statistics among different management components in a Fagaceae species. All other pertinent studies identified subtler disparities among populations. Clearly, more studies are needed in this area.

The genetic effects of coppice conversion to high forest have been insufficiently studied thus far and the few relevant studies reported non-concordant results. Valbuena-Carabaña et al. [53] advice against intensive conversion practices of *Q. pyrenaica* coppice into seedling (high) forest due to the anticipated significant reduction in genetic variation when unique genotypes were taken away. Ortego et al. [54] found an extensive clonal structure of *Quercus ilex* coppice with high number ramets constituting a single genet and reached similar conclusions. The arguments against intensive thinning of Valbuena-Carabaña et al. [53], may be regarded as somewhat in contrast to those made by Mattioni et al. [48], concerning their observation that the coppice management system may increase the total allelic correlation among loci and decrease effective population size, while thinning operations in (over-mature) coppice will reverse these trends. Indeed, fundamental genetic theory suggests that population inbreeding is a function of effective population size [55] and therefore thinning of old coppice may offset genetic drift, and assist ultimately in the preservation of the inherent genetic variation diversity [48].

Overall, genetic diversity between seedling high forest and coppice forest does not appear to differ significantly, while there is also absence of strong differences in mating system parameters. On the other hand, long-term management techniques have an impact as coppice populations, present high clonality and higher genetic uniformity for quantitative adaptive traits and appear to present a slower evolutionary response. From a genetics and ecology perspective coppice conversion to high forest is a favorable prospect, as coppice management may potentially reduce the evolutionary rate of local adaptive potential. Conversion to high forest may reinstate natural succession in forest ecological dynamics and restore a more typical course of evolution and adaptation in an era of rapid environmental change [47]. In addition, it may reduce genetic drift and increase effective population sizes. However, thinning intensity must be moderate in order not to result in unique genotypes being lost [53].

4. Forest Management: Fragmentation and Over-Exploitation

The genetic outcomes of forest fragmentation and overexploitation are not well known. Forest fragmentation is the emergence of spatial discontinuities that break large, contiguous, forested areas into smaller pieces of forest causing population fragmentation and ecosystem destruction. Forest exploitation refers to the long-term intensive mis-management of forests that result in forest ecosystem degradation and inability to provide forest products and ecosystem services. In theory, forest fragmentation may cause a reduction or loss of genetic connectivity. Both forest fragmentation and overexploitation may lead in a reduction of effective population size and inbreeding depression, which will eventually result in a decrease in species genetic variation and fitness.

Certain studies report a notable deficit of genetic diversity and enlarged genetic differentiation after forest fragmentation and reduction of tree density, while others fail to establish such effects [56,57]. Contrasting results may be expected for perennial organisms like forest tree species which exhibit vastly different life history characteristics, temporal scales of sampling and time-frame that has elapsed since fragmentation [57]. Recent studies of Mediterranean species, such as in *Taxus baccata* [58] and *Plantago brutia* [59], point towards the negative genetic consequences of forest fragmentation as an outcome of intensive management. Ortego et al. [54] suggested that intensive management leading to long-term population fragmentation induces negative genetic consequences, such as reduced pollen exchange, augmented genetic differentiation between different management units, erosion of resident genetic diversity and the prevalence of asexual over sexual reproduction. An ample volume of empirical data indicates that forest trees are sensitive to fragmentation [60], however, the available evidence has been a matter of debate, principally regarding temperate forests [61], since in numerous situations fragmentation might have been rather new and fragments may encompass notable remaining forests. Particular reports have even indicated that fragmentation will create open landscapes and facilitate pollen movement which may counterweigh anticipated adverse genetic consequences [57].

Over-exploitation effects in the genetic structure of forest tree populations were explored in a study of *Pinus brutia* in Turkey [42]. Over-exploited populations under intensive forest management were compared to natural populations that provided a background benchmark of genetic diversity.

No significant differences between the two population types were found in genetic polymorphism and heterozygosity [42]. However, a homozygosity excess was observed (about 6% higher) in over-exploited populations, suggesting the potential influence of inbreeding in these populations. Intensive forest management, fragmentation, reduction of standing wood, and reduction of crown closure did not seem to result in drastic changes in the amount of genetic diversity [42]. The authors warn though that genetic diversity may gradually diminish over subsequent generations, if over-exploitation pertains [42,57].

Overall, case studies do indicate that both fragmentation and overexploitation may have adverse effects on genetic diversity, but generalized conclusive statements cannot be made given the variety of forest ecological conditions, field situations during sampling, study approaches and species life history characteristics.

5. Genetic Impact of Forestry Plantations

Plantation forestry is a major form of forestry practice that is expected to increase in the future in terms of land use, wood volume produced, and socio-economic importance. Worldwide, planted forests constitute almost 7% of the total forest area [62], exceeding an area of 250 m ha worldwide [63], while at least 33% of the planet's industrial round-wood derived from plantations by 2012 [64]. Plantation forestry may also impact the genetic diversity of forest species [65]. Plantation germplasm is precisely selected for the optimal performance of economically important quantitative traits, a selection procedure that could result in a more restricted genetic base for ensuing generations and in the weakening of local adaptation in natural forests. Clearly in the choice between genetic gain and genetic diversity, the number of parental seed trees (e.g., from a seed orchard or arboretum) used to provide the FRM is crucial: a low number will cause inbreeding among progeny, while a large number will increase genetic diversity and reduce differentiation among plantations. Forestry plantations are the first step in forest tree domestication, a process where the choice, nursery production and transfer of FRM may affect genetic diversity and future local adaptation [36,37,66] (see also Section 2.3).

Plantation forestry is usually spatially proximal to natural forest and employs in many cases the same species or genera as the natural forest, a result of tuning species selection to local ecological, edaphic and climatic requirements. Plantation to natural forest gene flow, forms an essential (however, not yet well studied) threat of genetic introgression from exotic conspecifics into indigenous populations. Large scale plantation forestry is frequently associated with a mechanized plantlet production from seed sources unrelated to the planting site. From the population genetics perspective the most important harmful consequence is artificial genetic homogenization, namely the progressive upsurge in genetic similarity between introduced and native gene pools through gene flow [67]. In this respect, anthropogenically brought gene flow of exotic origin may ultimately overwhelm locally adapted genotypes in natural populations.

An excellent relevant case study is the work of Steinitz et al. [68] on the results of forest plantations on the genetic constitution of conspecific native *Pinus halepensis*, a highly resilient species to stressful environments. The study considered natural *Pinus halepensis* populations with diverse levels of spatial separation from neighboring conspecific plantations approximately 40 years after plantation establishment. Results showed that native populations in two locations were significantly different from their respective surrounding plantations. Allele frequency changes (presence of novel alleles, common cohort allele frequency alterations), among different age classes (mature tree age classes, natural regeneration) were observed. Changes in allele frequencies were due to considerable gene flow from the plantations as was an increase in the genetic variation of the young age class. A notable result of forest plantations on the genetic variance within and between native proximal conspecific populations has been detected, resulting in a modification in the genetic constitution of the natural population younger age classes. Steinitz et al. [68] demonstrated that even in cases of highly isolated natural populations, elevated aggregate rate of gene flow over a period of many years will result in a strong long-term genetic contamination from plantations. Changes in allele frequencies produced by incoming gene flow may have potentially adverse effects in native populations. Plantations comprised of faraway provenances exert a gene flow that could result in a migration load (outbreeding depression) in the recipient population [69]. Even a small genetic difference between native and non-indigenous populations may endanger widespread local adaptations [70] for multilocus quantitative traits [71].

An additional outcome of widespread gene introgression from plantations bearing exotic material is the genetic homogenization of the various native populations. The anthropogenically induced gene flow by plantation establishment may disrupt local adaptive complexes and induce genetic homogenization in natural populations, especially if they are small compared to plantation size. In *P. halepensis* it was found that genetic divergence among natural forests was lower at the younger age class which indicates a genetic homogenization process [67,68]. These results are concordant to other reports concerning brutia pine (*P. brutia* Ten.) [41] reported considerable genetic variation within and among *P. brutia* stands in Turkey, but no distinct diversity patterns according to elevation, geography, or breeding zones. It has been suggested that the extensive use of artificial regeneration from non-local sources homogenized genetic diversity patterns in natural populations [41,42]. Another characteristic example is with regards to *Pinus nigra* var. *salzmanni* in southern France, where natural populations cover about 5000 ha (non-native planted *Pinus nigra* covers >200,000 ha), putting the integrity of the indigenous genetic resources at risk of contamination [72].

Plantation genetic diversity was studied in Syria where plantations constitute most of the forest cover. *P. brutia* plantations exhibited a notable decrease in mean genetic variation parameters compared to natural populations and were more genetically differentiated [73]. The stronger genetic differentiation in plantations may reflect the low number of seed trees from only a few populations that contributed to planted progenies [73]. In the same study, plantations of *Cupressus sempervirens* did not manifest a significant decline in mean genetic diversity, but were also more differentiated. The former result may be associated to an elevated gene flow among natural source populations and an overall uniform genetic variation present in natural forests.

Overall, the reduction of genetic variation found in plantations, may be regarded as an expected outcome of the selection process during domestication. Moreover, the demonstrated genetic introgression from non-local plantations (as well as introgression from artificial regeneration based on non-local sources within the natural forest) and the ensuing genetic homogenization of natural indigenous forests calls for a thoughtful reassessment of silviculture, afforestation, and forest management strategies and guidelines. Genetic homogenization results in a reduction of spatial gene diversity and variation in quantitative traits, and could reduce the capacity for adaptation to environmental change, weakening the resilience of biological communities [67,68]. These results clearly show once more that the prerequisite for the preservation of the needed elevated levels of genetic variation among forest plantations, is a well-planned seed collection from a large number of trees originating from well-established populations.

6. Conclusions and Perspectives

Silviculture and forest management systems have an influence on the genetic variation and structure of the populations of forest trees. Silvicultural and management systems, such as those referring to the management of natural regeneration systems, artificial regeneration using plant material of various origin, seedling, coppice and under conversion forest, fragmentation and overexploitation, as well as the establishment of plantations, influence genetic parameters. Nevertheless, forest tree populations appear to be generally highly resilient to forest management practices. Averaging genetic variation parameters, do not present significant differences between populations exposed to specific forest management methods and "control" populations. Observed differences are rather subtler and regard primarily the structure of the genetic diversity. Potentially negative differences of forest management approaches for instance refer to gene flow, effective population size, rare allele frequencies, founder effects, and disruption of local adaptive complexes. An adverse influence on genetic diversity and structure appears to be more noticeable in intensive forest management situations.

Because of environmental change and anthropogenic impact, species and ecosystem resilience in Mediterranean forests needs to be high. Neutral and adaptive genetic diversity provides a mechanism reinforcing both population perseverance and persistence of ecosystem functions [74], and loss of genetic diversity may reduce resilience [75]. As the impact on genetic diversity of silviculture and forest management practices used in the Mediterranean does not generally appear to be substantial, resilience associated to genetic diversity is not largely expected to be compromised in the immediate future as an upfront result of management practices. To the level that genetic diversity affects species resilience [75,76], there is no direct cause of concern for species resilience by the silviculture and forest management practices used. However, some of the finer genetic changes found, especially those dealing with genetic connectivity and effective population size, may eventually decrease fitness [74,77] and exert some unfavorable influence on the long-term resilience of Mediterranean forests.

Overall, the silviculture and forest management genetic impacts on long-impacted ecosystems, call for cautious methodologies, practices and policies at the strategic and operational levels, under the scope of sustainable multi-purpose forestry.

Acknowledgments: Partial assistance by the ERANet FORESTERRA research project "Integrated research on forest resilience and management in the Mediterranean" (INFORMED), is gratefully acknowledged. The author thankfully acknowledges the input and constructive criticism received by two anonymous reviewers.

Conflicts of Interest: The author declares no conflict of interest.

References

1. Ratnam, W.; Rajora, O.P.; Finkeldey, R.; Aravanopoulos, F.; Bouvet, J.-M.; Vaillancourt, R.E.; Kanashiro, M.; Fady, B.; Tomita, M.; Vinson, C. Genetic effects of forest management practices: Global synthesis and perspectives. *For. Ecol. Manag.* **2014**, *333*, 52–65. [CrossRef]
2. Kavaliauskas, D.; Fussi, B.; Westergren, M.; Aravanopoulos, F.; Finzgar, D.; Baier, R.; Alizoti, P.; Bozic, G.; Avramidou, E.; Konnert, M. The interplay between forest management practices, genetic monitoring, and other long-term monitoring systems. *Forests* **2018**, *9*, 133. [CrossRef]
3. Savolainen, O.; Kärkkäinen, K. Effect of forest management on gene pools. In *Population Genetics of Forest Trees*; Springer: Berlin, Germany, 1992; pp. 329–345.
4. Finkeldey, R.; Ziehe, M. Genetic implications of silvicultural regimes. *For. Ecol. Manag.* **2004**, *197*, 231–244. [CrossRef]
5. Critical Ecosystem Partnership Fund (CEPF). Explore the Biodiversity Hotspots. Available online: https://www.cepf.net/our-work/biodiversity-hotspots (accessed on 19 February 2018).
6. Palahi, M.; Mavsar, R.; Gracia, C.; Birot, Y. Mediterranean forests under focus. *Int. For. Rev.* **2008**, *10*, 676–688. [CrossRef]
7. Barbéro, M.; Loisel, R.; Quezel, P.; Richardson, D.M.; Romane, F. Pines of the Mediterranean basin. In *Ecology and Biogeography of Pinus*; Cambridge University Press: Cambridge, UK, 1998; pp. 153–170.
8. Fady-Welterlen, B. Is there really more biodiversity in Mediterranean forest ecosystems? *Taxon* **2005**, *54*, 905–910. [CrossRef]
9. Aravanopoulos, F.; Bucci, G.; Akkak, A.; Blanco Silva, R.; Botta, R.; Buck, E.; Cherubini, M.; Drouzas, A.; Fernandez-Lopez, J.; Mattioni, C. Molecular population genetics and dynamics of chestnut (*Castanea sativa*) in Europe: Inferences for gene conservation and tree improvement. In *Proceedings of the III International Chestnut Congress*; Acta Horticulturae: Leiden, The Netherlands, 2005.
10. Scarascia-Mugnozza, G.; Oswald, H.; Piussi, P.; Radoglou, K. Forests of the Mediterranean region: Gaps in knowledge and research needs. *For. Ecol. Manag.* **2000**, *132*, 97–109. [CrossRef]
11. Bariteau, M.; Alptekin, U.; Aravanopoulos, F.; Asmar, F.; Bentouati, A.; Benzyane, M.; Derridj, A.; Ducci, F.; Isik, F.; Khaldi, A.; et al. Les ressources génétiques forestières dans le bassin Méditerranéen. *Forêt Méditerranéenne* **2003**, *24*, 148–158.
12. Walker, B.; Holling, C.S.; Carpenter, S.R.; Kinzig, A. Resilience, adaptability and transformability in social–ecological systems. *Ecol. Soc.* **2004**, *9*, 5. [CrossRef]

13. Sasaki, T.; Furukawa, T.; Iwasaki, Y.; Seto, M.; Mori, A. Perspectives for ecosystem management based on ecosystem resilience and ecological thresholds against multiple and stochastic disturbances. *Ecol. Indic.* **2015**, *57*, 395–408. [CrossRef]

14. Fabbio, G.; Merlo, M.; Tosi, V. Silvicultural management in maintaining biodiversity and resistance of forests in Europe—The Mediterranean region. *J. Environ. Manag.* **2003**, *67*, 67–76. [CrossRef]

15. Yakovlev, I.; Fossdal, C.G.; Skrøppa, T.; Olsen, J.E.; Jahren, A.H.; Johnsen, O. An adaptive epigenetic memory in conifers with important implications for seed production. *Seed Sci. Res.* **2012**, *22*, 63–76. [CrossRef]

16. Gomory, D.; Foffova, E.; Longauer, R.; Krajmerova, D. Memory effects associated with early-growth environment in Norway spruce and European larch. *Eur. J. For. Res.* **2015**, *134*, 89–97. [CrossRef]

17. Johnsen, O.; Fossdal, C.G.; Nagy, N.; Molmann, J.; Daehlen, O.G.; Skroppa, T. Climatic adaptation in *Picea abies* progenies is affected by the temperature during zygotic embryogenesis and seed maturation. *Plant Cell Environ.* **2005**, *28*, 1090–1102. [CrossRef]

18. Avramidou, E.V.; Doulis, A.G.; Aravanopoulos, F.A. Determination of epigenetic inheritance, genetic inheritance, and estimation of genome DNA methylation in a full-sib family of *Cupressus sempervirens* L. Gene **2015**, *562*, 180–187. [CrossRef] [PubMed]

19. Asuka, Y.; Tomaru, N.; Munehara, Y.; Tani, N.; Tsumura, Y.; Yamamoto, S. Half-sib family structure of *Fagus crenata* saplings in an old-growth beech–dwarf bamboo forest. *Mol. Ecol.* **2005**, *14*, 2565–2575. [CrossRef] [PubMed]

20. El-Kassaby, Y.; Dunsworth, B.; Krakowski, J. Genetic evaluation of alternative silvicultural systems in coastal montane forests: Western hemlock and *Amabilis fir. Theor. Appl. Genet.* **2003**, *107*, 598–610. [CrossRef] [PubMed]

21. Wickneswari, R.; Ho, W.; Lee, K.; Lee, C. Impact of disturbance on population and genetic structure of tropical forest trees. *For. Genet.* **2004**, *11*, 193–201.

22. Lourmas, M.; Kjellberg, F.; Dessard, H.; Joly, H.; Chevallier, M.-H. Reduced density due to logging and its consequences on mating system and pollen flow in the African mahogany *Entandrophragma cylindricum*. *Heredity* **2007**, *99*, 151–160. [CrossRef] [PubMed]

23. Savolainen, O. Guidelines for gene conservation based on population genetics. In *Forest and Society: The Role of Research, Proceedings of the XXI IUFRO World Congress, Kuala Lumpur, Malaysia, 7–12 August 2000*; Malaysian XXI IUFRO World Congress Organising Committee: Kuala Lumpur, Malaysia, 2000; pp. 100–109.

24. Robledo-Arnuncio, J.J.; Smouse, P.E.; Gil, L.; Alía, R. Pollen movement under alternative silvicultural practices in native populations of Scots pine (*Pinus sylvestris* L.) in central Spain. *For. Ecol. Manag.* **2004**, *197*, 245–255. [CrossRef]

25. Karlsson, C.; Örlander, G. Mineral nutrients in needles of *Pinus sylvestris* seed trees after release cutting and their correlations with cone production and seed weight. *For. Ecol. Manag.* **2002**, *166*, 183–191. [CrossRef]

26. Adams, W.T.; Zuo, J.; Shimizu, J.Y.; Tappeiner, J.C. Impact of alternative regeneration methods on genetic diversity in coastal Douglas-fir. *For. Sci.* **1998**, *44*, 390–396.

27. Morgante, M.; Vendramin, G.; Rossi, P. Effects of stand density on outcrossing rate in two Norway spruce (*Picea abies*) populations. *Can. J. Bot.* **1991**, *69*, 2704–2708. [CrossRef]

28. Stoehr, M.U. Seed production of western larch in seed-tree systems in the southern interior of British Columbia. *For. Ecol. Manag.* **2000**, *130*, 7–15. [CrossRef]

29. Perry, D.J.; Bousquet, J. Genetic diversity and mating system of post-fire and post-harvest black spruce: An investigation using codominant sequence-tagged-site (STS) markers. *Can. J. For. Res.* **2001**, *31*, 32–40. [CrossRef]

30. Aravanopoulos, F. Wild fires as a factor contributing to the erosion of the forest gene pool: Towards a genetic holocaust? In *Proceedings of the 14th Pan-Hellenic Forest Science Conference, Patra, Greece, 4–7 October 2009*; Geotechnical Chamber of Greece: Patra, Greece, 2009; pp. 853–865.

31. Krauss, S.L. Low genetic diversity in *Persoonia mollis* (Proteaceae), a fire-sensitive shrub occurring in a fire-prone habitat. *Heredity* **1997**, *78*, 41–49. [CrossRef] [PubMed]

32. England, P.R.; Usher, A.V.; Whelan, R.J.; Ayre, D.J. Microsatellite diversity and genetic structure of fragmented populations of the rare, fire-dependent shrub *Grevillea macleayana*. *Mol. Ecol.* **2002**, *11*, 967–977. [CrossRef] [PubMed]

33. Uchiyama, K.; Goto, S.; Tsuda, Y.; Takahashi, Y.; Ide, Y. Genetic diversity and genetic structure of adult and buried seed populations of *Betula maximowicziana* in mixed and post-fire stands. *For. Ecol. Manag.* **2006**, *237*, 119–126. [CrossRef]

34. Aravanopoulos, F.A.; Panetsos, K.P.; Skaltsoyiannes, A. Genetic structure of *Pinus brutia* stands exposed to wild fires. *Plant Ecol.* **2004**, *171*, 175–183. [CrossRef]
35. Rajora, O.; Pluhar, S. Genetic diversity impacts of forest fires, forest harvesting, and alternative reforestation practices in black spruce (*Picea mariana*). *Theor. Appl. Genet.* **2003**, *106*, 1203–1212. [CrossRef] [PubMed]
36. Konnert, M.; Fady, B.; Gomory, D.; A'Hara, S.; Wolter, F.; Ducci, F.; Koskela, J.; Bozzano, M.; Maaten, T.; Kowalczyk, J. *Use and Transfer of Forest Reproductive Material in Europe in the Context of Climate Change*; European Forest Genetic Resources Programme (EUFORGEN); Bioversity International: Rome, Italy, 2015; 86p.
37. Cambell, R.K.; Sorensen, F.C. Genetic Implications of Nursery Practices. In *Forest Nursery Manual: Production of Bareroot Seedlings*; Duryea, M.L., Landis, T.D., Eds.; Martinus Nilhoff/Dr W. Junk Publ.: The Hague, The Netherlands; Boston, MA, USA; Lancaster, UK, 1984; p. 386.
38. Ivetic, V.; Devetakovic, J.; Nonic, M.; Stankovic, D.; Sijacic-Nikolic, M. Genetic diversity and forest reproductive material—From seed source selection to planting. *iForest* **2016**, *9*, 801–812. [CrossRef]
39. Acácio, V.; Holmgren, M.; Moreira, F.; Mohren, G. Oak persistence in Mediterranean landscapes: The combined role of management, topography, and wildfires. *Ecol. Soc.* **2010**, *15*, 40. [CrossRef]
40. Almeida, M.; Sampaio, T.; Merouani, H.; Costa e Silva, F.; Nunes, A.; Chambel, M.; Branco, M.; Faria, C.; Varela, M.; Pereira, J. Influência da qualidade dos materiais de reprodução na reflorestação com sobreiro. In *Gestão Ambiental e Eeconómica do Ecossistema Montado na Península Ibérica*; Jornadas Técnicas: Madrid, Spain, 2005.
41. Kandedmir, G.E.; Kandemir, I.; Kaya, Z. Genetic variation in Turkish red pine (*Pinus brutia* Ten.) seed stands as determined by RAPD markers. *Silvae Genet.* **2004**, *53*, 169–175. [CrossRef]
42. Lise, Y.; Kaya, Z.; Isik, F.; Sabuncu, R.; Kandemir, I.; Onde, S. The impact of over-exploitation on the genetic structure of Turkish red pine (*Pinus brutia* Ten.) populations determined by RADP markers. *Silva Fenn.* **2007**, *41*, 211–220. [CrossRef]
43. Bouffier, L.; Raffin, A.; Kremer, A. Evolution of genetic variation for selected traits in successive breeding populations of maritime pine. *Heredity* **2008**, *101*, 156–165. [CrossRef] [PubMed]
44. Icgen, Y.; Kaya, Z.; Cengel, B.; Velioğlu, E.; Öztürk, H.; Önde, S. Potential impact of forest management and tree improvement on genetic diversity of Turkish red pine (*Pinus brutia* Ten.) plantations in Turkey. *For. Ecol. Manag.* **2006**, *225*, 328–336. [CrossRef]
45. Amorini, E.; Manetti, M.; Turchetti, T.; Sansotta, A.; Villani, F. Impact of silvicultural system on *Cryphonectria parasitica* incidence and on genetic variability in a chestnut coppice in central Italy. *For. Ecol. Manag.* **2001**, *142*, 19–31. [CrossRef]
46. Aravanopoulos, F.A.; Drouzas, A.D.; Alizoti, P.G. Electrophoretic and quantitative variation in chestnut (*Castanea sativa* Mill.) in Hellenic populations in old-growth natural and coppice stands. *For. Snow Landsc. Res.* **2001**, *76*, 429–434.
47. Aravanopoulos, F.; Drouzas, A. Does forest management influence genetic diversity in chestnut (*Castanea sativa* Mill) populations? In *Proceedings of the 11th Pan-Hellenic Forest Science Conference, Olympia, Greece, 1–3 October 2003*; Geotechnical Chamber of Greece: Olympia, Greece, 2003; pp. 329–337.
48. Mattioni, C.; Cherubini, M.; Micheli, E.; Villani, F.; Bucci, G. Role of domestication in shaping *Castanea sativa* genetic variation in Europe. *Tree Genet. Genomes* **2008**, *4*, 563–574. [CrossRef]
49. Hill, W.G. Estimation of effective population size from data on linkage disequilibrium. *Genet. Res.* **1981**, *38*, 209–216. [CrossRef]
50. Ducci, F.; Proietti, R.; Cantiani, P. Genetic and social structure within a Turkey oak coppice with standards. In *Selvicoltura Sostenibile Nei Boschi Cedui*; Annali CRA—Centro di Ricerca per la Selvicoltura: Arezzo, Italy, 2006; Volume 33, pp. 143–158.
51. Papadima, A.; Drouzas, A.; Aravanopoulos, F. A gene flow study in natural seedling and coppice populations of *Castanea sativa* Mill. In *Proceedings of the 13th Pan-Hellenic Forest Science Conference, Kastoria, Greece, 7–10 October 2007*; Geotechnical Chamber of Greece: Kastoria, Greece, 2007; pp. 444–452.
52. Alizoti, P.; Aravanopoulos, F.; Diamantis, S. Genetic variation of chestnut (*Castanea sativa* Mill) populations under different management practices for seedling quantitative traits. In Proceedings of the 9th Pan-Hellenic Conference of the Hellenic Plant Genetics and Breeding Scientific Society, Thermi, Greece, 30 October–2 November 2002; pp. 164–171.

53. Valbuena-Carabaña, M.; González-Martínez, S.; Gil, L. Coppice forests and genetic diversity: A case study in *Quercus pyrenaica* Willd. from central Spain. *For. Ecol. Manag.* **2008**, *254*, 225–232. [CrossRef]

54. Ortego, J.; Bonal, R.; Muñoz, A. Genetic consequences of habitat fragmentation in long-lived tree species: The case of the Mediterranean holm oak (*Quercus ilex*, L.). *J. Hered.* **2010**, *101*, 717–726. [CrossRef] [PubMed]

55. Falconer, D.S.; Mackay, T.F.C. *Introduction to Quantitative Genetics*; Pearson/Prenctice Hall: New York, NY, USA, 1996; 464p.

56. Young, A.; Boyle, T.; Brown, T. The population genetic consequences of habitat fragmentation for plants. *Trends Ecol. Evol.* **1996**, *11*, 413–418. [CrossRef]

57. Lowe, A.; Boshier, D.; Ward, M.; Bacles, C.; Navarro, C. Genetic resource impacts of habitat loss and degradation; reconciling empirical evidence and predicted theory for neotropical trees. *Heredity* **2005**, *95*, 255–273. [CrossRef] [PubMed]

58. Dubreuil, M.; Riba, M.; Gonzalez-Martinez, S.C.; Vendramin, G.G.; Sebastiani, F.; Mayol, M. Genetic effects of chronic habitat fragmentation revisited: Strong genetic structure in a temperate tree, *Taxus baccata* (Taxaceae), with great dispersal capability. *Am. J. Bot.* **2010**, *97*, 303–310. [CrossRef] [PubMed]

59. De Vita, A.; Bernardo, L.; Gargano, D.; Palermo, A.; Peruzzi, L.; Musacchio, A. Investigating genetic diversity and habitat dynamics in *Plantago brutia* (Plantaginaceae), implications for the management of narrow endemics in Mediterranean mountain pastures. *Plant Biol.* **2009**, *11*, 821–828. [CrossRef] [PubMed]

60. Pautasso, M. Geographical genetics and the conservation of forest trees. *Perspect. Plant Ecol. Evol. Syst.* **2009**, *11*, 157–189. [CrossRef]

61. Kramer, A.T.; Ison, J.L.; Ashley, M.V.; Howe, H.F. The paradox of forest fragmentation genetics. *Conserv. Biol.* **2008**, *22*, 878–885. [CrossRef] [PubMed]

62. Food and Agriculture Organization (FAO). *State of the World's Forests*; FAO: Rome, Italy, 2011.

63. Forest Stewardship Council (FSC). *Strategic Review on the Future of Forest Plantations*; FSC: Helsinki, Finland, 2012.

64. Jürgensen, C.; Kollert, W.; Lebedys, A. *Assessment of Industrial Roundwood Production from Planted Forests*; Planted Forests and Trees Working Papers; FP/48/E; Food and Agriculture Organization (FAO): Rome, Italy, 2014.

65. Lefèvre, F. Human impacts on forest genetic resources in the temperate zone: An updated review. *For. Ecol. Manag.* **2004**, *197*, 257–271. [CrossRef]

66. El-Kassaby, Y.A. Domestication and genetic diversity-should we be concerned. *For. Chron.* **1992**, *68*, 687–700. [CrossRef]

67. Olden, J.D.; Poff, N.L.; Douglas, M.R.; Douglas, M.E.; Fausch, K.D. Ecological and evolutionary consequences of biotic homogenization. *Trends Ecol. Evol.* **2004**, *19*, 18–24. [CrossRef] [PubMed]

68. Steinitz, O.; Robledo-Arnuncio, J.; Nathan, R. Effects of forest plantations on the genetic composition of conspecific native Aleppo pine populations. *Mol. Ecol.* **2012**, *21*, 300–313. [CrossRef] [PubMed]

69. Lenormand, T. Gene flow and the limits to natural selection. *Trends Ecol. Evol.* **2002**, *17*, 183–189. [CrossRef]

70. Savolainen, O.; Pyhäjärvi, T.; Knürr, T. Gene flow and local adaptation in trees. *Ann. Rev. Ecol. Evol. Syst.* **2007**, *38*, 595–619. [CrossRef]

71. Allendorf, F.W.; Leary, R.F.; Spruell, P.; Wenburg, J.K. The problems with hybrids: Setting conservation guidelines. *Trends Ecol. Evol.* **2001**, *16*, 613–622. [CrossRef]

72. Fady, B.; Brahic, P.; Cambon, D.; Gilg, O.; Rei, F.; Roig, A.; Royer, J.; Thévenet, J.; Turion, N. Valoriser et conserver le pin de salzmann en France. *Forêt Méditerranéenne* **2010**, *31*, 3–14.

73. Al-Hawija, B.N.; Wagner, V.; Hensen, I. Genetic comparison between natural and planted populations of *Pinus brutia* and *Cupressus sempervirens* in Syria. *Turk. J. Agric. For.* **2014**, *38*, 267–280. [CrossRef]

74. Oliver, T.H.; Heard, M.S.; Isaac, N.J.B.; Roy, D.B.; Procter, D.; Eigenbrod, F.; Freckleton, R.; Hector, A.; Orme, C.D.L.; Petchey, O.L.; et al. Biodiversity and resilience of ecosystem functions. *Trends Ecol. Evol.* **2015**, *30*, 673–684. [CrossRef] [PubMed]

75. Schaberg, P.G.; DeHayes, D.H.; Hawley, G.J.; Nijensohn, S.E. Anthropogenic alterations of genetic diversity within tree populations: Implications for forest ecosystem resilience. *For. Ecol. Manag.* **2008**, *256*, 855–862. [CrossRef]

76. Cavers, S.; Cortell, J.E. The basis of resilience in forest tree species and its use in adaptive forest management in Britain. *Forestry* **2015**, *88*, 13–26. [CrossRef]

77. Van Oppen, M.J.H.; Gates, R.D. Conservation genetics and the resilience of reef-building corals. *Mol. Ecol.* **2006**, *15*, 3863–3883. [CrossRef] [PubMed]

© 2018 by the author. Licensee MDPI, Basel, Switzerland. This article is an open access article distributed under the terms and conditions of the Creative Commons Attribution (CC BY) license (http://creativecommons.org/licenses/by/4.0/).

forests
MDPI

Article

Genetic Structure and Population Demographic History of a Widespread Mangrove Plant *Xylocarpus granatum* J. Koenig across the Indo-West Pacific Region

Yuki Tomizawa [1,†,‡], Yoshiaki Tsuda [2,†], Mohd Nazre Saleh [3,†], Alison K. S. Wee [4,5],
Koji Takayama [6], Takashi Yamamoto [4,7], Orlex Baylen Yllano [8], Severino G. Salmo III [9],
Sarawood Sungkaew [10], Bayu Adjie [11], Erwin Ardli [12], Monica Suleiman [13],
Nguyen Xuan Tung [14], Khin Khin Soe [15], Kathiresan Kandasamy [16], Takeshi Asakawa [1],
Yasuyuki Watano [1], Shigeyuki Baba [4] and Tadashi Kajita [4,7,*]

1   Department of Biology, Graduate School of Science, Chiba University, Chiba 263-8522, Japan;
    richstream.courage@gmail.com (Y.T.); asakawa@faculty.chiba-u.jp (T.A.); watano@faculty.chiba-u.jp (Y.W.)
2   Sugadaira Research Station, Mountain Science Center, University of Tsukuba, 1278-294 Sugadairakogen,
    Ueda, Nagano 386-2204, Japan; ytsuda.gt@gmail.com
3   Faculty of Forestry, Universiti Putra Malaysia, 43400 UPM Serdang, Selangor Darul Ehsan, Malaysia;
    mnazre@gmail.com
4   Iriomote Station, Tropical Biosphere Research Center, University of the Ryukyus, 870 Uehara, Taketomi-cho,
    Yaeyama-gun, Okinawa 907-1541, Japan; alisonwks@gmail.com (A.K.S.W.);
    t.yamamoto.vm@gmail.com (T.Y.); babasan@lab.u-ryukyu.ac.jp (S.B.)
5   Guangxi Key Laboratory of Forest Ecology and Coservation, College of Forestry, Guangxi University,
    Nanning 530000, China
6   Museum of Natural and Environmental History, Shizuoka, 5762 Oya, Suruga-ku, Shizuoka 422-8017, Japan;
    gen33takayama@gmail.com
7   United Graduate School of Agricultural Science, Kagoshima University, 1-21-24 Korimoto,
    Kagoshima 890-0065, Japan
8   Biology Department, College of Sciences and Technology, Adventist University of the Philippines,
    Putting Kahoy, 4118 Silang, Cavite, Philippines; obyupd@yahoo.com
9   Department of Environmental Science, School of Science and Engineering, Ateneo de Manila University,
    1108 Quezon City, Philippines; ssalmo@ateneo.edu
10 Forest Biology Department, Faculty of Forestry, Kasetsart University, 50 Ngamwongwan Rd., ChatuChak,
    Bangkok 10900, Thailand; sungkaes@tcd.ie
11 Bali Botanical Garden, Indonesian Institute of Sciences (LIPI), Candikuning, Baturiti, Tabanan, Bali 82191,
    Indonesia; biobayu@gmail.com
12 Faculty of Biology, Jenderal Soedirman University, Jalan Dr. Suparno 63, Purwokerto 53122, Indonesia;
    eriantoo@yahoo.com
13 Institute for Tropical Biology and Conservation, Universiti Malaysia Sabah, Jalan UMS, 88400 Kota Kinabalu,
    Sabah, Malaysia; monicas@ums.edu.my
14 Mangrove Ecosystem Research Centre, Hanoi National University of Education, 136 Xuan Thuy Street,
    Cau Giay District, Hanoi, Vietnam; xt_xuthanh@yahoo.com
15 Department of Botany, University of Yangon, Kamayut 11041, Yangon, Myanmar; khinkhinsoe9@gmail.com
16 Center of Advanced Study in Marine Biology, Faculty of Marine Science, Annamalai University,
    Parangipettai, 608 502 Tamil Nadu, India; kathiresan57@gmail.com
*   Correspondence: kajita@mail.ryudai.jp; Tel.: +81-98-085-6560; Fax: +81-98-085-6830
†   These authors contributed equally to this work.
‡   Present address: Seed Production Department, Kaneko Seeds Co. Ltd. 50-12, Furuichimachi 1-chome,
    Mabashi City 371-8503, Japan.

Received: 29 August 2017; Accepted: 29 November 2017; Published: 5 December 2017

Abstract: *Xylocarpus granatum* J. Koenig is one of the most widespread core component species of mangrove forests in the Indo-West Pacific (IWP) region, and as such is suitable for examining how genetic structure is generated across spatiotemporal scales. We evaluated the genetic structure of this species using maternally inherited chloroplast (cp) and bi-parentally inherited nuclear DNA markers, with samples collected across the species range. Both cp and nuclear DNA showed generally similar patterns, revealing three genetic groups in the Indian Ocean, South China Sea (with Palau), and Oceania, respectively. The genetic diversity of the Oceania group was significantly lower, and the level of population differentiation within the Oceania group was significantly higher, than in the South China Sea group. These results revealed that in addition to the Malay Peninsula—a common land barrier for mangroves—there is a genetic barrier in an oceanic region of the West Pacific that prevents gene flow among populations. Moreover, demographic inference suggested that these patterns were generated in relation to sea level changes during the last glacial period and the emergence of Sahul Shelf which lied northwest of Australia. We propose that the three genetic groups should be considered independent conservation units, and that the Oceania group has a higher conservation priority.

Keywords: demographic inference; genetic structure; mangrove; seed dispersal; ocean current; Indo-West Pacific region; *Xylocarpus*

1. Introduction

Mangrove is a unique forest ecosystem that connects different habitats: land, freshwater, and ocean. Mangrove forests form the center of mangrove ecosystems, and are distributed in tropical and subtropical coastal and riverine regions of the world [1]. The ecological services and economic contribution of mangrove forests are important to many tropical and subtropical countries and provide tremendous benefits to human beings [1,2]. However, because mangroves forests are distributed close to areas of high anthropogenic disturbance and land use change, they are rapidly declining and under destruction, which results in the severe fragmentation of forests [3,4]. Thus, it is well recognized that mangrove forests have a high priority in conservation biology and forest management in many countries, not only due to human activities, but also because of recent global warming [5].

Despite the wide distribution range of mangrove forests, the number of component species is relatively small compared with other tropical forest habitats [6]. Only 30–40 species are recognized as core mangrove component species [1,2]. Even though this may be a small number, most species are widespread, enabling mangrove forests to be distributed globally across the tropics. Core groups of mangroves, such as *Avicennia* spp., *Rhizophora* spp., *Bruguiera* spp., and *Xylocarpus* spp., are found across different continents and oceanic regions [1,3]. Each group is an important component of mangrove ecosystems, and contributes to forest formation and succession processes. One of the mechanisms by which the core mangrove species can maintain their wide distribution is due to their effective method of seed dispersal, that is, sea-dispersal. By dispersing buoyant propagules via ocean currents, mangrove species have established wide distribution ranges spanning over continents. However, the specific processes involved in the formation and maintenance of the wide distribution of core mangrove species are not well understood.

Recently, molecular studies have suggested the presence of genetically distinct units within widespread mangrove species that are formed by barriers to gene flow. A number of studies have been performed in the Indo-West Pacific (IWP) region where a higher level of mangrove species diversity exists, e.g., *Ceriops tagal* (Perr.) C.B.Rob. [7], *Bruguiera gymnorhiza* (L.) Lam. [8,9], and *Rhizophora* spp. [10–12]. These studies showed strong genetic structure in the IWP, especially across the Malay Peninsula, which is a significant biogeographic barrier that has shaped genetic structure between the Indian and Pacific oceans. An additional "cryptic barrier" [12] in the West Pacific is reported in a study of *Rhizophora stylosa* Griff. Distinct genetic structure between the South China Sea and Oceania has also been found in *Ceriops tagal* [13] and *Bruguiera gymnorhiza* [14]; this concordance in phylogeography may indicate shared biogeographic histories. To understand the genetic structure of mangroves across the entire IWP region, it is necessary to focus on species that are widespread across the region. Moreover, although coalescent-based population demographic inferences have started to provide a deeper understanding of the genetic diversity and structure of species and their history (e.g., [15–17]), this kind of approach has not yet been well conducted in mangrove species. Thus, further inferences of past demographic history of mangrove species can shed new light on not only the population genetics of mangrove species, but also mangrove forest management and conservation, as demonstrated in forest tree species [16,18–20].

Xylocarpus granatum J. Koenig is one of the most widespread core mangrove species in the IWP, and can provide an ideal system that allows us to test the presence of genetic structures across the region. The genus *Xylocarpus* is the only mangrove genus in the large mahogany family, Meliaceae. The genus occurs solely in the IWP region, from east Africa westward to Oceania, and is comprised of three accepted species, *X. granatum*, *X. moluccensis* M.Roem., and *X. rumphii* (Kostel.) Mabb. [21]. *Xylocarpus* species have small, white flowers that are most likely insect-pollinated. Unlike the iconic mangrove species in the Rhizophoraceae that disperse viviparous propagules, *Xylocarpus* produces large, cannonball-like, multi-seeded fruits, and their seeds are buoyant for dispersal via the ocean. Within the genus, *X. granatum* is the most widespread species, but the genetic diversity and structure of this species has not been examined. Therefore, this species can be a suitable model to evaluate in order to provide a deeper understanding of the genetic structure and past demographic history of the mangrove species across the IWP region.

The aims of this study are: (1) to examine the genetic structure of *X. granatum* over a wide distribution range across the IWP using maternally inherited cpDNA and bi-parentally inherited nuclear DNA markers; (2) to infer the past demographic history of the species to further understand any genetic structure and barriers detected; and (3) to propose conservation units to maintain local genetic diversity.

2. Materials and Methods

2.1. Sample Collection and DNA Extraction

Leaf samples of 111 individuals of *X. granatum* from 18 populations were collected, covering 14 countries across the entire distribution range of *X. granatum* in the Indo-West Pacific (Table 1, Figure 1). Total DNA was extracted from silica-dried leaves using a modified cetyltrimethylammonium bromide (CTAB) method [22]. The extracted DNA was eluted in Tris-EDTA Buffer (TE) and stored at −20 °C until sequencing and genotyping. *Xylocarpus moluccensis* and *X. rumphii* were also examined in this study to evaluate whether cpDNA variation was shared among closely related species. For *X. moluccensis*, which is also widespread in SE Asia, 17 individuals from five populations were examined.

Table 1. Information of population samples used in this study.

ID	Country	Locality	Coordinate	Chloroplast DNA		h	Nulcear SSRs	
				n	H		n	Voucher
Xylocarpus granatum J. Koenig								
1	Mozambique	Maputo	25.85° S, E32.696° E	9	2	0.198	-	TK10122702
2	Mozambique	Quelimane	17.883° S, 36.861° E	4	1	0.000	4	TK10122504
3	Seychelles	Mahe	4.654° S, 55.403° E	6	1	0.000	-	KT12022201
4	Myanmar	Ayeyarwady	15.95° N, 95.357° E	6	1	0.000	4	TK11100805
5	Thailand	Phuket	8.204° N, 98.442° E	4	2	0.375	4	KT09121701
6	Malaysia	Klang	2.988° N, 101.355° E	7	2	0.245	4	TK11121403
7	Malaysia	Kemaman	4.65° N, 103.427° E	6	1	0.000	4	TK11121603
8	Malaysia	Sabah	5.945° N, 118.022° E	6	1	0.000	4	TK11072202
9	Singapore	Sungei Buloh	1.448° N, 103.73° E	5	1	0.000	-	KT09111202
10	Indonesia	Java	7.709° S, 108.9° E	6	1	0.000	-	KT09111508
11	Indonesia	Bali	8.734° S, 115.197° E	8	1	0.000	4	KT09111704
12	Vietnam	Dong Rui	21.226° N, 107.375° E	3	2	0.444	-	TK10050104
13	Philippines	Panay	11.802° N, 122.207° E	11	1	0.000	4	TK11062404
14	Palau	Airai	7.368° N, 134.576° E	5	2	0.320	4	YT60917
15	Micronesia	Kosrae	5.279° N, 162.973° E	8	1	0.000	3	TK11121601
16	Vanuatu	Malatie	17.55° S, 168.341° E	3	1	0.000	4	KT10073101
17	New Caledonia	Baie de Taare	22.259° S, 167.013° E	7	1	0.000	4	KT13022607
18	Australia	Daintree River	16.258° S, 145.399° E	7	1	0.000	3	KT13031902
				Total 111			Total 50	
Xylocarpus moluccensis M.Roem.								
19	India	Pichavaram	11.431° N, 79.794° E	2	1	0.000	-	TK10112308
20	Thailand	Trat	12.165° N, 102.482° E	2	2	0.500	-	KT09121902
21	Malaysia	Klang	3.038° N, 101.264° E	7	1	0.000	-	TK11121404
22	Vietnam	Ca Mau	8.716° N, 104.814° E	1	1	0.000	-	TK10042705
23	Philippines	Panay	11.802° N, 122.2° E	4	1	0.000	-	TK11062405
24	Myanmar	Ayeyarwady	15.95° N, 95.357° E	1	1	0.000	-	KK09122608
				Total 17				
Xylocarpus rumphii (Kostel.) Mabb.								
25	Malaysia	Kilim	6.42° N, 99.417° E	1	1	0.000	-	KLG20130620

n: sample size; H: number of haplotypes; h: gene diversity. See text for detail. Voucher specimens are preserved at URO (University of the Ryukyus) or UPM (University Putra Malaysia); SSR: simple sequence repeat.

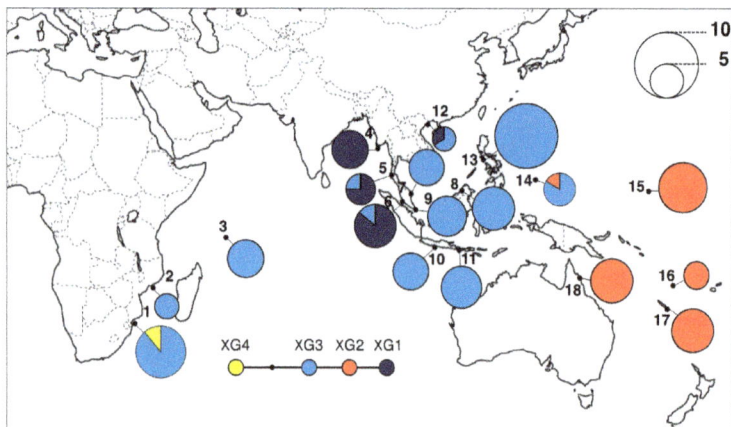

Figure 1. Geographical distributions of the four chloroplast (cp) DNA haplotypes of *Xylocarpus granatum* J. Koenig. The size of the pie charts indicates the sample size of the population. Population ID are defined in Table 1. The inset (bottom) denotes the haplotype network of the four haplotypes extracted from Figure 2.

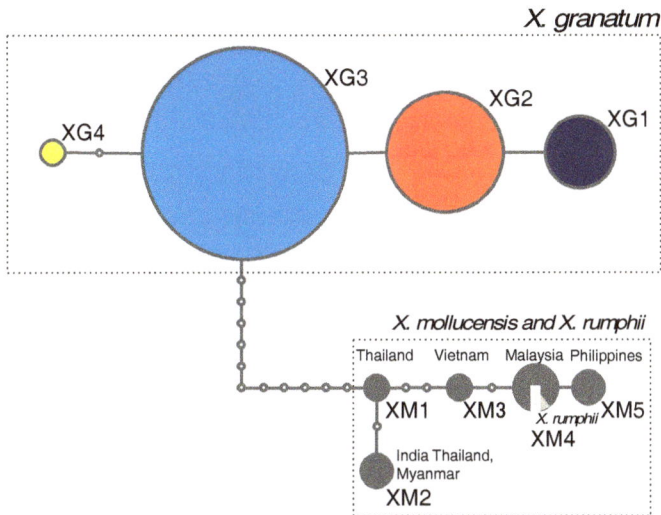

Figure 2. A statistical parsimony network constructed using the concatenated cpDNA sequences used in this study. XG1–4 designate the haplotypes of Xylocarpus granatum J. Koenig and XM1–5 of Xylocarpus moluccensis M.Roem. XM4 was also possessed by Xylocarpus rumphii (Kostel.) Mabb.

2.2. Chloroplast DNA Sequencing and Data Analysis

Three cpDNA intergenic regions, *trnD-trnT* [23], *trnL-trnF* [24], and *accD-psaI* [25], were sequenced for all of the samples using published universal primer pairs. Each polymerase chain reaction (PCR) was performed in a total volume of 10 μL with 1.0 μL of 10× Ex Taq buffer (Takara Bio Inc., Shiga Prefecture, Japan), 0.8 μL of each 10 mM deoxynucleotide (dNTP) mixture (Takara Bio Inc., Shiga Prefecture, Japan), 0.4 μL of each 10 pM primer pair, 0.05 μL of TaKaRa Ex Taq (Takara Bio Inc., Shiga Prefecture, Japan), and 6.35 μL of sterilized water mixed with 1.0 μL of template DNA. The PCR protocol used was initial denaturation at 95 °C for 60 s, followed by 35 cycles of 95 °C for 45 s, 55 °C for 45 s, and 72 °C for 60 s, and a final elongation at 72 °C for 10 min. The PCR products were purified using ExoSAP-IT (USB Corporation, Cleveland, OH, USA) and cycle sequencing was performed on the purified product using the BigDye cycle sequencing kit ver. 3.1 (Applied Biosystems, Foster City, CA, USA). Subsequently, the nucleotide sequence was determined with an ABI3500 Genetic Analyzer (Applied Biosystems, Foster City, CA, USA). Raw sequence data was imported into Auto Assembler ver. 2.1.1 (Applied Biosystems, Foster City, CA, USA), and peak calling was visually corrected. The sequence data of the three regions were concatenated and aligned using the ClustalW algorithm [26] incorporated within MEGA5 [27]. The haplotype of each individual was identified using the final concatenated sequence by considering only base substitution and indels. A statistical parsimony network was constructed, using the final concatenated sequences of 1702 bp by TCS 1.21 [28] to visualize the relationships among the cpDNA haplotypes. Genetic diversity (*h*) was calculated for each population. Genetic differentiation among populations was evaluated by calculating F_{ST} [29] and its standardized value, F'_{ST}, which always ranges from 0 to 1 [30], using GenAlEx 6.5 (hereafter, GenAlEx [31]).

2.3. Nuclear SSR Genotyping and Data Analysis of X. Granatum

We obtained Short Sequence Repeat (SSR) data for a total of 13 populations of *X. granatum* (Table 1). The SSR data of nine populations for 11 loci was obtained from Tomizawa et al. [32] (Table 1). The remaining four populations—5-Phuket, 14-Airai, 17-Baie de Tare, and 18-Daintree—were genotyped with the eleven SSR loci based on the PCR and fragment analysis protocol in Tomizawa et al. [32]. Allele calling was done using GeneMapper ver. 4.1 (Applied Biosystems, Foster City, CA, USA). The apportionment of genetic variation among group (F_{RT}), among populations within a group (F_{SR}), and among individuals within populations (F_{ST}) were evaluated based on a Bayesian model-based clustrering analysis results of $K = 2$ (Indian Ocean–South China Sea, and Oceania groups) and $K = 3$ (Indian Ocean, South China Sea, and Oceania groups) (see results for details) using an analysis of molecular variance (AMOVA, [33]) implemented in GenAlEx. The AMOVA was also conducted for each pair of the three groups at $K = 3$. The standardized values of F'_{RT} and F_{ST} were also calculated using GenAlEx. The genetic relationships among populations were evaluated by generating a neighbor-joining (NJ) tree based on the Nei's genetic distances (DA) [34] using Populations 1.2.30 beta software [35]. The statistical confidence in the topology of the tree was evaluated by 1000 bootstraps derived using the same software (Figure S1). The NJ tree was reconstructed on a topographic map using Mapmaker and GenGIS2 software [36]. Individual-level genetic relationships were analyzed with principal coordinates analysis (PCoA) using GenAlEx. Individual-based genetic structure was evaluated using a Bayesian model-based clustering algorithm implemented in the software STRUCTURE v. 2.3.4 [37,38]. STRUCTURE analysis was performed using the admixture model [38], with 20 runs for each number of subpopulations (K), from $K = 1$ to 13. Each run consisted of 30,000 replicates via the Markov chain Monte Carlo (MCMC) method after a burn-in of 20,000 replicates. STRUCTURE HARVESTER [39] was employed to evaluate the probability of the data (LnP (D)) for each K, and to calculate ΔK according to the method described by Evanno et al. [40]. Multimodality among runs and major clustering patterns at each K were evaluated using the CLUMPAK server [41].

2.4. Inference of Demographic History

The software DIYABC v2.0 [42,43] was used to analyze the nuclear SSR data of *X. granatum* to infer demographic history based on the approximate Bayesian computation (ABC) approach. To keep the scenarios in the ABC analysis simple, three groups were defined based on the result of the STRUCTURE analysis: Pop I (Pop IDs 2, 4, 5 and 6), Pop II (Pop IDs 7, 8, 11, 13 and 14), and Pop III (Pop IDs 15, 16, 17 and 18). The main aims were: 1) to infer population size change through time for each group, and 2) to infer the order of splitting into the three groups with their time scale, to understand how the modern genetic structure of *X. granatum*, as shown in the NJ tree and the STRUCTURE analysis, was generated (Figure 3a,b). Thus, we conducted two analyses: 'ABC1– Inference of population size change in each regional group', and 'ABC2– Inference of population demographic history among regional groups'. In all of the scenarios, t# represented the time scale, measured by generation time, and N# represented the effective population size of the corresponding populations (PopI, II, and III) during the relevant time period (e.g., 0–t1, t1–t2, and t2–t3).

ABC1—Inference of population size change in each regional group.

Four simple scenarios were tested for each group (Figure 4a). The scenarios were as follows:

- Scenario 1—Constant population size model: The effective population size of Pop# was constant at N1 from the past to the present.
- Scenario 2—Population shrinkage model: The effective population size of Pop# reduced from Na to N1 at t1.
- Scenario 3—Population expansion model: The effective population size of Pop# increased from Nb to N1 at t1.
- Scenario 4—Bottleneck model: The effective population size of Pop# experienced a bottleneck from Nd to Nc at t2 followed by population size recovery from Nc to N1 at t1.

ABC2—Inference of population demographic history among regional groups.

Six simple population demographic scenarios were tested with considering all the possible hierarchical population splits and populations to be traced back to the common ancestral population (Figure 4b). As population shrinkage was detected in Pop III in ABC1 (see results), population shrinkage for Pop III at t1 was added to all the scenarios, giving Na and N3 for the effective population size before and after the population size reduction, respectively. The change in the number of repeats followed a generalized stepwise mutation model (GSM; [44]) and single nucleotide indels (SNI) were also allowed. The mutation rate of the former was assumed to be higher than the mutation rate of the latter. The default values of the priors were used for all of the parameters (Table S1) except t2, according to our preliminary test runs (data not shown). The mean values of the expected heterozygosity (H_E), the number of alleles, the size of variance among the alleles, and Garza and Williamson's M [45] were used as summary statistics for each of the three populations in ABC1 and 2. As for summary statistics for each of the population pairs in ABC2, we used H_E, A, Garza and Williamson's M, F_{ST}, genotype likelihood, shared allele distance, and $(\delta\mu)^2$ distance [46]. One million simulations were performed for each scenario, and the most likely scenario was evaluated by comparing posterior probabilities with the logistic regression method. The goodness-of-fit of the scenarios was also assessed by principal component analysis (PCA) using the option "model checking" in DIYABC.

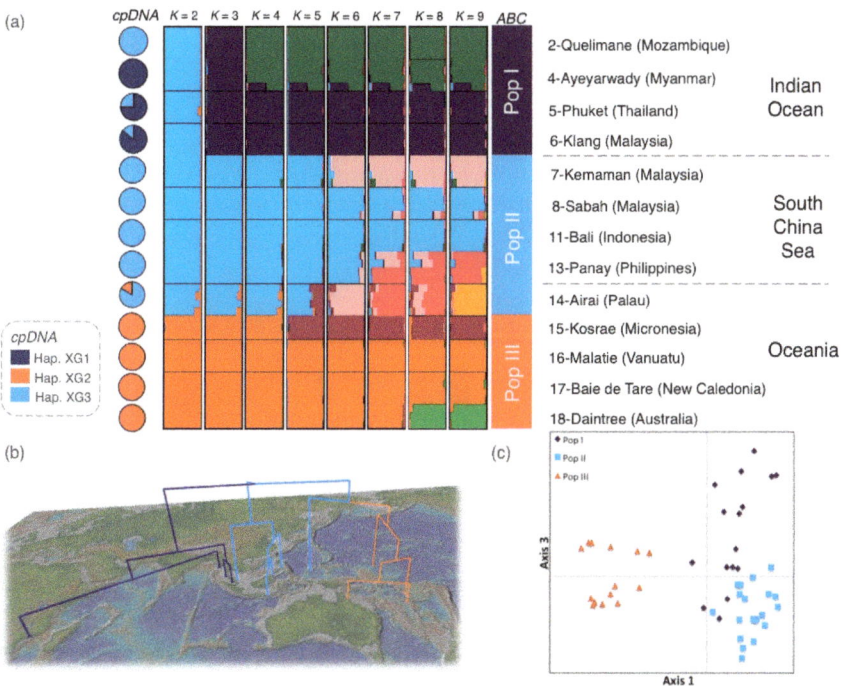

Figure 3. Bayesian clustering analyses of nuclear simple sequence repeat (SSR) data for the 18 populations of *Xylocarpus granatum* J. Koenig. (**a**) Bar plots indicating the putative genetic ancestry of each individual in comparison with the cpDNA haplotype composition of each population. (**b**) Population neighbor-joining (NJ) tree reconstructed on a topographic map. Each branch was colored based on the population subdivision used for approximate Bayesian computation (ABC) analyses (Pop I: orange, Pop II: blue, and Pop III: green). (**c**) Plot of the second two coordinates of principal coordinates analysis (PCoA) based on the SSR data. Samples were colored according to the subdivision used for ABC analyses.

(a)

ABC1

(b)

ABC2

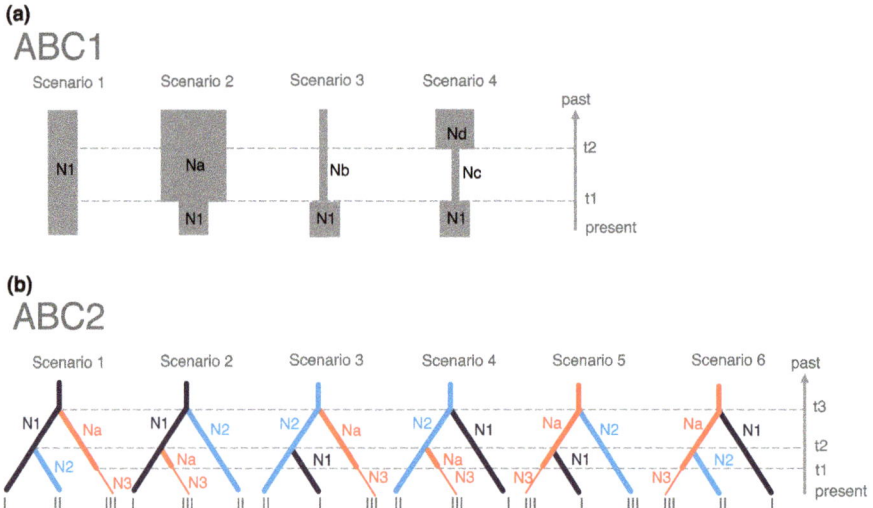

Figure 4. Scenarios tested using DIYABC in this study. (**a**) The four scenarios tested in ABC1. (**b**) The six scenarios tested in ABC2. See text for explanations. In all the scenarios, t# represents the time scale measured in number of generations, and N# represents the effective population size of the corresponding populations.

3. Results

3.1. Phylogenetic Relationships of the cpDNA Haplotypes in Xylocarpus

A total of 1702 bp of cpDNA sequence was used for analysis, including 687 bp of *trnD–trnT* Intergenic Spacer (IGS), 435 bp of *trnL–trnF* IGS, and 580 bp of *accD–psaI*. In *X. granatum*, four haplotypes (XG1–XG4) were characterized based on four nucleotide substitutions; while in *X. moluccensis*, five haplotypes (XM1–XM5) were characterized based on six nucleotide substitutions and one indel (Table 2). The two species did not share any haplotypes, but the haplotype of *X. rumphii* used in this study was the same as a haplotype of *X. moluccensis* (XM4).

The statistical parsimony network constructed by TCS showed that *X. granatum* and *X. moluccensis* are clearly differentiated. In *X. granatum*, haplotype XG2 and XG4 were derived from the major haplotype XG3, and haplotype XG1 was derived from XG2 (Figure 2). There were only one or two steps between neighboring haplotypes. In *X. moluccensis* and *X. rumphii*, nearby haplotypes were differentiated by one to three steps.

Table 2. Haplotypes identified by the concatenated sequences of three cpDNA intergenic regions. Polymorphic sites are designated by columns with the positions in the alignment of concatenated sequences.

Haplotype	trnD–trnT										trnL–trnF				accD–psaI				
	Accession No.	4	185	191	196–201	202	223	231	232	404	Accession No.	946	950	979	Accession No.	1222	1266	1325	1525
Xylocarpus granatum J. Koenig																			
XG1	LC217845	C	A	A	-	T	T	A	A	T	LC217851	C	C	C	LC217857	G	C	A	-
XG2	LC217846	C	A	C	-	T	T	A	A	T	LC217851	C	C	C	LC217857	G	C	A	-
XG3	LC217846	C	A	C	-	T	T	A	A	T	LC217851	C	C	C	LC217858	G	C	G	-
XG4	LC217846	C	A	C	-	T	T	A	A	T	LC217852	T	A	C	LC217858	G	C	G	-
Xylocarpus moluccensis M. Roem.																			
XM1	LC217847	T	C	C	AAAGGA	A	G	A	A	G	LC217853	T	C	G	LC217859	G	G	G	-
XM2	LC217847	T	C	C	AAAGGA	A	G	A	A	G	LC217854	C	C	C	LC217859	G	G	G	-
XM3	LC217848	T	C	C	AAAGGA	A	G	G	A	G	LC217854	C	C	C	LC217860	A	G	G	T
XM4	LC217848	T	C	C	AAAGGA	A	G	G	A	G	LC217855	T	A	C	LC217860	A	G	G	T
XM5	LC217849	T	C	C	AAAGGA	A	G	G	G	G	LC217855	T	A	C	LC217860	A	G	G	T
Xylocarpus rumphii (Kostel.) Mabb.																			
XR (XM4)	LC217850	T	C	C	AAAGGA	A	G	G	A	G	LC217856	T	A	C	LC217861	A	G	G	T

3.2. Geographical Distribution of cpDNA Haplotypes

The geographical distribution of the haplotypes for the 18 populations of *X. granatum* revealed that haplotype XG3 was widely distributed across the Indo-West Pacific (Table 1, Figure 1). The distribution of haplotypes demonstrated no clear genetic break between the Indian and Pacific Oceans (due to the presence of XG3 in both). Instead, we observed a localized distribution in the other three haplotypes, with XG1 and XG2 largely confined to the Andaman Sea (4-Ayeyarwady, 5-Phuket, 6-Klang, and 12-Dong Rui) and in Oceania (14-Airai, 15-Kosrae, 16-Malatie, 17-Baie de Tare, and 18-Daintree), respectively, while XG4 was found in only one individual from Mozambique (1-Maputo). Haplotypes XG1 and XG2 were fixed in one population (4-Ayeyarwady), and four populations (15-Kosrae, 16-Malatie, 17-Baie de Tare, and 18-Daintree), respectively. The values of F_{ST} and F'_{ST} calculated among populations by cpDNA data were 0.853 and 0.933, respectively, suggesting the level of population differentiation was high.

3.3. Genetic Structure Revealed by SSR Markers

The total number of alleles detected was five to 20 per locus. Although the highest ΔK was detected at $K = 2$, the Ln $P(D)$ values increased together with the number of K up to around $K = 9$ in the STRUCTURE analysis (Figure S2), and genetic clustering patterns were clear from $K = 2$ to 9. At $K = 2$, the geographic pattern of the two clusters was clear; cluster 1 (C1 in blue color) was distributed in the Indian Ocean and South China Sea, while cluster 2 (C2 in orange color) was distributed in Oceania, consisting of 15-Kosrae (Micronesia), 16-Malatie (Vanuatu), 17-Baie de Tare (New Caledonia), and 18-Daintree (Australia). One exception was the 14-Airai (Palau) population, which clustered into C1 despite its location in Oceania (Figure 3a). At $K = 3$, C1 at $K = 2$ was further divided into two clusters, which corresponded to the Indian Ocean (C3 in dark blue color), consisting of 2-Quelimane (Mozambique), 4-Ayeyarwady (Myanmar), 5-Phuket (Thailand), and 6-Klang (Malaysia); and the South China Sea, consisting of 7-Kemaman (Malaysia), 8-Sabah (Malaysia), 11-Bali (Indonesia), 13-Panay (Philippines), and 14-Airai (Palau). A similar pattern to the results of $K = 3$ was also detected by the NJ tree for 13 populations (Figure 3b). Although more local genetic clusters were detected up to $K = 9$, the NJ tree for nine clusters revealed the same three genetic groups detected at $K = 3$ and the NJ tree for 13 populations. An individual level PCoA analysis demonstrated three well-segregated groups in the plot of the second two axes (1 vs. 3 explained 26.56% of the total variation) (Figure 3c) and the first two axes (1 vs. 2 explained 27.92%). Thus, finally, the 13 populations were clustered into three groups: Pop I, II, and III, prior to conducting ABC analyses (as stated in the Materials and Methods). In concordance with the cpDNA data, a genetic break was detected at the boundary between the South China Sea and Oceania. The cpDNA haplotypes from both regions were detected in population 14-Airai, which is located between the two regions, and this population also revealed a small amount of admixture between the Indian Ocean–South China Sea cluster and the Oceania cluster in the nuclear DNA data.

The overall value of F_{ST} was 0.404, suggesting that the level of population differentiation among the 13 populations was high, especially when considering the standardized F'_{ST} value of 0.802 (Table 3). The value of F_{ST} among the four populations in Oceania was significantly higher than among the five populations in the South China Sea, while significant differences were neither detected among the three groups nor between other pairs of groups. There was relatively high genetic differentiation among the populations within groups (F'_{SR} values were 0.678 and 0.610 for $K = 2$ and 3, respectively), although larger differentiation was detected among groups (F'_{RT} values were 0.748 and 0.648 for $K = 2$ and 3, respectively) (Table 3). For all of the comparisons, F_{ST} values are always two to three times higher, which can be due to higher variation among individuals than among populations or groups. Moreover, the AMOVA for each group pair suggested that genetic differentiation between the South China Sea and Oceania ($F'_{RT} = 0.815$) was much higher than that between the Indian Ocean and the South China Sea ($F'_{RT} = 0.389$), and the Indian Ocean and Oceania ($F'_{RT} = 0.416$).

Table 3. Analysis of molecular variance (AMOVA) for 50 individuals of *X. granatum* based on 11 nuclear SSRs.

Source of Variation	Total Variance (%)	F-Statistics	p-Value	F'-Value
13 populations				
Among populations	40.4%	$F_{ST} = 0.404$	0.001	$F'_{ST} = 0.802$
Among individuals within populations	59.6%	-	-	-
K = 2				
Among group	19.8%	$F_{RT} = 0.198$	0.001	$F'_{RT} = 0.748$
Among population within groups	27.2%	$F_{SR} = 0.339$	0.001	$F'_{SR} = 0.678$
Among individuals within populations	53.0%	$F_{ST} = 0.470$	0.001	-
K = 3				
Among group	18.3%	$F_{RT} = 0.183$	0.001	$F'_{RT} = 0.648$
Among population within groups	24.9%	$F_{SR} = 0.302$	0.001	$F'_{SR} = 0.610$
Among individuals within populations	56.8%	$F_{ST} = 0.432$	0.001	-
Between the Indian Ocean (Pop 2,4,5 and 6) and the South China Sea (Pop 7,8,11,13 and 14)				
Among group	10.4%	$F_{RT} = 0.104$	0.001	$F'_{RT} = 0.389$
Among population within groups	23.9%	$F_{SR} = 0.266$	0.001	$F'_{SR} = 0.577$
Among individuals within populations	65.8%	$F_{ST} = 0.342$	0.001	-
Between the Indian Ocean (Pop 2,4,5 and 6) and Oceania (Pop 15,16,17 and 18)				
Among group	11.6%	$F_{RT} = 0.116$	0.001	$F'_{RT} = 0.416$
Among population within groups	23.5%	$F_{SR} = 0.266$	0.001	$F'_{SR} = 0.567$
Among individuals within populations	64.9%	$F_{ST} = 0.351$	0.001	-
Between the South China Sea (Pop 7,8,11,13 and 14) and Oceania (Pop 15,16,17 and 18)				
Among group	25%	$F_{RT} = 0.246$	0.001	$F'_{RT} = 0.815$
Among population within groups	21%	$F_{SR} = 0.283$	0.001	$F'_{SR} = 0.572$
Among individuals within populations	54%	$F_{ST} = 0.459$	0.001	-

SSR: simple sequence repeat.

3.4. Inferences of Demographic History

When considering the changes in the temporal population size in each regional group in ABC1, scenario 1 (constant population size model) showed the highest posterior probability in Pop1 (Indian Ocean) and 2 (South East Asia), and summary statistics and the PCA suggested that this scenario best described the data (Table 4, Figures S3b and S4b). However, the 95% confidence interval (CI) of scenario 1 overlapped with other scenarios in these two groups (Table 4). Regarding Pop3 (Oceania), scenario 2 (population shrinkage model) showed the highest posterior probability (0.3917) and its 95% CI (0.3858–0.3977) did not overlap with other scenarios. In this scenario, the effective population size after the shrinkage was estimated to be 2310 (95% hyper probability density (HPD), 806–6960), and the population shrinkage was inferred to have occurred 12800 (95% HPD, 888–29,200) generations ago (Table S3, Figure S5a). This time scale can be translated to 128,000 years Before Present (BP) (95% HPD, 8880–292,000) under the assumption of a generation time of 10 years for *X. granatum*. The median values of the mean mutation rate of SSR, mean *P* (the parameter of the geometric distribution to generate multiple stepwise mutations) and SNI were estimated to be 4.53×10^{-4} (95% CI: 1.30×10^{-4}–9.65×10^{-4}), 0.254 (95% CI: 0.128–0.300) and 2.54×10^{-7} (95% CI: 1.18×10^{-8}–7.72×10^{-6}), respectively. The summary statistics and the PCA suggested the good fit of the scenario (Table S4, Figure S5b). All four summary statistics in this scenario were not significantly differentiated from the observed value, and the PCA showed that the observed data point (yellow) was in the middle of a small cluster of datasets from the posterior predictive distribution (red, Figure S5b), suggesting scenario 2 fits the observed data well in Pop3.

Table 4. Posterior probability of each scenario in ABC1 and its 95% hyper probability density (HPD) based on the logistic estimate by DIYABC v2.0 [42,43]. The scenarios tested in ABC1 are shown in Figure 4.

ABC1	Posterior Probability (95% Confidence Interval (Lower–Upper))		
Scenario	Region 1	Region 2	Region 3
1	0.2640 (0.2586–0.2693)	0.2564 (0.2477–0.2651)	0.2554 (0.2500–0.2607)
2	0.2543 (0.2491–0.2596)	0.2645 (0.2482–0.2660)	0.3917 (0.3858–0.3977)
3	0.2340 (0.2288–0.2392)	0.2221 (0.2136–0.2305)	0.1624 (0.1579–0.1669)
4	0.2477 (0.2420–0.2533)	0.2571 (0.2482–0.2660)	0.1905 (0.1858–0.1952)

In the inference of population demographic history among regional groups (ABC2), the highest posterior probability was detected for scenario 1 (0.3794, 95% CI, 0.3712–0.3877) without overlapping with the 95% CI of other scenarios (Table 5). In this scenario, the median values of N1, N2, N3, and Na were estimated as 4090 (95% HPD, 1680–7610), 7880 (95% HPD, 4140–9850), 3090 (95% HPD, 822–8510) and 7840 (95% HPD, 3740–9910), respectively. However, Na was not well estimated (Table S5, Figure S6a). The median values of the time scales for the population shrinkage of Pop3 (t1), the splitting of Pop1 and Pop2 (t2), and the splitting of Pop1 and Pop3 (t3) were 4810 (95% HPD, 191–25,400), 4990 (95% HPD, 1460–18,500), and 15500 (95% HPD, 4590–44,300) generations ago, respectively, corresponding to 48,100 (95% HPD, 1910–254,000), 49,900 (95% HPD, 14,600–185,000), and 155,000 (95% HPD, 45,900–443,000) years BP, respectively, although t1 was not well estimated. The median values of the mean mutation rate of SSR, mean P (the parameter of the geometric distribution to generate multiple stepwise mutations), and SNI were estimated to be 3.47×10^{-4} (95% CI: 1.58×10^{-4}–7.82×10^{-4}), 0.248 (95% CI: 0.134–0.300), and 1.13×10^{-7} (95% CI: 1.11×10^{-8}–2.89×10^{-6}), respectively. All 36 summary statistics for this scenario were not significantly differentiated from the observed value, and the PCA suggested a good fit of the scenario (Table S6, Figure S6b). Moreover, the PCA showed that the observed data point (yellow) was in the middle of a small cluster of datasets from the posterior predictive distribution (light green, Figure S6b), suggesting that scenario 1 fits the observed data well.

Table 5. Posterior probability of each scenario in ABC2 and its 95% hyper probability density (HPD) based on the logistic estimate by DIYABC v2.0 [42,43]. The scenarios tested in ABC2 are shown in Figure 4.

ABC2	
Scenario	Posterior Probability (95% Confidence Interval: Lower–Upper)
1	0.3794 (0.3712–0.3877)
2	0.0839 (0.0784–0.0893)
3	0.3111 (0.3031–0.3190)
4	0.0877 (0.0821–0.0934)
5	0.0590 (0.0541–0.0639)
6	0.0789 (0.0734–0.0844)

4. Discussion

4.1. Genetic Structure of a Widespread Mangrove Plant, X. granatum, across the IWP Region

This study detected clear genetic structure with three genetic clusters across the distribution of *Xylocarpus granatum*, namely the Indian Ocean, the South China Sea, and Oceania clusters, in both cp and nuclear DNA datasets. In particular, a genetic break with high genetic differentiation ($F'_{RT} = 0.815$) was detected between the South China Sea and Oceania in the West Pacific, suggesting limited gene flow by sea-dispersed propagules between these two regions. This genetic break could be recognized as the "cryptic barrier" [12] that might have historically prevented gene flow between South East Asia and Oceania, as shown in *Rhizophora stylosa* [12], *Bruguiera gymnorhiza* [14], *Sonneratia alba* Griff. [47], and *Vigna marina* (Burm.) Merr. [48].

Although it is difficult to compare the exact location of genetic breaks among these mangrove species due to studies having different sampling locations, the presence of genetic breaks in a similar oceanic region is surprising, because all of the core mangrove species are sea-dispersed plants that can utilize oceans as corridors for migration. Sea dispersal is considered to be one of the most effective modes of seed dispersal for land plants, with dispersal ranges reaching over 100 km [49]. Indeed, long-distance dispersal by sea dispersal is well supported in population genetic studies in extremely widespread plants, especially for the pantropical plants with sea drifted seeds that have vast distribution ranges across the tropics, such as *Hibiscus tiliaceus* L. [50,51], *Ipomoea pes-caprae* (L.) R. Br. subsp. *brasiliensis* (L.) van Ooststr. [52] and the genus *Rhizophora* [10]. However, topographic barriers can act as genetic barriers even in these species. For example, Takayama et al. [10] detected the Central American Isthmus as a genetic barrier dividing the Pacific and Atlantic groups of *Rhizophora mangle* L. and *Rhizophora racemose* G. Mey. in both cp and nuclear DNA, while extended gene flow was suggested within those groups.

4.2. Inferences of the Demographic History of X. granatum

Our results suggest a clear genetic break between the South China Sea and Oceania, despite no obvious topographic barrier in this oceanic area. The ABC inference of past demographic history is helpful to try and understand this pattern. The inferred divergence time of 155,000 (95% HPD, 45,900–443,000) years BP between the South China Sea and Oceania groups suggests that the shallow sea between the Australian continent and New Guinea (the Torres Strait) might have acted as a significant land geographic and genetic barrier during and since the last glacial maximum (LGM, ca. 20,000 years BP), when the entire Sahul Shelf was exposed [53]. The past influence of the Sahul Shelf, compounded by the present large geographic distance among oceanic islands, may have resulted in the observed genetic differentiation of multiple core mangrove species over the West Pacific region. Furthermore, the westward Equatorial Current that bifurcates north and south as it approaches the Indo-Malay Archipelago [54] could be a strong barrier preventing the gene dispersal of haplotype XG2 beyond Palau. Thus, although we need to be cautious about uncertainties in the ABC inferences, such as the generation time of species, overlapping of generations, and wide 95% CI of inferred parameters in the assumed model [16], the present results suggest that genetic differentiation caused by a genetic barrier between the South China Sea and Oceania was generated by episodes of eustatic change in sea levels in relation to ice ages over the past several hundred thousand years. Moreover, population shrinkage was detected in the Oceania group (estimated 128,000 years BP (95% HPD, 8880–292,000)), suggesting that effective population size was reduced following genetic divergence from the South China Sea. In addition, as the mean value of F_{ST} among populations within Oceania (0.4090) was significantly higher than within the South China Sea (0.1800), higher genetic isolation within the Oceania group might have also contributed to a decrease in the regional effective population size in Oceania. Regarding the assumed model in the ABC, DIYABC does not consider gene flow after divergence, and it may bias the inferred temporal parameters [16,20]. However, as genetic differentiation between the South China Sea and Oceania is high enough with less gene flow ($F'_{RT} = 0.815$), this bias may be limited. In addition, since no consideration of gene flow after divergence may underestimate divergence times [16], the main discussion here would not be changed with the population split still occurring before the Last Glacial Maximum (LGM). Finally, although the ABC approach in this study revealed the demographic history of the species, the results obtained in this study might be based on the limited sample size in this study. Thus, further analysis would be required with more loci and more sample sizes in future studies in order to obtain a deeper understanding of the demography of the species.

The population in Palau (14-Airai) revealed a unique genetic composition, and the cpDNA showed a partial mixture of two haplotypes (XG2 and 3), which were dominant in the South China Sea and

Oceania, respectively. In addition, nuclear DNA revealed that this population was clustered with the South China Sea group, although it is located in Oceania. As Palau is geographically located close to the boundary between the South China Sea and Oceania, a mixture of cpDNA haplotypes might be reasonable when seed flow between them is considered. Indeed, a mixture of cpDNA halotypes from different lineages has been commonly detected around the boundary areas of lineages in many plant populations (e.g., [48,55,56]). However, nuclear DNA showed only a small amount of admixture between the South China Sea and Oceania clusters, and was basically grouped with the South China Sea cluster. Thus, although the limited sample size examined in this study should be considered, this pattern could be due to differential gene flow between pollen and seeds between the South China Sea and Oceania groups, or incomplete sorting of cpDNA haplotypes. These scenarios need to be further tested with more extended sample collections that cover the species distribution.

4.3. Malay Peninsula as a Common Barrier to Widespread Mangrove Plants

The results of the NJ tree, PCoA, and the STRUCTURE analysis of nuclear SSR data showed clear genetic structure between the Indian Ocean and South China Sea, suggesting that the Malay Peninsula also acts as a barrier for gene flow in *X. granatum*. The Malay Peninsula has been reported as a clear land barrier for several mangrove species, such as *B. gymnorhiza* [8,9,14] and *Ceriops tagal* [7]. This land barrier emerged in this region during the LGM, and has been routinely evoked as an explanation for the genetic structure of mangrove plants across the Malay Peninsula [57]. Indeed, the inferred divergence time of 49,900 (95% HPD, 14,600–185,000) years BP in the ABC generally supports this hypothesis. Although cpDNA generally showed similar patterns to the nuclear DNA, cpDNA did not suggest clear genetic structure across the Malay Peninsula, especially when considering the presence of the widespread haplotype, XG3, from East Africa to Palau (Figure 1). The presence of the widespread haplotype over this wide range implies either frequent long-distance migration, or less variation (resolution) of the molecular marker.

4.4. Relationships with Other Xylocarpus Species

Similar to other mangrove groups, *Xylocarpus granatum* has its closely related species, *X. moluccensis*, in a core group of mangroves that has overlapping distribution with *X. granatum* [21]. Another species, *X. rumphii*, is also recognized as a mangrove, but is not very common, and the distribution is scattered. In these results, *X. granatum* and *X. moluccensis* formed two clear clades, and *X. rumphii* shared the same haplotype with *X. moluccensis* (Figure 2, Table 2). Perhaps it is necessary to further study the taxonomic status of *X. rumphii* with extensive sampling from its distribution range. Only cpDNA data was collected for *X. moluccensis*, and the haplotype distribution map shows genetic structure across the Malay Peninsula (Figure S7), which also suggests that the Peninsula plays a role as a barrier for the seed dispersal of mangroves, as discussed above. It is interesting that no evidence of hybridization between *X. granatum* and *X. moluccensis* was detected, from field observations or with molecular data, despite collecting samples from the same area and population in some cases (Philippines and Malaysia). In general, core mangrove species tend to have a sister mangrove species, e.g., *Bruguiera gymnorhiza*—*B. sexangula*, *Rhizophora stylosa*—*R. mucronata*, and *Rhizophora mangle*—*R. racemosa* [58]. The distribution ranges of the sister species often overlap, and when they grow in the same location, hybrids are sometimes formed. Hybrid formation is quite common, especially for the mangrove species of Rhizophoraceae. Hybrid formation can be an important phenomena for plants to increase variation and promote speciation, but may not occur in *Xylocarpus*.

5. Conclusions

In this study, a clear genetic structure was detected within the distribution range of one of the most widespread and core component species of mangrove forests in the IWP. The genetic structure was shaped not only by the Malay Peninsula, a common land barrier for mangroves, but also by a cryptic barrier across the West Pacific. Moreover, the demographic inferences revealed that genetic divergence across the West Pacific was older than that across the Malay Peninsula, implying that a common factor such as the Sahul Shelf acted to shape the genetic structure of sea-dispersed mangrove plants. Other factors that generated genetic structure also need to be considered and evaluated in future studies, including adaptation to the local environment, ancient ocean currents, the location of refugia during the ice ages, and expansion routes during oscillation cycles. However, although further study would be needed, this study has provided tentative information on conservation genetics from a neutral genetic variation point of view, and the three groups detected in this study should be treated as independent conservation units in order to maintain local genetic diversity. In addition, conservation priority should be given to the Oceania region, as genetic diversity is lower and genetic isolation among populations is more prominent than in other areas.

Supplementary Materials: The following are available online at www.mdpi.com/1999-4907/8/12/480/s1, Figure S1: Neighbor-joining (NJ) tree for 18 populations of *Xylocarpus granatum* based on nuclear SSR dataset. Genetic distances were calculated by Populations 1.2.30. Statistical confidence of the topology was evaluated by 1000 bootstrap replicates. Figure S2: Plot of LnP(D) and ΔK calculated for K ranging from 1 to 13. See text for detail. Figure S3: (a) The prior and posterior distributions for each parameter and (b) Principal Component Analysis (PCA) of "model checking" for the scenario 1 in ABC1 obtained for Pop1 using DIYABC. Priors and posterior probability densities are shown in the Y-axis., Figure S4: (a) The prior and posterior distributions for each parameter and (b) Principal Component Analysis (PCA) of "model checking" for the scenario 1 in ABC1 obtained for Pop2 using DIYABC. Priors and posterior probability densities are shown in the Y-axis., Figure S5: (a) The prior and posterior distributions for each parameter and (b) Principal Component Analysis (PCA) of "model checking" for the scenario 2 in ABC1 obtained for Pop3 using DIYABC. Priors and posterior probability densities are shown in the Y-axis. Figure S6: (a) The prior and posterior distributions for each parameter and (b) Principal Component Analysis (PCA) of "model checking" for the scenario 1 in ABC2 using DIYABC. Priors and posterior probability densities are shown in the Y-axis. Figure S7: Geographical distributions of the five cpDNA haplotypes of *Xylocarpus moluccensis*. The size of the pie charts indicates sample size of the population. Population ID are defined in Table 1. The inset (bottom) denotes the haplotype network of XM1–5 extracted from Figure 2. Table S1: Prior distributions of the parameters used in ABC1, Table S2: Prior distributions of the parameters used in ABC2, Table S3: Demographic parameters of the most likely scenario in each region in ABC1 obtained by DIYABC, Table S4: Comparison of summary statistics for the observed data set and posterior simulated data sets in the most likely scenario in each region in ABC1, Table S5: Demographic parameters of the scenario 1 in ABC2 obtained by DIYABC, Table S6: Comparison of summary statistics for the observed data set and posterior simulated data sets in the scenario 2 in ABC2.

Acknowledgments: The authors thank Sankararamasubramanian Halasya Meenakshisundaram, Myint Aung, Ian Cowie, Sanjay Deshmukh, Norman Duke, Kyaw Kyaw Khaung, Jurgenne Primavera, Vando Márcio da Silva, Norhaslinda Malekal, Sengo Murakami, Hoho Takayama, Masaru Banba, Sabah Forestry Department (SFD), M. S. Swaminathan Research Foundation, and the Department of Environment and Natural Resources, Region VI for participation in fieldwork to collect materials. This work was supported by JSPS KAKENHI 22405005, 25290080 and 17H01414 to TK, JSPS JENESYS Programme 2009 and 2011 to the Graduate School of Science of Chiba University (coordinated by T.K.), TBRC Joint Usage Project Grant and the Collaborative Research Grant 2016 of the Tropical Biosphere Research Center, University of the Ryukyus (to T.K. and to Y.Ts.). This study is a part of the MSc dissertation of Y.To. at Chiba University. JSPS KAKENHI 17H01414 (to T.K.) covered the costs to publish this work in open access.

Author Contributions: Y.To., Y.Ts., K.T. and T.K. designed the study; Y.To., M.N.S. and K.T. conducted experiments; Y.To., Y.Ts., T.Y. and T.K. performed analyses; Y.To., M.N.S., A.K.S.W., K.T., O.B.Y., S.G.S., S.S., B.A., E.A., M.S., N.X.T., K.K.S., K.K., T.A., Y.W., S.B. and T.K. contributed to sampling; Y.Ts., M.N.S., A.K.S., K.T., T.Y. and T.K. wrote the paper.

Conflicts of Interest: The authors have no conflicts of interest to declare.

References

1. Spalding, M. *World Atlas of Mangroves*; Routledge: Abingdon, UK, 2010.
2. Duke, N.C.; Meynecke, J.-O.; Dittmann, S.; Ellison, A.M.; Anger, K.; Berger, U.; Cannicci, S.; Diele, K.; Ewel, K.C.; Field, C.D.; et al. A World without Mangroves? *Science* **2007**, *317*, 41–43. [CrossRef] [PubMed]
3. Polidoro, B.A.; Carpenter, K.E.; Collins, L.; Duke, N.C.; Ellison, A.M.; Ellison, J.C.; Farnsworth, E.J.; Fernando, E.S.; Kathiresan, K.; Koedam, N.E.; et al. The loss of species: Mangrove extinction risk and geographic areas of global concern. *PLoS ONE* **2010**, *5*, e10095. [CrossRef] [PubMed]

4. Richards, D.R.; Friess, D.A. Rates and drivers of mangrove deforestation in Southeast Asia, 2000–2012. *Proc. Natl. Acad. Sci. USA* **2016**, *113*, 344–349. [CrossRef] [PubMed]

5. Gilman, E.L.; Ellison, J.; Duke, N.C.; Field, C. Threats to mangroves from climate change and adaptation options: A review. *Aquat. Bot.* **2008**, *89*, 237–250. [CrossRef]

6. Duke, N.C.; Ball, M.C.; Ellison, J.C. Factors influencing biodiversity and distributional gradients in mangroves. *Glob. Ecol. Biogeogr. Lett.* **1998**, *7*, 27–47. [CrossRef]

7. Liao, P.-C.; Havanond, S.; Huang, S. Phylogeography of *Ceriops tagal* (Rhizophoraceae) in Southeast Asia: The land barrier of the Malay Peninsula has caused population differentiation between the Indian Ocean and South China Sea. *Conserv. Genet.* **2007**, *8*, 89–98. [CrossRef]

8. Minobe, S.; Fukui, S.; Saiki, R.; Kajita, T.; Changtragoon, S.; Ab Shukor, N.A.; Latiff, A.; Ramesh, B.R.; Koizumi, O.; Yamazaki, T. Highly differentiated population structure of a Mangrove species, *Bruguiera gymnorhiza* (Rhizophoraceae) revealed by one nuclear *GapCp* and one chloroplast intergenic spacer *trnF-trnL*. *Conserv. Genet.* **2010**, *11*, 301–310. [CrossRef]

9. Urashi, C.; Teshima, K.M.; Minobe, S.; Koizumi, O.; Inomata, N. Inferences of evolutionary history of a widely distributed mangrove species, *Bruguiera gymnorrhiza*, in the Indo-West Pacific region. *Ecol. Evol.* **2013**, *3*, 2251–2261. [CrossRef] [PubMed]

10. Takayama, K.; Tamura, M.; Tateishi, Y.; Webb, E.L.; Kajita, T. Strong genetic structure over the American continents and transoceanic dispersal in the mangrove genus *Rhizophora* (Rhizophoraceae) revealed by broad-scale nuclear and chloroplast DNA analysis. *Am. J. Bot.* **2013**, *100*, 1191–1201. [CrossRef] [PubMed]

11. Ng, W.L.; Onishi, Y.; Inomata, N.; Teshima, K.M.; Chan, H.T.; Baba, S.; Changtragoon, S.; Siregar, I.Z.; Szmidt, A.E. Closely related and sympatric but not all the same: Genetic variation of Indo-West Pacific *Rhizophora* mangroves across the Malay Peninsula. *Conserv. Genet.* **2015**, *16*, 137–150. [CrossRef]

12. Wee, A.K.S.; Takayama, K.; Chua, J.L.; Asakawa, T.; Meenakshisundaram, S.H.; Onrizal, B.A.; Ardli, E.R.; Sungkaew, S.; Malekal, N.B.; Tung, N.X.; et al. Genetic differentiation and phylogeography of partially sympatric species complex *Rhizophora mucronata* Lam. and *R. stylosa* Griff. using SSR markers. *BMC Evol. Biol.* **2015**, *15*. [CrossRef] [PubMed]

13. Huang, Y.; Tan, F.; Su, G.; Deng, S.; He, H.; Shi, S. Population genetic structure of three tree species in the mangrove genus *Ceriops* (Rhizophoraceae) from the Indo West Pacific. *Genetica* **2008**, *133*, 47–56. [CrossRef] [PubMed]

14. Ono, J.; Tsuda, Y.; Wee, A.K.S.; Takayama, K.; Onrizal, B.A.; Ardli, E.R.; Sungkaew, S.; Suleiman, M.; Tung, N.X.; Salmo, S.G., III; et al. Genetic structure of a widespread mangrove species *Bruguiera gymnorhiza* suggests limited gene flow by sea-dispersal in oceanic regions. Unpublished work. 2017.

15. Budde, K.B.; González-Martínez, S.C.; Hardy, O.J.; Heuertz, M. The ancient tropical rainforest tree *Symphonia globulifera* L. f. (Clusiaceae) was not restricted to postulated Pleistocene refugia in Atlantic Equatorial Africa. *Heredity* **2013**, *111*, 66–76. [CrossRef] [PubMed]

16. Tsuda, Y.; Nakao, K.; Ide, Y.; Tsumura, Y. The population demography of *Betula maximowicziana*, a cool-temperate tree species in Japan, in relation to the last glacial period: Its admixture-like genetic structure is the result of simple population splitting not admixing. *Mol. Ecol.* **2015**, *24*, 1403–1418. [CrossRef] [PubMed]

17. Gutiérrez-Ortega, J.S.; Yamamoto, T.; Vovides, A.P.; Pérez-Farrera, A.M.; Martínez, J.F.; Molina-Freaner, F.; Watano, Y.; Kajita, T. Aridification as a driver of biodiversity: A case study for the cycad genus *Dioon* (Zamiaceae). *Ann. Bot.* **2017**. [CrossRef] [PubMed]

18. Bagnoli, F.; Tsuda, Y.; Fineschi, S.; Bruschi, P.; Magri, D.; Zhelev, P.; Paule, L.; Simeone, M.C.; González-Martínez, S.C.; Vendramin, G.G. Combining molecular and fossil data to infer demographic history of *Quercus cerris*: Insights on European eastern glacial refugia. *J. Biogeogr.* **2016**, *43*, 679–690. [CrossRef]

19. Soliani, C.; Tsuda, Y.; Bagnoli, F.; Gallo, L.A.; Vendramin, G.G.; Marchelli, P. Halfway encounters: Meeting points of colonization routes among the southern beeches *Nothofagus pumilio* and *N. antarctica*. *Mol. Phylogenet. Evol.* **2015**, *85*, 197–207. [CrossRef] [PubMed]

20. Tsuda, Y.; Semerikov, V.; Sebastiani, F.; Vendramin, G.G.; Lascoux, M. Multispecies genetic structure and hybridization in the *Betula* genus across Eurasia. *Mol. Ecol.* **2017**, *26*, 589–605. [CrossRef] [PubMed]

21. Mabberley, D.J.; Pannell, C.M.; Sing, A.M. *Flora Malesiana: Series I. Spermatophyta Volume 12, Part 1, 1995. Meliaceae*; Rijksherbarium, Foundation Flora Malesiana: Leiden, The Netherlands, 1995; ISBN 9071236269.

22. Doyle, J.J.; Doyle, J.L. A rapid DNA isolation procedure for small quantities of fresh leaf tissue. *Phytochem. Bull.* **1987**, *19*, 11–15.

23. Demesure, B.; Sodzi, N.; Petit, R.J. A set of universal primers for amplification of polymorphic non-coding regions of mitochondrial and chloroplast DNA in plants. *Mol. Ecol.* **1995**, *4*, 129–131. [CrossRef] [PubMed]

24. Taberlet, P.; Gielly, L.; Pautou, G.; Bouvet, J. Universal primers for amplification of three non-coding regions of chloroplast DNA. *Plant Mol. Biol.* **1991**, *17*, 1105–1109. [CrossRef] [PubMed]

25. Small, R.L.; Ryburn, J.A.; Cronn, R.C.; Seelanan, T.; Wendel, J.F. The tortoise and the hare: Choosing between noncoding plastome and nuclear *Adh* sequences for phylogeny reconstruction in a recently diverged plant group. *Am. J. Bot.* **1998**, *85*, 1301–1315. [CrossRef] [PubMed]

26. Thompson, J.D.; Gibson, T.J.; Higgins, D.G. Multiple sequence alignment using ClustalW and ClustalX. *Curr. Protoc. Bioinform.* **2002**. [CrossRef]

27. Tamura, K.; Peterson, D.; Peterson, N.; Stecher, G.; Nei, M.; Kumar, S. MEGA5: Molecular evolutionary genetics analysis using maximum likelihood, evolutionary distance, and maximum parsimony methods. *Mol. Biol. Evol.* **2011**, *28*, 2731–2739. [CrossRef] [PubMed]

28. Clement, M.; Posada, D.; Crandall, K.A. TCS: A computer program to estimate gene genealogies. *Mol. Ecol.* **2000**, *9*, 1657–1659. [CrossRef] [PubMed]

29. Weir, B.S.; Cockerham, C.C. Estimating *F*-statistics for the analysis of population structure. *Evolution* **1984**, *38*, 1358–1370. [CrossRef] [PubMed]

30. Meirmans, P.G.; Hedrick, P.W. Assessing population structure: F_{ST} and related measures. *Mol. Ecol. Resour.* **2011**, *11*, 5–18. [CrossRef] [PubMed]

31. Peakall, R.; Smouse, P.E. GenAlEx 6.5: Genetic analysis in Excel. Population genetic software for teaching and research-an update. *Bioinformatics* **2012**, *28*, 2537–2539. [CrossRef] [PubMed]

32. Tomizawa, Y.; Shinmura, Y.; Wee, A.K.S.; Takayama, K.; Asakawa, T.; Yllano, O.B.; Salmo, S.G., III; Ardli, E.R.; Tung, N.X.; Binti Malekal, N.; et al. Development of 11 polymorphic microsatellite markers for *Xylocarpus granatum* (Meliaceae) using next-generation sequencing technology. *Conserv. Genet. Resour.* **2013**, *5*, 1159–1162. [CrossRef]

33. Excoffier, L.; Smouse, P.E.; Quattro, J.M. Analysis of molecular variance inferred from metric distances among DNA haplotypes: Application to human mitochondrial DNA restriction data. *Genetics* **1992**, *131*, 479–491. [CrossRef] [PubMed]

34. Nei, M.; Tajima, F.; Tateno, Y. Accuracy of estimated phylogenetic trees from molecular data. *J. Mol. Evol.* **1983**, *19*, 153–170. [CrossRef] [PubMed]

35. Langella, O. Populations 1.2.30: Population Genetic Software (Individuals or Population Distances, Phylogenetic Trees). 2007. Available online: http://bioinformatics.org/~tryphon/populations/ (accessed on 25 July 2017).

36. Parks, D.H.; Porter, M.; Churcher, S.; Wang, S.; Blouin, C.; Whalley, J.; Brooks, S.; Beiko, R.G. GenGIS: A geospatial information system for genomic data. *Genome Res.* **2009**, *19*, 1896–1904. [CrossRef] [PubMed]

37. Pritchard, J.K.; Stephens, M.; Donnelly, P. Inference of population structure using multilocus genotype data. *Genetics* **2000**, *155*, 945–959. [CrossRef] [PubMed]

38. Falush, D.; Stephens, M.; Pritchard, J.K. Inference of population structure using multilocus genotype data: Linked loci and correlated allele frequencies. *Genetics* **2003**, *164*, 1567–1587. [CrossRef] [PubMed]

39. Earl, D.A.; von Holdt, B.M. STRUCTURE HARVESTER: A website and program for visualizing STRUCTURE output and implementing the Evanno method. *Conserv. Genet. Resour.* **2012**, *4*, 359–361. [CrossRef]

40. Evanno, G.; Regnaut, S.; Goudet, J. Detecting the number of clusters of individuals using the software STRUCTURE: A simulation study. *Mol. Ecol.* **2005**, *14*, 2611–2620. [CrossRef] [PubMed]

41. Kopelman, N.M.; Mayzel, J.; Jakobsson, M.; Rosenberg, N.A.; Mayrose, I. CLUMPAK: A program for identifying clustering modes and packaging population structure inferences across K. *Mol. Ecol. Resour.* **2015**, *15*, 1179–1191. [CrossRef] [PubMed]

42. Cornuet, J.-M.; Santos, F.; Beaumont, M.A.; Robert, C.P.; Marin, J.-M.; Balding, D.J.; Guillemaud, T.; Estoup, A. Inferring population history with DIY ABC: A user-friendly approach to approximate Bayesian computation. *Bioinformatics* **2008**, *24*, 2713–2719. [CrossRef] [PubMed]

43. Cornuet, J.-M.; Pudlo, P.; Veyssier, J.; Dehne-Garcia, A.; Gautier, M.; Leblois, R.; Marin, J.-M.; Estoup, A. DIYABC v2.0: A software to make approximate Bayesian computation inferences about population history using single nucleotide polymorphism, DNA sequence and microsatellite data. *Bioinformatics* **2014**, *30*, 1187–1189. [CrossRef] [PubMed]

44. Estoup, A.; Jarne, P.; Cornuet, J.-M. Homoplasy and mutation model at microsatellite loci and their consequences for population genetics analysis. *Mol. Ecol.* **2002**, *11*, 1591–1604. [CrossRef] [PubMed]

45. Garza, J.C.; Williamson, E.G. Detection of reduction in population size using data from microsatellite loci. *Mol. Ecol.* **2001**, *10*, 305–318. [CrossRef] [PubMed]

46. Goldstein, D.B.; Ruiz Linares, A.; Cavalli-Sforza, L.L.; Feldman, M.W. Genetic absolute dating based on microsatellites and the origin of modern humans. *Proc. Natl. Acad. Sci. USA* **1995**, *92*, 6723–6727. [CrossRef] [PubMed]

47. Wee, A.K.S.; Teo, J.X.H.; Chua, J.L.; Takayama, K.; Asakawa, T.; Meenakshisundaram, S.H.; Onrizal, B.A.; Ardli, E.R.; Sungkaew, S.; Suleiman, M.; et al. Vicariance, oceanic barriers and geographic distance drive contemporary genetic structure of widespread mangrove species *Sonneratia alba* in the Indo-West Pacific. *Forests* **2017**, in press.

48. Yamamoto, T.; Tsuda, Y.; Takayama, K.; Nagashima, R.; Tateishi, Y.; Kajita, T. The presence of a cryptic barrier in the West Pacific Ocean suggests the effect of glacial climate changes on a widespread sea-dispersed plant, *Vigna marina* (Fabaceae). *Ecol. Evol.* **2017**. under review.

49. Harwell, M.C.; Orth, R.J. Long-distance dispersal potential in a marine macrophyte. *Ecology* **2002**, *83*, 3319–3330. [CrossRef]

50. Takayama, K.; Kajita, T.; Murata, J.; Tateishi, Y. Phylogeography and genetic structure of *Hibiscus tiliaceus*—speciation of a pantropical plant with sea-drifted seeds. *Mol. Ecol.* **2006**, *15*, 2871–2881. [CrossRef] [PubMed]

51. Takayama, K.; Tateishi, Y.; Murata, J.; Kajita, T. Gene flow and population subdivision in a pantropical plant with sea-drifted seeds *Hibiscus tiliaceus* and its allied species: Evidence from microsatellite analyses. *Mol. Ecol.* **2008**, *17*, 2730–2742. [CrossRef] [PubMed]

52. Miryeganeh, M.; Takayama, K.; Tateishi, Y.; Kajita, T. Long-distance dispersal by sea-drifted seeds has maintained the global distribution of *Ipomoea pes-caprae* subsp. *brasiliensis* (Convolvulaceae). *PLoS ONE* **2014**, *9*, e91836. [CrossRef] [PubMed]

53. Voris, H.K. Maps of Pleistocene sea levels in Southeast Asia: Shorelines, river systems and time durations. *J. Biogeogr.* **2000**, *27*, 1153–1167. [CrossRef]

54. Wyrtki, K. *Physical Oceanography of the Southeast Asian Waters*; Scientific Results of Marine Investigations of the South China Sea and the Gulf of Thailand; University of California, Scripps Institution of Oceanography: San Diego, CA, USA, 1961; Volume 2, p. 195. [CrossRef]

55. Magri, D.; Vendramin, G.G.; Comps, B.; Dupanloup, I.; Geburek, T.; Gömöry, D.; Latałowa, M.; Litt, T.; Paule, L.; Roure, J.M.; et al. A new scenario for the Quaternary history of European beech populations: Palaeobotanical evidence and genetic consequences. *New Phytol.* **2006**, *171*, 199–221. [CrossRef] [PubMed]

56. Tsuda, Y.; Ide, Y. Chloroplast DNA phylogeography of *Betula maximowicziana*, a long-lived pioneer tree species and noble hardwood in Japan. *J. Plant Res.* **2010**, *123*, 343–353. [CrossRef] [PubMed]

57. Triest, L. Molecular ecology and biogeography of mangrove trees towards conceptual insights on gene flow and barriers: A review. *Aquat. Bot.* **2008**, *89*, 138–154. [CrossRef]

58. Tomlinson, P.B. *The Botany of Mangroves*; Cambridge University Press: Cambridge, UK, 1986.

© 2017 by the authors. Licensee MDPI, Basel, Switzerland. This article is an open access article distributed under the terms and conditions of the Creative Commons Attribution (CC BY) license (http://creativecommons.org/licenses/by/4.0/).

![forests logo] *forests*

MDPI

Article

Intraspecific Variation in Pines from the Trans-Mexican Volcanic Belt Grown under Two Watering Regimes: Implications for Management of Genetic Resources

Andrés Flores [1,2], José Climent [2,3], Valentín Pando [2,4], Javier López-Upton [5] and Ricardo Alía [2,3,*]

[1] CENID-COMEF, National Institute for Forestry, Agriculture and Livestock Research, Progreso 5, Coyoacán 04010, Mexico; foga12@gmail.com
[2] Research Institute on Sustainable Forest Management, University of Valladolid, Av. Madrid s/n, 34004 Palencia, Spain; climent@inia.es (J.C.); vpando@uva.es (V.P.)
[3] INIA-CIFOR, Department of Ecology and Forest Genetics, Ctra. Coruña km 7.5, 28040 Madrid, Spain
[4] Statistics Department, University of Valladolid, Av. Madrid s/n, 34004 Palencia, Spain
[5] Colegio de Postgraduados, Postgrado en Ciencias Forestales, Km 36.5 Carr. Mexico-Texcoco, Montecillo 56230, Mexico; jlopezupton@gmail.com
* Correspondence: alia@inia.es; Tel.: +34-913-473-959

Received: 19 October 2017; Accepted: 26 January 2018; Published: 30 January 2018

Abstract: Management of forest genetic resources requires experimental data related to the genetic variation of the species and populations under different climatic conditions. Foresters also demand to know how the main selective drivers will influence the adaptability of the genetic resources. To assess the inter- and intraspecific variation and plasticity in seedling drought tolerance at a relevant genetic resource management scale, we tested the changes in growth and biomass allocation of seedlings of *Pinus oocarpa*, *P. patula* and *P. pseudostrobus* under two contrasting watering regimes. We found general significant intraspecific variation and intraspecific differences in plasticity, since both population and watering by population interaction were significant for all three species. All the species and populations share a common general avoidance mechanism (allometric adjustment of shoot/root biomass). However, the intraspecific variation and differences in phenotypic plasticity among populations modify the adaptation strategies of the species to drought. Some of the differences are related to the climatic conditions of the location of origin. We confirmed that even at reduced geographical scales, Mexican pines present differences in the response to water stress. The differences among species and populations are relevant in afforestation programs as well as in genetic conservation activities.

Keywords: drought stress; genetic variation; early testing; adaptive variation; genecology; phenotypic plasticity

1. Introduction

In the last decades, there has been an increasing concern about the consequences of climate change on the future distribution and productivity of forest species. Many forest areas have experienced a decrease in rainfall and a subsequent increase in drought severity. In particular, Mexico will experience, on average, an increase of 1.5 °C in mean annual temperature, and a decrease of 6.7% in annual precipitation by 2030 [1]. This is already posing practical problems in the management of many forest tree species, derived from the shifts in species distribution [2], and the future requirements in terms of adaptation and productivity.

We are far from having enough experimental data to address important aspects related to the adaptability, i.e., the potential or ability of a population to adapt to changes in environmental conditions through changes in its genetic structure [3]. For example we lack information about the roles of genetic variation, phenotypic plasticity [4], and of phenotype changes derived from the trade-offs among life-history traits, among others [5]. This information is essential at scales that are meaningful for forest management (i.e., at forest or forest-landscape scales), as it is necessary to make decisions when selecting the species and the basic material to use in afforestation and restoration programs (e.g., local vs. non-local), or to suggest changes in silvicultural systems (e.g., regeneration methods, selection of parent trees) to increase forest resilience. Therefore, the evaluation of local genetic resources at fine scales is essential for the management of local genetic resources, complementing information at larger scales.

Low water availability has been identified, particularly in conifers, as one of the major abiotic stressors, conducive to stomatal closure, reduced photosynthesis and death due to carbon starvation [6]. Tolerance to low water availability is an important selective factor, involving quite different traits, such as rooting depth, transpiration area of leaves and shoots, and size and number of shoots [7]. There are, therefore, important adaptive differences in the response at different levels, from species to individuals [8,9].

Intra-specific genetic variation is crucial in forest trees species, which must endure both abiotic and biotic stressors for long periods of time [10]. Particularly, it is necessary to develop management options for the genetic resources of target species, and to determine if genotypes would be able to grow efficiently under future stressful conditions. However, testing drought-tolerant genotypes amongst mature trees growing in the field is cumbersome, due to the previously mentioned complexity of plant responses to drought and the lack of control of watering treatments [11]. An alternative approach is to develop controlled experimental conditions to test genotypes at early stages [12]. Early developmental stages in plants are the most critical in the survival of forest trees, and are related to the future adaptability of the species [13] depending on the genetic intraspecific variation in these genetic traits. Evaluating morphological and physiological changes in response to low water availability at early ages is a recognized way to know their adaptive responses (i.e., leaf water potential and gas exchange [14] and changes in growth and survival [15]). Inter- and intraspecific variations among populations of different pines species, when cultivated under contrasting water availability, reveal high population divergence for phenotypic changes and marked allocational shifts, a plastic response [16]. Moreover, different works have addressed some of the features involved in the growth process that can skew the results of early testing in plants, e.g., pot size, water quality and salinity [17,18].

Pinus is the largest genus of the *Pinaceae* family, with 114 species widely distributed in the Northern Hemisphere [19]. Mexico presents the highest specific diversity (46 species), with contrasting geographical and intraspecific genetic patterns, as a result of adaptive responses to climate changes in the past [20]. Among them, *Pinus oocarpa* Schiede ex Schltdl., *P. patula* Schiede ex Schltdl. & Cham. and *P. pseudostrobus* Lindl. are three economically important Mexican pines, used in highly productive forest plantations established in the tropics and subtropics [21]. These pines occupy diverse habitats in the country, and present a variety of ecological roles and life histories. Specifically, the Trans-Mexican Volcanic Belt (TMVB) covers a wide range of environments differing in altitude, precipitation, temperature and soil. Thus, the Volcanic Belt constitutes a good model area to check for intraspecific differences in growth and performance to drought stress in Mexican pines, as a way to improve our recommendations for the management of genetic resources under climate change scenarios.

The objective of this study was to assess the inter- and intraspecific genetic variation in seedling drought tolerance in *Pinus oocarpa*, *P. patula* and *P. pseudostrobus* from the TMVB. We tested the seedlings under two contrasted controlled watering regimes and we measured different adaptive morphological and allocation traits. Our hypotheses were that at a fine geographical scale: (i) both seedling growth and biomass are affected by low water availability, a potential adaptive response and (ii) these responses differ due to species intraspecific variation.

This information is essential to implement breeding and conservation programs under climate change scenarios.

2. Materials and Methods

2.1. Plant Material

The natural distribution of the study species covers small and large areas throughout the north, center and south of Mexico. *Pinus oocarpa* (OC) and *P. pseudostrobus* (PS) are usually found in fragmented mixed stands, while *P. patula* (PA) occurs in pure stands. We sampled populations of the three species from the TMVB. The number of sampling sites (populations) was different for each species (Figure 1 and Table 1): *P. oocarpa*, two populations (OC01, OC02), *P. patula*, 10 populations (PA01, PA02, PA03, PA04, PA06, PA07, PA08, PA09, PA11, PA12), and *P. pseudostrobus*, five populations (PS01 to PS05). Seedlots were either samples provided by academic institutions (15 out of 17 samples) or commercial seedlots provided by a private seed supplier (two out of 17), and were composed of seeds from at least 20 mother trees per population. The sampling for *P. oocarpa* was limited to areas where the taxonomic identification of the species was clear, to avoid biases in the comparisons. This is particularly important in the eastern area of the study, where three new species have been recently described but assigned to *P. oocarpa* by the National Forest Inventory.

2.2. Experiment Description and Experimental Design

Three hundred seeds per population were sown in trays containing moistened rock wool and covered with plastic film (see Appendix A for details in the experimental set-up). Trays were placed inside a germination chamber at $25 \pm 1\ ^{\circ}C$, $60 \pm 5\%$ relative humidity and an eight-hour photoperiod. The germination was recorded three times a week and then used to calculate the germination curve parameters (total germination in %, speed) based on a sample of 60 seeds per population. Germination for the three species started at three days. *P. oocarpa* and *P. pseudostrobus* had a higher germination rate than *P. patula* (Supplementary information Figure S1).

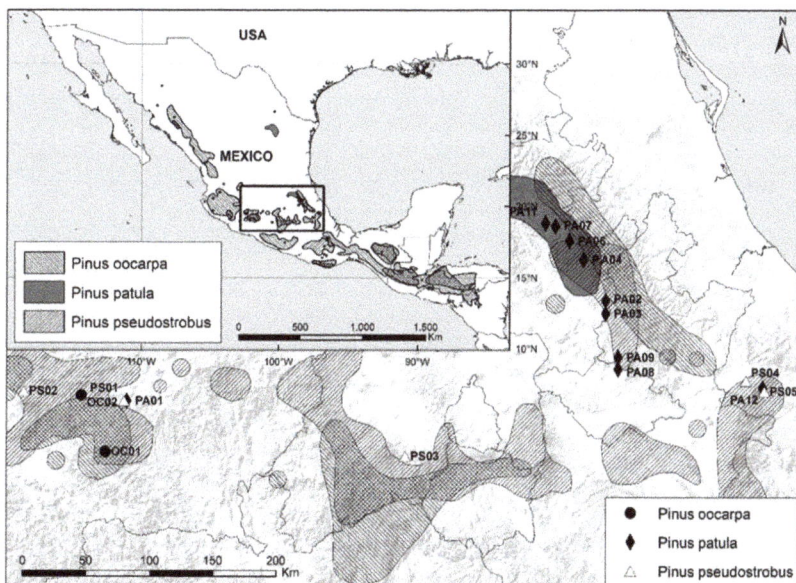

Figure 1. Location of sampled populations and distribution range of the species [19] for *Pinus oocarpa*, *P. patula* and *P. pseudostrobus*.

Table 1. Location and general characteristics of the *Pinus* spp. populations sampled in Mexico.

Code [1]	Population, State	Supplier [2]	Latitude and Longitude	Altitude (m)	MAT [3] (°C)	MAP [4] (mm)
OC01	Ario de Rosales, Mich.	INIFAP	19°04′/101°44′	1490	20.7	1112
OC02	San Ángel Zurumucapio, Mich.	INIFAP	19°27′/101°54′	1700	17.0	1299
	Range [5]				13.8 to 21.3	891 to 1422
PA01	Casas Blancas, Mich.	Colpos	19°25′/101°35′	2258	15.7	1060
PA02	Acaxochitlán, Hgo.	Colpos	20°06′/98°12′	2190	13.8	962
PA03	Ahuazotepec, Pue.	Colpos	20°01′/98°12′	2250	13.8	847
PA04	Apulco, Hgo.	Colpos	20°23′/98°22′	2200	15.2	909
PA06	Huayacocotla, Ver.	Colpos	20°31′/98°28′	2050	16.1	1099
PA07	Tlahuelompa, Hgo.	Colpos	20°37′/98°34′	2020	16.2	1234
PA08	Tlaxco, Tlax.	Colpos	19°38′/98°07′	2800	12.1	764
PA09	Villa Cuauhtémoc, Pue.	Colpos	19°43′/98°07′	2720	12.4	730
PA11	Zacualtipán, Hgo.	Colpos	20°38′/98°38′	2030	16.1	1199
PA12	Xico, Ver.	Asoc. For.	19°30′/97°05′	2839	11.5	1019
	Range				11.1 to 17.7	615 to 1223
PS01	Casas Blancas, Mich.	INIFAP	19°25′/101°36′	2244	15.7	1054
PS02	Nu. San Juan Parangaricutiro, Mich.	INIFAP	19°29′/102°19′	2245	15.2	1173
PS03	Tenango del Valle, Ver.	INIFAP	19°02′/99°37′	2990	11.3	1156
PS04	Perote, Ver.	Colpos	19°33′/97°12′	3200	9.5	1322
PS05	Xico, Veracruz.	Asoc. For.	19°30′/97°05′	2839	11.5	1019
	Range				9.0 to 16.9	717 to 1415

[1] OC: *Pinus oocarpa*; PA: *P. patula*; PS: *P. pseudostrobus*; [2] INIFAP: National Institute for Forestry, Agricultural and Livestock Research, Michoacán; Colpos.: Colegio de Postgraduados en Ciencias Agrícolas; Asoc. For.: Asociación Forestal Especializada AC.; [3] MAT = Mean annual temperature; [4] MAP = Mean annual precipitation; [5] Range: MAT and MAP ranges in the Trans-Mexican Volcanic Belt (TMVB) region. All values calculated with ANUSPLIN software [1,22].

We transplanted 50 seedlings into individual plastic containers, except for three *P. patula* populations (PA02, PA07 and PA08) that had a low germination rate, for which we transplanted, respectively, 38, 26 and 35 seeds. The total number of seedlings used was 786. We used individual plastic containers with a mixture of peat moss and vermiculite substrate (3:1 v/v) whose size was big enough (250 cm^3) to avoid root restriction, given the short duration of the experiment [17]. The trial was established in a greenhouse under controlled conditions (Appendix A). The trial was set up with a randomized complete block design, with five seedlings per experimental unit, and five blocks in each of the two watering treatments. Seedlings were maintained in a slow-growth phase over 135 days from November to March to allow the material to harden. Then plants were cultivated in a normal-growing phase (April to June). Fifty seedlings per population were submitted to two watering treatments during 90 days (25 seedlings per watering regime): Field Capacity (FC) and Drought-Stress (DS). For those populations with lower seed germination rate we set an equal number of seedlings per treatment (PA02, PA07 and PA08). The watering regimes were based on the mean saturation level of the substrate: 90–100% on FC and 35–45% on DS treatments. We determined the amount of water for each watering event every two days by weighing plants randomly chosen from each treatment.

2.3. Variables Measured

We periodically recorded the survival, height (mm) and ontogenetic stage of all seedlings [16]. Species were in the epicotyl elongation and formation of axillary buds phase at the beginning of the experimental phase, and had dwarf shoots by the end of it (Appendix A). We obtained the height growth increment (HG in mm) during the watering experiment as the difference between height at the beginning and the end of the watering experiment (H_t-H_0). At the end of the experiment (90 days of watering treatment, 225 days old), all plants were harvested and partitioned in roots, stems, and leaves. They were dried (65 °C/72 h) and weighed (g, ±0.01) [23] to assess total dry mass (TDM in mg) and that of its components: roots, stems and needles (RDM, SDM, and NDM, respectively, in mg). The root mass fraction (RMF, root dry mass to total dry mass), stem mass fraction (SMF, stem dry mass to total dry mass) and needle mass fraction (NMF, needle dry mass to total dry mass) were also computed. The specific leaf area (SLA in cm^2/g) was estimated from 10 needles randomly chosen from each plant [13].

2.4. Data Analysis

2.4.1. Seedling Survival

For seedling survival, we performed a logistic regression analysis using a maximum likelihood method:

$$p_{ik(j)} = 1/[1 + \exp(-z_{ik(j)})] \tag{1}$$

$$z_{ik(j)} = \log[p_{ik(j)}/(1 - p_{ik(j)})] = \mu + W_i + S_j + P_{k(j)} \tag{2}$$

where $p_{ik(l)}$ is the survival probability in the ith watering regime of the kth population within the jth species; $z_{ik(j)}$ is the logit estimation in the ith watering treatment of the kth population within the jth species; μ is the grand mean; W_i is the effect of the ith watering regime (1 to 2); S_j is the effect of the jth species (1 to 3); and $P_{k(j)}$ is the effect of the kth population within the jth species (1 to 10). The WS_{ij} interaction was not included in the model due to its lack of significance.

2.4.2. Mixed Model

For the other variables, we conducted an inter-species variance analysis according to the following mixed model:

$$y_{ijkl} = \mu + W_i + S_j + WS_{ij} + PS_{k(j)} + BW_{l(i)} + c\, x_{ijkl} + \varepsilon_{ijkl}, \tag{3}$$

where y_{ijkl} is the value of observation in the lth block of the kth population of the jth species in the ith watering treatment; μ is the general mean; W_i is the fixed effect of the ith watering treatment (1–2); S_j is the fixed effect of the jth species (1–3); WS_{ij} is the interaction fixed effect of the ith treatment with the jth species; $PS_{k(j)}$ is the random effect of the kth population within the jth species; $BW_{l(i)}$ is the random effect of the lth block (1–5) within the ith treatment; c is the lineal effect of the covariate x_{ijkl} (seedling height at the beginning of the watering regimen); and ε_{ijkl} is the experimental error.

In order to examine the intra-species variation, a variance analysis was performed for each species, using the following model:

$$y_{ijk} = \mu + W_i + P_j + WP_{ij} + BW_{k(i)} + c\, x_{ijkl} + \varepsilon_{ijk}, \tag{4}$$

where y_{ijk} is the value of observation in the kth block of the jth population of the ith watering regime; μ is the general mean; W_i is the fixed effect of the ith treatment (1-2); P_j is the fixed effect of the jth population (2–10); WP_{ij} is the interaction fixed effect of ith treatment with the jth population; $BW_{k(i)}$ is the random effect of the kth block (1–5) within the ith treatment; c is the lineal effect of the covariate x_{ijk} (seedling height at the beginning of the watering regimen); and ε_{ijk} is the experimental error.

We analyzed the variation of dry masses and mass fractions including the initial height as a covariate to correct the bias due to differences in the initial growth [24]. Consequently, the experimental error of the models was reduced in each case.

2.4.3. Phenotypic Plasticity

For each species, we calculated the plasticity index of a trait due to drought stress effect as [25]:

$$PI = (V_1 - V_2)/V_1 \times 100 \tag{5}$$

where V_1 is the trait mean under the FC treatment; and V_2 is the trait mean under the DS treatment.

In species with significant treatment x population interaction (WP), a plasticity analysis for each population was conducted, plotting the mean value trait by population on a dimensional plane where the x-axis was the drought stress treatment (DS) and the y-axis was the field capacity treatment (FC) [26].

2.4.4. Allometric Analysis

We further used allometric analysis based in log-transformed data to study the changes in root dry mass compared to the sum of stem and needle dry mass. Differences between the two watering regimes in slopes and intercepts for the three species with their populations were assessed by parallelism test using watering regimes [23,27].

2.4.5. Factor Analysis

In order to display the overall performance of the populations tested, we performed, for *P. patula* and *P. pseudostrobus* (species with more than two populations), a factor analysis using a maximum likelihood method and a Varimax rotation to maximize the variation of factor loadings and to facilitate the interpretation of the factors. We used variables with highly significant differences in the watering treatment (model 1): HG, RDM, SDM, SMF and SLA. A Biplot using the values of the factors for the FC and DS treatments was considered for each population. A correlation coefficient was computed for the mean values of the populations of the two axes, the plasticity (differences among FC/DS treatments), and the altitude and rainfall.

All the statistical analyses were performed using SAS software (SAS Institute, Cary, NC, USA) [28].

2.5. Data Access

The data are to be stored in the Zenodo repository, and DOIs are being provided for the various datasets in the Supplementary Material section.

3. Results

3.1. Response to Watering Regimes

Water stress treatment significantly affected species survival, irrespective of seed origin (Table S1). Mortality (from the beginning to the end of the drought experiment) ranged from 30% for *Pinus oocarpa* to 4% for *P. pseudostrobus*, with *P. patula* offering an intermediate value, 12% (Table S2).

Watering produced significant differences for all three pine species in seedling phenotypic changes (Table 2 and Table S3). Most traits, with the exception of relative biomass allocation to roots (RMF) and needle biomass (NDM), showed distinct phenotypic changes (i.e., plasticity) in response to drought, indicating the importance of the watering treatment. We also found differences in the plastic responses of the species (species by watering interactions). Moreover, data confirmed a general significant intraspecific variation and intraspecific differences in plasticity, since both population and watering x population interaction were significant for all the three species.

Table 2. Mean squares and level of significance [1] in the inter-specific analysis estimated for all species for different functional traits in three Mexican pines.

Trait [2]	W	S	WxS	c	P(S)	B(W)
df	1	2	2	1	14	8
HG	800,680 **	174,440 **	31,469 **	-	13,167 **	5394 **
RDM	588,837 *	122,126 *	103,861 **	4,820,150 **	31,919 **	148,087 **
SDM	1,611,569 **	112,624 *	88,819 **	3,649,888 **	24,332 **	47,780 **
NDM	1,936,624 ns	3,957,069 **	168,257 ns	26,807,653 **	240,005 **	1,031,535 **
TDM	11,731,955 *	4,778,621 **	936,486 **	86,216,460 **	461,308 **	2,379,984 **
RMF	0.003 ns	0.055 *	0.010 *	0.022 **	0.011 **	0.024 **
SMF	0.214 **	0.198 **	0.025 **	0.020 **	0.013 **	0.007 **
NMF	0.167 **	0.467 **	0.032 **	0.086 **	0.024 **	0.014 **
SLA	33,146 **	72,346 **	1693 ns	33,228 **	7498 **	3819 **

[1] Mean squares, and Level of significance: ** significant differences ($p < 0.01$); * significant differences ($p < 0.05$); ns, not significant ($p < 0.05$). [2] HG: height growth increment; RDM: root dry mass; SDM: stem dry mass; NDM: needle dry mass; TDM: total dry mass; RMF: root mass fraction; SMF: stem mass fraction; NMF: needle mass fraction; SLA: specific leaf area. W: Watering. S: Species. WxS: Watering x Species interaction. c: Covariate: Initial height for all traits except for HG. P(S): Population within species. B(W): Block within treatment. df: degrees of freedom.

3.2. Allocation Patterns

Overall, regression models between root dry mass with stem plus needles dry mass, representing relative allocation to roots, had a positive relationship with low watering regime. For FC and DS, regression lines did not share a common trajectory ($p < 0.0001$) for all three species. However, intercepts were different for *P. patula* and *P. pseudostrobus* ($p < 0.0001$) but not for *P. oocarpa* ($p = 0.344$) (Figure 2, Table S4).

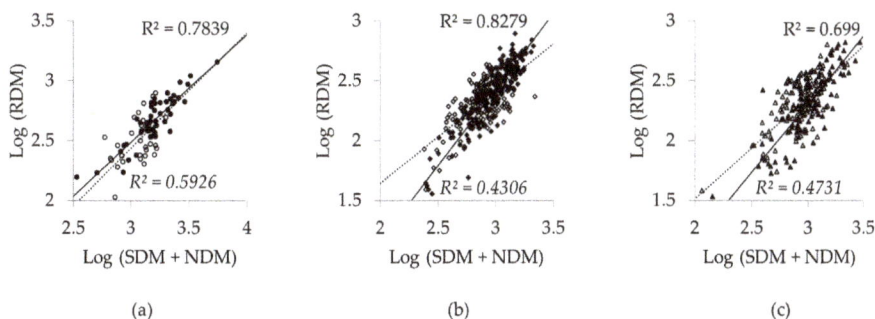

Figure 2. Allometric regression between RDM with SDM + NDM for field capacity (FC) and drought-stress (DS) regimes: (**a**) *P. oocarpa*, (**b**) *P. patula* and (**c**) *P. pseudostrobus*. Solid lines, filled symbols and R^2 are for FC treatment while dotted lines, empty symbols and *italics R^2* are for DS treatment.

3.3. Intraspecific Variation

Height growth increment significantly varied with watering treatment for all study species, the extent of the change significantly varying for *P. patula* and *P. pseudostrobus* populations, but not for *P. oocarpa*'s (Table 3). The more plastic traits were related to height growth increment, stem and needle biomass and specific leaf area (Table 4).

We found differences among populations in many of the analyzed traits, especially those related to the biomass components, but not for allocation fractions: stem and total biomass in *Pinus oocarpa*, height growth increment, total biomass and biomass components and specific leaf area in *P. patula*, and all the traits except stem biomass in *P. pseudostrobus*. For many of those traits that showed a significant population effect, significant differences in population phenotypic plasticity were detected, indicating differences among species and populations in response to drought, e.g., population phenotypic changes in stem and needle biomass in *Pinus oocarpa* and *P. patula*, or biomass allocation and specific leaf area in *P. pseudostrobus*.

The patterns of phenotypic plasticity among populations were quite contrasting depending on the trait (Figure 3). The height growth increment showed sharp differences in phenotypic plasticity for two of the species (*P. patula* and *P. pseudostrobus*), with a higher variation for the first species. It is interesting to notice that for SDM (Figure 3b), *Pinus patula* populations were quite homogeneous in allocating biomass to stems despite the differences in height, while the two *P. oocarpa* populations had quite different patterns. *P. pseudostrobus* populations showed differences in phenotypic plasticity for SMF and SLA, with populations PS05 and PS04 being the most interactive for the two traits (Figure 3d,e).

Table 3. Mean squares and level of significance [1] in the intra-specific analysis per species for different functional traits in three Mexican pines.

Trait [2]	W	P	WxP	c	B(W)
Pinus oocarpa					
df	1	1	1	1	8
HG	193,937 **	517 ns	3106 ns	-	4504 **
RDM	411,344 ns	122,364 *	71,980 ns	1,543,789 **	92,363 **
SDM	515,777 *	122,994 **	164,325 **	1,506,868 **	66,208 **
NDM	1,218,092 ns	385,868 ns	601,087 *	5,944,773 **	574,158 **
TDM	6,067,398 ns	1,747,874 *	2,099,484 **	24,086,763 **	1,575,038 **
RMF	0.008 ns	0.001 ns	0.004 ns	0.001 ns	0.010 ns
SMF	0.012 ns	0.002 ns	0.005 ns	0.007 *	0.003 *
NMF	0.040 *	0.005 ns	0.000 ns	0.013 ns	0.005 ns
SLA	2.156 ns	4 ns	19 ns	2077 ns	502 ns
P. patula					
df	1	9	9	1	8
HG	736,102 **	3460 **	5617 **	-	2882 *
RDM	329,833 ns	10,358 ns	16,031 *	1,319,554 **	94,360 **
SDM	1,239,102 **	14,035 **	16,812 **	691,007 **	16,274 **
NDM	584,235 ns	85,401 **	61,937 ns	9,253,456 **	436,834 **
TDM	6,153,728 *	214,427 *	225,448 **	25,668,546 **	1,086,400 **
RMF	0.006 ns	0.003 ns	0.002 ns	0.007 ns	0.019 **
SMF	0.372 **	0.004 *	0.003 ns	0.001 ns	0.007 **
NMF	0.280 **	0.004 ns	0.003 ns	0.003 ns	0.016 **
SLA	178,790 **	2949 **	676 ns	39,736 **	4007 **
P. pseudostrobus					
df	1	4	4	1	8
HG	173,685 **	34,691 **	3742 *	-	3135 *
RDM	25,352 ns	75,630 **	18,592 ns	1,940,741 **	18,583 ns
SDM	272,247 **	4556 ns	12,154 ns	1,109122 **	18,678 **
NDM	486,907 ns	665,570 **	175,298 ns	11,061,214 **	353,560 **
TDM	1,872,890 ns	924,259 **	304,728 ns	33,341,014 **	665,116 **
RMF	0.023 ns	0.035 **	0.007 *	0.054 **	0.006 **
SMF	0.075 **	0.022 **	0.006 *	0.037 **	0.003 ns
NMF	0.013 ns	0.057 **	0.008 ns	0.180 **	0.006 ns
SLA	64,286 **	11,042 **	2603 *	345 ns	1286 ns

[1] Mean Squares and level of significance: ** significant differences ($p < 0.01$); * significant differences ($p < 0.05$); ns, not significant ($p < 0.05$). [2] HG: height growth increment; RDM: root dry mass; SDM: stem dry mass; NDM: needle dry mass; TDM: total dry mass; RMF: root mass fraction; SMF: stem mass fraction; NMF: needle mass fraction; SLA: specific leaf area. W: Watering. P: Population. WxP: Watering x Population interaction. c: Covariate: Initial height for all traits except for HG. B(W): Block within treatment. df: degrees of freedom.

Table 4. Plasticity Index of the traits under the drought stress treatment at the species level in three Mexican pines (only variables with a Watering significant effect are included).

Trait [1]	*P. oocarpa*	*P. patula*	*P. pseudostrobus*
HG	73.74	57.60	60.08
SDM	43.03	48.94	35.04
TDM	-	24.11	-
SMF	-	32.60	26.88
NMF	−7.74	−10.09	-
SLA	-	20.30	19.92

[1] HG: height growth increment; RDM: root dry mass; SDM: stem dry mass; NDM: needle dry mass; TDM: total dry mass; RMF: root mass fraction; SMF: stem mass fraction; NMF: needle mass fraction; SLA: specific leaf area.

3.4. Phenotypic Variation of the Mexican Species Under Full Capacity and Drought Stress Treatments

The two first factors explained 86.09% of the total variation, with the first factor (PC1), related to stem growth and SLA, explaining 59.85% of the total variation, and the second factor, related to root and stem dry biomass, explaining 26.24% of it (Table S5). The two treatments clearly differed, all populations analyzed showing a similar pattern, mainly due to an increment in stem mass under the full capacity treatment, although they differed either in the extent of the variation or in the allocation pattern (expressed in the two axes). The differences were higher for *Pinus patula* than for *P. pseudostrobus*. *P. pseudostrobus* populations showed a similar performance, PS03 and PS04 behaving similarly under the two treatments (Figure S2). *Pinus patula* showed a significant correlation ($r = 0.634$ *) between rainfall of the origin and plasticity in PC2 (Figure 4a). In the case of *Pinus pseudostrobus*, the value was significant ($r = 0.823$ *) in the case of PC2 (Figure 4b).

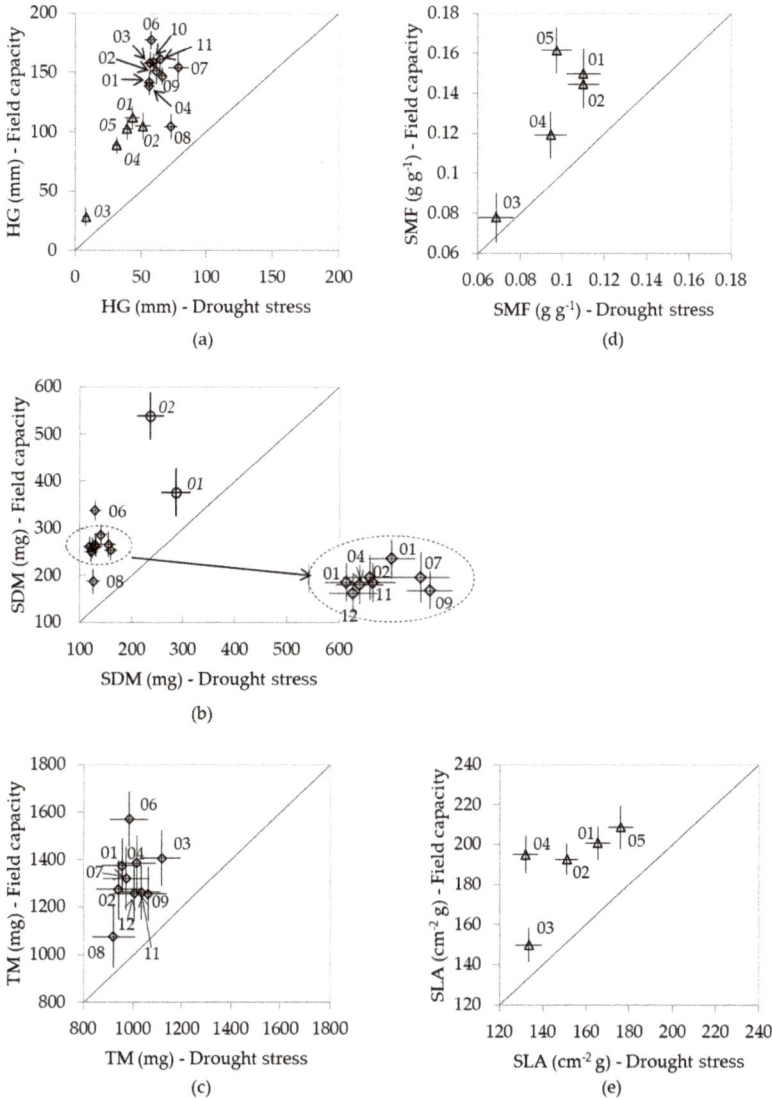

Figure 3. Analysis of plasticity by species for growth variables and biomass components with significant response to watering treatment in (**a**) HG: height growth increment; (**b**) SDM: stem dry mass; (**c**) TDM: total dry mass; (**d**) SMF: stem mass fraction; and (**e**) SLA: specific leaf area (*P. oocarpa* = ○, *P. patula* = ◊ and *P. pseudostrobus* = Δ). Bars indicate the standard errors in the two treatments.

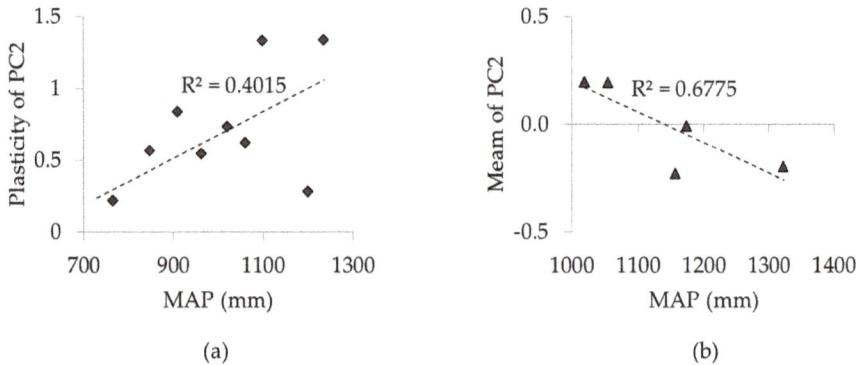

Figure 4. Relationships among rainfall (MAP, data from Table 1) with (**a**) plasticity of Principal Component 2 for *P. patula* (♦); and (**b**) mean of Principal Component 2 for *P. pseudostrobus*(▲).

4. Discussion

This paper evaluates the variation in growth and biomass allocation in seedlings of three Mexican pines grown under two contrasting watering regimes. The results showed inter- and intraspecific variation in seedling drought tolerance, which confirms our hypothesis that the watering regime had a significant effect in phenotypic changes for plants of *Pinus oocarpa*, *P. patula* and *P. pseudostrobus*. All species and populations shared a common general avoidance mechanism (increasing water uptake and reducing water loss [29]) in relation to changes in their allocation patterns, but the intraspecific variation and differences in phenotypic plasticity among populations modified the adaptation strategies of the species to drought. The sampling scheme allowed us to detect differences among geographically close populations, with strong implications for forest management.

Our study is limited to moderately stressful experimental conditions, as we were dealing with species and populations that differ in their tolerance to water stress, but in accordance to the climatic scenarios predicted by 2030 [1]. Our results evidenced the existence of an avoidance mechanism in the face of drought stress at the seedling stage, which is the most critical in both the natural and artificial regeneration methods. The existence of watering x population interaction in many traits implies differences in the genetic responses of the populations that are important for the in situ adaptation of the species, due to the possible selection of reaction norms. Experiments under more intense water stress, that is, more stressful conditions than those predicted for the next generation, could result in hidden reaction norms, i.e., responses of the populations not described previously [30]. Another caveat of the study is that maternal environmental effects at the seedling stage significantly modulate variability in the trees growing in the stressful environment [31]. However, we minimized the impact of these effects by using the initial height as a covariate. Finally, we focused our experiment in a restricted area, using a limited number of samples (in the case of *P. oocarpa*, only two, to avoid biases due to taxonomic errors in the identification, see Materials and Methods). The sampled populations, however, cover the range of mean temperature and rainfall of the study area (Table 1). We addressed the level and patterns of variation of close-together populations in the same region as a means to infer genetic resources management recommendations in the study area. We are not able, however, to provide estimates of the level of genetic variation of the species, which is largely dependent on the sampling scheme.

The adjustment to drought stress treatment in the Mexican pines analyzed mainly involved allometric changes by reduction of aerial biomass, although it is interesting to point out that root allocation was not significantly affected, and neither was needle dry biomass. Seedling allometric changes, linked to low water availability in the soil [32], are associated to particular physiological processes, including changes in photosynthetic and transpirational capacities, that depend on the level of stress [6]. A reduction in SLA, an important functional trait related to leaf assimilation capacity [33], was also observed. Such reduction in SLA under water-stress conditions has been repeatedly reported in seedling experiments (e.g., *P. canariensis* [34], *P. halepensis* [35]), under similar experimental conditions. SLA changes were due to shifts in the watering regime [36], and seedlings from drought-tolerant seed sources showed greater reductions in needle size, area per needle and stomata per needle than seedlings from non-tolerant sources [37].

The three species analyzed did not behave similarly, and presented significant differences in the level of intraspecific variation and phenotypic plasticity under water stress treatment. *P. oocarpa* showed the highest mortality, growth reduction and needle biomass fraction increment. *P. oocarpa* seemed the least tolerant to water stress, while *P. pseudostrobus* was the most tolerant. The climatic information from the sampled populations (and from the species in the area of study) is not exactly coherent with this behavior, since *P. oocarpa* lives under higher annual temperatures (18.8 °C) and rainfall (1205 mm) than the other two species (14.3 °C and 982 mm for *P. patula*, and 12.6 °C and 1145 mm for *P. pseudostrobus*). Therefore, climatic data (temperature and rainfall) cannot be solely relied upon in predicting drought tolerance in forest species, especially when dealing with populations from a restricted area, where other factors and climatic variables could have shaped local adaptation, determining the behavior of each species [38–40].

For the three species, several patterns have been described for the relationship between the ecological conditions and the performance in field or in greenhouse experiments of the species, indicating that these relationships depend both on the species and the experimental conditions (sampled material and site). In many cases there is a maximum (or minimum) of the performance at a given ecological (rainfall, altitude) value. For *P. oocarpa* seedlings, the occurrence of a seedling stage was high whenever the rainfall at the seed origin was less than 1250 mm. The ability to form a lignotuber (storage root typical from seedling-stage pines) is probably an adaptation to dry, fire-frequented environments [41]. Height growth was related to the altitude, rainfall and dry season of the seed origin [41], and the greatest growth would occur in populations originating from 1255 m a.s.l., with populations from either lower or higher altitudes having a lower growth [42]. *Pinus patula* provenances from lower altitudes showed higher growth and a larger number of shoots cycles than provenances from higher altitudes [43,44]. However, in a greenhouse-provenance trial, seedlings showed slightly higher growth potential in provenances from mid-altitude (2700 m a.s.l.) than those provenances originated in altitudinal extremes (2400 and 3000 m a.s.l.) [45]. *Pinus pseudostrobus* populations from lower altitudes (2300–2400 m a.s.l.) presented poorer health than populations from intermediate altitudes (2700 m a.s.l.), and those populations from altitudinal extremes (2300 and 2900 m a.s.l.) presented the lowest percentages of germination, while the highest germination rate corresponded to 2700 m a.s.l. [46].

Intraspecific variation will influence the strategies of the species at two main levels: genetic variation and differences in the plastic response of the populations. The three species showed significant levels of intraspecific variation within the sampled area, with *P. oocarpa*, for which only two populations were sampled, having a largest level of genetic variation, the two populations differing in phenotypic plasticity in response to drought stress. It has been reported that populations from low altitudes tend to show higher growth potential than trees from populations originating at higher altitudes [42], and that populations from altitudes of origin above 1000 m a.s.l. are less drought-tolerant than those of below 1000 m a.s.l. [47]. It is quite likely that populations (and species) from low altitudes have a more conservative growth strategy, related to the avoidance of drought stress [9,48]. However, at our sampling scale, the population from the high altitude (OC02) was more tolerant to drought stress, as indicated by its lower mortality, and its better stem biomass adjustment under our two watering regimes.

Pinus patula populations showed a significant among-population variation in most of the traits related to stem growth and SLA, and they also differed in levels of phenotypic plasticity for those traits, although not for SLA. It has been reported that *P. patula* provenances from lower altitudes have a higher growth [43]. In our study, there is a correlation among seed origin rainfall and the mean value of the first factor ($r = 0.65$), related to stem mass fraction, and SLA and the plasticity in the second factor ($r = 0.63$), related to root and stem dry biomass, indicating that even at local scales there is an adaptive pattern to climate of the integrated phenotypes.

P. pseudostrobus also showed intraspecific differences in traits related to stem biomass, and SLA, but also significant differences in phenotypic plasticity among populations. We found a correlation of the mean value of the populations in the factor 2 ($r = 0.82$ *) related to root and stem dry biomass. The low sampling size (5 populations), could influence the lack of significance of the factor 1 ($r = 0.65$ ns) and the plasticity of the factor 2 ($r = 0.68$ ns). The linear relationships described in this study can also be caused by the sampling area, as we cannot discard a more complex performance (as the ones described in the studies previously mentioned), when expanding the study area. It is interesting to notice that populations from western Mexico did not have significant genotype-environment interaction [49,50], when tested in close-by test sites. Therefore, estimating intraspecific differences in terms of adaptability at local scales will require an estimate of among-population genetic differences in terms of genetic phenotypic plasticity [51] in a larger number of populations.

The implications for forest genetic resources management are related to the natural and artificial regeneration of the species and conservation of genetic resources. In the TMVB, the populations of the species differ in adaptability to drought stress, and our ability to predict the responses requires a sufficient sample size, that is, at spatial scales significant for forest management we can detect differences in genetic variation and patterns of performance related to the climate of origin. In the case of *P. oocarpa*, even two very close populations performed differently and, for the other two species, the existence of intraspecific variation (population and drought-by-population interaction) justifies the use of local material in afforestation programs [52]. More productive allochthonous basic materials could be used in the region, ensuring that native populations were not introgressed with this potentially non-adapted material [53]. This study also shows the importance of the area for the genetic conservation of the species, as some conservation units can be selected having differential value in terms of adaptation for the future climatic conditions [54]. Also, our results show that, at early developmental stages, genetic differences in survival are important, depending on the species, and therefore silvicultural treatments must be taken into consideration to favor different biomass allocation (e.g., by reducing competition or light) [55]. Managing the genetic resources within a region, therefore, needs not only information at the species level, but a more precise information about major variation patterns of their populations, as the effects will affect the future adaptation and performance of the species in the area considered.

5. Conclusions

We confirmed that even at reduced geographical scales, Mexican pines present differences in the response to water stress. The responses differed among species, including the allometric phenotypic changes in biomass allocation (plasticity), the genetic differences among populations, and the differences in phenotypic plasticity among populations. Testing three different species that presented differences in water stress tolerance, allowed us to detect different strategies of avoidance (mainly changes in allometry, but also changes in needle structure for some of the populations), and some patterns of species response. These differences are relevant not only in afforestation programs, but also in genetic conservation activities.

Supplementary Materials: The following are available online at http://www.mdpi.com/1999-4907/9/2/71/s1: Table S1, Analysis of survival in the drought experiment for the three species. Table S2, Values of survival and ontogenic stages recorded per population. Table S3, Mean (± standard errors) for growth variables and biomass fractions for the two watering treatments (FC/DS). Table S4, Analysis of unequal slope and intercept estimated among watering regimes by species. Table S5, Percentage of the total variance explained by components and weights obtained among their variables. Figure S1, Germination speeds for *P. oocarpa* (OC), *P. patula* (PA) and *P. pseudostrobus* (PS) and among their populations (OC01-02; PA01-04, 06-09, 11-12; PS01-05). Figure S2, Biplot off the variables (**X**) and populations on the plane defined by the two Principal Components, for *P. patula* (♦) and *P.*

pseudostrobus (▲). Filled symbols and population's number represent FC treatment, while empty symbols and underlined population's number represent DS treatment. Also, the database of the four Pines species is available at: https://doi.org/10.5281/zenodo.1162044.

Acknowledgments: We are grateful to F. del Caño and S. Sansegundo for technical support during the drought stress treatment, and to O. Trejo, J.O. Romero, A. López and H.J. Muñoz for seed supply. We acknowledge funding from the Spanish National Research Plan (RTA2017-00063-C04-01, AGL2015-68274-C3) and AEG 17-041 (financed by FEADER -75%- and MAPAMA -25%- to implement the National Program for Rural Development). A.F. was supported by a grant from the Mexico's National Council for Science and Technology (CONACyT) and National Institute for Forestry, Agricultural and Livestock Research (INIFAP). We also thank P.C. Grant, the professional in English grammar and style, who has revised the manuscript.

Author Contributions: A.F. and R.A. conceived and designed the experiments; A.F. performed the experiments; A.F., J.C., V.P. and R.A. analyzed the data; J.L.-U. contributed materials; A.F., J.L., J.C., V.P. and R.A. wrote the paper.

Conflicts of Interest: The authors declare no conflict of interest.

Appendix A

Experimental Details

Populations: These are described in Table 1 and located in Figure 1.

Germination: Three hundred seeds per population were soaked in hydrogen peroxide (H_2O_2, 1:5 v/v) for 15 min, then rinsed twice and soaked in distilled water for 24 h. Seeds were then sown in trays containing moistened rock wool and covered with plastic film. Trays were placed inside a germination chamber at 25 ± 1 °C and $60 \pm 5\%$ relative humidity and 8-h photoperiod. The germination was recorded three times a week and then used to calculate the germination curve parameters (total germination in %, speed) based on a sample of 60 seeds per population. Germination for three species started at 3 days, *P. oocarpa* and *P. pseudostrobus* having a higher germination rate than *P. patula*.

Container characteristics: Fifty seedlings from each population were transplanted into individual plastic containers, except from *P. patula* populations (PA02, PA07 and PA08), which had a low germination rate. We used 250 cm^3 individual plastic containers with a mixture of peat moss and vermiculite substrate (3:1 v/v). All the containers were equally filled. This container size was big enough to avoid root restriction, given the short duration of the experiment [17].

Experiment design: The seedlings were installed in a greenhouse with temperature control (see next section for details). Plants were arranged in a randomized complete block design, with 5 seedlings per block, and 5 blocks in each of the two watering treatments (25 seedlings per treatment). Two different growing phases were established to hasten ontogenetic changes, differing in temperature and photoperiod. Seedlings were maintained in a slow-growth phase over 135 days from November to March (8 ± 2 °C and $60 \pm 5\%$ relative humidity), and then cultivated in a normal-growing phase from April to June (24 ± 2 °C and $80 \pm 5\%$ relative humidity). Both growth phases were implemented at the greenhouse. During this second growing phase, seedlings were submitted to two watering treatments during 90 days.

Greenhouse controlled conditions: See Table A1.

Table A1. Phases and greenhouse conditions for species studied.

	Slow-Growth Phase	Normal-Growth Phase
Duration	135 days, November-March	90 days, April to June
Position of the trays with respect to the solar angle	west to east	west to east
Temperature	8 ± 2 °C	24 ± 2 °C
Photoperiod	Short days	Long days
Irrigation system	misting nozzle	misting nozzle
Watering amount	Full capacity	Two watering regimes
Relative humidity	$60 \pm 5\%$	$80 \pm 5\%$
Heat shield	40–50% reflecting radiant heat	40–50% reflecting radiant heat
Fertilization	Peter's 20-20-20 -N-P-K	Peter's 20-20-20 -N-P-K

Watering was determined by weighing every second day 25 pots randomly chosen from each treatment (50 in the first phase). The two watering regimes were established based on the mean saturation level of the substrate: 90–100% on FC and 35–45% on DS treatments.

References

1. Sáenz-Romero, C.; Rehfeldt, G.E.; Crookston, N.L.; Duval, P.; St-Amant, R.; Beaulieu, J.; Richardson, B.A. Spline models of contemporary, 2030, 2060 and 2090 climates for Mexico and their use in understanding climate-change impacts on the vegetation. *Clim. Chang.* **2010**, *102*, 595–623. [CrossRef]
2. Thuiller, W.; Lavorel, S.; Araujo, M.B.; Sykes, M.T.; Prentice, I.C. Climate change threats to plant diversity in Europe. *Proc. Natl. Acad. Sci. USA* **2005**, *102*, 8245–8250. [CrossRef] [PubMed]
3. Hubert, J.; Cottrell, J. *The Role of Forest Genetic Resources in Helping British Forests*; Forestry Commission: Edinburgh, UK, 2007.
4. Chevin, L.-M.; Lande, R.; Mace, G.M. Adaptation, plasticity, and extinction in a changing environment: Towards a predictive theory. *PLoS Biol.* **2010**, *8*, e1000357. [CrossRef] [PubMed]
5. Santos-del-Blanco, L.; Alía, R.; González-Martínez, S.C.; Sampedro, L.; Lario, F.; Climent, J. Correlated genetic effects on reproduction define a domestication syndrome in a forest tree. *Evol. Appl.* **2015**, *8*, 403–410. [CrossRef] [PubMed]
6. Pallardy, S.G. *Physiology of Woody Plants*, 3rd ed.; Academic Press: San Diego, CA, USA, 2008; ISBN 978-0-12-088765-1.
7. Stebbins, G.L. *Variation and Evolution in Plants*, 2nd ed.; Columbia University Press: New York, NY, USA, 1950; ISBN 0069-6285.
8. Valladares, F.; Sánchez-Gómez, D. Ecophysiological traits associated with drought in Mediterranean tree seedlings: Individual responses versus interspecific trends in eleven species. *Plant Biol.* **2006**, *8*, 688–697. [CrossRef] [PubMed]
9. Barton, A.M.; Teeri, J.A. The ecology of elevational positions in plants: Drought resistance in five montane pine species in Southeastern Arizona. *Am. J. Bot.* **1993**, *80*, 15–25. [CrossRef]
10. Holt, R.D. The microevolutionary consequences of climate change. *Trends Ecol. Evol.* **1990**, *5*, 311–315. [CrossRef]
11. Jones, H.G. Monitoring plant and soil water status: Established and novel methods revisited and their relevance to studies of drought tolerance. *J. Exp. Bot.* **2007**, *58*, 119–130. [CrossRef] [PubMed]
12. López, R.; Rodríguez-Calcerrada, J.; Gil, L. Physiological and morphological response to water deficit in seedlings of five provenances of *Pinus canariensis*: Potential to detect variation in drought-tolerance. *Trees Struct. Funct.* **2009**, *23*, 509–519. [CrossRef]
13. Alía, R.; Chambel, R.; Notivol, E.; Climent, J.; González-Martínez, S.C. Environment-dependent microevolution in a Mediterranean pine (*Pinus pinaster* Aiton). *BMC Evol. Biol.* **2014**, *14*, 200. [CrossRef] [PubMed]
14. Wright, S.J.; Machado, J.L.; Mulkey, S.S.; Smith, A.P. Drought acclimation among tropical forest shrubs (Psychotria, Rubiaceae). *Oecologia* **1992**, *89*, 457–463. [CrossRef] [PubMed]
15. Engelbrecht, B.M.J.; Kursar, T.A. Comparative drought-resistance of seedlings of 28 species of co-occurring tropical woody plants. *Oecologia* **2003**, *136*, 383–393. [CrossRef] [PubMed]
16. Chambel, M.R.; Climent, J.; Alía, R. Divergence among species and populations of Mediterranean pines in biomass allocation of seedlings grown under two watering regimes. *Ann. For. Sci.* **2007**, *64*, 87–97. [CrossRef]
17. Poorter, H.; Bühler, J.; van Dusschoten, D.; Climent, J.; Postma, J.A. Pot size matters: A meta-analysis of the effects of rooting volume on plant growth. *Funct. Plant Biol.* **2012**, *39*, 839–850. [CrossRef]
18. Levy, Y.; Syvertsen, J. Irrigation water quality and salinity effects in citrus trees. In *Horticultural Reviews*; Janick, J., Ed.; John Wiley & Sons, Inc.: Hoboken, NJ, USA, 2004; Volume 30, pp. 37–82, ISBN 0-471-35420-1.
19. Farjon, A.; Filer, D. *An Atlas of the World's Conifers: An Analysis of Their Distribution, Biogeography, Diversity and Conservation Status*; Brill: Leiden, The Netherlands, 2013; ISBN 978-90-04-21180-3.
20. Perry, J.P. *The Pines of Mexico and Central America*; Timber Press, Inc.: Portland, OR, USA, 1991; ISBN 0881921742.
21. Cambrón-Sandoval, V.H.; Sánchez-Vargas, N.M.; Sáenz-Romero, C.; Vargas-Hernández, J.J.; España-Boquera, M.L.; Herrerías-Diego, Y. Genetic parameters for seedling growth in *Pinus pseudostrobus* families under different competitive environments. *New For.* **2012**, *44*, 219–232. [CrossRef]

22. Crookston, N.L. Research on Forest Climate Change: Potential Effects of Global Warming on Forests and Plant Climate Relationships in Western North America and Mexico. Available online: http://www.webcitation. org/6uJ6DepQ9 (accessed on 18 October 2017).

23. Poorter, H.; Nagel, O. The role of biomass allocation in the growth response of plants to different levels of light, CO2, nutrients and water: A quantitative review. *Aust. J. Plant Physiol.* **2000**, *27*, 595–607. [CrossRef]

24. South, D.; Larsen, H.S. Use of seedling size as a covariate for root growth potential studies. In *5th Biennial Southern Silvicultural Research Conference*; USDA Forest Service: Memphis, TN, USA, 1988; pp. 89–93.

25. Hernández-Pérez, C.; Vargas-Hernández, J.J.; Ramírez-Herrera, C.; Muñoz-Orozco, A. Variación geográfica en la respuesta a la sequía en plántulas de *Pinus greggii* Engelm. *Rev. Cienc. For. México* **2001**, *26*, 61–79.

26. Pigliucci, M.; Schlichting, C.D. Reaction norms of Arabidopsis. IV. Relationships between plasticity and fitness. *Heredity* **1996**, *76*, 427–436. [CrossRef] [PubMed]

27. Poorter, H.; Niklas, K.J.; Reich, P.B.; Oleksyn, J.; Poot, P.; Mommer, L. Biomass allocation to leaves, stems and roots: Meta-analyses of interspecific variation and environmental control. *New Phytol.* **2012**, *193*, 30–50. [CrossRef] [PubMed]

28. Institute, S.A.S. *SAS/GRAPH 9.1 Reference*; SAS Institute: Cary, NC, USA, 2004; ISBN 1590471954.

29. Poorter, L.; Markesteijn, L. Seedling traits determine drought tolerance of tropical tree species. *Biotropica* **2008**, *40*, 321–331. [CrossRef]

30. Schlichting, C.D. Hidden reaction norms, cryptic variation and evolvability. *Ann. N. Y. Acad. Sci.* **2008**, *1133*, 187–203. [CrossRef] [PubMed]

31. Zas, R.; Cendán, C.; Sampedro, L. Mediation of seed provisioning in the transmission of environmental maternal effects in Maritime pine (*Pinus pinaster* Aiton). *Heredity* **2013**, *111*, 248–255. [CrossRef] [PubMed]

32. Sáenz-Romero, C.; Rehfeldt, G.E.; Soto-Correa, J.C.; Aguilar-Aguilar, S.; Zamarripa-Morales, V.; López-Upton, J. Altitudinal genetic variation among *Pinus pseudostrobus* populations from Michoacán, México. Two location shadehouse test results. *Rev. Fitotec. Mex.* **2012**, *35*, 111–120.

33. Niinemets, Ü. Components of leaf dry mass per area—Thickness and density—Alter leaf photosynthetic capacity in reverse directions in woody plants. *New Phytol.* **1999**, *144*, 35–47. [CrossRef]

34. Climent, J.M.; Aranda, I.; Alonso, J.; Pardos, J.A.; Gil, L. Developmental constraints limit the response of Canary Island pine seedlings to combined shade and drought. *For. Ecol. Manag.* **2006**, *231*, 164–168. [CrossRef]

35. Baquedano, F.J.; Castillo, F.J. Comparative ecophysiological effects of drought on seedlings of the Mediterranean water-saver *Pinus halepensis* and water-spenders *Quercus coccifera* and *Quercus ilex*. *Trees Struct. Funct.* **2006**, *20*, 689–700. [CrossRef]

36. Reich, P.B.; Wright, I.J.; Cavender-Bares, J.; Craine, J.M.; Oleksyn, J.; Westoby, M.; Walters, M.B. The evolution of plant functional variation: Traits, spectra, and strategies. *Int. J. Plant Sci.* **2003**, *164*, S143–S164. [CrossRef]

37. Cregg, B.M. Carbon allocation, gas exchange, and needle morphology of *Pinus ponderosa* genotypes known to differ in growth and survival under imposed drought. *Tree Physiol.* **1994**, *14*, 883–898. [CrossRef] [PubMed]

38. Leimu, R.; Fischer, M. A meta-analysis of local adaptation in plants. *PLoS ONE* **2008**, *3*, e4010. [CrossRef] [PubMed]

39. Bansal, S.; Harrington, C.A.; Gould, P.J.; St.Clair, J.B. Climate-related genetic variation in drought-resistance of Douglas-fir (*Pseudotsuga menziesii*). *Glob. Chang. Biol.* **2015**, *21*, 947–958. [CrossRef] [PubMed]

40. Warwell, M.V.; Shaw, R.G. Climate-related genetic variation in a threatened tree species, *Pinus albicaulis*. *Am. J. Bot.* **2017**, *104*, 1205–1218. [CrossRef]

41. Greaves, A. *Review of the Pinus caribaea Mor. and Pinus oocarpa Schiede International Provenance*; CFI Occasional Papers; University of Oxford: Oxford, UK, 1980.

42. Sáenz-Romero, C.; Guzmán-Reyna, R.R.; Rehfeldt, G.E. Altitudinal genetic variation among *Pinus oocarpa* populations in Michoacán, Mexico. *For. Ecol. Manag.* **2006**, *229*, 340–350. [CrossRef]

43. Salazar-García, J.G.; Vargas-Hernández, J.J.; Jasso-Mata, J.; Molina-Galán, J.D.; Ramírez-Herrera, C.; López-Upton, J. Variación en el patrón de crecimiento en altura de cuatro especies de Pinus en edades tempranas. *Madera Bosques* **1999**, *5*, 19–34. [CrossRef]

44. Sáenz-Romero, C.; Beaulieu, J.; Rehfeldt, G.E. Altitudinal genetic variation among *Pinus patula* populations from Oaxaca, México, in growth chambers simulating global warming temperatures. *Agrociencia* **2011**, *45*, 399–411.

45. Sáenz-Romero, C.; Ruiz-Talonia, L.F.; Beaulieu, J.; Sánchez-Vargas, N.M.; Rehfeldt, G.E. Genetic variation among *Pinus patula* populations along an altitudinal gradient. Two environment nursery tests. *Rev. Fitotec. Mex.* **2011**, *34*, 19–25.
46. Lopez-Toledo, L.; Heredia-Hernández, M.; Castellanos-Acuña, D.; Blanco-García, A.; Saénz-Romero, C. Reproductive investment of *Pinus pseudostrobus* along an altitudinal gradient in Western Mexico: Implications of climate change. *New For.* **2017**, *48*, 867. [CrossRef]
47. Masuka, A.J.; Gumbie, C.M. Susceptibility of *Pinus oocarpa* to Armillaria root disease in Zimbabwe. *J. Appl. Sci. S. Afr.* **1998**, *4*, 43–48.
48. Poulos, H.M.M.; Berlyn, G.P.P. Variability in needle morphology and water status of *Pinus cembroides* across an elevational gradient in the Davis Mountains of west Texas, USA. *J. Torrey Bot. Soc.* **2007**, *134*, 281–288. [CrossRef]
49. Viveros-Viveros, H.; Sáenz-Romero, C.; López-Upton, J.; Vargas-Hernández, J.J. Variación genética altitudinal en el crecimiento de plantas de *Pinus pseudostrobus* Lindl. en campo. *Agrociencia* **2005**, *39*, 575–587.
50. Castellanos-Acuña, D.; Sáenz-Romero, C.; Lindig-Cisneros, R.A.; Sánchez-Vargas, N.M.; Lobbit, P.; Montero-Castro, J.C. Variación altitudinal entre especies y procedencias de *Pinus pseudostrobus, P. devoniana* y *P. leiophylla*. Ensayo de vivero. *Rev. Chapingo Serie Cienc. For. Ambient.* **2013**, *19*, 399–411. [CrossRef]
51. Chambel, M.R.; Climent, J.; Alía, R.; Valladares, F. Phenotypic plasticity: A useful framework for understanding adaptation in forest species. *Investig. Agrar. Sist. Recur. For.* **2005**, *14*, 334–344. [CrossRef]
52. McKay, J.K.; Christian, C.E.; Harrison, S.; Rice, K.J. "How local is local?"—A review of practical and conceptual issues in the genetics of restoration. *Restor. Ecol.* **2005**, *13*, 432–440. [CrossRef]
53. IUCN. *Afforestation and Reforestation for Climate Change Mitigation: Potentials for Pan-European Action*; IUCN Programme Office for Central Europe: Warsaw, Poland, 2004; ISBN 2-8317-0723-4.
54. Rodriguez-Quilon, I.; Santos-del-Blanco, L.; Serra-Varela, M.J.; Koskela, J.; Gonzalez-Martinez, S.C.; Alia, R. Capturing neutral and adaptive genetic diversity for conservation in a highly structured tree species. *Ecol. Appl.* **2016**, *26*, 2254–2266. [CrossRef] [PubMed]
55. Nocentini, S.; Buttoud, G.; Ciancio, O.; Corona, P. Managing forests in a changing world: The need for a systemic approach. A review. *For. Syst.* **2017**, *26*, 1–15. [CrossRef]

© 2018 by the authors. Licensee MDPI, Basel, Switzerland. This article is an open access article distributed under the terms and conditions of the Creative Commons Attribution (CC BY) license (http://creativecommons.org/licenses/by/4.0/).

forests

MDPI

Article

Mediterranean Islands Hosting Marginal and Peripheral Forest Tree Populations: The Case of *Pinus brutia* Ten. in Cyprus

Nicolas-George H. Eliades [1,2,†], Filippos (Phil) A. Aravanopoulos [3,†] and Andreas K. Christou [1,4,*,†]

1   Cyprus Association of Professional Foresters, P.O. Box 24258, 1703 Nicosia, Cyprus; niceliades@gmail.com or res.en@frederick.ac.cy
2   Nature Conservation Unit, Frederick University, P.O. Box 24729, 1303 Nicosia, Cyprus
3   Laboratory of Forest Genetics and Tree Breeding, School of Forestry and Natural Environment, Aristotle University of Thessaloniki, P.O. Box 238, 54124 Thessaloniki, Greece; aravanop@for.auth.gr
4   Department of Forests, Ministry of Agriculture, Rural Development and Environment, 1414 Nicosia, Cyprus
*   Correspondence: achristou@fd.moa.gov.cy; Tel.: +357-22805503
†   The authors contributed equally to this article.

Received: 20 July 2018; Accepted: 21 August 2018; Published: 24 August 2018

Abstract: Mediterranean islands have served as important Tertiary and glacial refuges, hosting important peripheral and ecologically marginal forest tree populations. These populations, presumably harboring unique gene complexes, are particularly interesting in the context of climate change. *Pinus brutia* Ten. is widespread in the eastern Mediterranean Basin and in Cyprus in particular it is the most common tree species. This study evaluated genetic patterns and morphoanatomical local adaptation along the species geographical distribution and altitudinal range in Cyprus. Analysis showed that the Cyprus population of *P. brutia* is a peripheral population with high genetic diversity, comprised of different subpopulations. Evidence suggests the presence of ongoing dynamic evolutionary processes among the different subpopulations, while the most relic and isolated subpopulations exhibited a decreased genetic diversity compared to the most compact subpopulations in the central area of the island. These results could be the consequence of the small size and prolonged isolation of the former. Comparing populations along an altitude gradient, higher genetic diversity was detected at the middle level. The phenotypic plasticity observed is particularly important for the adaptive potential of *P. brutia* in an island environment, since it allows rapid change in local environmental conditions.

Keywords: Mediterranean; island; isoenzymes; marginal and peripheral forests; forest tree genetics; genetic structure

1. Introduction

The islands of the Mediterranean basin comprise one of the 36 terrestrial biodiversity hot spots of the world and are characterized by high diversity of landscape and vegetation types due to the complex historical biogeography and the profound environmental heterogeneity [1,2]. Mediterranean islands contain a significant component of Mediterranean biodiversity, notably a number of range-restricted species and unusual vegetation types [3,4]. The vegetation types usually considered as "typically Mediterranean" are the evergreen and sclerophyllous shrublands and forests under semi-arid or subhumid bioclimates, corresponding to thermo-mediterranean and meso-mediterranean vegetation belts [4]. Although the majority of Mediterranean islands are "continental islands" (that is, they became progressively isolated from the mainland and from each other by a complex combination of tectonic and glacio-eustatic processes), there are also cases of Mediterranean "oceanic islands" in the geological sense which were part of mainland (such as islands that emerged from the bottom of the sea) (see previous papers [3,5]). Thus, "oceanic islands" often present lower richness of biodiversity elements compared to "continental islands" [6], while the genetic background of their wild populations could be restricted, due to isolation, small population, founder effects, bottlenecks, low effective population sizes, and genetic drift [7].

The Mediterranean islands have served as important Tertiary and glacial refuges, and, hence, Mediterranean islands possess highly polymorphic species and vicariant endemic plants which emerged from more or less recent speciation events [6,7]. The geographic isolation and the environmental heterogeneity of the Mediterranean basin have favored diverse evolutionary processes of gradual speciation of plants, such as genetic drift or adaptive radiation (see previous papers [4,8,9]). These features indicate the role of the wild populations of flora and fauna species in the Mediterranean islands as important peripheral (and marginal) populations. Currently, the value of peripheral forest tree populations is of particular interest in the context of climate change [10]. These populations may concurrently be those where the most significant evolutionary changes will occur; those with increasing extinction risk; the source of migrants for the colonization of new habitats at leading edges; or the source of genetic variation for reinforcing existing genetic variation in various parts of the range [10].

Several authors argued that demographic and evolutionary processes shape peripheral populations differently, compared to populations at the core of the distribution, depending on their situation in the geographic space [11–13]. Thus, whether leading edge populations are diverse enough to efficiently contribute to colonization will depend on their interpopulation gene flow and the amount of gene flow from core populations [10]. The Mediterranean island forest tree populations are identified as geographically peripheral populations, since they are found at the rear edge of distribution areas; indeed an increasingly unfavorable climate may lead to their ecological marginalization, with drastic consequences for their survival [10].

Brutia pine (*Pinus brutia* Ten.), along with Aleppo pine (*Pinus halepensis* Mill.) form a distinct group (Group Halepensis) within the Eurasian hard pines; their combined geographic distribution reflects their prominence among low-elevation Mediterranean forest species [14]. *Pinus brutia* Ten. is a coniferous species confined mainly to the Eastern Mediterranean region (incl. Greece, Turkey, Cyprus, Syria, and Lebanon), and can also be found in Iraq and Iran [15,16]. *P. brutia* grows under several variations of the Mediterranean climate [15], on a wide range of soil types [17,18], while it is recognized for its adaptation to drought and alkaline soils [19,20]. In addition, *P. brutia* is able to form stable vegetation associations with broad-leaved species [21,22], a characteristic which has led to an increased interest in the species for commercial plantations, illustrated by breeding and provenance trails carried out in various countries in the Mediterranean region [14,23,24] and even in wider geographical ranges [20,25,26]. The *P. brutia* forests correspond to habitat type "9450 Mediterranean Pine Forests with Endemic Mesogean pine" according to the European Directive 93/42/EC. Despite the fact that the largest area covered by *P. brutia* occurs in Turkey (3.8–5.4 million ha) [16,27,28], the conservation interest nowadays focuses on the geographically peripheral and ecologically marginal populations of this species.

The present study examines the genetic and ecological processes acting on the peripheral population of *P. brutia* in Cyprus, an oceanic island as defined above, located at the southern edge of the species' distribution. Thus, the current study examines possible genetic and morphoanatomical responses of *P. brutia* in relation to (i) its geographical distribution within the island of Cyprus and (ii) its distribution in different altitudes. Therefore, this study is an attempt to pursue a genetic and morphoanatomic analysis at a landscape scale, following a refined longitudinal, latitudinal, and altitudinal sampling within an island environment. Its outcomes will be invaluable in delineating a conservation strategy for mitigating adverse effects of global warming on the potential growth and survival of *P. brutia* forests in Cyprus and their genetic resources. The outcomes from this study will also contribute in obtaining more knowledge on the evolutionary capacity of geographically peripheral and potentially ecologically marginal populations of coniferous species, under the island environment (biogeography) and their adaptability to changing conditions, as well as the impact of altitude gradients on population genetic structure within the island ecosystems.

2. Material and Methods

2.1. The P. brutia Forest in Cyprus

In Cyprus the thermophilous pine forests with *P. brutia* is the most extensive and widespread forest type, occurring in all mountainous areas from dry to subhumid climates (0–1.400 m), covering 66% of island-wide forest land (~88,790 ha) [29]. The Troodos range is well covered with dense pine forests, which attain their best development in Pafos forest (60,159 ha), where the largest unfragmented and best conserved *P. brutia* forests are found. According to the Rivas–Martínez bioclimatic classification, Cyprus has a Mediterranean Mesophytic to Xerophytic–Oceania bioclimate with zones ranging from Thermo-Mediterranean–semi-arid (lowlands) to supra-Mediterranean-humid (Troodos) [30].

2.2. Sampling Design

For the purposes of the current study, sampling was carried out in the central and main mountain range of the island, namely the "Troodos Mountain range". Sampling covered a distance of 90 km longitudinal and 45 km latitudinal (Figure 1) and was implemented in the six forests, as these are defined by the Department of Forests, namely: Akamas forest, Pafos forest, Troodos forest, Adelphi forest, Limassol forest, and Macheras forest (referred to as "subpopulations", see Table 1 & Figure 1). Plant tissues were collected from adult trees 50–70 years old (see Table 1) at a distance of 200 m in order to avoid genetic kinship. Within Pafos forest, sampling adopted ecological parameters, namely: (i) altitudinal subpopulations (altitude zones of 400 m, 800 m, and 1200 m—referred to as "altitudinal subpopulations") and (ii) different aspects, namely the northeast and the southwest orientation (referred to as "aspect subpopulations"). *P. brutia* forest is the dominant vegetation in Pafos forest (size: 60,159 ha), shaping the best growing *P. brutia* forest on the island, covering a large area of forest from near-sea-level up to the peak of Tripylos at 1352 m. These characteristics allow a detailed assessment of genetic structure reflecting the dynamic effect of differential adaptation within Pafos forest (which is distinguished in three altitudinal subpopulations and two aspect subpopulations). The altitudinal subpopulations within Pafos forest were defined based on the Rivas–Martínez bioclimatic classification of Cyprus' bioclimatic zones [31], where: (i) 0–400 m a.s.l. corresponding to the Hot Arid to Mild Arid bioclimatic belt, (ii) 400–800 m a.s.l. corresponding to the semi wet mild bioclimatic belt and (iii) 800–1300 m a.s.l. corresponding to the semi wet cool to cool wet bioclimatic belt.

Table 1. Geographic location of samples along the Troodos mountain range and sample size per sampled subpopulation.

Location (Abbr.)	Distribution Area + (ha/km)	Elevation Range (m)	Sample Size					
			Genetic Analysis		Morphoanatomical Analysis			
			No. Trees	No. Mega-Gametophytes	No. Trees	No. Cones	No. Seeds	No. Needles
Macheras (Mach)	5583 ha	400–1200	60	50	60	180	540	180
Limassol (Lim)	7896 ha	800–1200	60	50	60	180	540	180
Akamas (Aka)	5743 ha	400	25	150 *	25	75	225	180
Adelphi (Ade)	12,826 ha	400–1200	60	50	60	180	540	180
Troodos (Tro)	8843 ha	400–1200	60	50	60	180	540	180
Pafos North (PaN)	37,794 ha	400–1200	60	172	60	180	540	180
Pafos South (PaS)	22,365 ha	400–1200	60	178	60	180	540	180
Pafos z.400 (PaZ.400)	320 km	400	60	100	60	180	540	180
Pafos z.800 (PaZ.800)	438 km	800	25	150 *	60	180	540	180
Pafos z.1200 (PaZ.1200)	35 km	1200	60	100	60	180	540	180
Total			580	1050	565	1695	5085	1695

+ Distribution area corresponds to the area that is covered by sampled "subpopulation" and estimated either in ha in case of extended forest area or in km in case of specific altitude zone. * For the Aka and PaZ.800 the 150 megagametophytes resulted by the collection of six (6) seeds per tree from 25 different trees. For the rest "subpopulations" the sampling of seeds for genetic analysis was done as described in Section 2.3.
Abbr.: Abbreviation.

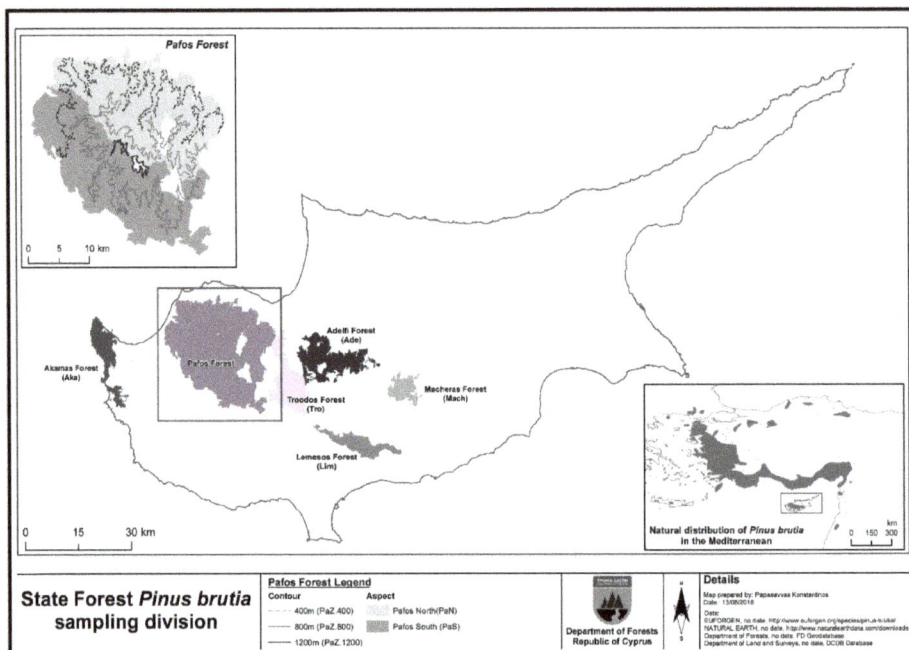

Figure 1. The distribution of sampled subpopulations of *Pinus brutia* in Cyprus.

2.3. Assessment of Genetic Diversity and Structure

For assessing the patterns of genetic diversity at both intra- and intersampling levels, three wind-pollinated cones were collected from the middle of the trees' crown (Table 1), based on the assertion that at the middle-to-high range of a tree's crown the possibility of autogamy (self-fertilization) is much lower (practically absent) than at the low level of the crown [32]. In continuation all sampled seeds from each "subpopulation" were bulked, and, hence, bulked seed material was obtained for further genetic analyses (Table 1). The genetic diversity in *P. brutia* from Cyprus was assessed based on haploid megagametophytes from germinated seeds assayed by isoenzyme horizontal starch gel electrophoresis. Isoenzyme analysis provides a nonrandom sampling of expressed genomic sequences and has proven invaluable in population genetic analysis over many decades of implementation. The protocols of Conkle et al. [33] and Cheliak and Pitel [34] were used to study the following enzyme systems: aspartate aminotransferase (AAT; EC 2.6.1.1), glutamate dehydrogenase (GDH; E.C.1.4.1.2), leucine aminopeptidase (LAP; E.C.3.4.11.1), and phosphoglucose isomerase (PGI; E.C.5.3.1.9), menadione reductase (MNR; E.C.1.6.99.2), isocitrate dehydrogenase (IDH; E.C.1.1.1.42), malate dehydrogenase (MDH; E.C.1.1.1.37), and 6-phosphogluconate dehydrogenase (6PGD; E.C.1.1.1.44).

In order to assess Mendelian inheritance in *P. brutia* in Cyprus, six megagametophytes from each sampled tree of the Akamas (Aka) subpopulation and the altitudinal subpopulation *PaZ.800* were used to derive the genotype of the (maternal) tree. Furthermore, the observed heterozygosity (*Ho*) and the expected heterozygosity (*He*) for these subpopulations were also estimated, based on the formulas by Nei [35,36]. In addition, the Hardy–Weinberg equilibrium (HWE) was calculated for these two subpopulations, as well as their genetic heterozygosity. The significance level of HWE was estimated based on the differentiation between observed and expected frequency of genotypes (as this was observed in each of the two subpopulations) using the chi-squared (X^2) statistic test. These subpopulations were chosen since both showed specific ecological and demographic features according to the national forest inventories of the Department of Forests (1981; 1991; 2001; 2011): The Aka subpopulation represents the driest and most degrading pine forest in Cyprus, while the subpopulation PaZ.800 corresponds to the altitude zone which is classified as the best ecological niche of the species growing in Cyprus.

For all sampling levels (subpopulations, altitude subpopulations, and orientation subpopulations) the multilocus intra-level genetic variation was assessed by the: percentage of polymorphic loci (PPL), observed number of allelic (*Na*), effective number of allelic (*Ne*), Shannon's index (*I*), and genetic diversity (*H_E*). Regarding the Pafos forest, the above measures were calculated as the mean values from the different subpopulations (i.e., altitude subpopulations and orientation subpopulations). The software GenAlEx 6.5 [37] was used for the calculation of the above interpopulation measures.

The intersampling level genetic diversity was assessed at three levels: (i) all sampled subpopulations, (ii) range-wide subpopulations (including all sampled subpopulations but without the altitude subpopulations), and (iii) altitudinal subpopulations. Furthermore, the subpopulations genetic structure was investigated using a Bayesian model-based clustering analysis [38], as implemented in the Structure v 2.3.4 software [38]. Bayesian analysis was performed using the admixture and the frequency-independent allele models with 50,000 Markov chain Monte Carlo (MCMC) steps and 10,000 burn-in periods. The number of *K* was set (i) from 1 to 15 when all subpopulations were included in the analysis, (ii) from 1 to 9 when the range-wide subpopulations were included, and (iii) from 1 to 5 when the three altitudinal subpopulations were included; for all runs each value of *K* for each case was run by three replicates. Post-processing of Structure software's results, for selecting the optimum number of clusters (*K*) based on the Evanno method [39] and producing the graphical output, was implemented using the software CLUMPAK [40].

The genetic structure was also examined using a hierarchical analysis of molecular variance (AMOVA) as this is applied in software GenAlEx 6.5 [37], while its significance level was computed using 999 permutations. As in the previous investigation, the AMOVA was performed at three different levels (all sampled subpopulations, range-wide subpopulations and altitudinal subpopulations). At the range-wide subpopulations level AMOVA was implemented by subdividing the sampling locations into six "Groups" (i.e., each group corresponding to each sampling forest), while for the range-wide subpopulation the analysis was performed into two hierarchical levels. In addition, the hierarchical levels of genetic structure were also investigated at the marginal subpopulations, where two separate AMOVA runs were carried out: one for the three altitude gradient subpopulations, and one for the two orientation subpopulations.

The genetic relationships among sampled subpopulations (all sampled subpopulations, range-wide subpopulations, and altitude gradient subpopulations), were analyzed by means of a cluster analysis based on the unweighted pair group method with arithmetic mean (UPGMA) and the genetic distance of Nei [41], using the software TFPGA (version 1.3) [42]. Bootstrap values for the dendrogram were generated using the same software, after 10,000 replications over individuals. Visualization of the genetic structure, at multivariate space was carried out based on a Principal Coordinate Analysis (PCoA) using the GenAlEx 6.5 [37].

Finally, the correlation between the three altitudinal subpopulations (PaZ.400, PaZ.800, and PaZ.1200) and their allele frequencies was investigated by the Spearman's rank correlation coefficient; using software SPSS 20.0 (SPSS Inc.®, New York, NY, USA).

2.4. Assessment of Morphoanatomical Diversity and Structure

The morphoanatomical variation of *P. brutia* in Cyprus was investigated using needles and cones. Three branches and three cones were collected from each sampled tree. For cones the following morphological traits were measured: cone length (CLen), cone width (CWid), and the ratio of cone length/width (CLen/CWid). From each of the measured cones three seeds (from the middle-part of the cone) were selected for measuring the morphological traits: seed length (SLen), seed width (SWid), length of seed's wing (SWing), and the total length of seed and wing (SLenWing). Needles were collected from the north side and middle part of tree crowns and three needles (two-years-old) from each sampled tree were used to measure 13 morphoanatomical traits: length of needle sheath (NShLen), needle length (NLen), needle width (NWid), needle thickness (NThic), number of resin ducts (internal side—NResIn), number of resin ducts (dorsal side—NResDo), total number of resin ducts per needle (NResTot), number of stomata rows (dorsal side—NStomDo), number of stomata rows (internal side—NStomIn), total number of stomata rows per needle (NStoRow), number of stomata per row (NSto/Row), total number of stomata per 1 cm^2 of needle (NStom), and number of needle teeth (NTeh). For 10 out of the 13 of the needle morphoanatomical traits, that is, apart from NLenSh, NLen, and NWin, measurements were carried out at the middle of the needle length, while the anatomic traits were measured using a stereoscope (magnification: 2 × 40).

The software SPSS 20.0 (SPSS Inc.®) was used to assess morphoanatomical trait variation from all sampled subpopulations at the intra- and intersampling location levels. For each trait and in each sampled subpopulation the following descriptive statistics were calculated: mean (μ), standard deviation (SD), and coefficient of variation (CV). The Spearman test was used to assess the correlation between the morphological and anatomical traits and to evaluate the correlation between traits and altitude within the Pafos forest. Furthermore, to assess morphoanatomical variation at the multivariate space level a principal component analysis (PCA) was used. The new independent components that were formed with eigenvalues above unity (>1) were used for the estimation of Euclidean distances among the sampled subpopulations. In order to visualize the classification patterns, in each of the three levels of sampling (all sampled subpopulations, range-wide sampled subpopulations, and altitudinal subpopulation) an unweighted pair group method with arithmetic average (UPGMA) dendrograms was constructed based on the morphological Euclidean distances (morphoanatomical distance), using the software NTSYS-pc 2.0 [43]. In addition, the hypothesis that trees belonging to their original sampled subpopulation are morphologically and anatomically similar was tested, using back-grouping, a nonparametric classification method analogous to discriminant analysis [44,45]. Finally, a Mantel test [46,47] was performed in order to investigate the possible relationship between morphoanatomical distance and genetic distance using also NTSYS-pc 2.0 [43].

3. Results

3.1. Genetic Diversity and Subpopulations Structure of P. brutia in Cyprus

Eight enzyme systems encoded by 10 loci (enzyme systems AAT and MDH encoded two loci each: AAT-1, AAT-2, MDH-1, and MDH-2) were analyzed. Nine out of 10 loci were found to be polymorphic. Six loci (AAT-1, AAT-2, LAP, GDH, MDH-1, and IDH) presented two alleles, two loci (PGI and MNR) displayed three alleles, and one locus (6PGD) showed four alleles. Remarkably, one out of the three allelic detected in locus MNR was found in subpopulation PaN and in the altitudinal subpopulation PaZ.800, and one out of the two alleles of locus IDH was found in subpopulation PaN and in the altitudinal subpopulation PaZ.1200. In the locus-by-locus analysis the genetic diversity (H_E) ranged from 0 (for IDH in PaZ.400 & PaZ.800 & the monomorphic MDH-2) to 0.66 (for 6PGD in PaZ.800); the latter exhibited the highest mean genetic diversity (H_E = 0.631) (Table S1). However, the highest average number of allelic (Na = 2.667) was detected in locus PGI (Table S1).

Mendelian inheritance was verified for seven out of the nine polymorphic loci (see Table S2 for more details), where X^2 and $p > 0.05$ at a 99% CI were tested. The outcomes from this analysis support the absence of segregation distortion of the tested loci (except for LAP where segregation distortion was found) (see Table S2). The Hardy–Weinberg equilibrium (HWE) was also calculated for the two subpopulations (Akamas and PaZ.800) where the Mendelian inheritance was investigated. Both subpopulations were under HWE for most of the study loci, since loci MNR-1 and LAP-1 showed nonsignificant HWE for the Aka subpopulation and the locus LAP-1 was not significant in PaZ.800. Based on individuals' genotype, the overall genetic diversity of Aka subpopulation was estimated: observed heterozygosity (H_o) = 0.164, expected heterozygosity (H_e) = 0.216, and the value of inbreeding

(F_{is}) = 0.241. The level of genetic diversity for PaZ.800 was slightly higher (H_o = 0.216) and the H_e = 0.242, while the F_{is} was almost half (F_{is} = 0.107).

Assessment of multilocus genetic diversity within each of the sampled subpopulations (Table 2) showed that PaN and PaZ.1200 have the highest PPL value (90%), whereas Mach, Lim, and Tro the lowest PPL value (60%). PaN was the subpopulation with the highest number of allelic per locus (Na = 2.300) and Lim the one with the lowest (Na = 1.700). The effective number of alleles (Ne) was the lowest in Lim (Ne = 1.300), but relatively similar for two out of the ten subpopulations (PaN, Ne = 1.415 and Paz.800, Ne = 1.464). In overview, PaZ.800 presented the highest genetic diversity (I = 0.418 and H_E = 0.244) and Lim the lowest (I = 0.269 and H_E = 0.170). Notably, in Pafos forest (PaN, PaS, PaZ.400, PaZ.800, and PaZ.1200) the overall mean genetic diversity is PPL = 84%, Na = 2.160, Ne = 1.397, I = 0.373, and H_E = 0.217. The present study detected the total genetic diversity of *P. brutia* in Cyprus as PPL = 90%, Na = 2.300, Ne = 1.421, I = 0.381, and H_E = 0.220.

Table 2. Patterns of genetic variation at multilocus level over subpopulations in *P. brutia* in Cyprus.

Subpopulation	Genetic Parameters				
	PPL	**Na**	**Ne**	**I**	**H_E**
Mach	60%	1.800	1.435	0.331	0.203
Lim	60%	1.700	1.300	0.269	0.170
Aka	70%	2.000	1.442	0.371	0.217
Ade	80%	2.100	1.473	0.393	0.242
Tro	60%	1.800	1.430	0.335	0.214
PaN	90%	2.300	1.415	0.393	0.228
PaS	80%	2.100	1.415	0.375	0.220
PaZ.400	80%	2.000	1.375	0.350	0.209
PaZ.800	80%	2.200	1.464	0.418	0.244
PaZ.1200	90%	2.200	1.315	0.327	0.186
Overall	90%	2.300	1.421	0.381	0.220

PPL: % of polymorphic loci; *Na*: Number of allelic per locus; *Ne*: Effective number of allelic; *I*: Shannon's index; H_E: Genetic diversity.

Furthermore, when only the samples of altitudinal subpopulations were considered, PaZ.800 presented the highest genetic diversity and PaZ.1200 the lowest, although the latter presented the highest presentence of polymorphic loci and a relatively high number of alleles per locus (Table 2).

The Bayesian clustering analyses were performed at three levels, nevertheless results were not explicitly clear. Analysis of all subpopulations, showed the highest Evanno's ΔK index for K = 2 (with relatively high statistical support; ΔK = 55.64), while two more K values demonstrated a trend of grouping, but with low statistical support of ΔK index, K = 5 (ΔK = 13.01) a number corresponding to the geographic origins of sampled subpopulations and K = 10 (ΔK = 10.05) a number equal to the sampled subpopulations (Figure 2a(i)). Similar outcomes were recorded when the Bayesian analysis was performed including the range-wide subpopulations, by recording the highest Evanno's ΔK index for K = 2 (with high statistical support; ΔK = 81.81) and a further peak on the graphical illustration of ΔK in K = 7 (ΔK = 30.62), a number equal to the range-wide subpopulations unit (all sampled subpopulation without the PaZ.400, PaZ.800, and PaZ1200) (Figure 2a(ii)). Bayesian clustering analysis was implemented in the altitudinal subpopulations, revealing three distinct clusters (K = 3; ΔK = 56.25) (Figure 2a(iii)).

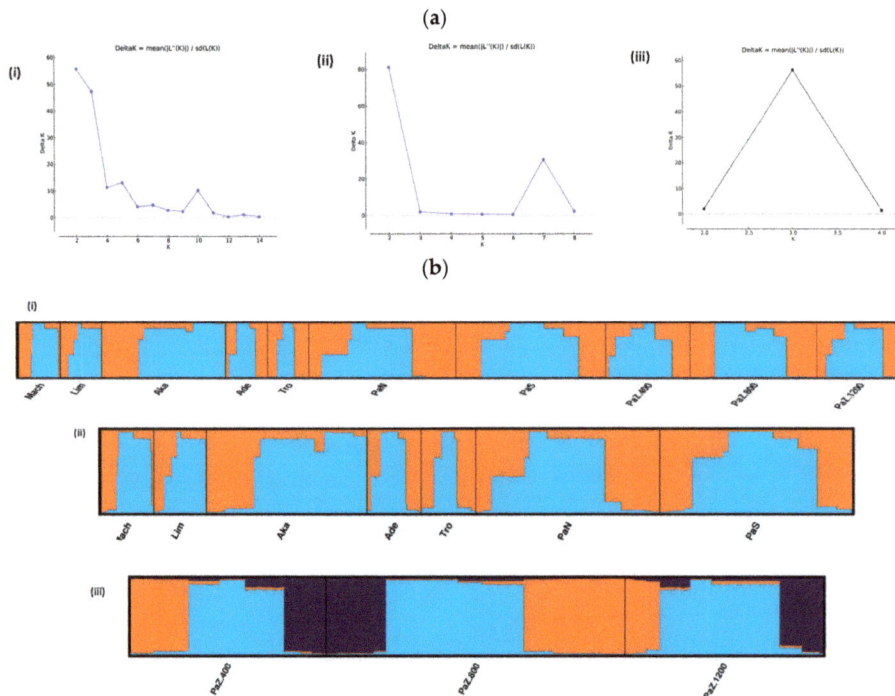

Figure 2. Bayesian cluster analysis of the optimum K clusters. (**a**) The optimal number of clusters (*K*) based on the Evanno method: (**i**) all sampled subpopulations, (**ii**) range-wide subpopulations, and (**iii**) altitudinal subpopulations. (**b**) Bar plot with individual mosquitoes represented as vertical bars colored in proportion to their assignment to clusters inferred at (**i**) all sampled subpopulations—*K* = 2, (**ii**) range-wide subpopulations—*K* = 2, and (**iii**) altitudinal subpopulation—*K* = 3.

The AMOVA showed that most of the genetic variation occurs within the sampled subpopulations (Table 3a,b) and the hierarchical analysis based on the subpopulations origin (grouping based on forest origin) revealed significant genetic differentiation (Phi_{RT} = 0.118 ***) (Table 3a). The same analysis did not detect significant genetic differentiation between groups that consist of the core (central) area subpopulations and the peripheral subpopulations (forests). Quantification of genetic differentiation among all sampled subpopulations using AMOVA found a significant Phi_{ST} = 0.129 *** (Table 3a), while low but significant genetic differentiation was also detected among the altitudinal populations (Phi_{ST} = 0.018 **, Table 3b) and between the two subpopulations with different orientation (Phi_{ST} = 0.012 **, Table 3b).

Table 3. Summary of the hierarchical analysis of molecular variance (AMOVA). (**a**) Hierarchical AMOVA based on the sampled subpopulation patterns, (i) all sampled subpopulations and (ii) range-wide subpopulations. (**b**) Hierarchical AMOVA based on marginality level, (i) altitude gradient subpopulations and (ii) different orientation subpopulations.

(a)

Source of Variation	d.f.	All Sampled Subpopulations [§]			d.f.	Range-Wide Subpopulations [†]		
		Variance Components	Percentage Variation	Fixation Indices [†]		Variance Components	Percentage Variation	Fixation Indices [†]
Among Groups	5	0.137	12%	$Phi_{KT} = 0.118$ ***	3	0.005	0%	$Phi_{KT} = 0.004$ n.s
Among subpopulations within Groups	4	0.013	1%	$Phi_{PR} = 0.012$ ***	3	0.011	1%	$Phi_{PR} = 0.010$ *
Within subpopulations within Groups	1040	1.017	87%	$Phi_{PT} = 0.129$ ***	693	1.088	99%	$Phi_{PT} = 0.014$ ***
Total	1049	1.167	100%		693	1.104	100%	

(b)

Source of Variation	d.f.	Altitudinal Subpopulations			d.f.	Aspect Subpopulations		
		Variance Components	Percentage Variation	Fixation Indices [†]		Variance Components	Percentage Variation	Fixation Indices [†]
Among subpopulations	2	0.020	2%	$Phi_{PT} = 0.018$ **	1	0.014	1%	$Phi_{PT} = 0.012$ **
Within subpopulations	347	1.087	98%		348	1.118	99%	
Total	349	1.107	100%		349	1.132	100%	

§ Sampled subpopulations grouped in six "Groups" (based on their geographical origin) for AMOVA —Group #1: Mach; Group #2: Lim; Group #3: Aka; Group #4: Ade; Group #5: Tro; Group #6: PaN, PaS, PaZ.400, PaZ.800, PaZ.1200. † The sampled location grouped in three "Groups" for AMOVA. Group #1: Mach; Group #2: Lim; Group #3: Aka; Group #4: Ade, Tro, PaN, and PaS. † Phi_{KT}: proportion of genetic differentiation due to differences between groups; Phi_{PR}: proportion of genetic differentiation due to different populations within groups; Phi_{PT}: proportion of genetic differentiation among populations among groups. d.f., degrees of Freedom. Significant level of genetic differentiation: n.s., not significant; *** , $p < 0.001$; **, $p < 0.01$; *, $p < 0.05$.

Pairwise genetic distances (Nei's minimum genetic distance—Table S3) among sampled subpopulations were depicted using UPGMA (Figure 3). The genetic similarities among all subpopulations reflect significant subdivisions among two major groups (Figure 3a). One group includes subpopulations Lim, PaZ.400, and PaZ.1200 and the second group the rest. However, the clades in the latter group were shown to be unimportant due to the low values of bootstraps. A similar observation was made in the range-wide subpopulations UPGMA. In this case, Lim formed a separate group (Figure 3b). Contrary to the previous dendrograms, the UPMGA on altitudinal subpopulations reflects significant subdivisions among them, since PaZ.800 seems clearly subdivided from the other two subpopulations (PaZ.400 and PaZ.1200) with the highest bootstrap value (Figure 3c).

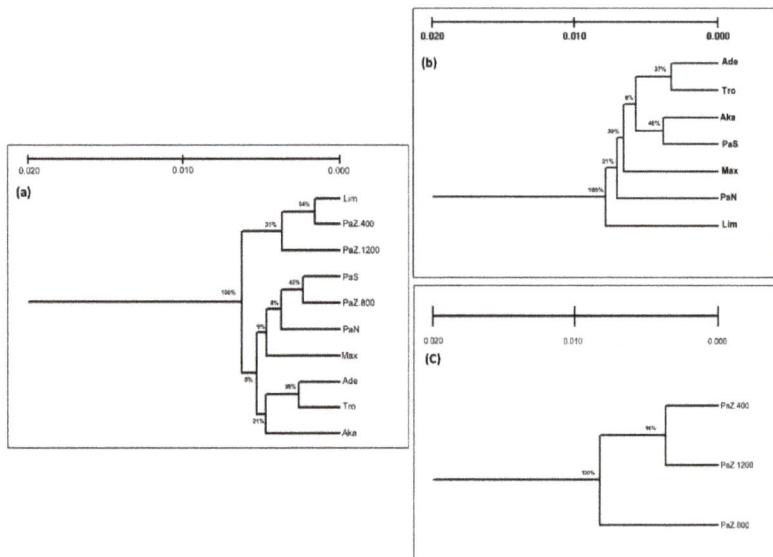

Figure 3. The genetic similarities in *Pinus brutia* illustrated by UPGMA dendrogram based on Nei's minimum genetic distance. (**a**) Sampling location level; (**b**) subpopulation level; (**c**) altitude level.

A PCoA was used to discover and depict the major patterns within a multivariate dataset, by detecting the relationship between the distance matrix elements in a two-dimensional space. When all sampled subpopulations were considered, PCoA revealed four loosely formed groups that do not correspond well to subpopulation geographic origin (Figure 4a). On the other hand, when analysis was performed at the range-wide subpopulations, Ade and Tro were completely isolated from the remaining populations, while the orientation subpopulations (PaN and PaS) grouped with the Mach population (Figure 4b). Contrary to the above analyses that considered the latitudinal and longitudinal sampling of populations, the visualization of the genetic structure of the altitudinal populations showed a clear disjunction among the populations of the three zones sampled (Figure 4c).

Given the significant influence of altitude in subpopulations genetic structure, the Spearman's correlation coefficient analysis examined the relation between allele frequency across loci and altitudinal subpopulations. It revealed a positive significant correlation between the altitudinal subpopulations in a single case, namely regarding one allelic in locus 6PGD.

(a)

(b)

(c)

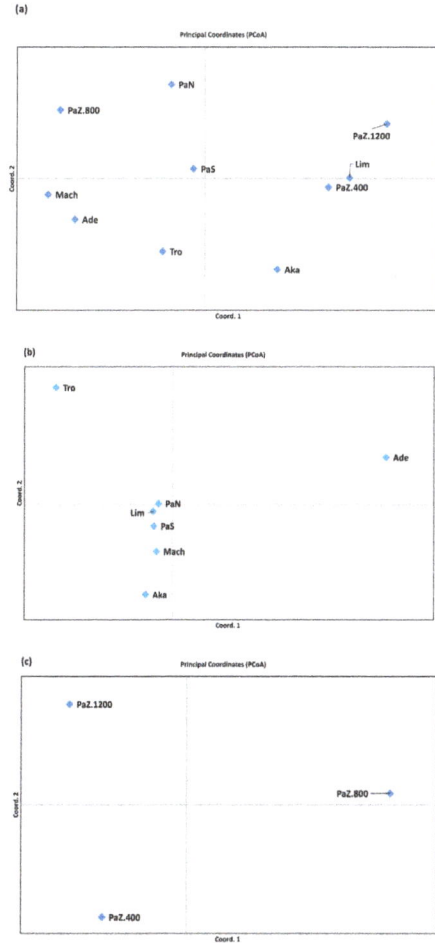

Figure 4. Two dimensional principal component analysis (PCA) plot of the genetic distance with regard to the first two principal components. (**a**) All sampled subpopulations; (**b**) wide-range subpopulation level; (**c**) altitudinal subpopulations.

3.2. The Patterns of Morphoanatomical Diversity and Structure in P. brutia

Table 4 presents the statistical description of morphoanatomic traits of *P. brutia* in Cyprus. Aka and Lim presented the lowest values of morphological traits for cones and seeds, while cone and seed length and width increased from low to high elevation in the altitudinal analysis. In addition, needle morphoanatomical traits showed that needle length (NLen) is variable among sampled subpopulations, since Lim presented the lowest values (106 mm), while Ade and PaN presented the largest (131 mm and 128 mm, respectively). Aka showed the highest values of NWid (1.43 mm) and NThic (0.88 mm), while PaZ.800 showed the lowest values NWid (1.25 mm) and NThic (0.77 mm). Concerning needle anatomical traits, Aka presented the highest values for resin ducts (NResIn: 4.23; NResDO: 6.72; NResTot: 10.95) and PaS showed the lowest values (NResIn: 3.15; NResDO: 5.21; NResTot: 8.36). In addition, resin ducts appear to increase with the altitude zonation. Regarding the five anatomical traits of stomata, Aka, Lim, and Tro, presented lower values (measurements relative to stomata traits) than the other subpopulations.

Table 4. Statistical measures (average value, standard deviation, and coefficients of variance) of morphological and anatomical traits in *P. brutia* from Cyprus: (**a**) the values in each sampled subpopulations and (**b**) the overall value of morphoanatomical traits in *P. brutia*.

(a)

Traits	Mach			Lim			Aka			Ade			Tro		
	μ	SD	CV	μ	SD	CV	μ	SD	CV	μ	SD	CV	μ	SD	CV
Clen (mm)	69.47	8.50	0.12	68.85	7.80	0.11	63.27	7.16	0.11	68.23	9.61	0.14	67.39	8.38	0.12
Cwid (mm)	40.40	3.17	0.08	37.77	3.31	0.08	35.335	3.60	0.10	38.56	3.65	0.34	38.53	3.11	0.08
Clen/Cwid	1.72	0.18	0.10	1.77	0.12	0.06	1.79	0.13	0.07	1.77	0.17	0.09	1.75	0.17	0.09
Slen (mm)	8.15	0.61	0.07	7.76	0.57	0.07	7.33	0.55	0.07	8.09	0.72	0.09	7.88	0.57	0.06
SWid (mm)	5.01	0.45	0.09	4.63	0.43	0.09	4.44	0.35	0.08	4.88	0.46	0.09	4.71	0.39	0.08
SWing (mm)	18.18	1.90	0.10	17.02	2.17	0.12	15.97	2.52	0.16	17.69	2.29	0.13	16.98	1.86	0.11
SLenWing (mm)	26.33	4.80	0.08	24.78	2.45	0.09	23.30	2.74	0.12	25.78	2.68	0.10	24.86	2.12	0.08
NShLen (mm)	6.82	1.48	0.21	6.40	0.99	0.15	6.81	0.87	0.13	6.57	0.96	0.14	6.46	1.03	0.16
NLen (mm)	126.76	15.62	0.12	106.10	13.01	0.12	115.72	12.83	0.11	131.05	15.63	0.12	119.33	14.91	0.12
NWid (mm)	1.33	0.09	0.06	1.31	0.06	0.05	1.43	0.09	0.06	1.32	0.08	0.06	1.28	0.08	0.06
NThic (mm)	0.82	0.05	0.06	0.79	0.04	0.05	0.88	0.06	0.07	0.82	0.06	0.07	0.78	0.06	0.07
NResIn	3.73	0.89	0.23	3.43	0.98	0.28	4.23	0.48	0.11	3.64	0.79	0.21	3.34	0.88	0.26
NResDo	6.05	0.98	0.16	5.78	0.99	0.17	6.72	0.95	0.14	6.16	1.06	0.17	5.64	1.24	0.22
NResTot	9.78	1.48	0.15	9.21	1.69	0.18	10.95	1.10	0.10	9.80	1.60	0.16	8.99	1.85	0.20
NStomDo	8.98	1.17	0.13	9.00	0.83	0.09	8.92	0.87	0.10	8.90	1.07	0.12	9.06	1.21	0.13
NStomIn	3.75	0.76	0.21	3.87	0.65	0.16	3.49	0.74	0.21	3.66	0.60	0.16	3.73	0.68	0.18
NStoRow	12.74	1.63	0.12	12.71	1.64	0.13	12.41	1.26	0.10	12.56	1.39	0.11	12.78	1.57	0.12
NSto/Row	83.37	8.69	0.10	83.37	9.75	0.11	84.25	10.68	0.13	87.47	8.14	0.09	81.65	9.57	0.11
Nstom	13,471	2917	0.21	11,268	2472	0.21	12,062	2171.60	0.18	14411	2730.80	0.19	12,465	2654.40	0.21
NTeh	70.23	9.43	0.13	71.43	12.08	0.17	75.07	8.24	0.11	70.73	10.61	0.15	67.21	13.50	0.20

Traits	PaN			PaS			PaZ.400			PaZ.800			PaZ.1200		
	μ	SD	CV	μ	SD	CV	μ	SD	CV	μ	SD	CV	μ	SD	CV
CLen (mm)	69.96	10.26	0.14	67.69	10.05	0.14	68.75	9.08	0.13	68.75	10.09	0.14	67.02	8.79	0.13
CWid (mm)	39.04	3.84	0.09	38.64	4.15	0.10	38.26	3.58	0.09	39.4	4.09	0.1	39.49	3.59	0.09
CLen/CWid	1.74	0.16	0.09	1.74	0.17	0.09	1.79	0.17	0.09	1.75	0.16	0.09	1.69	0.15	0.08
SLen (mm)	8.12	0.61	0.07	8.05	0.66	0.08	7.95	0.61	0.07	8.17	0.67	0.08	8.23	0.62	0.07
SWid (mm)	4.83	0.40	0.08	4.81	0.42	0.08	4.73	0.37	0.08	4.86	0.41	0.08	4.96	0.43	0.08
SWing (mm)	17.71	2.55	0.14	17.51	2.51	0.14	17.44	2.32	0.13	17.89	2.51	0.14	17.66	2.08	0.11
SLenWing (mm)	25.83	2.95	0.11	25.56	2.98	0.11	25.39	2.72	0.1	26.06	2.96	0.11	25.89	2.42	0.09
NShLen (mm)	6.54	1.07	0.16	6.33	0.89	0.14	6.11	0.77	0.12	6.39	0.91	0.14	7.04	1.1	0.16
NLen (mm)	128.47	19.12	0.15	114.50	14.14	0.12	121.7	21.62	0.17	127.4	15.86	0.12	116.49	15.55	0.13
NWid (mm)	1.27	0.09	0.07	1.28	0.07	0.05	1.29	0.09	0.07	1.25	0.08	0.06	1.29	0.07	0.05
NThic (mm)	0.78	0.05	0.06	0.77	0.05	0.06	0.79	0.06	0.07	0.77	0.05	0.06	0.78	0.05	0.06

Table 4. *Cont.*

	μ	SD	CV	μ	SD	CV	μ	SD	CV	μ	SD	CV	μ	SD	CV
NResIn	3.55	0.90	0.25	3.15	0.87	0.27	3.24	1.08	0.33	3.37	0.8	0.23	3.46	0.79	0.22
NResDo	5.84	1.20	0.20	5.21	1.21	0.23	5.21	1.55	0.3	5.58	1.16	0.2	5.8	1.11	0.19
NResTot	9.39	1.90	0.20	8.36	1.84	0.22	8.46	2.34	0.27	8.94	1.79	0.2	9.26	1.66	0.18
NStomDo	8.92	1.06	0.12	9.18	1.02	0.11	8.91	1.06	0.12	8.58	1.01	0.11	9.48	1.1	0.11
NStomIn	3.67	0.61	0.16	3.69	0.62	0.16	3.69	0.6	0.16	3.56	0.59	0.16	3.79	0.69	0.18
NStoRow	12.59	1.38	0.11	12.88	1.32	0.10	12.61	1.41	0.11	12.14	1.22	0.1	13.27	1.54	0.11
NSto/Row	84.44	7.33	0.09	86.64	8.31	0.09	85.81	8.32	0.09	85.09	8.07	0.09	85.88	8.82	0.1
Nstom	13,678	2837.13	0.20	12,777	2391.70	0.18	13,138	2854.60	0.21	13,163	2430.70	0.18	13,270.18	2676.90	0.2
NTeh	69.97	9.98	0.14	71.06	10.22	0.14	71.45	9.85	0.13	70.79	10.09	0.14	71.98	10.82	0.15

(b)

	CLen	CWid	CLen/CWid	SLen	SWid	SWing	SLenWing
μ	67.60	38.74	1.75	8.02	4.81	17.50	25.51
SD	0.22	0.09	0.004	0.01	0.01	0.05	0.06
CV	0.14	0.10	0.10	0.08	0.09	0.13	0.11

	NShLen	NLen	NWid	NThic	NResIn	NResDo	NResTot	NStomDo	NStomIn	NStoRow	NSto/Row	Nstom	NTeh
μ	6.53	121.07	1.29	0.79	3.47	5.74	9.21	9.00	3.71	12.69	84.83	13026	70.74
SD	1.06	17.83	0.09	0.06	0.91	1.23	1.88	1.08	0.66	1.48	8.83	2769.6	10.79
CV	0.16	0.15	0.07	0.66	0.26	0.21	0.20	0.12	0.18	0.12	0.10	0.21	0.15

The Spearman correlation among the investigated morphological and anatomical traits showed that 52 out of the 190 paired correlations (matrix table of 20 traits) are statistically significant for *p*-value > 95% (Table S4). Interestingly, significant correlations between needle size and the stomata rows, between cones traits as well as between the cone and seed traits, were found. Investigation on the association (Spearman correlation) between morphoanatomical traits and altitude, detected positive correlations for 12 out of the 20 traits; namely 10 correlations were positively significant and two negatively significant (Table 5). Despite the fact that the correlation coefficient is relatively low, the analysis showed a significant increase of the size of specific traits (e.g., morphological traits for cone and seed; morphoanatomical traits: NShLen, NResTot, and NStom) as the altitudinal subpopulations increased from the PaZ.400 to PaZ.1200. The identification of the traits that contribute more significantly in the overall phenotypic variation observed was investigated by PCA. The use of the eingenvalues (e.g., Kaiser's criterion), reduced the dimension of the 20 morphoanatomical traits to nine axes (for components see Table 6), of which the first six explain 97% of the total variance (Table 6). The first axis presents strong correlations with the initial variable expression of NWid, NThic, and SLen, explaining 40.60% of the total morphological variance. The second axis, accounting for 27.90% of the total variation, was associated with traits NLen and NStom, while the third axis interpreted 14% of the total variation, with NShLen and SWid to be the associated traits for this axis. The next three axes accounted for 14.6% of the overall variation, and were associated with the anatomical traits (stomata and resin ducts) and with the morphological traits of cones and seeds (Table 6). Furthermore, a morphoanatomical Euclidean distance matrix (Table S3) among the sampled subpopulations was produced using the PCA first six axes. The subpopulation analysis showed that the highest morphoanatomical distance was recorded between Aka and PaZ.800 (3.188) and the lowest between PaS and PaZ.800 (0.292). On the other hand, based on the altitudinal subpopulations, the morphoanatomical distance between PaZ.400 and PaZ.1200 was the highest (1.482) distance occurring compared to the other morphoanatomical distances recorded between the PaZ.800 (middle range) and the other altitudinal subpopulations (Table S2). The illustration of these distances in a UPGMA dendrogram indicated a specific grouping pattern (Figure 5). The subpopulations from Pafos forest (PaN, PaS, PaZ.400, PaZ.800, and PaZ.1200) and Tro shaped a clear geographically defined group and formed a common clade. The other clade was built by the Mach and Ade subpopulations. The geographically isolated subpopulations of Lim and Aka were incorporated in two different (separate) clades on the dendrogram. With regards to the altitude gradient subpopulations analysis, and contrary to the UPGMA based on genetic distances, the morphoanatomical Euclidean distances were lower between PaZ.400 and PaZ.800, whereas the PaZ.1200 was grouped in a single clade (Figure 5).

Table 5. Results of Pearson Correlation coefficient between investigated morphoanatomical traits and the sampled altitudinal subpopulations (PaZ.400, PaZ.800, and PaZ.1200).

	CLen	CWid	CLen/CWid	SLen	SLenWing	SWid
Altitude zone	−0.79 n.s.	0.13 **	−0.24 **	0.20 **	0.09 **	0.23 **

	NShLen	NLen	NThic	NWid	SWing	NResIn	NResDo	NResTot	NStomDo	NStomIn	NStoRow	NSto/Row	Nstom	NTeh
Altitude zone	0.36 **	−0.10 *	−0.05 n.s.	−0.03 n.s.	0.04 n.s.	0.09 *	0.16 **	0.14 **	0.20 **	0.05 n.s.	0.16 n.s.	−0.001 n.s.	0.01 n.s.	0.03

Significant level: * $p < 0.05$; **, $p < 0.01$; n.s.: nonsignificant.

Table 6. Principal component analysis (PCA) of the 20 morphoanatomical traits of *Pinus brutia* in Cyprus.

Variable	Axis 1	Axis 2	Axis 3	Axis 4	Axis 5	Axis 6	Axis 7	Axis 8	Axis 9
% of variance explained	40.6	27.9	14.0	7.2	4.4	3.0	2.0	0.6	0.2
NWid	−0.628 *	0.277	0.216	0.191	0.199	−0.126	−0.185	−0.273	0.136
NThic	−0.578 *	0.557	0.159	0.103	0.296	−0.074	−0.268	−0.126	0.086
SLen	0.446 *	0.145	0.281	0.149	0.185	0.166	−0.055	0.033	0.188
NLen	0.328	0.742 *	−0.273	0.023	−0.052	−0.181	−0.033	−0.214	0.258
NStoRow	0.250	0.439 *	0.011	0.409	0.060	−0.154	−0.408	−0.269	0.224
NShLen	−0.047	0.171	0.579 *	0.009	−0.288	−0.218	0.046	−0.063	0.293
SWid	0.348	0.259	0.475 *	0.109	0.419	0.108	−0.039	0.113	−0.019
NStom	0.068	0.060	−0.028	0.639 *	0.012	0.334	−0.193	0.262	0.018
NResIn	−0.263	0.375	0.282	−0.182	−0.553 *	0.128	−0.287	0.011	0.001
NResTot	−0.296	0.415	0.278	−0.168	−0.530 *	0.118	−0.131	−0.216	−0.078
CWid	0.337	0.136	0.325	−0.218	0.432 *	0.073	0.156	−0.019	−0.023
CLen	0.153	0.076	−0.040	−0.093	0.363 *	0.166	−0.018	−0.060	0.231
NStomDo	−0.019	−0.159	0.394	0.246	0.087	−0.452 *	−0.388	−0.424	−0.002
NTeh	−0.110	0.002	0.063	0.341	−0.102	0.361 *	0.354	−0.277	0.263
SWing	0.253	0.154	0.148	0.008	0.329	0.344 *	0.079	−0.122	−0.167
SLenWing	0.324	0.167	0.195	0.042	0.328	0.337 *	0.055	−0.098	−0.099
NStomIn	−0.005	−0.136	0.190	−0.154	0.213	0.139	−0.454 *	−0.411	0.376
NStomDo	0.004	−0.141	0.364	0.158	0.168	−0.387	−0.428	−0.472 *	0.134
NResDo	−0.250	0.343	0.189	−0.097	−0.339	0.069	0.117	−0.454 *	−0.160
CLen/CWid	−0.128	−0.022	−0.380	0.083	0.105	0.159	−0.136	−0.007	0.424 *

* Variables that showed strong correlation with each component.

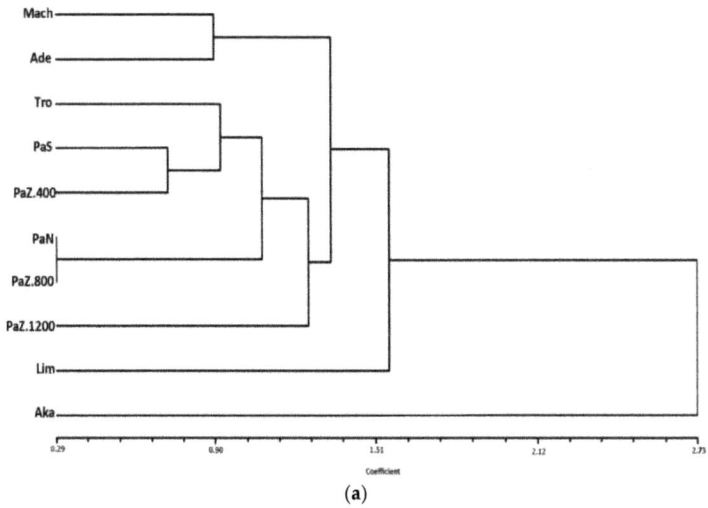

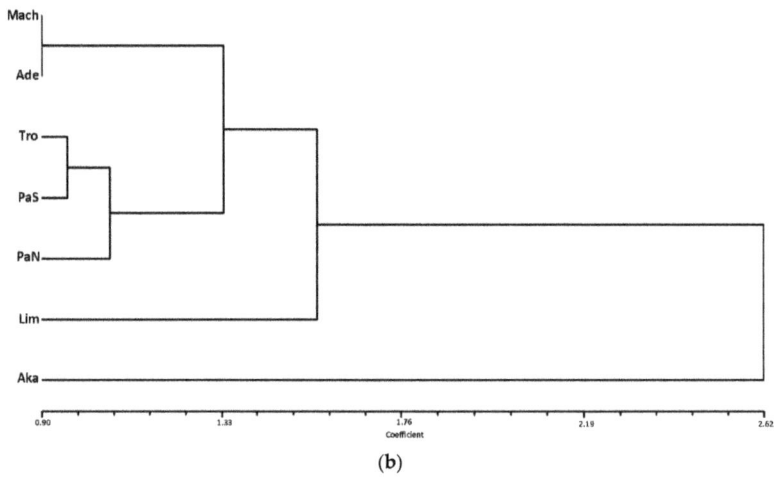

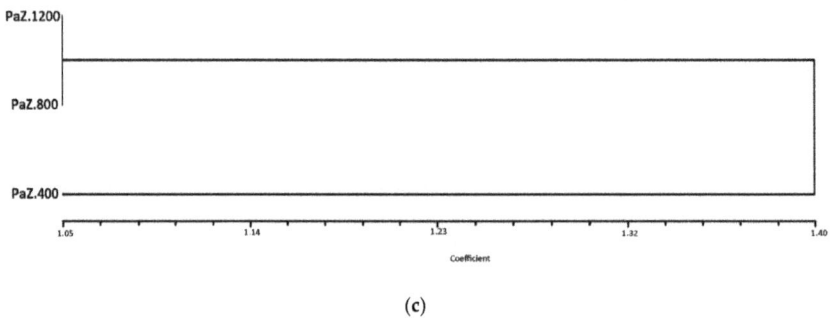

Figure 5. Unweighted pair group method wit arithmetic average (UPGMA) dendrogram based on Euclidean distances for morphoanatomical traits. (**a**) All sampled subpopulations; (**b**) wide-range subpopulation level; (**c**) altitudinal subpopulations.

In the alternative grouping pattern (using the back-grouping method), 37.2% of the initial observations (sampled trees) were back-grouped in the original sampled subpopulations. Despite the overall relatively low percentage of classification, Aka was the subpopulation with the highest value of back-grouping (82.7%), while in the other geographically disjunct subpopulations of Lim and Mach the percentage of return was 51.7% and 48.3%, respectively. On the contrary, the subpopulations PaN and PaS presented a low percentage of back-grouping (Figure 6).

Finally, in the present study there does not seem to be a significant correlation ($p > 0.05$) between the genetic distance (Nei's minimum distance) and morphoanatomical Euclidean distance, for any of the three levels of analysis, as these are classified as (i) all sampled subpopulations, (ii) range-wide subpopulations, and (iii) altitudinal subpopulations.

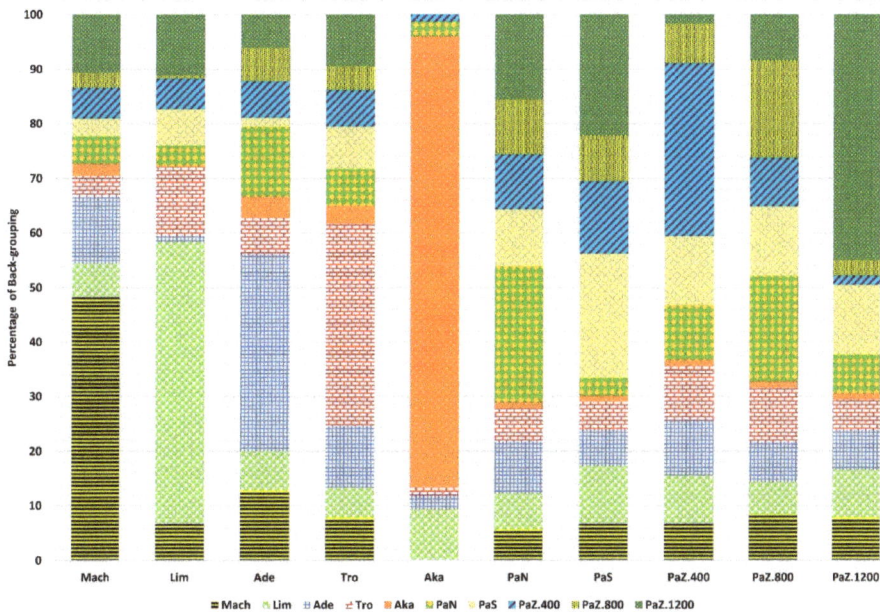

Figure 6. Back-grouping (classification) results of the morphoanatomical traits, using the sampling locations as a dependent variable.

4. Discussion

4.1. P. brutia in Cyprus as a Source of Peripheral and Marginal forest Genetic Resources

To the authors' knowledge, this study represents one of the few cases of genetic and morphoanatomic analysis, considering a refined longitudinal, latitudinal, and altitudinal sampling in an island environment. In such a geographically peripheral and ecologically marginal environment, an overview argument supports that numerous factors may influence the patterns of genetic diversity and population structure in a peripheral population, including phylogeographical constraints, meta-population dynamics, refugium, and autochthonous origin [48,49].

The loci used conform to Mendelian inheritance in agreement with previous studies on *Pinus* sp. for the same isoenzyme loci [50,51]. The observed total genetic diversity (H_E) was $H_E = 0.220$; a value relatively higher than the mean genetic diversity ($H_E = 0.118$) found at the *P. brutia* subsp *brutia* in the east Mediterranean Basin (incl. populations from Greece, Cyprus, and Asia Minor) [52] and slightly lower than the genetic diversity of the neighboring population in the mainland in the northern part of

Asia Minor where H_E = 0.265 [53] using isoenzyme loci. The higher number of rare alleles recorded in the Cyprus population of *P. brutia*, for specific loci (i.e., loci: AAT, PGI, LAP, 6PGD, IDH, and MNR), compared to the populations from its continuous range [52,54,55], implies that the *P. brutia* peripheral population in Cyprus contains unique genetic variants due to its distinctive evolutionary history. Thus, this population can be appreciated as a living gene bank for forest genetic resources for this species.

The nonsignificant inbreeding coefficient values observed in the two populations (Aka and PaZ.800), in conjunction to the presence of Hardy–Weinberg equilibrium, support the general argument, as in most conifers, of a random mating system. These results indicate that these two subpopulations (which showed reverse ecological and demographic features) are not characterized by a founder effect and have not been affected by strong bottleneck events. Hence, the *P. brutia* forest in Cyprus may have been somewhat preserved from genetic bottlenecks; especially during displacement and confinement of populations in Pleistocene glaciations, or by raised water levels during intervening warm periods as this event has been observed for several species in Europe [12,56].

The above arguments are also supported by the comparison of several morphoanatomical traits from the present study and other studies on *P. brutia* populations from other islands (i.e., Rodos, Crete, etc.) and/or populations of larger origins (continuous populations). The coefficient of variation (CV) [57,58] and the mean values [57–59] for numerous morphoanatomical traits show similarities between the populations of Cyprus and those from other origins, which implies that the morphological and anatomic traits of the former are not affected by restriction of genetic diversity and genetic drift (or strong inbreeding events). A remarkable observation on Cypriot *P. brutia* is the significant correlation between morphological and anatomical traits (Table S4). This morphoanatomic association in this rear-edge population could be a consequence of macro-environmental differentiation from the continuous distribution range of the species. Such adaptive needle traits have been mentioned for other coniferous species [60]. Often the acclimation to their environment leads to the development of specific adaptive phenotypic responses by altering their length, width, number of stomata, their angle towards the shoot, or by forming needle clumping [61–64]. In the current study, the correlation between NLen and Nstom, indicates that long needles are associated with a high number of stomata, an observation which could be a relative advantage for Cypriot *P. brutia*, providing effective resilience and adaptability for this peripheral population in different environmental conditions.

4.2. Patterns of Genetic Diversity of P. brutia in Cyprus

A nonuniform genetic diversity distribution across longitudinal, latitudinal, and altitudinal distributions was detected. Genetic variation among subpopulations is most likely a consequence of different demographic and evolutionary events. The intensive and negative impact of human activities on forests resilience and the deforestations in Cyprus, since the first human presence (11th millennium B.C.), is well-known [65,66]. During the Ottoman period (1570–1878 A.C.) reported goat stocking rates [67] are far above the forest carrying capacity, while the Cypriot forests were repeatedly logged in order to cover the energy needs of Bronze Age copper production (c. 3300–1200 BC) [68]. These historical facts may reasonably be linked to the fragmentation of *P. brutia* forest and lead to negative pressures on genetic variability within specific relic subpopulations, and consequently to different genetic variation patterns. Thus, the divergence of genetic variation (H_E) and effective number of alleles (N_e) between subpopulations, seems to be linked to high past anthropogenic pressure, particularly for subpopulations located in peripheral and more isolated forests (Aka, Lim, and Mach) where the pressure was higher (see Figure 1 and Table 2).

The clustering of subpopulations in two groups (Structure analysis; K = 2) and the detection of significant genetic differentiation among the subpopulations (AMOVA, see Table 3), alongside with the unclear geographic clustering of subpopulations from UPMGA and PCoA, could be attributed either to the fragmentation of an earlier larger and uniform population, or to the fact that the present forests are relics of previously differentiated populations of *P. brutia*. This question could not be directly answered by the present study, as more powerful molecular markers may be needed, for a more clear estimation of the best *K*-clustering of *P. brutia* in Cyprus. Meanwhile, the existence of landscape barriers to effective gene flow among the current subpopulations, shape genetic differentiation through generations.

Comparison of the genetic differentiation values ($Phi_{PT} = G_{ST} = 0.129$) between this and other studies reveals that a notably higher level of differentiation was observed in the case of *P. brutia* in Cyprus. In particular, populations of *P. brutia* originating from islands of the north-eastern Aegean sea showed $G_{ST} = 0.021$ [55], whereas populations from south Asia Minor ($G_{ST} = 0.053$) support the previous argument that rear-edge populations shape disproportionately high levels of genetic differentiation present even between geographically proximal ones, leading to exceptionally high levels of regional genetic diversity, than the populations of the continuous range [12,69–71].

Fingerprinting of the genetic diversity in Pafos forest allowed assessing the impact of local landscape on genetic patterns. Despite the fact that the comparison of northeast and southwest (aspect) subpopulations detected equal expected heterozygosity ($H_E = 0.228$ and $H_E = 0.220$, respectively), a low but significant genetic differentiation was detected ($Phi_{PT} = 0.012$). Moreover, the three altitude gradient subpopulations showed significant but low differentiation ($Phi_{PT} = 0.018$) and structure ($K = 3$); while the middle altitude gradient subpopulation (PaZ.800) recorded the highest genetic diversity ($H_E = 0.244$) (Table 2). The highest value of H_E was found in PaZ.800, being also the highest value among all subpopulations in Cyprus. This result is in concordance with other studies regarding *P. brutia*, particularly in the Taurus mountains, where the middle altitude zone recorded a higher genetic diversity than other zones [53]. The significant differentiation among the three altitude gradient subpopulations could be attributed to a combination of anthropogenic activities and small scale disturbance, or to the processing of different genetic evolutionary factors within each subpopulation (at local microscale), after their last postglaciation separation. An alternative explanation, could be altitudinal movements amplified by local topography (upward and downward movement within a single mountain region) during Pleistocene glaciations and interglaciations. Such an option has been presented for Euro-Mediterranean ecosystems [70], and is supported for instance by the findings on *Cedrus brevifolia*, a narrow endemic tree in Cyprus [72]. Therefore, this shift at different altitudinal gradient zones (i.e., ecological niche) during interglaciation and postglaciation periods is potentially the reason for the formation of an admixture zone. Alternatively, the significant differentiation among the altitude gradient subpopulations could be attributed to the processing of different evolutionary factors at microscale, after their last postglaciation partition.

The hypothesis of local (microenvironmental) adaptation dynamics is supported by relevant literature which demonstrates several examples of wild populations on ecologically marginal sites (i.e., altitudinal gradients, and different ecological aspects) [73–76] and further reinforced by the presence of altitudinal clinal variation in *P. brutia* in the various morphoanatomical traits, since significant positive correlation between traits and altitudinal variation was detected (discussed below).

4.3. Patterns of Morphoanatomical Traits of P. brutia in Cyprus

The sampled subpopulations exhibit varying degrees of diversity in the 20 morphoanatomical traits studied, in relation to the longitudinal or the altitudinal gradient. In particular, morphological and anatomical traits constituted powerful tools for describing the phenotypic diversity and structure in the present study. The interpretation of the morphoanatomical trait diversity patterns, suggests that the observed diversity and structure, possibly are the result of the manifestation of phenotypic plasticity at different micro-environments. High phenotypic plasticity, leads to the species being distributed in a wider geographical and ecological range. The morphoanatomical traits detected a similar range of values between the populations of Cyprus and those from other origins, which bring the notion of the connection of phenotypic plasticity to genetics (see a previous paper [77]) and to the underlying quantitative trait loci variation that shapes phenotypic patterns. Phenotypic plasticity is reflected in: (i) the clustering of sampled subpopulations in a morphoanatomical dendrogram (Figure 5) where clustering of the subpopulations is concordant to their geographical origin, a relation not seen in the genetic data; and (ii) the back-grouping method results (Figure 6), in which the isolated subpopulations illustrated the highest value of back-grouping (Aka, Lim, and PaZ.1200). Phenotypic plasticity among subpopulations could be a consequence of different reaction norms, where a set of phenotypes can be produced by an individual genotype when exposed to different environmental conditions, such as different soil type and meteorological conditions. Furthermore, phenotypic plasticity is manifested in the positive significant clinal variation of the morphoanatomical traits in the altitudinal range subpopulations (PaZ.400, PaZ.800, PaZ.1200). This clinal variation across the altitudinal range may be due to trade-offs between plasticity and stress tolerance in harsh environments, as indicated for other species [78–80]. Thus, *P. brutia* from Cyprus showed an increased number of resin ducts and stomata per needle, as well as the largest size of cones and seeds as the altitude increased within the mountain elevation gradient.

The morphoanatomical trait patterns observed could be the consequence of isolation and phenotypic plasticity seen over different environmental gradients. A similar interpretation was also invoked in analyses of other peripheral populations of the same species in Crete [58], in other tree species, such as *Abies cephalonica* in Mt. Parnitha [81] and *Fagus* sp. in western Eurasia [82]. Therefore, this study further supports the notion that the evolution of plasticity increases the response to selection, thus reducing maladaptation induced by gene flow (see previous works [83,84]). The same studies support that, in interaction with local population growth, the evolving plasticity allows a species to occupy a larger geographical range, presenting higher plasticity in marginal than in central habitats [84].

The morphological and anatomical diversity of *P. brutia* in Cyprus is greatly dependent on a number of needle (NWid, NThic, NLen, NStomRow, and NShLen) and seed (SLen and Swid) traits, with high loadings in the first three axes of PCA where 82.5% of the total variance measured is explained (Table 6). In addition, the significant correlation between specific traits, such as needle size (i.e., between length and width), needle morphology, and anatomy (i.e., between length or width of needle and resin ducts or stomata), and cone size with seed size (i.e., between cone width and total length of seed and wing), probably implies that the species developed mechanisms and characters that permit different morphological and anatomical harvesting strategies. Peripheral subpopulations (i.e., Lim, Aka, and PaS) appear to develop different mechanisms to survive in different micro-environments. They are characterized by short, wide, thick needles (especially Aka), which function as an adaptation mechanism (morphological features) in dry sites [85]. The Aka and the PaZ.1200 (the subpopulation at the highest altitude) present the highest number of resin ducts. This feature is linked to a plant species mechanism under extreme environment conditions (dry or cold), since resin ducts seem to protect vascular tissues and ground tissues in most species [85,86]. Alternatively stomata numbers are low at the driest micro-environments (subpopulations: Aka, Lim, and Tro), which implies a development of adaptation mechanisms against water loss in the summer. Similar adaptation mechanisms to dry environments were recorded in other studies [81,85]. Cone and seed size was smaller in peripheral, drier, and ecologically degraded subpopulations (Aka, Lim), while the opposite results were observed in subpopulations of higher altitude, where the same traits present clinal variation. These patterns were also observed in other studies of *P. brutia* and have been attributed to ecological factors, in particular to the brief sprouting period which is coupled to low temperatures [58,85]. Such factors are apparently present in this study, particularly at a higher altitude.

5. Conclusions and Perspectives of Conservation Action

5.1. Inferences from the Study of a Mediterranean Oceanic Island's Peripheral Tree Population

The population of *P. brutia* in Cyprus is an example of an isolated (both geographically and ecologically) island peripheral population, whose patterns of genetic diversity were shaped by past demographic and ecological stochasticity. The detection of high genetic diversity of *P. brutia* in Cyprus may respond to more stable effective population sizes in eastern Mediterranean forest tree populations as argued for Mediterranean conifers by Fady [8]. This study likely constitutes an example of a peripheral forest tree population that does not tend towards genetic erosion, neither towards increased inbreeding or genetic drift, when compared to core populations. These findings are contrary to earlier observations that peripheral tree populations showed genetic erosion and elevated inbreeding leading to genetic drift, in comparison to core populations [87–89]. The detection of significant genetic structure (Bayesian and AMOVA) among geographically sampled subpopulations shows that the population of *P. brutia* in Cyprus is not genetically homogeneous, comprising of genetically different subpopulations. Thus, ongoing genetic evolution dynamic processes seem to occur within the study population, in spite of its small geographical distribution within an island.

In the present study, the peripheral population of *P. brutia* presented a similar value range regarding morphoanatomical traits, to populations from other origins in the Mediterranean Basin (island and mainland). This is in contrast to earlier findings suggesting that in peripheral populations' phenotypic trait descriptive statistics, show different values from those found in core populations, indeed for a limited range of traits usually related to growth [90]. The phenotypic plasticity that was observed in the present study is of particular importance for the adaptive potential of the targeted population in an island environment. Such phenotypic plasticity, especially for long-lived plants, is particularly important since it allows wild organisms to accommodate rapid change in local environmental conditions [91,92].

5.2. Future Conservation Actions

Nowadays the argument that peripheral forest tree populations need to be managed under an evolution-oriented forestry [10,93,94] is gaining ground. In the case of Cyprus, where the consequences of climate change will lead to the increase of the mean annual temperature and to the decrease of the mean annual precipitation [95], it is crucial to develop and adopt a rational and sustainable management plan for the *P. brutia* forest, characterized by compatible management of the habitat, together with genetic conservation. Thus, dynamic conservation of the target species need to be ensured at two levels: (i) in situ conservation by establishing a conservation unit within each of the sampled subpopulations where the germplasm will be protected and the natural regeneration will be safeguarded and (ii) ex situ conservation by maintaining (or establishing) seed orchards. In addition to these conservation measures, the sampled subpopulations from this study could be delineated as seed zones or provenance regions, whereas seedlots from each subpopulation need to be sampled and stored in seed banks. Also, the existing provenance trials should be maintained and evaluated, while new ones should be established at different phytosociological associations within the island according to the present study's outcomes. Critical for the sustainable management of *P. brutia* genetic resources is the collection and use of local genetic material for seed sowing and for seedling production in postfire restoration programs. Especially for subpopulations that show low genetic diversity (i.e., Aka, and PaZ.1200) the seed material must be collected from the whole range of their distribution; collected bulked seed materials should then be used towards implementing postfire management plans.

Supplementary Materials: The following are available online at http://www.mdpi.com/1999-4907/9/9/514/s1, Table S1: Patterns of genetic variation at single-locus level for each sampled subpopulation; Table S2: Assessment of Mendelian inheritance in *P. brutia* from Cyprus. (For loci where the analysed megagametophytes were lower than 18, the X^2 was not calculated.); Table S3: Pairwise genetic distance among sampled subpopulations based on Nei's minimum distance (below the diagonal) and pairwise morphoanatomical Euclidean distance (above diagonal); Table S4: Assessment of correlation (Spearman correlation test) signal among the investigated morphological and anatomical traits.

Author Contributions: The manuscript was collectively written and edited by N.-G.H.E., F.(P.)A.A. and A.K.C. and is a part of the doctorate research of A.K.C., referenced as: "A.K.C. Assessment of genetic variation of *Pinus brutia* in Cyprus. Ph.D. Thesis, Faculty of Agriculture, Forestry and Natural Environment, Aristotle University of Thessaloniki, Greece, 2000".

Funding: This research received no external funding.

Acknowledgments: This manuscript was elaborated under the framework of research tasks of projects: V4-1438, V4-1614, COST Action FP1202 "Strengthening conservation: a key issue for adaptation of marginal/peripheral

Forests **2018**, *9*, 514

populations of forest trees to climate change in Europe (MaP-FGR)". The authors wish to thank the Cyprus Department of Forests for the valuable help in sampling. In addition, the authors would like to dedicate this paper to Prof. Kostas Panetsos. Prof. Kostas Panetsos has been a mentor for A.K.C. and F.(P.)A.A. and a prominent figure for forest genetics research in the south-eastern Mediterranean region. He was among the first who stressed the importance of the conservation of forest genetic resources and conceived studies on local adaptation, the topic of this work.

Conflicts of Interest: The authors declare no conflicts of interest.

References

1. Médail, F.; Myers, N. Mediterranean Basin. In *Hotspots Revisited: Earth's Biologically Richest and Most Endangered Terrestrial Ecoregions*; Mittermeier, R.A., Robles-Gil, P., Hoffmann, M., Pilgrim, J., Brooks, T., Mittermeier, C.G., Lamoreux, J., da Fonseca, G.A.B., Eds.; CEMEX: Monterrey, Mexico; Conservation International: Washington, DC, USA; Agrupación Sierra Madre: Mexico City, Mexico, 2004; pp. 144–147.
2. Noss, R.F.; Platt, W.J.; Sorrie, B.A.; Weakley, A.S.; Means, D.B.; Costanza, J.; Pee, R.K. How global biodiversity hotspots may go unrecognized: Lessons from the North American Coastal Plain. *Divers. Distrib.* **2015**, *21*, 236–244. [CrossRef]
3. Vogiatzakis, I.N.; Pungetti, G.; Mannion, A.M. *Mediterranean Island Landscapes: Natural and Cultural Approaches, Landscape Series*; Springer Science & Business Media: New York, NY, USA, 2008; Volume 9.
4. Médail, F. The specific vulnerability of plant biodiversity and vegetation on Mediterranean islands in the face of global change. *Reg. Environ. Chang.* **2017**, *17*, 1775–1790. [CrossRef]
5. Whittaker, R.J.; Fernández-Palacios, J.M. *Island Biogeography, Ecology, Evolution, and Conservation*, 1st ed.; Oxford University Press: New York, NY, USA, 2007.
6. Mansion, G.; Rosenbaum, G.; Schoenenberger, N.; Bacchetta, G.; Rosselló, J.A.; Conti, E. Phylogenetic analysis informed by geological history supports multiple, sequential invasions of the Mediterranean Basin by the angiosperm family Araceae. *Syst. Biol.* **2008**, *57*, 269–285. [CrossRef] [PubMed]
7. Rosselló, J. A perspective of plant microevolution in the Western Mediterranean islands as assessed by molecular markers. In *Proceedings of the Islands and Plants: Preservation and Understanding of Flora on Mediterranean Islands, 2nd Botanical conference in Menorca, Maó, Spain, 26–30 April 2011*; Cardona Pons, E., Estaún Clarisó, I., Comas Casademont, M., Fraga i Arguimbau, P., Eds.; Institut Menorquı́ d'Estudis: Illes Balears, Spain, 2013; pp. 21–34.
8. Fady, B. Is there really more biodiversity in Mediterranean forest ecosystems? *Taxon* **2005**, *54*, 905–910. [CrossRef]
9. Fady, B.; Conord, C. Macroecological patterns of species and genetic diversity in vascular plants of the Mediterranean basin. *Divers. Distrib.* **2010**, *16*, 53–64. [CrossRef]
10. Fady, B.; Aravanopoulos, F.A.; Alizoti, P.; Mátyás, C.; von Wühlisch, G.; Westergren, M.; Belletti, P.; Cvjetkovic, B.; Ducci, F.; Huber, G.; et al. Evolution-based approach needed for the conservation and silviculture of peripheral forest tree populations. *For. Ecol. Manag.* **2016**, *375*, 66–75. [CrossRef]
11. Alleaume-Benharira, M.; Pen, I.R.; Ronce, O. Geographical patterns of adaptation within a species' range: Interactions between drift and gene flow. *J. Evol. Biol.* **2006**, *19*, 203–215. [CrossRef] [PubMed]
12. Hampe, A.; Petit, R.J. Conserving biodiversity under climate change: The rear edge matters. *Ecol. Lett.* **2005**, *8*, 461–467. [CrossRef] [PubMed]
13. Ohsawa, T.; Ide, Y. Global patterns of genetic variation in plant species along vertical and horizontal gradients on mountains. *Glob. Ecol. Biogeogr.* **2008**, *17*, 152–163. [CrossRef]
14. Panetsos, C.P. Monograph of *Pinus halepensis* (Mill.) and *Pinus brutia* (Ten.). *Annu. For.* **1981**, *9*, 39–77.
15. Emberger, L.; Gaussen, H.; Kassa, A.; De Phillips, A. *Bioclimatic Map of the Mediterranean Zone: Ecological Study of the Mediterranean Zone*; UNESCO-FAO: Paris, France, 1963; 58p.
16. Fady, B.; Semerci, H.; Vendramin, G.G. *EUFORGEN Technical Guidelines for Genetic Conservation and Use for Aleppo pine (Pinus halepensis) and Brutia pine (Pinus brutia)*; Bioversity International: Rome, Italy, 2003.
17. Selik, M. Botanical investigation on *Pinus brutia* especially in comparison with *P. halepensis*, Istanbul Univ. *Fac. For. J.* **1958**, *8*, 161–198.
18. Mirov, N.T. *The Genus Pinus*; The Roland Press Company: New York, NY, USA, 1967; pp. 252–254.

19. Oppenheimer, H.R. *Mechanisms of Drought Resistance in Conifers of the Mediterranean Zone and the Arid West of the U.S.A. Part 1. Physiological and Anatomical Investigations*; Final Report on Project No. A 10-FS 7, Grant No. FG-Is-119; Hebrew University of Jerusalem: Rehovot, Israel, 1967; 73p.

20. Spencer, D. *Conifers in the Dry Countries*; A Report for the RIRDC/L&W Australia/FWPRDC Joint Venture Agroforestry Program, RIRDC Publication No. 01/46; Rural Industries Research and Development Corporation: Kingston, Australia, 2001.

21. Zohary, M. *Geobotanical Foundation of the Middle East*; Gustav Fischer Verlag: Stuttgart, Germany, 1973; 740p.

22. Barbéro, M.; Loisel, R.; Quézel, P.; Richardson, D.M.; Romane, F. Pines of the Mediterranean Basin. In *Ecology and Biogeography of Pinus*; Richardson, D.M., Ed.; Cambridge University Press: Cambridge, UK, 1998; pp. 153–170.

23. Isik, K. Altitudinal variation in *Pinus brutia* Ten.: Seed and seedling characteristics. *Silvae Genet.* **1986**, *35*, 58–67.

24. Bariteau, M. Variabilité géographique et adaptation aux contraintes du milieu méditerranéen des pins de la section *halepensis*: Resultants (provisoires) d'ún essai en plantations comparatives en France. *Annu. Sci. For.* **1992**, *49*, 261–276. [CrossRef]

25. Souvannavong, O.; Malagnoux, M.; Palmberglerche, C. Nations join to conserve forests and woodlands of the Mediterranean region. *Diversity* **1995**, *11*, 19–20.

26. Spencer, D.J. Dry country pines: Provenance evaluation of the *Pinus halepensis-P. brutia* complex in the semi-arid region of south-east Australia. *Aust. For. Res.* **1985**, *15*, 263–279.

27. Boydak, M. Silvicultural characteristics and natural regeneration of *Pinus brutia* Ten.—A review. *Plant Ecol.* **2004**, *171*, 153–163. [CrossRef]

28. Kurt, Y.; González-Martínez, S.C.; Alía, R.; Isik, K. Genetic differentiation in *Pinus brutia* Ten. using molecular markers and quantitative traits: The role of altitude. *Annu. For. Sci.* **2012**, *69*, 345–351. [CrossRef]

29. Department of Forests. *Criteria and Indicators—For the Sustainable Forest Management in Cyprus*; Ministry of Agriculture, Natural Resources and Environment: Nicosia, Cyprus, 2006.

30. Barber i Valles, A. *Contribution to the Knowledge of the Bioclimate and Vegetation of the Island of Cyprus*; Cyprus Forestry College: Nicosia, Cyprus, 1995.

31. Pantelas, V. *The Bioclima and Phytosociology in Cyprus*; Report in Department of Forests, Ministry of Agriculture, Natural Resources and Environment: Nicosia, Cyprus, 1996.

32. Shen, H.; Rudin, D.; Lindgren, D. Study of the pollination pattern in a scots pine seed orchard by means of Isozyme Analysis. *Silvae Genet.* **1981**, *30*, 7–15.

33. Conkle, M.T.; Hodgkiss, P.D.; Nunnally, L.B.; Hunter, S.C. *Starch Gel Electrophoresis of Conifer Seeds, A Laboratory Mannual*; Gen. Techn. Rep. PSW-64; Pacific Southwest Forest and Range Experiment Station: Berkeley, CA, USA, 1982.

34. Cheliak, W.M.; Pitel, J.A. Genetic control of allozyme variants in mature tissues of white spruce trees. *J. Hered.* **1984**, *75*, 34–40. [CrossRef]

35. Nei, M. F-statistics and analysis of gene diversity in subdivided populations. *Annu. Hum. Genet.* **1977**, *41*, 225–233. [CrossRef]

36. Nei, M. Estimation of average heterozygosity and genetic distance from a small number of individuals. *Genetics* **1978**, *89*, 583–590. [PubMed]

37. Peakall, R.; Smouse, P.E. GenAlEx 6.5: Genetic analysis in Excel. Population genetic software for teaching and research-an update. *Bioinformatics* **2012**, *28*, 2537–2539. [CrossRef] [PubMed]

38. Pritchard, J.K.; Stephens, M.; Donnelly, P. Genetics Society of America Inference of Population Structure Using Multilocus Genotype Data. *Genetics* **2000**, *155*, 945–959. [PubMed]

39. Evanno, G.; Regnaut, S.; Goudet, J. Detecting the number of clusters of individuals using the software STRUCTURE: A simulation study. *Mol. Ecol.* **2005**, *14*, 2611–2620. [CrossRef] [PubMed]

40. Kopelman, N.M.; Mayzel, J.; Jakobsson, M.; Rosenberg, N.A.; Mayrose, I. CLUMPAK: A program for identifying clustering modes and packaging population structure inferences across K. *Mol. Ecol. Res.* **2015**, *15*, 1179–1191. [CrossRef] [PubMed]

41. Nei, M. Genetic distance between populations. *Am. Nat.* **1972**, *106*, 283–292. [CrossRef]

42. Miller, M.P. *Tools for Population Genetic Analyses (TFPGA) Version 1.3—A Windows Program for the Analysis of Allozyme and Molecular Population Genetic Data*; Northern Arizona University: Flagstaff, AZ, USA, 1997.

43. Rohlf, F.J. *NTSyS-p.c: Numerical Taxonomy and Multivariate Analysis System*, version 2.0; Exeter Software Publishers Ltd.: Setauket, NY, USA, 1998.

44. Breiman, L.; Friedman, J.; Olshen, R.; Stone, C. *Classification and Regression Trees*; Wadsworth International Group: Belmont, CA, USA, 1984.

45. De'ath, G.; Fabricius, K.E. Classification and regression trees: A powerful yet simple technique for the analysis of complex ecological data. *Ecology* **2000**, *81*, 3178–3192. [CrossRef]

46. Mantel, N.A. The detection of disease clustering and a generalized regression approach. *Can. Res.* **1967**, *27*, 209–220.

47. Sokal, R. Testing statistical significance of geographic variation patterns. *Syst. Zool.* **1979**, *28*, 227–232. [CrossRef]

48. Pujol, B.; Pannell, J.R. Reduced responses to selection after species range expansion. *Science* **2008**, *321*, 96. [CrossRef] [PubMed]

49. Holt, R.D.; Keitt, T.H.; Lewis, M.A.; Maurer, B.A.; Taper, M.L. Theoretical models of species' borders: Single species approaches. *Oikos* **2005**, *108*, 18–27. [CrossRef]

50. Adams, W.T.; Joly, R.J. Genetics of allozyme variants in loblolly pine. *J. Hered.* **1980**, *71*, 33–40. [CrossRef]

51. Morgante, M.; Vendramin, G.G.; Giannini, R. Inheritance and linkage relationships of isozymes variants of *Pinus leucodermis* Ant. *Silvae Genet.* **1993**, *42*, 231–237.

52. Conkle, M.T.; Schiller, G.; Grunwald, C. Electrophoretic analysis of diversity and phylogeny of *Pinus brutia* and closely related taxa. *Syst. Bot.* **1988**, *13*, 411–424. [CrossRef]

53. Kara, N.; Korol, L.; Isik, K.; Schiller, G. Genetic Diversity in *Pinus brutia* TEN.: Altitudinal Variation. *Silvae Genet.* **1997**, *46*, 2–3.

54. Schiller, G.; Conkle, M.T.; Grunwald, C. Local differentiation among Mediterranean populations of Aleppo pine in their isoenzymes. *Silvae Genet.* **1985**, *35*, 11–19.

55. Panetsos, K.P.; Aravanopoulos, F.; Scaltsoyiannes, A. Genetic Variation of *Pinus brutia* from Islands of the Northeastern Aegean Sea. *Silvae Genet.* **1998**, *47*, 2–3.

56. van Staaden, M.J.; Michener, G.R.; Chesser, R.K. Spatial analysis of microgeographic genetic structure in Richardson's ground squirrels. *Can. J. Zool.* **1996**, *74*, 1187–1195. [CrossRef]

57. Panetsos, K.; Scaltsoyiannes, A.; Aravanopoulos, F.A.; Dounavi, K.; Demetrakopoulos, A. Identification of *Pinus brutia* TEN., *P. halepensis* MILL. and Their Putative Hybrids. *Silvae Genet.* **1997**, *46*, 253–257.

58. Dangasuk, O.G.; Panetsos, K.P. Altitudinal and longitudinal variations in *Pinus brutia* (Ten.) of Crete Island, Greece: Some needle, cone and seed traits under natural habitats. *New For.* **2004**, *27*, 269–284. [CrossRef]

59. Calamassi, R.; Puglisi, S.R.; Vendramin, G.G. Genetic variation in morphological and anatomical needle characteristics in *Pinus brutia* Ten. *Silvae Genet.* **1988**, *37*, 5–6.

60. Tian, M.; Yu, G.; He, N.; Hou, J. Leaf morphological and anatomical traits from tropical to temperate coniferous forests: Mechanisms and influencing factors. *Sci. Rep.* **2016**, *6*, 19703. [CrossRef] [PubMed]

61. Niinemets, Ü.; Cescatti, A.; Lukjanova, A.; Tobias, M.; Truus, L. Modification of light-acclimation of *Pinus sylvestris* shoot architecture by site fertility. *Agric. For Meteorol.* **2002**, *111*, 121–140. [CrossRef]

62. Smolander, S.; Stenberg, P. A method to account for shoot scale clumping in coniferous canopy reflectance models. *Remote Sens. Environ.* **2003**, *88*, 363–373. [CrossRef]

63. Niinemets, Ü. A review of light interception in plant stands from leaf to canopy in different plant functional types and in species with varying shade tolerance. *Ecol. Res.* **2010**, *25*, 693–714. [CrossRef]

64. Gebauer, R.; Volařík, D.; Urban, J.; Børja, I.; Nagy, N.E.; Eldhuset, T.D.; Krokene, P. Effect of thinning on the anatomical adaptation of Norway spruce needles. *Tree Physiol.* **2011**, *31*, 1103–1113. [CrossRef] [PubMed]

65. Knapp, B. Cyprus's earliest prehistory: Seafarers, Forgers and Settlers. *J. World Prehist.* **2010**, *23*, 79–120. [CrossRef]

66. Vigne, J.-D.; Briois, F.; Zazzo, A.; Willcox, G.; Cucchi, T.; Thiébault, S.; Carrère, I.; Franel, Y.; Touquet, R.; Martin, C.; et al. First wave of cultivators spread to Cyprus at least 10,600 y ago. *Proc. Natl. Acad. Sci. USA* **2012**, *109*, 8445–8449. [CrossRef] [PubMed]

67. Thirgood, J.V. *Cyprus, a Chronicle of Its Forest, Land, and People*; University of British Columbia Press: Vancouver, BC, Canada, 1987.

68. Geological Survey Department. *The Geology of Cyprus*; Article 10; Ministry of Agriculture, Natural Resources and Environment: Nicosia, Cyprus, 2002.

69. Hampe, A.; Arroyo, J.; Jordano, P.; Petit, R.J. Range wide phylogeography of a bird-dispersed Eurasian shrub: Contrasting Mediterranean and temperate glacial refugia. *Mol. Ecol.* **2003**, *12*, 3415–3426. [CrossRef] [PubMed]

70. Petit, R.J.; Aguinagalde, I.; de Beaulieu, J.L.; Bittkau, C.; Brewer, S.; Cheddadi, R.; Ennos, R.; Fineschi, S.; Grivet, D.; Lascoux, M.; et al. Glacial refugia: Hotspots but not melting pots of genetic diversity. *Science* **2003**, *300*, 1563–1565. [CrossRef] [PubMed]

71. Martin, P.R.; McKay, J.K. Latitudinal variation in genetic divergence of populations and the potential for future speciation. *Evolution* **2004**, *58*, 938–945. [CrossRef] [PubMed]

72. Eliades, N.-G.; Gailing, O.; Leinemann, L.; Fady, B.; Finkeldey, R. High genetic diversity and significant population structure in *Cedrus brevifolia* Henry, a narrow Mediterranean endemic tree from Cyprus. *Plant Syst. Evol.* **2011**, *294*, 185–198. [CrossRef]

73. Gonzalo-Turpin, H.; Hazard, L. Local adaptation occurs along altitudinal gradient despite the existence of gene flow in the alpine plant species *Festuca eskia*. *J. Ecol.* **2009**, *97*, 742–751. [CrossRef]

74. Sáenz-Romero, C.; Guzmán-Reyna, R.R.; Rehfeldt, G.E. Altitudinal genetic variation among *Pinus oocarpa* populations in Michoacán, Mexico Implications for seed zoning, conservation, tree breeding and global warming. *For. Ecol. Manag.* **2006**, *229*, 340–350. [CrossRef]

75. Byars, S.G.; Papst, W.; Hoffmann, A.A. Local adaptation and cogradient selection in the Alpine plant, *Poa hiemata*, along a narrow altitudinal gradient. *Evolution* **2007**, *61*, 2925–2941. [CrossRef] [PubMed]

76. Brousseau, L.; Postolache, D.; Lascoux, M.; Drouzas, A.D.; Källman, T.; Leonarduzzi, C.; Liepelt, S.; Piotti, A.; Popescu, F.; Roschanski, A.M.; et al. Local Adaptation in European Firs Assessed through Extensive Sampling across Altitudinal Gradients in Southern Europe. *PLoS ONE* **2016**, *11*, e0158216. [CrossRef] [PubMed]

77. Forsman, A. Rethinking phenotypic plasticity and its consequences for individuals, populations and species. *Heredity* **2015**, *115*, 276–284. [CrossRef] [PubMed]

78. Matyas, C.; Yeatman, C.W. Effect of geographical transfer on growth and survival of jack pine (*Pinus banksiana* Lamb.) populations. *Silvae Genet.* **1992**, *41*, 370–376.

79. Körner, C. The use of "altitude" in ecological research. *Trends Ecol. Evol.* **2007**, *22*, 569–574. [CrossRef] [PubMed]

80. Corcuera, L.; Cochard, H.; Gil-Pelegrin, E.; Notivol, E. Phenotypic plasticity in mesic populations of *Pinus pinaster* improves resistance to xylem embolism (P50) under severe drought. *Trees* **2011**, *25*, 1033–1042. [CrossRef]

81. Papageorgiou, A.C.; Kostoudi, C.; Sorotos, I.; Varsamis, G.; Korakis, G.; Drouzas, A.D. Diversity in needle morphology and genetic markers in a marginal *Abies cephalonica* (Pinaceae) population. *Annu. For. Res.* **2015**, *58*, 217–234. [CrossRef]

82. Denk, T.; Grimm, G.; Stögerer, K.; Langer, M.; Hemleben, V. The evolutionary history of *Fagus* in western Eurasia: Evidence from genes, morphology and the fossil record. *Plant Syst. Evol.* **2002**, *232*, 213–236. [CrossRef]

83. Chevin, L.-M.; Collins, S.; Lefèvre, F. Phenotypic plasticity and evolutionary demographic responses to climate change: Taking theory out to the field. *Funct. Ecol.* **2013**, *27*, 966–979. [CrossRef]

84. Chevin, L.M.; Lande, R. Adaptation to marginal habitats by evolution of increased phenotypic plasticity. *J. Evol. Biol.* **2011**, *24*, 1462–1476. [CrossRef] [PubMed]

85. Tiwari, S.P.; Kumar, P.; Yadav, D.; Chauhan, D.K. Comparative morphological, epidermal, and anatomical studies of *Pinus roxburghii* needles at different altitudes in the North-West Indian Himalayas. *Turk. J. Bot.* **2013**, *37*, 65–73.

86. Sheue, C.R.; Yang, Y.P.; Huang, L.L.K. Altitudinal variation of resin ducts in *Pinus taiwanensis* Hayata (Pinaceae) needles. *Bot. Bull. Acad. Sin.* **2003**, *44*, 305–313.

87. Brown, J.H.; Stevens, G.C.; Kaufman, D.M. The geographic range: Size, shape, boundaries, and internal structure. *Annu. Rev. Ecol. Syst.* **1996**, *27*, 597–623. [CrossRef]

88. Sagarin, R.D.; Gaines, S.D.; Gaylord, B. Moving beyond assumptions to understand abundance distributions across the ranges of species. *Trends Ecol. Evol.* **2006**, *21*, 524–530. [CrossRef] [PubMed]

89. Eckert, C.G.; Samis, K.E.; Lougheed, S.C. Genetic variation across species' geographical ranges: The centralmarginal hypothesis and beyond. *Mol. Ecol.* **2008**, *17*, 1170–1188. [CrossRef] [PubMed]

90. Rehfeldt, G.E.; Tchebakova, N.M.; Parfenova, Y.I.; Wykoff, W.R.; Kuzmina, N.A.; Milyutin, L.I. Intraspecific responses to climate in *Pinus sylvestris*. *Glob. Chang. Biol.* **2002**, *8*, 912–929. [CrossRef]

91. Schlichting, C.; Pigliucci, M. *Phenotypic Evolution: A Reaction Norm Perspective*; Sinauer Associates Incorporated: Sunderland, MA, USA, 1998.

92. Sultan, S.E. Phenotypic plasticity for plant development, function and life-history. *Trends Plant Sci.* **2000**, *5*, 537–542. [CrossRef]

93. Lefèvre, F.; Boivin, T.; Bontemps, A.; Courbet, F.; Davi, H.; Durand-Gillmann, M.; Fady, B.; Gaüzere, J.; Gidoin, C.; Karam, M.J.; et al. Considering evolutionary processes in adaptive forestry. *Annu. For. Sci.* **2014**, *71*, 723–739. [CrossRef]

94. Aravanopoulos, F.A. Do silviculture and forest management affect the genetic diversity and structure of long-impacted forest tree populations? *Forests* **2018**, *9*, 355. [CrossRef]

95. Hadjinicolaou, P.; Giannakopoulos, C.; Zerefos, C.; Lange, M.A.; Pashiardis, S.; Lelieveld, J. Mid-21st century climate and weather extremes in Cyprus as projected by six regional climate models. *Reg. Environ. Chang.* **2011**, *11*, 441–457. [CrossRef]

© 2018 by the authors. Licensee MDPI, Basel, Switzerland. This article is an open access article distributed under the terms and conditions of the Creative Commons Attribution (CC BY) license (http://creativecommons.org/licenses/by/4.0/).

forests

MDPI

Article

Effective Seed Dispersal and Fecundity Variation in a Small and Marginal Population of *Pinus pinaster* Ait. Growing in a Harsh Environment: Implications for Conservation of Forest Genetic Resources

Jesús Charco [1], Martin Venturas [2], Luis Gil [1] and Nikos Nanos [1,3,*]

[1] Departamento de Sistemas y Recursos Naturales, Universidad Politécnica de Madrid,
   Ciudad Universitaria s/n, 28040 Madrid, Spain; jesus.charco@gmail.com (J.C.); luis.gil@upm.es (L.G.)
[2] Department of Biology, University of Utah, 257 S 1400 E Salt Lake, UT 84112, USA;
   martin.venturas@utah.edu
[3] Forest Research Institute, Greek Agricultural Organization-Dimitra, 57006 Vassilika-Thessaloniki, Greece
* Correspondence: nikolaos.nanos@upm.es; Tel.: +34-91-336-7113

Received: 3 July 2017; Accepted: 23 August 2017; Published: 26 August 2017

Abstract: Small-size, relict and marginal tree-species populations are a priority for conservation of forest genetic resources. In-situ conservation of these populations relies on adequate forest management planning based on knowledge and understanding of both ecological (i.e., recruitment or dispersal dynamics) and population-genetic processes (i.e., female reproductive success, gene flow or inbreeding). Here, we estimate the fecundity (or female reproductive success) of adult trees (i.e., the number of successfully established offspring/adult tree) and the effective dispersal distance distribution in the pine forest of Fuencaliente (southern Spain), a small-sized, marginal and relict population of maritime pine growing on a steep, craggy hill with just 312 reproductively active individuals. Previous studies have shown the population to present reduced allelic richness and suffer from genetic introgression from nearby exotic plantations of unknown origin. Between 2003 and 2004, we surveyed all adults and recruits and we measured several adult-specific covariates, including the number of cones of all adults. The population was found to be distributed into two nuclei with 268 (Stand 1) and 44 adults (Stand 2). We used inverse modeling to adjust several dispersal-and-fecundity models including a model with random variation in fecundity among adults (Unrestricted Fecundity or UF model). Results show that: (i) the average fecundity is 2.5–3.2 recruits/adult; (ii) the mean effective dispersal distance is restricted to 13–24 m and (iii) fecundity is most likely controlled by the spatial location of adult trees in Stand ,1 but it should be considered randomly distributed in Stand 2 (in this stand five adults mothered 80% of recruits). We conclude that the low fecundity in Stand 1 and the unequal fecundity in Stand 2 may decrease the population genetic diversity and lead to lower effective population size while the low average dispersal distance may reduce the probability of this population expanding to adjacent areas. In light of the results, we define the management priorities for in-situ conservation of this population.

Keywords: dispersal kernel; forest management; regeneration; seed shadow model

1. Introduction

Small-sized and isolated populations of tree species located in the rear-edge of the range distribution are considered long-term stores of genetic diversity and their conservation is crucial for adaptation of tree species to climate change [1,2]. In situ conservation of genetic resources in such populations relies greatly on adequate forest management aiming at maintaining ecological and population-genetic processes while simultaneously reducing the risk of random genetic drift and allele fixation [3]. Determination, however, of specific management actions to be taken is very challenging without prior information and understanding of some relevant processes governing recruitment by adults, including the adult seed dispersal potential (i.e., the dispersal distance distribution between parents and their successfully established offspring) or the factors affecting adult reproduction success or fecundity (i.e., the number of successfully established offspring produced by an adult tree).

The fecundity of adult trees and the dispersal-distance distribution has often been studied via inverse modeling (IM), a method that estimates simultaneously the number of successfully established offspring and the dispersal-distance distribution using the seed-shadow model [4,5]. In addition, estimation of both parameters through the use of molecular markers for parental assignment has further enhanced the accuracy and precision of these models [6].

The average fecundity and its variation across adults of a population are crucial parameters in population dynamics, especially in small-sized and isolated populations. Fecundity variation among adults will determine the population´s effective size or the within-population spatial genetic structure [7,8]. In addition, depending on pollen flow dynamics, unequal fecundity across adults may likely lead to biparental inbreeding and to a higher risk of random allele fixation and genetic drift that will shape the within-population genetic diversity and its adaptive potential.

For marginal, small-sized and isolated populations suffering from frequent human-induced changes in their population size seed dispersal at short and long distances, determine their probability to recover their initial population size or to occupy new territories [9,10]. Thus, seed dispersal in these populations is intimately related to the probability of the population to survive and/or migrate to new territories, leading to local extinction in cases with limited seed dispersal and/or restricted fertility of adults [9]. Therefore, the study of seed dispersal dynamics is necessary not only for studying species' responses to climate change but also for developing realistic management plans aiming at conserving species genetic resources in situ [11].

Maritime pine, the focal species of this study, is a wind-dispersed species of the western Mediterranean. In Spain, its natural distribution has been divided into 27 provenances, five of which are considered restricted due to their small population size [12]. This study focuses on one of the restricted provenances of maritime pine, the relict and marginal pine forest of Fuencaliente (Ciudad Real, Spain; Figure 1) located on the southern edge of the species natural distribution range. The Fuencaliente population is considered the unique representative of natural maritime pine forests in the Sierra Morena mountain range (central-southern Spain) and is highly isolated from other natural populations of the species (see the species distribution map in [12]). Historic [13] and palaeobotanical studies suggest that maritime pine grew throughout Sierra Morena mountain range and its surroundings since the Pliocene until the Late Holocene [14–16]. The maritime pine range-size reduction, that led Fuencaliente to become a relict, resulted from anthropogenic landscape transformation during the last 4000 years [17]. The main transformation drivers were wood and charcoal over-exploitation for mining and recurrent fires to favor pasture for livestock [18].

The small population of Fuencaliente managed to survive in a marginal habitat (a steep craggy hill) presumably because fires of the last centuries (either natural or human-induced) could not propagate easily in this rocky site. The evolution of this fragmented population depends largely on a well-known process taking place all over the Mediterranean basin: extensive livestock management until the 1970s (approximately) has ceased and the associated activities (frequent pasture burning to induce resprouting that assures livestock feeding) have disappeared. As a result, tree species' populations are expanding to areas previously used as pastureland [19].

Several studies using allozymes, chloroplast or microsatellite markers have highlighted the importance of Fuencaliente population (as well as other relict populations) to the total species diversity in the Iberian Peninsula [20,21]. This population has a marked reduction in allelic richness, compared to other populations of maritime pine in Spain [22]. In order to aid Fuencaliente's conservation, recent investigations have studied genetic introgression patterns from nearby exotic plantations of unknown origin [23–25]. Another recent study has evaluated the damage that deer (*Cervus elaphus* L.) cause to this population by rubbing [26]. However, information on recruitment dynamics and its dispersal potential is still scarce, and this data would serve well to establish an adequate conservation plan, which has yet to be reported.

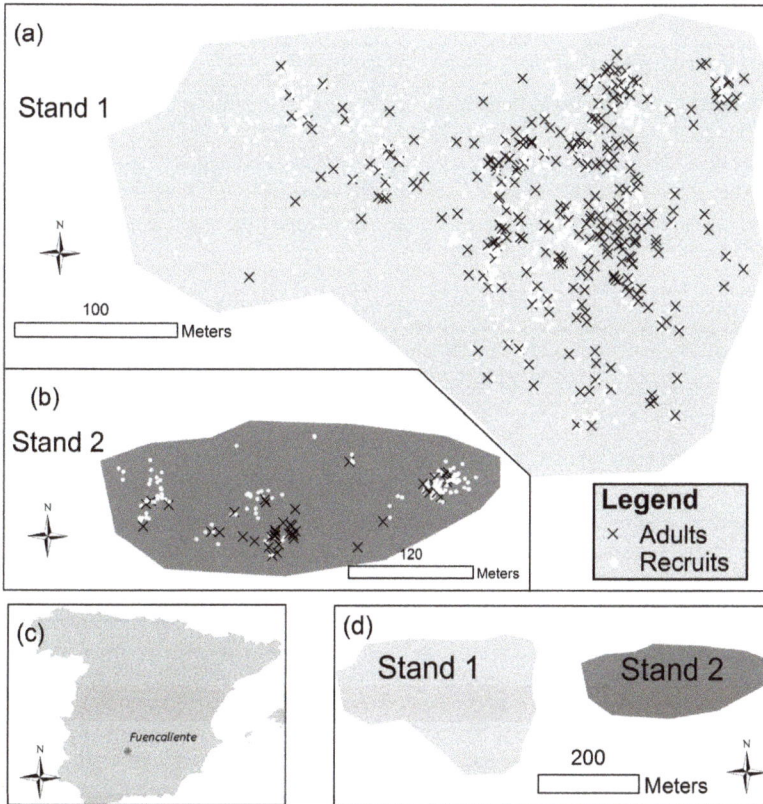

Figure 1. Maps of the maritime pine population of Fuencaliente. (**a**) Location of adults and recruits in Stand 1 and (**b**) in Stand 2; (**c**) Location of Fuencaliente population within Spain; (**d**) Location of both stands in relation to each other.

The general objective of this study was to determine recruitment patterns in the population of Fuencaliente and propose specific forest-management actions aiming at conservation of forest genetic resources. We produced accurate estimates of the effective dispersal distance distribution and of adult fecundity using IM. In addition, we determined the most likely tree-specific traits controlling fecundity variation among adults.

2. Materials and Methods

2.1. Study Area

The relict population of *Pinus pinaster* Ait. is located in the northern part of Sierra Morena mountain range and, more specifically, in Sierra Madrona (38°25′ N, 4°15′ W; Figure 1) close to Fuencaliente village (Ciudad Real). The site is protected under the Natura 2000 network. The climate in this area is Mediterranean-continental, with 14.5 °C mean annual temperature and 680 mm mean annual precipitation (59 mm in the summer months; [26]). The average altitude of the site is 1011 m a.s.l. (ranging from 907 to 1114 a.s.l.). Soils are rocky, poor in nutrients and acid with a main lithological substrate of quartzites [27]. The steepness (40% average slope), the southern exposure, and the eroded terrain greatly reduce the site quality, which can be considered as the most limiting factor for growth and regeneration, conferring this site its marginal characteristics.

2.2. Forest Inventory

In 2003 we performed a full forest survey within the area occupied by the pine population by tagging all individuals with a basal diameter larger than 1.5 mm (only lignified individuals were recorded). The spatial location of surveyed individuals was defined using a GPS (positioning errors were smaller than 1 m). In 2004, we measured tree total height, diameter at the base of the tree and diameter at breast height (DBH) of all trees tagged in 2003 and we determined the number of open cones and the number of serotinous cones on the crown of all trees by counting all visible cones. For the following analyses, trees were classified into two categories:

-      Adults: trees with at least one cone on their crown
-      Recruits: the rest of the individuals

In 2015 we selected 20 recruits to provide a rough description of the age of the recruitment cohort at the time when the forest inventory was realized. Sampling was based on a systematic rectangular grid superimposed on the spatial distribution map of recruits; individuals closest to the grid nodes were selected for sampling. The age of the sampled recruits was determined by growth-ring counting from cores extracted from their base with a Pressler drill (when their basal diameter was sufficiently large). Small-sized recruits were transported to the laboratory for age measurements via visual counting of tree rings at the tree base.

2.3. Modeling Effective Dispersal

Effective dispersal was modeled using the seed-shadow approach [28]. Quadrat counts (2 m × 2 m in size) of recruits were computed using their spatial coordinates. The observed sample $(n_1, \ldots, n_M)$ consists of the locations and the number of recruits in the j-th quadrat ($j = 1, \ldots, M$) as well as the locations of N adult trees (tree is indexed by i, $i = 1, \ldots, N$) along with several covariates measured on each adult (i.e., DBH, number of cones, etc.). We assumed that the number of recruits λ_{ij} originating from tree i and dispersed to quadrat (or site) j is Poisson distributed with expected value:

$$\lambda_{ij} = S_i \, P(\text{a recruit from tree } i \text{ lands upon site } j) \approx S_i A_j f_R(r) / 2\pi \, r_{ij} \tag{1}$$

Being S_i the fecundity of tree i (i.e., the number of recruits originating from the specific adult), r_{ij} the distance separating adult i form quadrat j and A_j the area of site j. Under this model, the total number of recruits on site j from the N trees is also Poisson distributed with expected value:

$$\lambda_j = \sum_{i=1}^{N} \lambda_{ij} \tag{2}$$

The $f_R(r)$ term of Equation (1) designates the probability density for the random dispersal distance (r) assumed to follow a lognormal distribution with probability density:

$$f_R(r) = \frac{1}{r\sigma\sqrt{2\pi}} \exp\left(-\frac{(\ln r - \mu)^2}{2\sigma^2}\right) \quad r > 0 \tag{3}$$

and μ and σ^2 the scale and shape parameters, respectively.

2.4. Models for Fecundity

We modeled the fecundity term of the seed-shadow model using three approaches. First, fecundities were allowed to vary among adults without any restriction by assigning a parameter to every adult [29]. This model (UF model) has a large number of parameters to be estimated and assumes that variability among the adult´s reproductive success is purely random (Table 1).

Table 1. Alternative models for the fecundity term of the seed shadow model.

Type of Model	Model Name	Abbr.	Formula	Number of Parameters	
				Stand 1	Stand 2
Full model	Unrestricted Fecundity	UF	S_i	268	44
Null model	Mean fecundity	MF	$\overline{S}$	1	1
Tree size covariates	Basal area	BA	$S_i = \beta \times Ba_i$	1	1
	Height	H	$S_i = \beta \times h_i$	1	1
Cone number covariates	Total cones	Tc	$S_i = \beta \times Tc_i$	1	1
	Open cones	Oc	$S_i = \beta \times Oc_i$	1	1
	Serotinous cones	Sc	$S_i = \beta \times Sc_i$	1	1
Spatial covariates	X coordinate of adult	Xco	$S_i = \beta \times Xco_i$	1	1
	Y coordinate of adult	Yco	$S_i = \beta \times Yco_i$	1	1

S_i: Fecundity of tree i; $\overline{S}$: average fecundity (over all trees of the stand); Ba_i: basal area; h_i: tree height; Tc_i: total cones; Oc_i: number of open cones; Sc_i: Number of serotinous cones; Xco_i and Yco_i: spatial coordinates of tree i and β a parameter.

The second model (MF) assumes that adult fecundities may be modeled using the average fecundity of adults. Under this model, adults are assumed to have produced the same number of recruits so that a unique estimated parameter (i.e., the average fecundity) is sufficient to describe between-adult variation in reproductive success. Finally, we used adult-specific covariates to model fecundity (as in classical seed-shadow modeling). Several covariates were used for this purpose (Table 1) that can be grouped into three categories depending on the nature of the covariate:

- Tree size covariates (basal area, BA, and tree height, H). From an ecological perspective, these models inherently assume that reproductive success is a linear function of tree size.
- Cone number covariates (total cones, Tc, open cones, Oc, and serotinous cones, Sc). Models using covariates related to the cone number assume that the number of seeds and number of recruits produced by adults is linearly related.
- Spatial covariates (the east-west, Xco, and the north-south, Yco, coordinates of adults). Inherently, these models assume the reproductive success has some relation to the microhabitat conditions surrounding the adult tree.

2.5. Parameter Estimation

Parameters of the UF model were estimated through maximization of the incomplete-data log-likelihood function of the Poisson distribution:

$$l_c(\mu, \sigma, S_1, \ldots, S_N) = \sum_{i=1}^{N} \sum_{j=1}^{M} \left(-\lambda_{ij} + n_{ij} \ln(\lambda_{ij}) - \ln(n_{ij}!) \right) \tag{4}$$

through the EM-algorithm [30] using the procedure described in [29]. Parameters of the rest of the models were estimated via maximization of the complete-data log-likelihood:

$$\sum_{j=1}^{M} \left(-\lambda_j + n_j \ln(\lambda_j) - \ln(n_j!) \right) \tag{5}$$

as described in several publications on seed dispersal modeling (see, for instance, [28] or [4]). Maximization was achieved through numerical optimization using the nlminb function of R [31]. The negative binomial distribution that has been employed in other studies [32] could not be used since the UF model parameters may be estimated analytically only through the Poisson likelihood.

2.6. Model Comparison

The best model to describe fecundity and dispersal was selected using the corrected Akaike´s Information Criterion (AICc) [33]. Model selection was facilitated by the computation of the following measures/indices:

- The difference between the AICc for the k-th model and the one with the smallest AICc (AICc$_{min}$):

$$\Delta_k = AICc_k - AICc_{min} \qquad (6)$$

- The correlation coefficient between observed and predicted counts in quadrats of the k-th model

When the UF model resulted in a valid model, we computed the correlation coefficient between fecundities estimated by this model with the adult-specific covariates. A standard t-test was used to test the hypothesis of the correlation coefficient being larger than zero.

3. Results

3.1. Descriptive Results

The population was found to be divided into two nuclei occupying an area of 7 ha (western stand) and 3.5 ha (eastern stand; we will use the term Stand 1 and Stand 2 for future reference of the western and the eastern stand, respectively). The spatial distribution of adults and recruits in the two stands can be seen in Figure 1. In Stand 1, the average fecundity was 2.56 recruits/adult (268 adults and 686 recruits). In Stand 2, the average fecundity was slightly higher, 3.25 recruits/adult (44 adults and 143 recruits). The average number of cones/adult of Stand 1 (58.2 cones/adult) was twice as large as for Stand 2 (33.9 cones/adult). In addition, a remarkable among-tree variation was found for both stands in the number of cones/adult, which varied between 1 and 587 in Stand 1, and between 1 and 193 in Stand 2. Both distributions were skewed with 25% of trees bearing less than 5 and 3.8 cones/adult for Stand 1 and 2, respectively.

Both stands had a considerably low adult density (38.2 trees/ha and 12.5 trees/ha for Stand 1 and 2, respectively). Both stands were uneven-aged in their structure with several individuals occupying the lower diameter classes (see the DBH column in Table 2). The diameter at the base of the tree-trunk (a variable used as a substitute to DBH in this multi- cohort stand) was very similar for the two stands (27.3 cm and 28.1 cm for Stands 1 and 2, respectively). Finally, the average adult-tree height was 5.4 m for both stands, and 25% of adult trees had a height smaller than 3.5 m.

The mean height of recruits (i.e., individuals without visible cones on their crown) was 0.8 m for both stands, while the mean DBH was 0.5 and 0.3 cm for Stand 1 and Stand 2, respectively. Some large sized individuals exhibited no cones in their crown and, therefore, were classified as recruits (the maximum height of recruits was 6.6 and 4.3 m for Stand 1 and 2, respectively; Table 2). Large-sized recruits without any cones on their crown may have been dispersing some seeds during the previous years, but the total number of seeds produced must have been small (cones of maritime pine, especially serotinous ones, persist on the crown during several years), thus expected errors from misclassification in the recruitment cohort may have no practical importance.

Tree age measurements performed on a subsample of recruits showed that this cohort consisted of individuals that germinated during the time interval from 1973 to 1995 (in 2004 recruits were between 9 and 31 years old, the average age being 22 years).

Table 2. Descriptive statistics for adults and recruits of both stands.

		Dbase (cm)		DBH (cm)		Height (m)		Total Cones	
		St1	St2	St1	St2	St1	St2	St1	St2
Adults	Min.	5.7	7.6	0.0	0.0	0.5	1.2	1.0	1.0
	1st Qu.	16.5	18.9	8.8	13.1	3.5	3.5	5.0	3.8
	Median	23.2	27.4	17.2	19.8	5.0	5.8	17.0	11.5
	Mean	27.3	28.1	19.2	20.1	5.4	5.4	58.2	33.9
	3rd Qu.	35.7	35.5	28.7	27.5	7.0	6.8	67.5	52.0
	Max.	68.8	65.6	56.3	44.6	16.5	10.0	587.0	193.0
Recruits	Min.	0.1	0.3	0.0	0.0	0.04	0.07	0.0	0.0
	1st Qu.	2.1	4.1	0.0	0.0	0.3	0.5	0.0	0.0
	Median	4.7	7.6	0.0	0.0	0.6	0.8	0.0	0.0
	Mean	5.7	7.7	0.5	0.3	0.8	0.8	0.0	0.0
	3rd Qu.	8.2	11.3	0.0	0.0	1.0	1.0	0.0	0.0
	Max.	34.3	31.5	19.0	11.1	6.6	4.3	0.0	0.0

St1: Stand 1; St2: Stand 2.

3.2. Choosing the Best Model for Fecundity

Not surprisingly, the UF model showed the largest log-likelihood in both stands (Table 3). Given the high number of free parameters, this model has a very flexible structure that permits local adjustment due to tree-to-tree differences in fecundities. However, the use of AICc (that punishes models having too many parameters) showed that the best-fit model differed depending on the stand considered.

Table 3. Comparison statistics for different dispersal-and-fecundity models for the two stands (models are ordered, within each stand, according to smaller AICc).

	Model	Ln(L)	AICc	Δ_k	cor
Stand 1	Yco	−2753.7	5513.4	0.0	0.20
	Tc	−2762.5	5531.0	17.6	0.20
	Sc	−2764.1	5534.3	20.9	0.19
	MF-null	−2771.3	5548.7	35.3	0.18
	BA	−2823.4	5652.8	Nc	0.18
	UF	−2431.8	5749.5	Nc	0.30
	H	−2883.6	5773.2	Nc	0.15
	Xco	−2911.9	5829.8	Nc	0.16
	Oc	−2982.4	5970.8	Nc	0.16
Stand 2	UF	−546.2	1220.9	0.0	0.34
	Yco	−628.1	1262.3	41.4	0.22
	Xco	−642.0	1290.2	69.3	0.20
	MF-null	−642.8	1291.8	70.9	0.18
	Tc	−661.8	1329.8	Nc	0.18
	Sc	−663.8	1333.8	Nc	0.18
	Oc	−667.5	1341.1	Nc	0.17
	BA	−691.0	1388.1	Nc	0.11
	H	−692.0	1390.2	Nc	0.12

Ln(L): log-likelihood; AICc: Bias-corrected Akaike´s Information Criterion; Δk: Delta AIC (with respect to the model with smaller AICc); Nc: indicates that the corresponding model was not considered (models exhibiting an AICc larger than the MF model were not considered in comparisons); cor: correlation coefficient between observed and predicted counts in quadrats.

In Stand 1, the model using one of the spatial coordinates of adults (Yco, the north-south coordinate) was the best in terms of AICc (for this model $\Delta_k = 0$). The model that assumes an average fecundity for adults (MF model) had a $\Delta_k = 35.3$, substantially larger than the best model. In addition, five models (BA, UF, H, Xco and Oc) had a Δ_k larger than the MF model (Table 3) and therefore were not considered in ulterior analyses. Two models, however, including total cones (Tc) or serotinous cones (Sc) as covariates showed an Δ_k smaller than the MF model (17.6 and 20.9, respectively) and could be considered as alternatives. Notwithstanding, their relative efficiency showed that both were poor approximations of the process generating the dispersal data of Stand 1 as values of $\Delta_k > 10$ are indicative of no empirical support for the corresponding model [34]. Therefore, the results indicated that there was sufficient experimental evidence to conclude that the model using the north-south coordinate of adults (Yco model) was the best model to describe the process generating the dispersal pattern in Stand 1. However, the Yco model showed a rather poor fit considering the low correlation coefficient between observed and predicted counts ($r = 0.2$, Table 3). In addition, note that the best correlation coefficient was much higher for the UF model ($r = 0.3$, Table 3).

The results obtained in Stand 2 were quite different. The smallest AICc was obtained for the UF model (AICc = 1220.9), that, by definition, was assigned $\Delta_k = 0$. The null model (MF model), on the other hand, exhibited an $\Delta_k = 70.9$ with respect to the UF model. Five other models showed a Δ_k larger than the MF model and were not considered in ulterior analyses (Table 3). These were the two tree-size covariates models (BA and H models) and the three seed-set covariate models (Tc, Oc and Sc). Thus, the only models exhibiting an Δ_k lower than the one obtained for the null model (apart from the best in terms of AICc, i.e., the UF) were the ones incorporating the spatial coordinates as proxies to fecundity (Yco and Xco). Though, the evidence of the UF model against the two "spatial" models was very strong ($\Delta_k = 41.4$ and 69.3 for the Yco and the Xco models, respectively). Conclusively, our results supported the hypothesis that the UF model was the best model to describe the process generating the dispersal pattern in Stand 2. Finally, the UF model was the best in terms of minimizing the information loss as it exhibited a rather acceptable fit given the high correlation coefficient obtained between observed and predicted counts ($r = 0.34$, Table 3).

3.3. Dispersal and Fecundity Parameter Estimates

Table 4 shows the estimated parameters for the two stands and the nine postulated models. In addition, it presents the average dispersal distances (mean, mode and median) only for models showing a Δ_k smaller than the null model of the corresponding stand. Within the same stand, mean dispersal distances estimated with different models were very similar. For Stand 1, the mean dispersal distance estimates varied between 20.7 and 25.9 m, whereas for Stand 2 they were slightly smaller as they ranged between 12.9 and 15.5 m. The mean dispersal distance in Stand 1 (24 m) was twice as large as in Stand 2 (12.9 m) according to the best model of each stand (i.e., Yco in Stand 1 and UF in Stand 2).

Table 4. Parameter estimates and average dispersal distances for the two stands and nine tested models (models are ordered according to increasing Δ_k as in Table 3).

	Model	μ	σ^2	β	Median	Mean	Mode
	Yco	2.66	1.02	0.01	14.30	24.05	5.05
	Tc	2.87	0.77	0.04	17.67	23.74	9.79
	Sc	2.95	0.78	0.08	19.11	25.90	10.40
	MF-null	2.57	0.96	2.58	13.07	20.71	5.20
Stand 1	BA	2.71	0.86	35.33	Nc	Nc	Nc
	UF	2.26	0.87	-	Nc	Nc	Nc
	H	2.72	0.99	0.00	Nc	Nc	Nc
	Xco	2.77	1.16	0.01	Nc	Nc	Nc
	Oc	4.25	1.45	0.14	Nc	Nc	Nc
	UF	2.27	0.76	-	9.69	12.94	5.43
	Yco	2.31	0.89	0.02	10.07	14.97	4.56
	Xco	2.32	0.92	0.004	10.18	15.54	4.36
	MF-null	2.32	0.91	3.32	10.18	15.40	4.45
Stand 2	Tc	2.67	0.80	0.10	Nc	Nc	Nc
	Sc	2.67	0.80	0.24	Nc	Nc	Nc
	Oc	2.69	0.81	0.16	Nc	Nc	Nc
	BA	2.46	0.86	44.50	Nc	Nc	Nc
	H	2.38	0.97	0.01	Nc	Nc	Nc

μ,σ^2: Dispersal kernel parameters, β: parameter of the fecundity model; Nc: Not considered (average dispersal distances (median, mean and mode) are estimated only for models showing a Δ_k smaller than the one for the null model of the corresponding stand.

Fecundity estimates in both stands are presented in Table 5. The best model for Stand 1 (Yco model) predicted an average fecundity of 2.6 recruits/adult; this is very similar to the value obtained from the inventory results. Variability of the estimated fecundities was, however, very small for this model (fecundities ranged from 0.1 to 4.3 recruits/adult). Figure 2 shows, in addition, that the relative contribution of each adult to the regeneration cohort, estimated through the fecundities of the Yco model is very homogeneous in Stand 1. Notwithstanding, the variance of (estimated) fecundities is much larger for other models in this stand (according to model Tc, for instance, fecundities of adults varied between 0.04 and 26.08 recruits/adult).

Table 5. Descriptive statistics for estimated fecundities for adjusted models in two stands.

	Model	Ave	Min	Max	Var	Sum
	Yco	2.6	0.1	4.3	0.9	699
Stand 1	Tc	2.5	0.04	26.0	18.3	692
	Sc	2.5	0.0	47.6	24.3	693
	MF-null	2.5	2.5	2.5	0.0	693
	UF	3.2	0.0	36.8	63.1	143
Stand 2	Yco	3.3	2.5	4.5	0.3	145
	Xco	3.3	2.5	4.0	0.2	146
	MF-null	3.3	3.3	3.3	0.0	146

Ave: Average fecundity; Max, Min: Maximum and minimum estimated fecundity; Sum: Summation of all the estimated fecundities in the stand; Var: Among-adults variance of the estimated fecundities. Note: Models are ordered according to increasing Δ_k However, only models with Δ_k larger than the null model are shown in each stand.

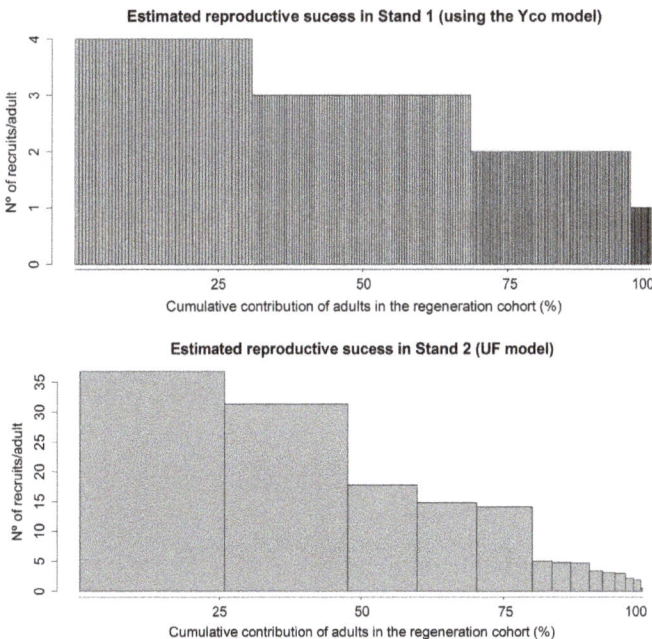

Figure 2. Adult reproductive success estimated through the best model in each stand. The models using the spatial coordinate (north-south) as a proxy to fecundity (Yco model) is used in Stand 1 (**upper** panel). In Stand 2 (**lower** panel) the UF model is used to estimate fecundities. Bar heights represent the estimated number of recruits produced by each adult. Lines represent the estimated relative contribution of each adult (cumulative) to the total number of recruits in each stand.

In Stand 2, where the best model was the UF model, we found a much larger variability among estimated fecundities (ranging between 0 and 36.8 recruits/adult) while the variance of the estimated fecundities (63.1) was the largest of all models and stands (Table 5). The rest of the candidate models (Yco, Xco and MF) showed very similar average fecundity estimates but a much smaller among-tree variance in fecundity estimates. Finally, according to the UF model predictions, the fecundity distribution in Stand 2 was very skewed, with three trees having produced 60% of the regeneration (Figure 2). In addition, 30 adults exhibited an estimated fecundity equal to zero and may be considered as being reproductively inactive. Notwithstanding, the number of reproductively inactive adults may be overestimated by the UF model when adults are spatially clustered [35].

Finally, a *t*-test for the correlation coefficient between the fecundities estimated via the UF model and other adult covariates (i.e., basal area, height, cone numbers and spatial coordinates) showed that correlation was significantly larger than zero only in the case of the y-coordinate of the adult tree ($r = 0.34$, $t = 2.31$ with 42 degrees of freedom, $p = 0.02$). This seems to confirm the results of Table 4 (Stand 2), where we showed that the best model (after the UF model) in terms of AICc reduction was the one using the y-coordinate of adults as a proxy for fecundity. A similar statistical test for Stand 1 was not performed since the UF was not judged appropriate to model these data.

4. Discussion

4.1. Fecundity Dynamics

Cone production by adult trees does not seem to be a limiting factor to regeneration and expansion of this population given that 268 trees in Stand 1 and 44 in Stand 2 are seen with cones in their crowns. Furthermore, some of them (the most vigorous) may be classified as high seed producers having up to 587 and 193 cones in Stand 1 and 2, respectively, while some really short individuals (just 0.5 m height) are seen with cones in their small crowns. Nevertheless, the distribution in reproductive success (in the seed stage) that was shown to be unequal among adults (25% of adults have less than 5 visible cones on their crown in Stand 1) should be a factor that reduces the regeneration potential of this stand.

On the other hand, fecundity is dramatically reduced in both stands. Recruitment in this population started to be effective only after 1973 (the maximum age in the sample of recruits was 31 years when sampling was performed in 2004). This time period coincides with the ceasing of extensive livestock management and the disappearance of the associated activities, such as frequent pasture burning. However, the average number of recruits/adult (3.25 recruits/adult in Stand 2) is a clear indication of low recruitment potential. Several factors may be responsible; among others is the high density of deer living in the area, which by browsing and fraying can cause severe damage and mortality to the regeneration cohort [26]. Deer can also affect fecundity of mature pines since those trees highly damaged by fraying produce fewer cones [26]. In addition, severe summer drought effects in conjunction with the poor edaphic conditions (complete absence of soil in many areas due to erosion) may be important factors contributing to massive mortality during the seed-to-sapling transition. A recent study using seed from this population [23] showed that seedlings obtained from seeds of Stand 1 exhibited higher mortality rates in water-stress experiments as compared to seedlings grown from seeds from nearby plantations.

Our results for Stand 2 also indicate that the distribution of fecundity across adults is highly skewed with the three most successful individuals (6.8% of adults) mothering 60% of the regeneration cohort. Unequal reproductive success in the seedling and sapling stages is a common phenomenon in forest stands and it has been reported previously in maritime pine in Spain [36] and in other tree species [35]. Should this pattern repeat itself during the subsequent years, we may predict an additional loss of genetic variation in a population that has been shown to be poor in terms of allelic composition of the adult cohort [22]. In addition, the unequal fecundity of adults in small populations can increase the genetic bottlenecks and further exacerbate the risk of random genetic drift. These results can aid conservation measures by helping identify individuals from which seed should be collected in order to minimize diversity losses, i.e., from those trees that the model shows that are not mothering recruits.

Results on variance of reproductive success are not reported in Stand 1 because the UF model that allows for a covariate-free estimation of fecundities had to be rejected due to its Δ_k being smaller than the null model. Interestingly, the UF model was the best model in Stand 2 (with 3.25 recruits/adults) but could not be informative enough in Stand 1 with just 2.56 recruits/adult given the high number of parameters to be estimated.

The results of this study may further enhance our knowledge on the factors governing fecundity and dispersal. According to our results, the hypothesis that fecundity is greater for bigger individuals should be rejected (see also [8,36]). Models based on tree size (basal area and tree height) to estimate fecundity were rejected for their performance being smaller than the null model. The same conclusion may be reached for covariates related to cone counts in Stand 2. However, in Stand 1, the use of the number of cones as a covariate improved the null model's performance so, in the absence of other more informative covariates, we may accept the hypothesis that effective fecundity is positively correlated to seed set variables (total number of cones/adult). Nevertheless, these results have to be interpreted with caution since this species has serotinous cones [35] and, therefore, it may be possible that results differ if regeneration is evaluated after a fire since, in that case, we may expect higher fecundity for trees with a higher number of serotinous cones.

Additionally, future studies using molecular markers to identify parentage relationships between parents and successfully established offspring may be used to confirm the patterns revealed by inverse modeling ([37,38]). Previous studies have shown that results for sapling dispersal kernels estimated via classical inverse modeling were remarkably similar to the ones obtained by genetic methods [36,39].

Our results highlight the importance of spatial covariates in estimating the effective adult fecundity and further support the hypothesis that in this relict population of maritime pine the most important factors in shaping the reproductive success of adults (evaluated as the number of seedlings and saplings they disperse) are related to their spatial location. According to model parameter estimates, adults with higher reproductive success are located in the northern part of both stands (note that the β parameter for the Yco models is positive in both stands, Table 4), where we may find better site qualities that allow for higher success in seed germination and establishment. It should be noted that the northern part of these stands corresponds to higher elevations in this population; therefore, Yco covariate might be capturing an environmental gradient, (i.e., a factor causing a direct effect on seed germination and establishment). Another possible explanation is that population expansion from south to north is causing higher reproductive success in the northern part because more sites free of intraspecific competition are available in this previously unoccupied area.

4.2. Effective Dispersal Distances

Average effective dispersal distances were reduced in this population (24 m in Stand 1 and 13 m in Stand 2). In other *Pinus pinaster* stands of central range [14], an average effective dispersal distance of 40 to 60 m was reported (depending on the model used). The most comprehensive study of maritime pine primary seed dispersal (i.e., before Janzen-Connell effects and secondary dispersal) performed in Spain [40] reported average primary seed dispersal distances of 14–25 m. Therefore, our findings suggest that the average effective dispersal of Fuencaliente population is in the same order of magnitude as primary seed dispersal in other populations.

Reduced dispersal distances may be due to the small height of trees in this population. Indeed, the average tree height of adult individuals was only 5.4 m, and the shortest mature tree in our data-set, bearing two cones on its crown, was just 45 cm in height. As seed dispersal distances are strongly correlated with plant height [41], dispersal distances may have been drastically reduced by tree height (seed release height), which is directly related to site quality. Other factors that could also account for the strong dispersal limitation observed, such as the absence of animal-mediated dispersal or density-dependent mortality [34,42], should also be considered but our data-set does not allow for more specific conclusions to be drawn.

In addition, the fact that the population grows in low stand densities does not seem to have favored effective dispersal at larger distances by offering sites free of conspecific competition [19]. This should be expected in this stand since the low stem density does not imply the existence of adequate sites for seedling establishment. Indeed, favorable microsites for establishment are very limited due to complete soil loss (in some cases) or because they have been already occupied by adult conspecifics. Under these conditions, it seems very unlikely that new recruits will establish in this population in the mid-term. Seemingly, dispersal potential and population expansion are drastically reduced in poor environments due to both decreasing adult tree height and safe-site limitation for seedling establishment.

4.3. Management Implications

In situ conservation of this population is a great priority for forest management and conservation, especially taking into account the results of recent studies on genetic diversity [26,28]. Due to its relict nature, the population of Fuencaliente has been included as one of the provenances of *Pinus pinaster* and is also considered as a model population for studying genetic introgression by exotic plantations [23]. However, no specific measures have been implemented for the protection, conservation, and restoration of Fuencalinente pine population other than being part of a Natura 2000 Site.

Measures to support the conservation and recovery of this population should include both legislative initiatives towards generating a special legal status that will recognize the uniqueness of the population as well as actions with local institutions and society to achieve higher awareness for this forest genetic resource.

Managerial actions may also enhance the viability of the population. Seeds from the individuals that show lower fecundity, mainly located in the southern part of both stands, should be used for population augmentation. Seedlings grown from these seeds could be used in small plantation programs within the population targeting favorable microhabitats with limitations in seed arrival. This action would increase or at least maintain stand genetic diversity by enhancing recruitment of reproductively inactive individuals. Additionally, as we report in a parallel study [26], deer density in the area should be reduced to avoid excessive damage caused by these animals to regeneration and adult individuals.

On the other hand, silvicultural treatments should be applied in Fuencaliente's surroundings to reduce fire risk as currently, the population might not be able to recover if it burns due to its small size. Moreover, canopy seed bank contained in closed cones has been severely reduced since the arrival of squirrels (*Sciurus vulgaris* L.) in this area in 2005 [26]. Finally, seeds from this local population should be used in plantation programs in the nearby areas in accordance with [23] proposal after studying introgression dynamics between the local and the introduced gene pools.

5. Conclusions

Fecundity is very low in the pine forest of Fuencaliente and managerial actions seem to be necessary for the conservation of its genetic resources. In addition, unequal fecundity among adults in one of the stands suggests that further loss of genetic diversity may be expected from reproductive inactivity of the majority of adult trees in Stand 2. Furthermore, we found no evidence that adult tree size is related to higher adult fecundity. However, we found evidence that adult fecundity is mostly related to the spatial location of adults, a finding suggesting that the most limiting factor to regeneration is the lack of microsites favorable for seed emergence and survival. Finally, we report very low dispersal distances for successfully established offspring, a fact that further reduces the probability of the population to expand to adjacent and unoccupied areas. Forest management should aim at reinforcing the reproductive potential of less fecund adults by designing small plantation programs within the population targeting favorable microhabitats with limitations in seed arrival.

Acknowledgments: This work is part of an agreement on conservation and improvement of conifer genetic resources between the Directorate General for Nature Conservation (Ministry of Environment, Spain) and the Technical University of Madrid. It has also received financial support from the Regional Government of Castile-La Mancha. We also thank Fuencaliente Municipality for giving us permission and logistic support for accessing the site to collect the data.

Author Contributions: J.C., L.G. and N.N. conceived and designed the experiments; J.C. performed the experiments; J.C., L.G., M.V. and N.N. analyzed the data; L.G. contributed reagents/materials/analysis tools; N.N. led the writing of the paper. N.N. and L.G. supervised the doctoral thesis of J.C.

Conflicts of Interest: The authors declare no conflict of interest

References

1. Hampe, A.; Petit, R.J. Conserving biodiversity under climate change: The rear edge matters. *Ecol. Lett.* **2005**, *8*, 461–467. [CrossRef] [PubMed]
2. Ganopoulos, I.; Aravanopoulos, F.A.; Argiriou, A.; Kalivas, A.; Tsaftaris, A. Is the genetic diversity of small scattered forest tree populations at the southern limits of their range more prone to stochastic events? A wild cherry case study by microsatellite-based markers. *Tree Genet. Genom.* **2011**, *7*, 1299–1313. [CrossRef]
3. Lowe, A.J.; Boshier, D.; Ward, M.; Bacles, C.F.E.; Navarro, C. Genetic resource impacts of habitat loss and degradation; reconciling empirical evidence and predicted theory for neotropical trees. *Heredity* **2005**, *95*, 255–273. [CrossRef] [PubMed]
4. Ribbens, E.; Silander, J.A.; Pacala, S.W. Seedling recruitment in forests—Calibrating models to predict patterns of tree seedling dispersion. *Ecology* **1994**, *75*, 1794–1806. [CrossRef]
5. Nathan, R.; Muller-Landau, H.C. Spatial patterns of seed dispersal, their determinants and consequences for recruitment. *Trends Ecol. Evol.* **2000**, *15*, 278–285. [CrossRef]
6. Moran, E.V.; Clark, J.S. Estimating seed and pollen movement in a monoecious plant: A hierarchical Bayesian approach integrating genetic and ecological data. *Mol. Ecol.* **2011**, *20*, 1248–1262. [CrossRef] [PubMed]
7. Petit, R.J.; Bialozyt, R.; Garnier-Géré, P.; Hampe, A. Ecology and genetics of tree invasions: From recent introductions to Quaternary migrations. *For. Ecol. Manag.* **2004**, *197*, 117–137. [CrossRef]
8. Gerzabek, G.; Oddou-Muratorio, S.; Hampe, A. Temporal change and determinants of maternal reproductive success in an expanding oak forest stand. *J. Ecol.* **2017**, *105*, 39–48. [CrossRef]
9. Hampe, A.; Arroyo, J. Recruitment and regeneration in populations of an endangered South Iberian Tertiary relict tree. *Biol. Conserv.* **2002**, *107*, 263–271. [CrossRef]
10. Nathan, R.; Safriel, U.N.; Noy-Meir, I.; Schiller, G. Spatiotemporal variation in seed dispersal and recruitment near and far from *Pinus halepensis* trees. *Ecology* **2000**, *81*, 2156–2169. [CrossRef]
11. Lefèvre, F.; Boivin, T.; Bontemps, A.; Courbet, F.; Davi, H.; Durand-Gillmann, M.; Fady, B.; Gauzere, J.; Gidoin, C.; Karam, M.-J. Considering evolutionary processes in adaptive forestry. *Ann. For. Sci.* **2014**, *71*, 723–739. [CrossRef]
12. Alía, R.; Martín, S.; de Miguel, J.; Galera, R.; Agúndez, D.; Gordo, J.; Salvador, L.; Catalán, G.; Gil, L. *Las Regiones de Procedencia de Pinus pinaster Aiton en España*; Organismo Autónomo Parques Nacionales-ETSI de Montes: Madrid, Spain, 1996.
13. Gil, L. Consideraciones históricas sobre *Pinus pinaster* Aiton en el paisaje vegetal de la Península Ibérica. *Estudios Geogr.* **1991**, *202*, 5–27.
14. Díaz-Fernández, P.M. Relations between modern pollen rain and Mediterranean vegetation in Sierra Madrona (Spain). *Rev. Palaeobot. Palynol.* **1994**, *82*, 113–125. [CrossRef]
15. Rubiales, J.M.; García-Amorena, I.; Álvarez, S.G.; Morla, C. Anthracological evidence suggests naturalness of *Pinus pinaster* in inland southwestern Iberia. *Plant Ecol.* **2009**, *200*, 155–160. [CrossRef]
16. Ruiz-Zapata, B.; Gil-García, M.J.; de Bustamante, I. Paleoenvironmental reconstruction of Las Tablas de Daimiel and its evolution during the Quaternary period. In *Ecology of Threatened Semi-Arid Wetlands: Long-Term Research in Las Tablas de Daimiel*; Sánchez-Carrillo, S., Angeler, D.G., Eds.; Springer: Dordrecht, The Netherlands; Heidelberg, Germany; London, UK; New York, NY, USA, 2010; pp. 23–43.
17. Valbuena-Carabaña, M.; López de Heredia, U.; Fuentes-Utrilla, P.; González-Doncel, I.; Gil, L. Historical and recent changes in the Spanish forests: A socio-economic process. *Rev. Palaeobot. Palinol.* **2010**, *162*, 492–506. [CrossRef]

18. Charco, J. Evolución Histórica de los Bosques en Sierra Madrona y Valle de Alcudia (Ciudad Real) y Dinámica del Pinar Relicto de Navalmanzano. Ph.D. Thesis, E.T.S.I. de Montes, Universidad Politécnica de Madrid, Madrid, Spain, 2016.
19. Chauchard, S.; Carcaillet, C.; Guibal, F. Patterns of land-use abandonment control tree-recruitment and forest dynamics in Mediterranean mountains. *Ecosystems* **2007**, *10*, 936–948. [CrossRef]
20. González-Martínez, S.; Salvador, L.; Agúndez, D.; Alía, R.; Gil, L. Geographical variation of gene diversity of *Pinus pinaster* Ait. in the Iberian Peninsula. In *Genetic Response of Forest Systems to Changing Environmental Conditions*; Muller-Starck, G., Schubert, R., Eds.; Springer: Dordrecht, the Netherlands, 2001; pp. 161–171.
21. Salvador, L.; Alía, R.; Agúndez, D.; Gil, L. Genetic variation and migration pathways of maritime pine (*Pinus pinaster* Ait) in the Iberian Peninsula. *Theor. Appl. Genet.* **2000**, *100*, 89–95. [CrossRef]
22. González-Martínez, S.G.; Gil, L.; Alía, R. Genetic diversity estimates of *Pinus pinaster* in the Iberian Peninsula: A comparison of allozymes and quantitative traits. *For. Syst.* **2005**, *14*, 3–12. [CrossRef]
23. Ramírez-Valiente, J.A.; Robledo-Arnuncio, J.J. Adaptive consequences of human-mediated introgression for indigenous tree species: The case of a relict *Pinus pinaster* population. *Tree Physiol.* **2014**, *34*, 1376–1387. [CrossRef] [PubMed]
24. Unger, G.M.; Vendramin, G.G.; Robledo-Arnuncio, J.J. Estimating exotic gene flow into native pine stands: Zygotic vs. gametic components. *Mol. Ecol.* **2014**, *23*, 5435–5447. [CrossRef] [PubMed]
25. Unger, G.M.; Heuertz, M.; Vendramin, G.G.; Robledo-Arnuncio, J.J. Assessing early fitness consequences of exotic gene flow in the wild: A field study with Iberian pine relicts. *Evol. Appl.* **2016**, *9*, 367–380. [CrossRef] [PubMed]
26. Charco, J.; Perea, R.; Gil, L.; Nanos, N. Impact of deer rubbing on pine forests: Implications for conservation and management of *Pinus pinaster* populations. *Eur. J. For. Res.* **2016**, *135*, 719–729. [CrossRef]
27. García-Rayego, J.L. Modelados de detalle en roquedos cuarcíticos de áreas de montaña media apalachense de la Meseta sur y Sierra Morena oriental. *Ería* **2006**, *71*, 269–282.
28. Clark, J.S.; Silman, M.; Kern, R.; Macklin, E.; HilleRisLambers, J. Seed dispersal near and far: Patterns across temperate and tropical forests. *Ecology* **1999**, *80*, 1475–1494. [CrossRef]
29. Nanos, N.; Larson, K.; Millerón, M.; Sjöstedt-deLuna, S. Inverse modeling for effective dispersal: Do we need tree size to estimate fecundity? *Ecol. Model.* **2010**, *221*, 2415–2424. [CrossRef]
30. Dempster, A.P.; Laird, N.M.; Rubin, D.B. Maximum likelihood from incomplete data via EM algorithm. *J. R. Stat. Soc. B* **1977**, *39*, 1–38.
31. R Development Core Team. 2008 R: A Language and Environment for Statistical Computing. R Foundation for Statistical Computing. Available online: http://www.R-project.org (accessed on 12 January 2017).
32. Muller-Landau, H.C.; Wright, S.J.; Calderón, O.; Condit, R.; Hubbell, S.P. Interspecific variation in primary seed dispersal in a tropical forest. *J. Ecol.* **2008**, *96*, 653–667. [CrossRef]
33. Burnham, K.P.; Anderson, D.R. *Model Selection and Multimodel Inference: A Practical Information-Theoretic Approach*, 2nd ed.; Springer: New York, NY, USA, 2002; p. 488.
34. Janzen, D.H. Herbivores and the number of tree species in tropical forests. *Am. Nat.* **1970**, *104*, 501–528. [CrossRef]
35. Tapias, R.; Gil, L.; Fuentes-Utrilla, P.; Pardos, J.A. Canopy seed banks in Mediterranean pines of south-eastern Spain: A comparison between *Pinus halepensis* Mill., *P. pinaster* Ait., *P. nigra* Arn. and *P. pinea* L. *J. Ecol.* **2001**, *89*, 629–638. [CrossRef]
36. González-Martínez, S.C.; Burczyk, J.; Nathan, R.; Nanos, N.; Gil, L.; Alía, R. Effective gene dispersal and female reproductive success in Mediterranean maritime pine (*Pinus pinaster* Aiton). *Mol. Ecol.* **2006**, *15*, 4577–4588. [CrossRef] [PubMed]
37. De Heredia, U.L.; Nanos, N.; García-del-Rey, E.; Guzmán, P.; López, R.; Venturas, M.; Gil, L. High seed dispersal ability of *Pinus canariensis* in stands of contrasting density inferred from genotypic data. *For. Syst.* **2015**, *24*, 015. [CrossRef]
38. Millerón, M.; de Heredia, U.L.; Lorenzo, Z.; Perea, R.; Dounavi, A.; Alonso, J.; Nanos, N. Effect of canopy closure on pollen dispersal in a wind-pollinated species (*Fagus sylvatica* L.). *Plant Ecol.* **2012**, *213*, 1715–1728. [CrossRef]
39. Millerón, M.; Lopez de Heredia, U.; Lorenzo, Z.; Alonso, J.; Dounavi, A.; Gil, L.; Nanos, N. Assessment of spatial discordance of primary and effective seed dispersal of European beech (*Fagus sylvatica* L.) by ecological and genetic methods. *Mol. Ecol.* **2013**, *22*, 1531–1545. [CrossRef] [PubMed]

40. Juez, L.; González-Martínez, S.C.; Nanos, N.; de-Lucas, A.I.; Ordóñez, C.; del Peso, C.; Bravo, F. Can seed production and restricted dispersal limit recruitment in *Pinus pinaster* Aiton from the Spanish Northern Plateau? *For. Ecol. Manag.* **2014**, *313*, 329–339. [CrossRef]

41. Thomson, F.J.; Moles, A.T.; Auld, T.D.; Kingsford, R.T. Seed dispersal distance is more strongly correlated with plant height than with seed mass. *J. Ecol.* **2011**, *99*, 1299–1307. [CrossRef]

42. Sagnard, F.; Pichot, C.; Dreyfus, P.; Jordano, P.; Fady, B. Modelling seed dispersal to predict seedling recruitment: Recolonization dynamics in a plantation forest. *Ecol. Model.* **2007**, *203*, 464–474. [CrossRef]

© 2017 by the authors. Licensee MDPI, Basel, Switzerland. This article is an open access article distributed under the terms and conditions of the Creative Commons Attribution (CC BY) license (http://creativecommons.org/licenses/by/4.0/).

forests

MDPI

Article

Use of Nuclear Microsatellite Loci for Evaluating Genetic Diversity of Selected Populations of *Picea abies* (L.) Karsten in the Czech Republic

Pavlína Máchová, Olga Trčková and Helena Cvrčková *

Forestry and Game Management Research Institute, Strnady 136, 252 02 Jíloviště, Czech Republic;
machova@vulhm.cz (P.M.); trckova@vulhm.cz (O.T.)
* Correspondence: cvrckova@vulhm.cz; Tel.: +420-257-892-268

Received: 8 December 2017; Accepted: 13 February 2018; Published: 15 February 2018

Abstract: DNA polymorphism at nine nuclear microsatellites of nine selected naturally-regenerated Norway spruce populations growing mainly within gene conservation units in different parts of the Czech Republic was studied. To verify the genetic quality of the selected gene conservation unit, we analyzed nine Norway spruce subpopulations from gene conservation unit GZ 102–Orlické hory. Genetic parameters can be used in state administrative decision making on including stands into gene conservation units. The level of genetic diversity within 17 investigated Czech Norway spruce units was relatively high. Mean values for the number of different alleles ranged from 12.2 (population SM 08) to 16.2 (subpopulation SM T4). The values of observed heterozygosity (H_o) ranged from 0.65 to 0.80 and expected heterozygosity (H_e) from 0.74 to 0.81. Pairwise population F_{ST} values ranging from 0.006 to 0.027 indicated low genetic differentiation between units, and values of Nei's genetic distance among Norway spruce units ranged from 0.046 to 0.168, thus structuring of the investigated Norway spruce units was confirmed. Closer genetic similarity was seen in subpopulations from the gene conservation unit in Orlické hory than in the studied populations from other genetic conservation units. Additionally, the populations SM 01 and SM 05, both of Hurst ecotypes, were the closest to one another and the populations of mountain and alpine ecotypes were assembled into another group.

Keywords: *Picea abies*; nuclear microsatellites; genetic diversity; gene conservation unit

1. Introduction

Norway spruce (*Picea abies* (L.) Karsten) is a coniferous species belonging to the family Pinaceae. It is one of the most widespread tree species in Europe, where it is located mainly in Northern and Northeastern Europe and in the mountains of Central and Southern Europe. The natural distribution of this species within the territory of the Czech Republic is in the Hercyno-Carpathian region and *Picea abies* represents the only autochthonous species of the genus *Picea* growing in the Czech Republic. *Picea abies* belongs to the most variable taxon of its genus with a relatively large area of distribution [1]. There are three main important ecotypes in the Czech Republic differing morphologically and physiologically and growing at different altitudes. The alpine ecotype grows at an altitude above 1050 m. It is very resistant to wind, snow, and ice. Its crown is slim and dense, and the needles are stiff and short. The mountain ecotype grows at an altitude from 700 to 1050 m, and its crown is short and sparse. The Hurst ecotype grows at an altitude below 700 m. Its crown is broad and elliptical, and the needles are long, relatively sparse, and its cones are long [2]. Norway spruce is economically the most important tree species in the Czech Republic, where it is used for pulp and timber production. This species is very demanding of soil moisture and requires higher relative humidity. In the past 200 years, the spruce has been secondarily extended to everywhere in Central Europe, thereby pushing out most of the original trees [3]. Recently, the main coniferous area, such as that of spruce, pine, and larch, has gradually decreased in the Czech Republic, while the proportions of silver fir and deciduous trees have been increasing in order to achieve optimal species composition of forests [4]. A significant reduction today in the occurrence of spruce in some localities is due to the natural increase in drought caused by climate change [5]. One of the priority tasks under state forest policies and an international obligation of the Czech Republic is to conserve biodiversity in forest ecosystems while promoting the principles of sustainable management. Therefore, it is essential to determine the genetic variability in economically and ecologically valuable stands and populations. It is widely assumed that populations characterized by narrow genetic diversity could be more sensitive to environmental changes or disease, thereby leading to a decrease in productivity [6]. The genetic diversity of trees is crucial for the adaptation of forests to climate change [7,8] and for sustaining forest ecosystems [9]. Knowledge based on DNA analyses regarding the variability of genetic resources will contribute to the quality of the reproduction material and to creating an optimal species composition in forests. Norway spruce has been the subject of numerous genetic surveys using isozymes [10,11], expressed sequence tags markers [12,13], mitochondrial DNA [14,15], sequence tagged site markers [16,17], amplified fragment length polymorphism [18], single nucleotide polymorphisms [19,20], and microsatellite markers [21–25]. Nuclear simple sequence repeats (SSRs, or microsatellites) are widely used for assessing genetic diversity in forestry populations [6]. With their high degree of polymorphism, they provide an ideal tool for gene flow studies [26]. As co-dominant markers, they allow the assessment of heterozygosity. Nuclear SSR markers for Norway spruce have been developed from genomic dinucleotide and trinucleotide sequences and from expressed sequence tags (EST)-derived stretches [27]. The application of SSR markers developed from genomic DNA for Norway spruce is limited due to the tree's large (ca. 20-gigabase) genome [28] and high proportion of repetitive DNA [21], which frequently produce complex multi-locus amplification products. To obtain a single locus, EST-SSRs markers derived from expressed regions have been developed [22,24]. These markers combine the advantages of microsatellite variability with the information content potentially carried by expressed sequences. Variable microsatellite markers relative to the coding regions are useful in forest population genetics (for example, in assessing adaptive variation) [22] and may be useful for association with phenotypic traits [24].

To provide insight into the levels of genetic variation and differentiation of selected naturally-regenerated Norway spruce populations and subpopulations growing mainly within the gene conservation units in different parts of the Czech Republic, eight EST-SSRs and one genomic-derived microsatellite markers were used. The gene conservation units are natural or man-made tree populations which are managed for maintaining evolutionary processes and adaptive potential across generations [29]. They present a set of forest stands with a significant share of valuable regional forest tree stocks in sufficient area to maintain the biological diversity of the population, which is able to reproduce on its own in case of suitable management methods. Each unit should have a designated status and a management plan, and one or more tree species recognized as target species for genetic conservation [29]. In the Czech Republic, the gene conservation units are proclaimed by the state administration in accordance with the valid legal regulation (Act No. 149/2003 Coll.) in order to save, conserve, and reproduce genetic resources of forest trees. According to Czech legislation, the gene conservation unit can be declared for one or more tree species in one or more separate parts and the size of one gene conservation unit should not be less than 100 ha. Natural regeneration should be preferred as a regeneration method. If artificial restoration is required, the reproductive material should originate from the same gene conservation unit.

Europe is an example of a complex where the distribution ranges of tree species extend across large geographical areas with profound environmental differences, and with ranges often overlapping across many countries. Conservation of forest genetic diversity through the use of gene conservation units is also conducted at the European level by European Forest Genetic Resources Programme (EUFORGEN), with which the Czech Republic is involved. This international cooperation supports the countries in their efforts to conserve forest genetic resources as part of sustainable forest management, as agreed in the context of Forest Europe, and contributes to developing genetic conservation strategies for forest trees at the pan-European level. Pan-European minimum requirements for dynamic conservation units of forest genetic diversity were developed as part of the EUFGIS project (Establishment of a European Information System on Forest Genetic Resources, April 2007–September 2010) which is one of the 17 actions co-funded by the European Commission through the Council Regulation (EC No 870/2004) on genetic resources in agriculture. Silvicultural interventions in gene conservation units should be allowed to enhance genetic processes, as needed, and field inventories carried out to monitor regeneration and the population size. These minimum requirements are now used by 36 countries to improve the management of forest genetic diversity.

Verification of the genetic quality of selected Czech genetic conservation units was carried out at the request of the state administration in order to develop procedures for acquiring knowledge about the genetic quality of other units, and these procedures will be used for subsequent monitoring strategies of the gene conservation units. Knowledge from genetic investigation is also important from the point of view of international cooperation (EUFORGEN) and it also contributes to fulfilling one of the priority tasks of the state forest policies and an international obligation of the Czech Republic to conserve biodiversity in forest ecosystems and promote the principles of sustainable management. In order to verify genetic quality of the selected gene conservation unit, we analyzed nine Norway spruce subpopulations from gene conservation unit GZ 102–Orlické hory. The results relating to genetic parameters can be used for state administrative decision making as to which stands to include into a gene conservation unit.

2. Materials and Methods

Eight Czech Norway spruce populations and nine subpopulations from the Orlické hory population (a total of 17 units) were genetically screened by nuclear microsatellites. Sampling of young needles was carried out during 2012–2016 from individual trees growing in gene conservation units, national parks, a protected landscape area, and national nature reserves. A more detailed investigation was carried out in the Orlické hory locality, where genetic parameters were studied within nine subpopulations of one gene conservation unit (GZ 102 Trčkov–Šerlišský kotel–Vrchmezí) and compared to the other populations. The locality designations, geographic coordinates, altitudes, and natural origins are presented in Table 1, sampling locations in Figure 1, and mean sample size of studied units was 34.7. The distance between randomly-sampled adult trees was approximately 80–100 m within gene conservation units in different parts of the Czech Republic, and 25–30 m within nine subpopulations of one gene conservation unit (GZ 102 Trčkov–Šerlišský kotel–Vrchmezí). Investigated Norway spruce populations or subpopulations represent three main different ecotypes (Table 1).

Table 1. Geographic coordinates of the *Picea abies* units.

Units	Geographic Coordinates		Altitude	Natural Origin
	N from–to	E from–to	m.	
SM 01: Hurst ecotype population (Středočeská pahorkatina)	49°56′43″–49°58′43″	14°46′10″–14°48′39	400–500	National nature reserve
SM 05: Hurst ecotype population (Středočeská pahorkatina)	49°51′12″–49°51′54″	14°35′31″–14°35′7″	300–400	Gene conservation unit
SM 07: Mountain ecotype of Beskydy population	49°26′48″–49°31′38″	18°26′0″–18°29′22″	700–800	Gene conservation unit
SM 08: Autochthonous highland ecotype population (Českomoravská vrchovina)	49°30′41″–49°31′50″	15°22′5″–15°23′56″	550–600	Gene conservation unit
SM 09: Alpine ecotype population (Hrubý Jeseník)	50°4′13″–50°4′44″	17°14′10″–17°15′35″	1100–1350	National nature reserve protected landscape area
SM 10: Alpine ecotype of Beskydy population (Moravskoslezské Beskydy)	49°32′23″–49°32′58″	18°26′42″–18°50′20″	530–1200	Gene conservation unit
SM 11: Alpine ecotype population (Šumava)	49°4′35″–49°4′48″	13°28′26″–13°28′49″	825–840	National park
SM 12: Alpine ecotype population (Krkonoše)	50°44′28″–50°46′17″	15°32′47″–15°36′19″	980–1280	National park
SM S1: Mountain ecotype (Orlické hory–Šerlich) subpopulation	50°19′28″–50°19′46″	16°22′21″–16°22′37″	850–980	Gene conservation unit
SM S2: Mountain ecotype (Orlické hory–Šerlich) subpopulation	50°20′21″–50°20′33″	16°21′36″–16°22′6″	970–1020	Gene conservation unit
SM S4: Mountain ecotype (Orlické hory–Šerlich) subpopulation	50°19′39″–50°19′46″	16°22′20″–16°22′41″	860–970	Gene conservation unit
SM T1: Mountain ecotype (Orlické hory–Trčkov) subpopulation	50°18′47″–50°18′55″	16°24′51″–16°25′7″	780–830	Gene conservation unit
SM T2: Mountain ecotype (Orlické hory–Trčkov) subpopulation	50°19′3″–50°19′12″	16°24′46″–16°24′59″	780–900	Gene conservation unit
SM T4: Mountain ecotype (Orlické hory–Trčkov) subpopulation	50°18′43″–50°18′51″	16°24′53″–16°25′9″	780–870	Gene conservation unit
SM V1: Mountain ecotype (Orlické hory–Vrchmezí) subpopulation	50°21′25″–50°21′32″	16°21′1″–16°21′47″	900–950	Gene conservation unit
SM V2: Mountain ecotype (Orlické hory–Vrchmezí) subpopulation	50°21′7″–50°21′10″	16°20′44″–16°21′3″	820–880	Gene conservation unit
SM V4: Mountain ecotype (Orlické hory–Vrchmezí) subpopulation	50°21′26″–50°21′32″	16°21′34″–16°21′45″	920–960	Gene conservation unit

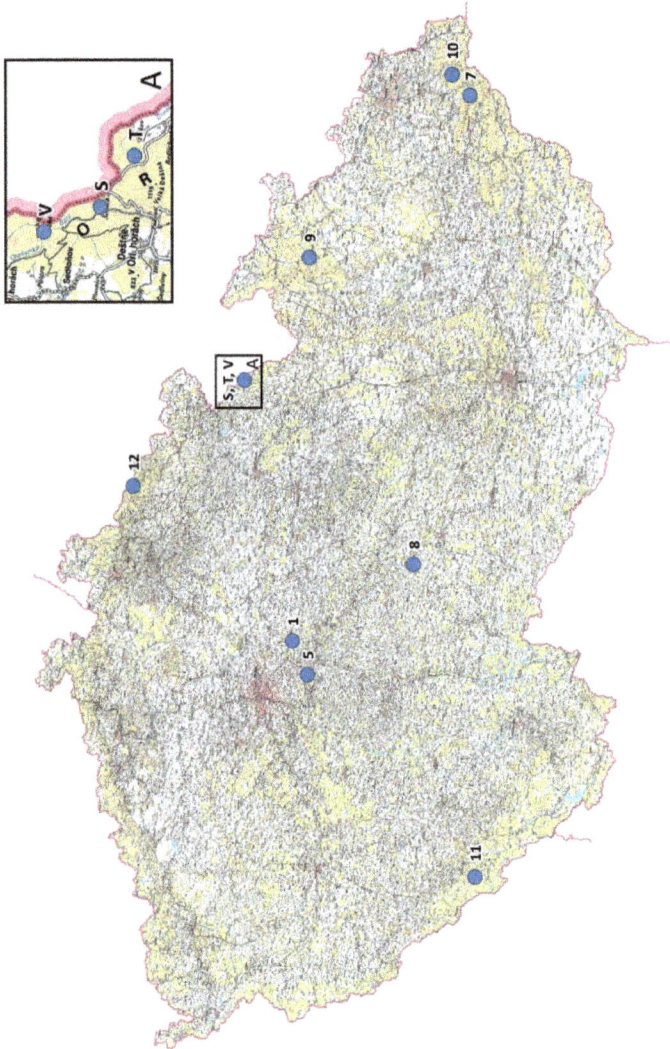

Figure 1. Geographical locations of the *Picea abies* units.

Total genomic DNA was extracted from 20 mg of dry or 100 mg of frozen young needles collected from 591 *Picea abies* individuals of the investigated localities using a DNeasy Plant Mini Kit (Qiagen, Hilden, Germany) while following the manufacturer's instructions. Liquid nitrogen was used for disrupting the plant material. DNA concentrations and purity were measured spectrophotometrically using a NanoPhotometer (Implen, Munich, Germany). SSR markers were used as molecular genetic markers for detecting genetic variation in the selected Norway spruce units. The following markers with dinucleotide motifs, whose PCR products provided single, clear, reproducible patterns were used: PAAC23, PAAC19 [22], WS00716.F13, WS0092.A19, WS0022.B15, WS0073.H08, WS00111.K13, WS0023.B03 [24] derived from expressed sequence tags (EST-SSRs), and SpAGD$_1$ [23] developed from genomic libraries of Norway spruce [21]. The nine nuclear microsatellite markers were assembled into three multiplexes from the viewpoint of the targeted allele sizes and amplification conditions. Specific primers were labelled fluorescently using FAM, VIC, and NED dyes. The amplification reaction conditions of the first multiplex with loci PAAC23, PAAC19, and SpAGD$_1$ were for each sample in a final volume of 15 μL and contained 1 μL of template DNA ($\approx$10–50 ng/μL), 1.5 μL of 10× PCR buffer (Mg-free), 0.2 mM of dNTP mixture (Takara Bio Inc., Otsu, Shiga, Japan), 2 mM MgCl$_2$, 0.37 U of Platinum® Taq DNA polymerase (Invitrogen, Carlsbad, CA, USA), and primer combinations of the forward and reverse primers. The primer concentrations of loci PAAC19 and SpAGD$_1$ were 0.1 μM and for locus PAAC23 0.2 μM. The reaction mixtures were supplemented with sterile water for molecular biology (Sigma-Aldrich, St. Louis, MO, USA). The PCR profile was as follows: initial denaturation at 94 °C for 3 min followed by 37 cycles of denaturation at 94 °C for 45 s, an annealing temperature of 57 °C for 45 s, and extension at 72 °C for 45 s, with a final extension step at 72 °C for 20 min. PCR for the second multiplex with loci WS00716.F13, WS0092.A19, and WS0022.B15 was performed using the Type-it® Microsatellite PCR Kit (Qiagen, Hilden, Germany). The concentration of each primer was 0.1 μM. PCR cycling conditions consisted of an initial denaturation at 95 °C for 15 min followed by 26 cycles of denaturation at 94 °C for 30 s, an annealing temperature at 53 °C for 90 s and extension at 72 °C for 30 s, with a final extension step at 60 °C for 30 min. PCR conditions for the third multiplex with loci WS0073.H08, WS00111.K13, and WS0023.B03 were the same as for the second multiplex, again with concentrations of each primer at 0.1 μM, except that the annealing temperature was 55 °C. Amplifications were carried out in a Veriti thermal cycler (Applied Biosystems, Foster City, CA, USA).

PCR products were separated by capillary electrophoresis using the Applied Biosystems 3500 genetic analyzer (Applied Biosystems, Foster City, CA, USA). As size standard, we used GeneScan^{TM} 600LIZ® (Applied Biosystems, Foster City, CA, USA). Alleles were sized using GeneMapper® 4.1 software (Applied Biosystems, Foster City, CA, USA). Micro-Checker software was used for identifying and correcting genotyping errors in microsatellite data and for estimation of null allele frequencies [30]. The majority of genetic diversity parameters—number of alleles, Shannon's information index, observed heterozygosity, expected heterozygosity, fixation index (F), pairwise population F$_{ST}$ values of the genetic divergence, Nei's genetic distance, and a principal coordinate analysis (PCoA) were calculated using the statistical program GenAlEx 6.501 [31,32]. The fixation index was calculated as $F = 1 - (H_o/H_e)$ according to Wright [33]. Deviations from Hardy-Weinberg equilibrium (HWE) for studied loci were assessed using CERVUS 3.0.7 with Bonferroni correction for evaluating the deviation significance [34]. Global Hardy-Weinberg tests across the studied populations were performed by GENEPOP 4.2 [35,36] using Markov chain Monte Carlo simulations with 10,000 dememorizations, 100 batches, and 10,000 iterations to detect significant heterozygote excess or deficiency. The genetic divergence F$_{ST}$ and Nei's genetic distance [37] between populations were estimated by computing a pairwise population matrix. Total F$_{ST}$ was calculated as the analysis of molecular variance (AMOVA) based on 999 permutations with the software GenAlEx 6.501 [31,32]. A chi-squared test was used for evaluating the Hardy–Weinberg equilibrium (HWE) for co-dominant genotypes at a single locus and for a single population. The number of private alleles is the number of alleles unique to a single population. It was calculated as the mean value from all studied loci. The Bayesian clustering method implemented in STRUCTURE 2.3.4 software [38–41] was used to infer the population structure. The admixture model and correlated allele frequencies were used. We used a Length of Burn-in Period of 10,000 and 100,000 Markov chain Monte Carlo (MCMC) Repeats after Burn-in. Multiple runs were performed by setting the number of populations (*K*) from *K* = 1 to *K* = 15. Each run was repeated ten times. The best estimate of *K* values was calculated using the web-based STRUCTURE HARVESTER program [42].

3. Results

Genetic diversity parameters were studied in the 17 Czech units (populations, subpopulations) using nine nuclear microsatellite markers. The nine subpopulations were taken from just one gene conservation unit (GZ 102 Trčkov–Šerlišský kotel–Vrchmezí) in order to compare the genetic distances among them and with the other studied populations situated in different localities of the Czech Republic.

The selected markers that generated simple patterns detected 23 (PAAC23), 36 (PAAC19), 38 (SPAGD1), 25 (WS00716.F13), seven (WS0092.A19), 24 (WS0022.B15), nine (WS0073.H08), 34 (WS00111.K13), and 37 (WS0023.B03) different alleles over 591 tested Norway spruce trees (Table 2). All SSR markers had dinucleotide repeats and their PCR products provided variable expected sizes. With one exception in subpopulation SM S4, the Micro-Checker software [30] found no evidence at any locus of scoring error due to stuttering and no evidence of large allele dropout for any unit. In SM S4, stuttering at locus PAAC23 might have resulted in scoring errors, as was indicated by the highly significant shortage of heterozygote genotypes with alleles of one repeat unit difference. Analyses indicated homozygote excess at loci PAAC19 and SpAGD1 in all units; at locus WS00716.F13 in units SM S1, SM V2, SM 05, and SM 11; at locus WS0022.B15 in units SM S1, SM S4, SM T1, and SM V2; at locus PAAC23 in subpopulation SM S4; at locus WS00111.K13 in subpopulation SM T1; at locus WS0073.H08 in population SM 01; and at locus WS0092.A19 in population SM 08. The Micro-Checker software indicated that these units are possibly in HWE while showing signs of a null allele. The evaluations of null allele frequencies in accordance with Oosterhout [30] are recorded in Table 2. Some significant deviations (*p* < 0.001) from HWE based on the chi-square test for HWE [31,32] were detected at locus PAAC23 in two units (SM T4, SM 05), at locus PAAC19 in 14 units (all except units SM V4, SM 01, and SM 12), at locus SpAGD1 (all except population SM 11), at locus WS0092.A19 (with significant deviation occurring in units SM T2, SM T4, SM V1, SM 01, and SM 08), at loci WS00716.F13 and WS0023.B03 (with significant deviation occurring only in population SM 05), at locus WS0022.B15 (in the SM S4 and SM V2 subpopulations), and at locus WS0073.H08 (in population SM 01). Significant deviations were not observed at locus WS00111.K13.

The genetic diversity parameters with the primer sequences of the studied markers are reported in Table 2. There were 234 different alleles detected at the nine loci in the 591 Norway spruce individuals. Expected heterozygosity (H_e) ranged from 0.20 (WS0092.A19) to 0.93 (SpAGD1, WS00111.K13), with a mean value of 0.78. The mean value of observed heterozygosity (H_o) was 0.71 and ranged from 0.22 (WS0092.A19) to 0.97 (WS0023.B03). Shannon's information index calculated for allelic and genetic diversity also depends on the evenness of allele frequencies, which ranged from 0.41 at locus WS0092.A19 to 2.86 at locus SpAGD1. Fixation index values varied from −0.075 (WS0023.B03) to 0.41 (PAAC19). Most of the loci exhibited homozygote excess with positive F values. The loci WS0073.H08, WS00111.K13, and WS0023.B03 with negative fixation indices reflect excesses of heterozygotes in comparison to their expected frequencies. According to the program CERVUS 3.0.7, significant deviations from HWE (*p* < 0.001) were detected at three loci (PAAC19; SpAGD1—deficiency of heterozygotes; WS0023.B03—excess of heterozygotes) across 17 investigated Norway spruce units.

Table 2. Characteristics of selected nuclear microsatellite loci across 17 investigated Norway spruce units.

Locus	Primer Sequence (5'–3')	PCR Product Size Range (bp)	Na	I	H_o	H_e	F	F (Null)
PAAC23	F: TGTGGCCCCACTTACTAATATCAG R: CGGGCATTGGTTTACAAGAGTTGC	266–314	23	1.66	0.67	0.71	0.04	0.0322
PAAC19	F: ATGGGCTCAAGGATGAATG R: AACTCCAAACGATTGATTTCC	141–237	37	2.52	0.53	0.90	0.41 ***	0.2129
SpAGD1	F: GTCAACCAACTGTAAAGCCA R: ACTTGTTTGGCATTTTCCC	110–188	38	2.86	0.65	0.93	0.31 ***	0.1555
WS00716.F13	F: tcaagtaatggacaaacgataca R: tttccaatagaatggtggattt	206–288	25	2.60	0.86	0.91	0.06 *	0.0397
WS0092.A19	F: gatgttgcaggcattcagag R: gcaccagcatcgattgacta	207–247	7	0.41	0.22	0.20	0.02	−0.0116
WS0022.B15	F: tttgtaggtgctgcagagatg R: tggttttattccagcaaga	166–214	24	2.27	0.84	0.86	0.02	0.0194
WS0073.H08	F: tgctctctattcgggcttc R: aagaacaaggcttcccaatg	182–216	9	1.24	0.69	0.67	−0.03	−0.0034
WS00111.K13	F: gactgaagatgccgatatgc R: ggccatatcatctcaaaataaagaa	209–271	34	2.82	0.94	0.93	−0.013	0.0074
WS0023.B03	F: agcagctg5g5gtcaaagtt R: aaagaaagcatgcatatgactcag	162–236	37	2.75	0.97	0.91	−0.075 ***	−0.022

Na: number of different alleles; I: Shannon's information index; H_o: observed heterozygosity; H_e: expected heterozygosity; F: fixation index; significant deviation from Hardy–Weinberg equilibrium (HWE) (* $p < 0.05$; *** $p < 0.001$), F (Null): estimated null allele frequency.

Genetic diversity characteristics of the 17 investigated Norway spruce units are given in Table 3. The mean values of the different alleles ranged from 12.2 (SM 08) to 16.2 (SM T4). The highest number of different alleles, at 28, occurred in subpopulation SM T4 at locus SpAGD1. The lowest, at only two alleles, was observed in the subpopulations SM V1 and SM V4 at locus WS0092.A19. The mean effective number of alleles ranged from 6.5 (SM 08) to 9.9 (SM T4). Mean values of population genetic diversity according to Shannon's information index (I) ranged from 1.93 to 2.19. The values of observed heterozygosity (H_o) ranged from 0.65 to 0.80 and of expected heterozygosity (H_e) from 0.74 to 0.81. Slightly negative fixation index (F) values occurred in two populations (SM 09 and SM 12). Other Norway spruce units showed positive values of fixation index (0.011–0.199), indicating heterozygote deficiencies relative to the expected fraction. Global Hardy-Weinberg test detected high significant deviations from HWE for all 17 Norway spruce units. The result "highly significant" is reported when at least one of the individual tests being combined yielded a zero *p*-value estimation.

Table 3. Mean values for genetic characteristics of 17 investigated Norway spruce units from nine selected nuclear microsatellite loci.

Characteristic/Populations	N	Na	Ne	I	Priv. Alleles	H_o	H_e	F
SM S1	35	14.6	9.1	2.15	0.33	0.68	0.79	0.109 ***
SM S2	35	15.1	9.3	2.16	0.11	0.69	0.78	0.101 ***
SM S4	35	15.4	9.6	2.17	0	0.68	0.78	0.104 ***
SM T1	35	15.3	9.5	2.18	0.44	0.67	0.78	0.116 ***
SM T2	35	15.4	9.4	2.18	0	0.69	0.78	0.142 ***
SM T4	35	16.2	9.9	2.19	0	0.71	0.78	0.091 ***
SM V1	35	13.4	8.9	2.09	0	0.65	0.77	0.199 ***
SM V2	35	14.9	9.5	2.14	0	0.68	0.77	0.095 ***
SM V4	35	14.7	8.6	2.11	0.44	0.73	0.77	0.038 ***
SM 01	32	14.2	7.9	2.09	0.67	0.68	0.78	0.121 ***
SM 05	40	14.7	7.9	2.10	0.44	0.72	0.78	0.059 ***
SM 07	30	14.3	8.5	2.10	0.11	0.74	0.77	0.011 ***
SM 08	24	12.2	6.5	1.93	0.11	0.72	0.74	0.051 ***
SM 09	60	15.8	8.8	2.19	0.33	0.79	0.81	−0.002 ***
SM 10	30	14	8.9	2.16	0.22	0.73	0.80	0.054 ***
SM 11	30	13.2	7.8	2.04	0.11	0.67	0.77	0.096 ***
SM 12	30	13.6	9.0	2.16	0	0.80	0.81	−0.004 ***
Mean	34.7	14.5	8.8	2.13	0.19	0.71	0.78	0.081 ***

N: sample size; Na: number of different alleles; Ne: number of effective alleles; I: Shannon's information index; Priv. alleles: number of private alleles; H_o: observed heterozygosity; H_e: expected heterozygosity; F: fixation index; significant deviation from HWE (*** *p* < 0.001).

The highest number (six) of private alleles was found in population SM 01, where one allele appeared at loci WS00716.F13, WS0073.H08, and WS00111.K13, and three alleles at locus WS0092.A19. In many units (SM V1, SM V2, SM T2, SM S4, SM T4, and SM 12), there were no private alleles. Significant allelic frequencies differences were found across the 17 studied units at most loci. For example, a 280 bp allele at locus PAAC23 varied in frequencies from 34% to 60%, a 165 bp allele at locus PAAC19 varied from 0% to 26%, and a 196 bp allele at WS0022.B15 from 1% to 17%.

Estimates of differentiation expressed by F_{ST} values ranging from 0.006 to 0.027 indicated low genetic differentiation between units. Differentiations between the nine subpopulations from the gene conservation unit GZ 102 Trčkov–Šerlišský kotel–Vrchmezí showed lower values (0.006–0.011). F_{ST} values were greater than zero and so confirm the structuring of the studied Norway spruce units. After calculation of total F_{ST} based on 999 permutations, the overall level of genetic differentiation was found to be very low (0.011). Genetic distances between units were calculated based on Nei's standard genetic distance [37]. The longest Nei's genetic distance (0.168) appeared between the SM S4 (Orlické hory–Šerlich) and SM 08 (Českomoravská vrchovina) units. The closest Nei's genetic distance (0.046) was between the SM S2 and SM T2 subpopulations, both from the Orlické hory locality. Values of Nei's genetic distance among subpopulations of the gene conservation unit ranged from 0.046 to 0.092. The closest pairwise Nei's genetic distance within other observed localities was 0.062 between SM 01 (Středočeská pahorkatina) and SM 05 (Středočeská pahorkatina), both of the Hurst ecotype.

Nei's standard genetic distances among the units are graphically illustrated in Figure 2, constructed on the basis of principal coordinate analysis, which shows the genetic similarity of the subpopulations from the gene conservation unit GZ 102 Trčkov–Šerlišský kotel–Vrchmezí. An interesting result was the grouping of the same ecotypes (Hurst, alpine, and mountain). The spatial differentiation was not confirmed by the STRUCTURE analysis. The Bayesian analysis identified $K = 2$ as the most relevant number of clusters. The admixture of clusters was very similar for studied Norway spruce units.

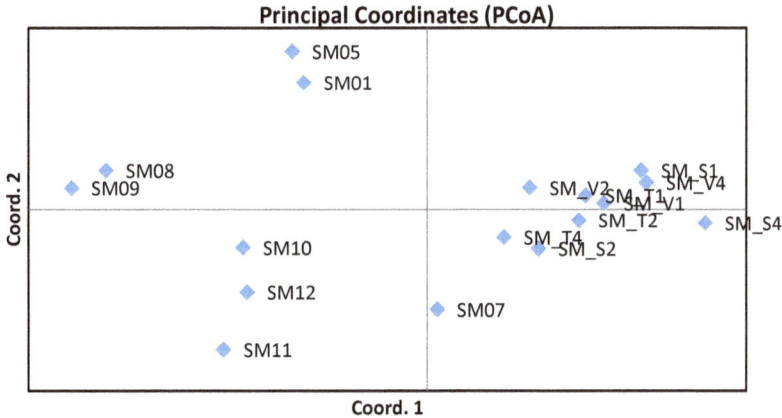

Figure 2. Results of the principal component analysis.

4. Discussion

The aim of this work was to determine the genetic parameters of selected Norway spruce units growing in different parts of the Czech Republic, where their regeneration is a naturally ongoing process, and to compare the genetic similarities of nine subpopulations from one gene conservation unit (GZ 102—Orlické hory) to other populations distributed across the Czech Republic. Nuclear microsatellite markers were tested in genetic studies of spruce populations. Nine nuclear microsatellite loci were chosen with clear, reproducible PCR products of expected sizes and sufficient polymorphism.

The highest variability was found in locus $SpAGD_1$ developed from genomic DNA. This was in accordance with Rungis et al. [24], who had found that the EST-SSRs showed significantly less variation than did the genomic-derived SSRs, as was the result also in his study, where the 25 EST-SSRs had approximately 9% less heterozygosity than did the 17 genomic-derived SSRs. However, the EST-SSR markers associated with the coding regions of the adaptive part of the spruce genome may be more useful in investigating defense against stress. At present, threats to this species are increasing at many localities of the Czech Republic, in particular due to climate changes and increasing water deficiency. It is generally presumed that populations with greater genetic diversity will have a stronger ability to adapt to changes in environmental conditions. Evaluation of genetic variation in genes controlling adaptive traits using genetic markers [8,13,20,43] could contribute to conserving the stability of long-lived forest trees. Analyzed Czech Norway spruce units have shown high levels of genetic diversity, so it can be said that these populations are appropriately included in the gene conservation unit. A prerequisite for these forests is their higher ability to adapt to changing environmental conditions. This knowledge is very important for forest management from the standpoint of utilizing the analyzed populations as important forest genetic resources of reproductive materials and for establishing genetic conservation strategies of gene conservation units. An interesting result was the grouping of studied populations according to three different ecotypes, as is illustrated by PCoA analysis. Already in a previous study of our workplace, similar results were found using six nuclear microsatellite loci, when populations from the České Švýcarsko National Park and the locality of Křivoklátsko, both Hurst ecotype, were distinguished from populations of mountain ecotype from locations Fláje (Krušné hory) and Trčkov–Šerlišský kotel (Orlické hory) based on PCoA analysis as well.

When we compared data based on nuclear microsatellites, we saw that the level of genetic diversity within the 17 investigated Czech Norway spruce units was relatively high and comparable with population diversities in such other countries as Austria [25], Poland [44], Italy [27], and Germany [45]. Diversity in Bosnia and Herzegovina populations was slightly lower in terms of the mean values of different alleles per population. Whereas in Czech populations, this ranged from 12.2 to 16.2, in Bosnia and Herzegovina populations, these means ranged from 7.8 to 15 [46]. The mean number of alleles at the same five loci (WS00716.F13, WS0022.B15, WS0073.H08, WS00111.K13, and WS0023.B03) over three Tyrol populations at different altitudes was, on average, 19.2, and across the 17 Czech units it was 25.8 alleles. The mean expected heterozygosity over these loci was much more similar in Tyrol populations (0.87) and in Czech units (0.86). A higher level of expected heterozygosity $H_e = 0.934$ was determined in Poland populations using three nuclear SSR markers [44]. Lower values were observed in Bosnia and Herzegovina populations (H_e ranged from 0.63 to 0.71).

Occurrences of null alleles were seen especially in loci PAAC19 and SpAGD1. A similar failure of amplifications for locus PAAC19 has been observed by Rungis et al. [24] and for locus SpAGD1 by Melnikova et al. [23]. Due to their recessive behavior, null alleles cause a shortage of heterozygote genotypes and result in incorrect estimates of allele frequencies. Experimental verification of null alleles is possible when new primers can be designed in microsatellite flanking regions, thereby resulting in a decrease of homozygotes [45]. With regard to the high number of different alleles (37 at locus PAAC19 and 38 at locus SpAGD1) across the studied Norway spruce units, we decided not to discard them from the evaluation; our main aim was to study the population diversities and, thus, it is better to use a greater number of polymorphic markers. Gömöry et al. [47] reported that null alleles could contribute to the overestimation of genetic differentiation among populations. We also calculated F_{ST} while excluding these two loci. In that case, we obtained slightly larger intervals of pairwise differentiation values (ranging from 0.005 to 0.031). The total level of genetic differentiation was 0.012 when these two loci were excluded. This compares to the previous result of 0.011 from nine loci. Among the geographically-closest subpopulations from the GZ 102–Orlické hory gene conservation unit, F_{ST} ranged in the lower values (0.005–0.012). Some results of other studies have shown that genetic differentiation could depend on the geographical distribution of sampling. Lower values of total F_{ST} estimation were found among three Austrian populations using SSR markers ($F_{ST} = 0.002$), thus indicating very small population differentiation. The spatial distance of these populations distributed at three different altitudes was only approximately 1000 m [25].

A higher total level of genetic differentiation ($F_{ST} = 0.026$) was observed among populations from Bosnia and Herzegovina that were spatially more distant [46]. In contrast to these results, however, in populations from the western Alps geographically located at shorter distances (only 40 km apart), meaningfully greater genetic differentiation was found (as high as $F_{ST} = 0.089$). The relatively high differentiation between geographically-close populations could be explained by their having originated from two distinct homogenous sources [48].

Low levels of genetic differentiation were observed among the Czech units, and the high genetic variation and heterozygosity within them is in conformity with similar results reported by other authors in previous studies for natural woody species populations, especially conifers. This could be explained by a mating system and high rate of migration provided by pollen and seed dispersal by wind or animals [6,48,49]. Another explanation may be associated with the changed geographic location of the populations due to extensive artificial forestation using Norway spruce since approximately the middle of the 19th century [50]. Genetic monitoring has been recognized on several international agreements and documents and can be an important tool for the protection of biodiversity. The use of genetic markers should be envisaged as a necessary complementary tool to demographic indicators for the complete assessment of a genetic resource. Genetic monitoring should concentrate on gene conservation units of such species, which should be advanced in a dynamic gene conservation scheme [51]. As the development of more powerful and affordable molecular markers and novel statistical and modelling tools is making genetic monitoring more feasible and cost-effective, it is reasonable to expect that an operational genetic monitoring system can be established for the dynamic conservation units in the near future [29].

5. Conclusions

The use of nuclear microsatellites has proven to be successful in detecting genetic variation in the studied units. In each naturally-regenerated population investigated, the allele number was high, thus indicating a higher level of genetic diversity that is comparable with that seen in Norway spruce populations of other European countries. Significant differences in allelic frequencies were found at single loci, but mean values from all loci showed low genetic differentiation between units. The shares of observed heterozygosity ranged from 0.65 to 0.80 and expected heterozygosity from 0.74 to 0.81. Significant knowledge for forestry management is that subpopulations from the gene conservation unit located in Orlické hory showed closer genetic similarity compared to that seen in other populations. Pairwise population F_{ST} values were greater than zero and so confirm the structuring of the investigated Czech populations. Structuring of populations was confirmed also by the subsequent finding in accordance with the results of Nei's standard genetic distances, where the populations SM 01 and SM 05, both of Hurst ecotypes, were the closest to one another and the populations of mountain and alpine ecotypes were assembled into another group. The relation between the genetic differences observed in Norway spruce ecotypes and local adaptation needs to be further investigated. Acquiring new knowledge about the genetic structure of coniferous species populations, especially in relation to valuable ecotypes of Norway spruce, is very important in order to both maintain the ecological stability of forests and for the optimization of timber production. The developed procedures of genetic monitoring with DNA markers will be used in the amendment of forestry legislation and in state subsidy policy in the area of protection and reproduction of forest tree gene resources. These procedures for verifying the genetic quality of selected Czech gene conservation units will be used for subsequent monitoring strategies of other gene conservation units.

Supplementary Materials: The results from CERVUS 3.0.7, the results from GENEPOP 4.2, results from GenAlEx 6.501 and from STRUCTURE 2.3.4 are available online at www.mdpi.com/1999-4907/9/2/92/s1.

Acknowledgments: This work was supported by the projects of the Ministry of Agriculture of the Czech Republic—Resolution RO0117 (reference number 6779/2017-MZE-14151.) and no. NAZV QJ1530294.

Author Contributions: P.M. and H.C. conceived and designed the experiments, and discussed the results; P.M., H.C., and O.T. performed the experiments and analyzed the data; and H.C. wrote the paper.

Conflicts of Interest: The authors declare no conflict of interest. The founding sponsors had no role in the design of the study; in the collection, analyses, or interpretation of data; in the writing of the manuscript, or in the decision to publish the results.

References

1. Musil, I.; Hamerník, J.; Leugnerová, G. *Lesnická Dendrologie 1. Jehličnaté Dřeviny. [Forest Dendrology 1. Coniferous Trees]*; The Czech Univerzity of Agriculture Prague: Praha, Czech Republic, 2003.
2. Směrnice pro uznávání a zabezpečení zdrojů reprodukčního materiálu lesních dřevin a pro jeho přenos. *[Directives for the Recognition and Security of Forest Resources and the Transfer of Forest Resources]*; Ministry of Forestry and Water Management and Woodworking Industry of the Czechoslovak Republic: Praha, Czech Republic, 1988.
3. Úradníček, L.; Maděra, P.; Tichá, S.; Koblížek, J. *Dřeviny České Republiky [Woody Species of the Czech Republic]*; Nakladatelství a vydavatelství Lesnická práce, s.r.o.: Kostelec nad Černými lesy, Czech Republic, 2009.

4. Ministry of Agriculture of the Czech Republic. Information on Forests and Forestry in the Czech Republic by 2014. Available online: http://eagri.cz/public/web/file/433136/ZZ2014AJ_16112015.pdf (accessed on 27 June 2016).

5. Šrámek, V.; Neudertová Hellebrandová, K. Mapy ohrožení smrkových porostů suchem jako nástroj identifikace rizikových oblastí [Maps of drought risk for Norway spruce stands as a decision tool indicating threatened regions in the Czech Republic: Short communication]. *Rep. For. Res.* **2016**, *61*, 305–309.

6. Maghuly, F.; Pinsker, W.; Praznik, W.; Fluch, S. Genetic diversity in managed subpopulations of Norway spruce [*Picea abies* (L). Karst.]. *For. Ecol. Manag.* **2006**, *222*, 266–271.

7. Hampe, A.; Petit, R.J. Conserving biodiversity under climate change: The rear edge matters. *Ecol. Lett.* **2005**, *8*, 461–467. [CrossRef] [PubMed]

8. Neale, D.B.; Kremer, A. Forest tree genomics: Growing resources and applications. *Nat. Rev. Genet.* **2011**, *12*, 111–122. [CrossRef] [PubMed]

9. Whitham, T.G.; Bailey, J.K.; Schweitzer, J.A.; Shuster, S.M.; Bangert, R.K.; LeRoy, C.J.; Lonsdorf, E.V.; Allan, G.J.; DiFazio, S.P.; Potts, B.M.; et al. A framework for community and ecosystem genetics from genes to ecosystems. *Nat. Rev. Genet.* **2006**, *7*, 510–523. [CrossRef] [PubMed]

10. Geburek, T. Genetic variation of Norway spruce (*Picea abies* [L.] Karst.) populations in Austria. III. Macrospatial allozyme patterns of high elevation populations. *For. Genet.* **1999**, *6*, 201–211.

11. Konnert, M. Genetic variation of *Picea abies* in southern Germany as determined using isozyme and STS markers. *Dendrobiology* **2009**, *61*, 131–136.

12. Schubert, R.; Mueller-Starck, G.; Riegel, R. Development of EST-PCR markers and monitoring their intrapopulational genetic variation in *Picea abies* (L.) Karst. *Theor. Appl. Genet.* **2001**, *103*, 1223–1231.

13. Bozhko, M.; Riegel, R.; Schubert, R.; Müller-Starck, G. A cyclophilin gene marker confirming geographical differentiation of Norway spruce populations and indicating viability response on excess soil-born salinity. *Mol. Ecol.* **2003**, *12*, 3147–3155. [CrossRef] [PubMed]

14. Maghuly, F.; Burg, K.; Pinsker, W.; Nittinger, F.; Praznik, W.; Fluch, S. Short Note: Development of mitochondrial markers for population genetics of Norway Spruce [*Picea abies* (L). Karst]. *Silvae Genet.* **2008**, *57*, 41–44. [CrossRef]

15. Tollefsrud, M.M.; Sønstebø, J.H.; Brochmann, C.; Johnsen, Ø.; Skroppa, T.; Vendramin, G.G. Combined analysis of nuclear and mitochondrial markers provide new insight into the genetic structure of North European. *Picea abies. Heredity* **2009**, *102*, 549–562.

16. Paglia, G.P.; Olivieri, A.M.; Morgante, M. Towards second-generation STS (sequence-tagged sites) linkage maps in conifers: A genetic map of Norway spruce (*Picea abies* K.). *Mol. Gen. Genet.* **1998**, *258*, 466–478.

17. Perry, D.J.; Isabel, N.; Bousquet, J. Sequence-tagged-site (STS) markers of arbitrary genes: The amount and nature of variation revealed in Norway spruce. *Heredity* **1999**, *83*, 239–248. [CrossRef] [PubMed]

18. Acheré, V.; Favre, J.M.; Besnard, G.; Jeandroz, S. Genomic organization of molecular differentiation in Norway spruce (*Picea abies*). *Mol. Ecol.* **2005**, *14*, 3191–3201. [CrossRef] [PubMed]

19. Chen, J.; Uebbing, S.; Gyllenstrand, N.; Lagercrantz, U.; Lascoux, M.; Källman, T. Sequencing of the needle transcriptome from Norway spruce (*Picea abies* Karst. L.) reveals lower substitution rates, but similar selective constraints in gymnosperms and angiosperms. *BMC Genom.* **2012**, *13*. [CrossRef] [PubMed]

20. Romšáková, I.; Foffová, E.; Kmeť, J.; Longauer, R.; Pacalaj, M.; Gömöry, D. Nucleotide polymorphisms related to altitude and physiological traits in contrasting provenances of Norway spruce (*Picea abies*). *Biologia* **2012**, *67*, 909–916. [CrossRef]

21. Pfeiffer, A.M.; Oliviery, A.M.; Morgante, M. Identification and characterization of microsatellites in Norway spruce (*Picea abies* K.). *Genome* **1997**, *40*, 411–419. [CrossRef] [PubMed]

22. Scotti, I.; Magni, F.; Fink, R.; Powell, W.; Binelli, G.; Hedley, P.E. Microsatellite repeats are not randomly distributed within Norway spruce (*Picea abies* K.) expressed sequences. *Genome* **2000**, *43*, 41–46. [CrossRef] [PubMed]

23. Melnikova, M.N.; Petrov, N.B.; Lomov, A.A.; la Porta, N.; Politov, D.V. Testing of Microsatellite Primers with Different Populations of Eurasian Spruces *Picea abies* (L.) Karst. and *Picea obovata* Ledeb. *Rus. J. Genet.* **2012**, *48*, 562–566. [CrossRef]

24. Rungis, D.; Bérubé, Y.; Zhang, J.; Ralph, S.; Ritland, C.E.; Ellis, B.E.; Douglas, C.; Bohlmann, J.; Ritland, K. Robust simple sequence repeat markers for spruce (*Picea* spp.) from expressed sequence tags. *Theor. Appl. Genet.* **2004**, *109*, 1283–1294. [CrossRef] [PubMed]

25. Unger, G.M.; Konrad, H.; Geburek, T. Does spatial genetic structure increase with altitude? An answer from *Picea abies* in Tyrol, Austria. *Plant Syst. Evol.* **2011**, *292*, 133–141.

26. Pastorelli, R.; Smulders, M.J.M.; VAN'T Westende, W.P.C.; Vosman, B.; Giannini, R.; Vettori, C.; Vendramin, G.G. Characterization of microsatellite markers in *Fagus sylvatica* L. and *Fagus orientalis* Lipsky. *Mol. Ecol.* **2003**, *3*, 76–78. [CrossRef]

27. Scotti, I.; Paglia, G.; Magni, F.; Morgante, M. Population genetics (*Picea abies* Karst.) at regional scale: Sensitivity of different microsatellite motif classes in detecting differentiation. *Ann. For. Sci.* **2006**, *63*, 485–491. [CrossRef]

28. Nystedt, B.; Street, N.R.; Wetterbom, A.; Zuccolo, A.; Lin, Y.; Scofield, D.G.; Vezzi, F.; Delhomme, N.; Giacomello, S.; Alexeyenko, A.; et al. The Norway spruce genome sequence and conifer genome evolution. *Nature* **2013**, *497*, 579–584. [CrossRef] [PubMed]

29. Koskela, J.; Lefèvre, F.; Schueler, S. Translating conservation genetics into management: Pan-European minimum requirements for dynamic conservation units of forest tree genetic diversity. *Biol. Conserv.* **2013**, *157*, 39–49. [CrossRef]

30. Van Oosterhout, C.V.; Hutchinson, W.F.; Wills, D.P.M.; Shipley, P. Micro-Checker: Software for identifying and correcting genotyping errors in microsatellite data. *Mol. Ecol.* **2004**, *4*, 535–538. [CrossRef]

31. Peakall, R.; Smouse, P.E. GENALEX 6: Genetic analysis in Excel. Population genetic software for teaching and research. *Mol. Ecol. Notes* **2006**, *6*, 288–295. [CrossRef]

32. Peakall, R.; Smouse, P.E. GenAlEx 6.5: Genetic analysis in Excel. Population genetic software for teaching and research—An update. *Bioinformatics* **2012**, *28*, 2537–2539. [CrossRef] [PubMed]

33. Wright, S. The interpretation of population structure by F-statistics with special regard to systems of mating. *Evolution* **1965**, *19*, 395–420. [CrossRef]

34. Kalinowski, S.T.; Taper, M.L.; Marshall, T.C. Revising how the computer program CERVUS accommodates genotyping error increases success in paternity assignment. *Mol. Ecol.* **2007**, *16*, 1099–1106. [CrossRef] [PubMed]

35. Raymond, M.; Rousset, F. GENEPOP (version 1.2): Population genetics software for exact tests and ecumenicism. *J. Hered.* **1995**, *86*, 248–249. [CrossRef]

36. Rousset, F. Genepop'007: A complete reimplementation of the Genepop software for Windows and Linux. *Mol. Ecol. Resour.* **2008**, *8*, 103–106. [CrossRef] [PubMed]

37. Nei, M. Genetic distance between populations. *Am. Nat.* **1972**, *106*, 283–392. [CrossRef]

38. Pritchard, J.K.; Stephens, M.; Donnelly, P. Inference of population structure using multilocus genotype data. *Genetics* **2000**, *155*, 945–959. [PubMed]

39. Falush, D.; Stephens, M.; Pritchard, J.K. Inference of population structure using multilocus genotype data: Linked loci and correlated allele frequencies. *Genetics* **2003**, *164*, 1567–1587. [PubMed]

40. Falush, D.; Stephens, M.; Pritchard, J.K. Inference of population structure using multilocus genotype data: Dominant markers and null alleles. *Mol. Ecol.* **2007**, *7*, 574–578. [CrossRef] [PubMed]

41. Hubisz, M.J.; Falush, D.; Stephens, M.; Pritchard, J.K. Inferring weak population structure with the assistance of sample group information. *Mol. Ecol. Resour.* **2009**, *9*, 1322–1332. [CrossRef] [PubMed]

42. Earl, D.A.; von Holdt, B.M. STRUCTURE HARVESTER: A website and program for visualizing STRUCTURE output and implementing the Evanno method. *Conserv. Genet. Resour.* **2012**, *4*, 359–361. [CrossRef]

43. Harfouche, A.; Meilan, R.; Altman, A. Molecular and physiological responses to abiotic stress in forest trees and their relevance to tree improvement. *Tree Physiol.* **2014**, *34*, 1181–1198.

44. Nowakowska, J.A. Mitochondrial and nuclear DNA differentiation of *Picea abies* populations in Poland. *Dendrobiology* **2009**, *61*, 119–129.

45. Nascimento de Sousa, S.; Finkeldey, R.; Gailing, O. Experimental verification of microsatellite null alleles in Norway spruce (*Picea abies* [L.] Karst.): Implications for population genetic studies. *Plant Mol. Biol. Rep.* **2005**, *23*, 113–119. [CrossRef]

46. Cvjetković, B.; Konnert, M.; Fussi, B.; Mataruga, M.; Šijačić-Nikolić, M.; Daničić, V.; Lučić, A. Norway spruce (*Picea abies* Karst.) variability in progeny tests in Bosnia and Herzegovina. *Genetika* **2017**, *49*, 259–272. [CrossRef]

47. Gömöry, D.; Ditmarová, L.; Hrivnák, M.; Jamnická, G.; Kmet, J.; Krajmerová, D.; Kurjak, D. Differentiation in phenological and physiological traits in European beech (*Fagus sylvatica* L.). *Eur. J. For. Res.* **2015**, *134*, 1075–1085. [CrossRef]

48. Meloni, M.; Perini, D.; Binelli, G. The distribution of genetic variation in Norway spruce (*Picea abies* [L.] Karst.) populations in the western Alps. *J. Biogeogr.* **2007**, *34*, 929–938. [CrossRef]
49. Hamrick, J.L.; Godt, M.J.W.; Sherman-Broyles, S.L. Factors influencing levels of genetic diversity in woody plant species. *New For.* **1992**, *6*, 95–124. [CrossRef]
50. Svoboda, P. *Lesní dřeviny a jejich porosty. Část I. [Forest Tree Species and Their Stands. Part I]*; Státní zemědělské nakladatelství: Prague, Czech Republic, 1953.
51. Aravanopoulos, F.A. Genetic monitoring in natural perennial plant populations. *Botany* **2011**, *89*, 75–81. [CrossRef]

© 2018 by the authors. Licensee MDPI, Basel, Switzerland. This article is an open access article distributed under the terms and conditions of the Creative Commons Attribution (CC BY) license (http://creativecommons.org/licenses/by/4.0/).

forests

MDPI

Article

Genetic Variation in *Quercus acutissima* Carruth., in Traditional Japanese Rural Forests and Agricultural Landscapes, Revealed by Chloroplast Microsatellite Markers

Yoko Saito [1,*], Yoshiaki Tsuda [2], Kentaro Uchiyama [3], Tomohide Fukuda [4], Yasuhiro Seto [5], Pan-Gi Kim [6], Hai-Long Shen [7] and Yuji Ide [1]

[1] Graduate School of Agricultural and Life Sciences, University of Tokyo, 1-1-1, Yayoi, Bunkyo, Tokyo 113-8657, Japan; ide@es.a.u-tokyo.ac.jp
[2] Sugadaira Research Station, Mountain Science Center, University of Tsukuba, 1278-294 Sugadairakogen, Ueda, Nagano 386-2204, Japan; tsuda.yoshiaki.ge@u.tsukuba.ac.jp
[3] Department of Forest Molecular Genetics and Biotechnology, Forestry and Forest Products Research Institute, Matsunosato 1, Tsukuba, Ibaraki 305-8687, Japan; kruchiyama@affrc.go.jp
[4] NTT DATA Intellink Corporation, Tsukishima 1-15-7, Chuo, Tokyo 104-0052, Japan; ckhut9343@gmail.com
[5] Mynavi Corporation, Chiyoda, Tokyo 100-0003, Japan; seto.yasuhiro@mynavi.jp
[6] School of Ecological and Environmental System, Kyungpook National University, Sangju 742-711, Korea; pgkim@knu.ac.kr
[7] State Key Laboratory of Forest Genetics and Breeding, Northeast Forestry University, Hexing Road 26, Harbin 150040, China; shenhl-cf@nefu.edu.cn
* Correspondence: yoko@es.a.u-tokyo.ac.jp; Tel.: +81-3-5841-8259

Received: 23 August 2017; Accepted: 13 November 2017; Published: 17 November 2017

Abstract: *Quercus acutissima* Carruth. is an economically important species that has long been cultivated in Japan, so is a valuable subject for investigating the impact of human activities on genetic variation in trees. In total, 2152 samples from 18 naturally regenerated populations and 28 planted populations in Japan and 13 populations from the northeastern part of Eurasia, near Japan, were analyzed using six maternally inherited chloroplast (cpDNA) simple sequence repeat (SSR) markers. Although 23 haplotypes were detected in total, both the Japanese natural and artificial populations exhibited much lower genetic diversity than the continental populations. The level of genetic differentiation among natural populations in Japan was also much lower ($G'_{ST} = 0.261$) than that on the continent ($G'_{ST} = 0.856$). These results suggest that human activities, such as historical seed transfer, have reduced genetic diversity within and among populations and resulted in a homogeneous genetic structure in Japan. The genetic characteristics of natural and artificial populations of *Quercus acutissima* in Japan are almost the same and it is likely that most of the natural populations are thought to have originated from individuals that escaped from plantations.

Keywords: genetic structure; human impact; seed transportation; artificial forest

1. Introduction

Genetic diversity and genetic structure reflect the evolutionary process, including aspects such as colonization history and adaptation to the environment over long periods. Ecological characteristics, such as the mating system, seed dispersal and hybridization with related species, are also reflected in the genetic variation of tree species. However, historical human activities, such as the transfer of seeds and saplings, breeding, domestication and environmental impacts may also be reflected in the genetic structure of tree species. Relatively few studies have examined tree species strongly affected by

Forests **2017**, *8*, 451

human activities over a long period [1–5]. Appropriate and sustainable use and management of forest systems, however, requires an understanding of the effects of human activities on genetic structure and population demography. For example, olive (*Olea europaea* L.) and cork oak (*Quercus suber* L.) are species that were considered to have been greatly influenced by human beings in Europe. Besnard et al. [6] reported five clades in a chlorotype consensus phylogram of olive populations in Africa and the Mediterranean, with each located in a specific geographic zone. However, within the Mediterranean area, the chlorotype originated in the east and had spread west, indicating seed transportation by humans. On the other hand, cork oak retains clear phylogeographic structure in the western Mediterranean and much less human impact than expected was detected [7]. This may be because cork oak has approximately the same distribution today as it did prior to the Neolithic [7] and seed transportation by humans has been rare. These studies suggest that the relationship between historical human activities and genetic structure of valuable tree species is complicated and their pattern is not always as expected.

Molecular makers are very useful tools to clarify the genetic structure patterns exhibited by tree species. Among the many molecular makers available, chloroplast DNA (cpDNA) is particularly useful for determining colonization routes, because (i) it does not recombine, therefore haplotypes remain mostly unchanged when passed to the next generation; and (ii) in angiosperms it is generally transmitted through seed only [8–10] because of its maternal inherited mode, so colonization patterns which derive from seed dispersal are not obscured by pollen flow [11]. Thus, levels of among-population genetic differentiation are expected to be much higher for cpDNA markers than for nuclear DNA markers [12]. However, detecting useful polymorphisms at the population level is often difficult because of the low level of substitutions in the chloroplast genome; chloroplast microsatellites represent potentially useful markers to circumvent this problem and, to date, studies have demonstrated high levels of inter- and intra-specific variability [7,13–15]. There are many studies which employed chloroplast microsatellites, so called cpSSR markers, to clarify the phylogeography of oak species widely in the world (e.g., [16–18]).

Quercus acutissima Carruth., Lepidobalanus Sect. Cerris, is one of the most economically and ecologically important deciduous tree species growing in Satoyama, a traditional Japanese rural forest and agricultural landscape. It is an anemophilous species and its seeds are dispersed by gravity and animals such as rodents [19]. This species is found in the warm-temperate zone from East Asia to the Himalayas [20], but the range is considered to be heavily influenced by human activities because of its long cultivation history, like that of European chestnut, *Castanea sativa* Mill. The origin of Japanese populations of *Q. acutissima* is unknown. Unlike other Japanese oak species, *Q. acutissima* is seldom observed in mountainous areas, but is found in and around human settlements in Japan. Kurata [21] questioned whether *Q. acutissima* is native to Japan. Fukamachi et al. [22] also suggested that *Q. acutissima* is an introduction from China, because it is less common than the native oaks *Quercus serrata* Murray and *Quercus crispla* Blume in woodlands, and it is usually pollarded, whereas the native oaks are coppiced in the Satoyama landscape. Although there is no available pollen fossil data specific to *Q. acutissima* in Japan, during the last glacial, warm-temperate evergreen broadleaved forest, in which *Q. acutissima* grows, was restricted to the Paleo-Yaku Peninsula, around the southern island of Kyushu, based upon pollen and plant macrofossil data [23]. Moreover, a recent study based on ecological niche model by Zhang et al. [24] revealed that *Q. acutissima* could not distribute in the main archipelago of Japan during the last glacial maximum (ca. 21,000 years BP). About 6000 years ago, when the climate was warmest, the evergreen forest extended to Kanto area, central Japan [25]. The plant residue of *Q. acutissima* or *Quercus variabillis* Blume (Lepidobalanus Sect. Cerris) has been found in the Japanese archipelago as fossil wood from the Jomon era, from approximately 16,500 to 3000 years ago [26]. The timber was used as a building material [26] and the seeds as food [27]. In recent years, especially since the 1960s, the demand for fresh Shiitake mushrooms has grown and *Q. acutissima* logs for cultivation have become more scarce, so the species has been planted widely in Japan [28].

In Japan, there are 15 native *Quercus* species [20] and they are widely distributed from south to north and from the coasts to the mountains. They are familiar to forest researchers and many ecological studies have been conducted on them, especially in the temperate zone (e.g., [29,30]). However, only a few reports have examined the phylogeographic characteristics of oak trees (e.g., [31–33]) and the genetic variation in *Q. acutissima* in Japan has not been investigated.

The aims of the study reported here were (1) to clarify the genetic diversity and structure of *Q. acutissima* using chloroplast microsatellite; (2) to detect evidence of the impact of human activity on genetic diversity and structure; and (3) to compare genetic variation in naturally regenerated and planted populations in Japan and continental populations in eastern Eurasia (South Korea and Northeast China).

2. Materials and Methods

2.1. Sampling of Plant Populations

Leaves of *Q. acutissima* of adult individuals were collected from 18 populations that had regenerated naturally along rivers (except for two populations, GO and SM) and two Gene Conservation Forests (Japanese natural populations) and 28 planted populations in Japan, covering the species' entire distribution (Table 1, Figure 1). Naturally regenerated populations were seldom found except for in sunny river valleys; they were small populations containing several to several dozen individuals. We selected the individuals separated by at least 10 m from the next tree in river side areas and 30 m in mountain areas. Every planted population was a subcompartment of a private forest without information of seed source and all of them were older than 40 years according to the forest registers. In order to compare Japanese populations with natural continental populations in mountainous areas or hillside, thirteen populations from northeastern Eurasia (referred to as continental populations, hereafter) (Table 1, Figure 1) were sampled, where possible, separated by at least 30 m from the next tree. In total, 59 populations, consisting of a total of 2152 individuals were sampled. Leaves were dried with silica gel in plastic bags before DNA extraction.

2.2. Chloroplast SSR Analysis

Total DNA was extracted from dried material using the modified CTAB method [34]. Each extract was then amplified by polymerase chain reaction (PCR) using a multiplex PCR Kit (Qiagen, Hilden, German) with six maternally inherited cpSSR primer pairs—µcd4, µcd5, µdt1, µdt3, µdt4, and µkk4 [35]—developed for *Quercus*. All primers were included in the same reaction mixture. Each 5.0-µL amplification reaction mixture contained 2.5 µL of MasterMix solution (Qiagen, Hilden, German), 1.5 µL of RNase-free water, 0.5 µL of the primer mix solution (including the six pairs of primers, each at 0.5 pmol/µL), and 0.5 µL of the extracted DNA (5–100 µg/mL). The reaction program was as follows: 15 min denaturation at 95 °C followed by 25 cycles of 30 s, denaturation at 94 °C, annealing for 90 s at 48 °C, and extension for 60 s at 72 °C, followed by a final extension at 60 °C for 30 min. CpSSR PCR products were analyzed using an ABI3100 Genetic Analyser (Thermo Fisher Scientific Inc., Waltham, MA, USA), and the sizes of the amplified alleles were estimated using GeneMapper software (Thermo Fisher Scientific Inc., Waltham, MA, USA).

2.3. Genetic Data Analysis

Haplotypes were determined based on the fragment length of the six cpSSR markers. The level of polymorphism within populations was estimated using haplotypic diversity based on unordered (hs) or ordered haplotypes (vs) following Pons and Petit [36], taking the number of differences in the repeats of cpSSR haplotypes into account. In the overall sample, total haplotypic diversity statistics based on unordered alleles which assumes that new allele is generated randomly, and ordered alleles assuming a stepwise mutation model, for the mutations of the SSR (ht and vt, respectively) [37], were also calculated following Pons and Petit [36]. Haplotypic richness was calculated using RAREFAC [38] software.

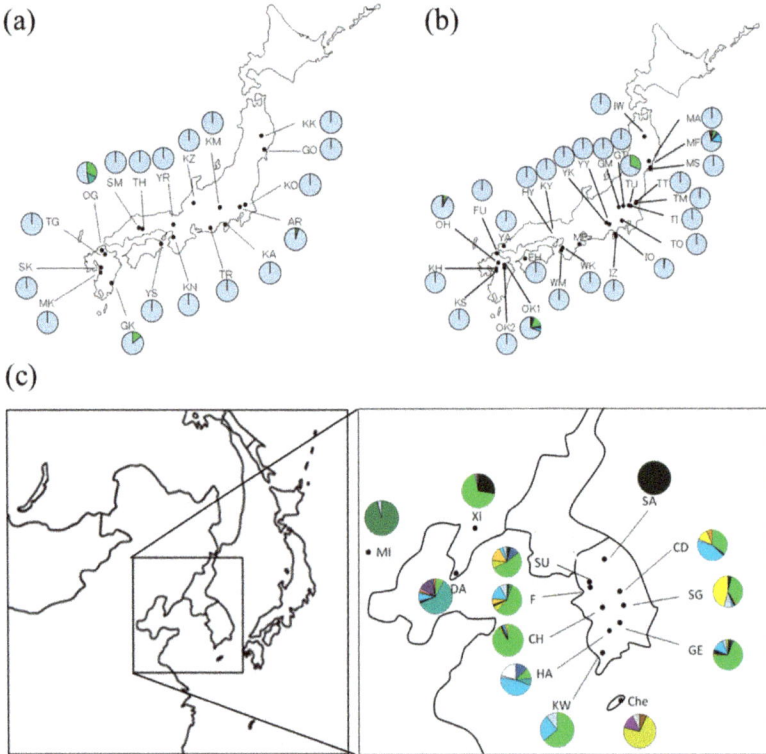

Figure 1. Chloroplast haplotype distribution of *Quercus acutissima* Carruth. for (**a**) 18 natural populations in Japan, (**b**) 28 artificial populations in Japan and (**c**) 13 continental populations. The color of haplotype is same as Figure 2.

Population differentiation at chloroplast loci was then evaluated in terms of unordered alleles (G_{ST}) [39] and ordered alleles, which assumes a stepwise mutation model of the SSR (R_{ST} [40]). Moreover, N_{ST} [36] was also employed to evaluate population differentiation. If a bias in homoplasy or multistep mutation of the SSR is large enough, it is better to employ N_{ST}, which considers the number of shared cpSSR loci with an identical allele as genetic distance among haplotypes. Following Pons and Petit [36], 1000 random haplotype permutations among populations were used to test whether the R_{ST} and N_{ST} values were significantly higher than the G_{ST} values. These calculations and tests were conducted using software PERMUT/CpSSR ver 2.0 [36]. Because G_{ST}, the index of the genetic differentiation values, is dependent on the level of genetic variation, standardized values of G_{ST} and G'_{ST} [41], which can range from 0 (no differentiation) to 1 (complete differentiation) regardless of polymorphism of examined markers, were also calculated. To determine phylogenetic relationships among the chloroplast DNA haplotypes, a Neighbor-net [42] was constructed with the software SplitsTree4 [43]. To clarify the genetic relationships between populations, a neighbor joining tree was constructed based on the genetic distance of $(\delta\mu)^2$ [44] using Populations ver.1.2.30 [45] and visualized on a hypometric map using Mapmaker and GenGIS2 software [46].

Table 1. Populations and sample size of *Quercus acutissima* Carruth. forest and detected haplotypes.

Region	Forest Type	Code	Latitude (N)	Longitude (E)	Sample Size	Detected Haplotype
Japan	Natural	KK	39°16′	141°07′	53	20
		GO	38°45′	141°23′	16	20
		ARD	36°07′	139°19′	41	6, 20
		KO	36°02′	140°01′	32	20
		KM	35°39′	138°30′	13	20
		KA	35°10′	138°90′	4	20
		TR	34°48′	137°49′	14	20
		KN	34°14′	135°09′	15	20
		YR	35°17′	135°19′	13	20
		KZ	35°54′	136°38′	15	20
		YS	34°06′	134°30′	13	20
		TH	34°52′	133°33′	6	20
		SM	34°53′	133°25′	12	20
		OG	33°35′	130°39′	19	7, 9, 20
		TG	33°20′	130°50′	8	20
		SK	32°46′	130°36′	14	20
		MK	32°42′	130°36′	13	20
		GK	32°39′	131°25′	21	7, 20
	Artifical	IW	39°46′	141°08′	37	20
		MF	38°20′	140°59′	33	6, 7, 19, 20, 38
		MA	38°38′	140°59′	49	20
		MS	38°21′	141°00′	22	20
		TT	36°31′	140°13′	46	20
		TM	36°30′	140°12′	46	20
		TI	36°29′	139°59′	48	20
		TU	36°29′	139°55′	48	7, 20
		GT	36°17′	139°04′	40	20
		GM	36°16′	138°45′	48	20
		TO	35°38′	139°18′	50	20
		IZ	34°53′	139°06′	40	20
		IO	35°01′	138°59′	50	4, 20
		YY	35°30′	138°28′	50	20
		YK	35°28′	138°30′	50	20
		ME	34°21′	136°29′	50	20
		KY	34°57′	135°32′	50	20
		HY	34°55′	135°27′	46	20
		WK	34°11′	135°17′	50	20
		WM	34°02′	135°13′	31	20
		YA	34°16′	131°18′	50	20
		EH	33°37′	132°49′	50	20
		FU	33°47′	130°53′	42	20
		OH	33°19′	130°57′	48	3, 7, 16, 19, 20
		OK1	33°11′	131°16′	50	3, 5, 7, 10, 15, 20
		OK2	33°08′	131°16′	50	20
		KS	33°02′	130°50′	48	20
		KH	33°04′	130°53′	47	20
Continent	Natural	SA	37°36′	128°01′	41	3
		SU	37°19′	127°01′	47	3, 5, 7, 12, 16, 19, 20
		F	37°12′	126°59′	50	3, 6, 7, 12, 15, 16, 19, 20
		CD	36°50′	127°57′	49	7, 15, 19, 20, 21, 26
		CH	36°24′	127°15′	49	7, 12, 15, 19, 20, 21
		SG	36°22′	128°08′	46	3, 7, 12, 15, 19, 20, 21
		GE	35°44′	127°47′	46	3, 5, 7, 12, 15, 19, 20, 21
		HA	35°38′	127°17′	39	5, 7, 9, 19, 20, 32
		KW	35°09′	126°56′	28	7, 19, 20
		Che	33°31′	126°32′	14	1, 26, 39, 41
		XI	40°23′	123°18′	54	3, 7, 20, 26
		MI	39°55′	116°24′	47	6, 20
		DA	38°55′	121°21′	51	7, 9, 15, 19, 20, 26, 34, 36, 37, 40

3. Results

3.1. Haplotypes

The six chloroplast microsatellite primers assayed for 2152 individuals in 59 populations (Table 1) gave 22 different alleles: μcd4, three alleles; μcd5, two alleles; μdt1, four alleles; μdt3, five alleles; μdt4, five alleles; and μkk4, three alleles. Thus, 23 haplotypes were identified (Table 2) and the haplotypes were divided into three groups on the Neighbor-net (Figure 2). Seven, eight and eight haplotypes belonged to Group I, II and III, respectively. In the total sample, haplotype 20 of group III and haplotype 7 of group I were the most common, with frequencies of 72.2% and 11.9%, respectively. Haplotype 20 was dominant in Japan and haplotype 7 was dominant on the continent (Table 2, Figure 1). In total, there were four haplotypes in the Japanese natural populations, twelve in the plantation populations, and twenty in the continental populations (Table 2). The number of haplotypes unique to each population group was three (haplotypes 4, 10, 38) in the Japanese plantations and ten (haplotypes 1, 12, 21, 32, 34, 36, 37, 39, 40, 41) in the continental populations. There were no haplotypes that only occurred in Japanese natural populations.

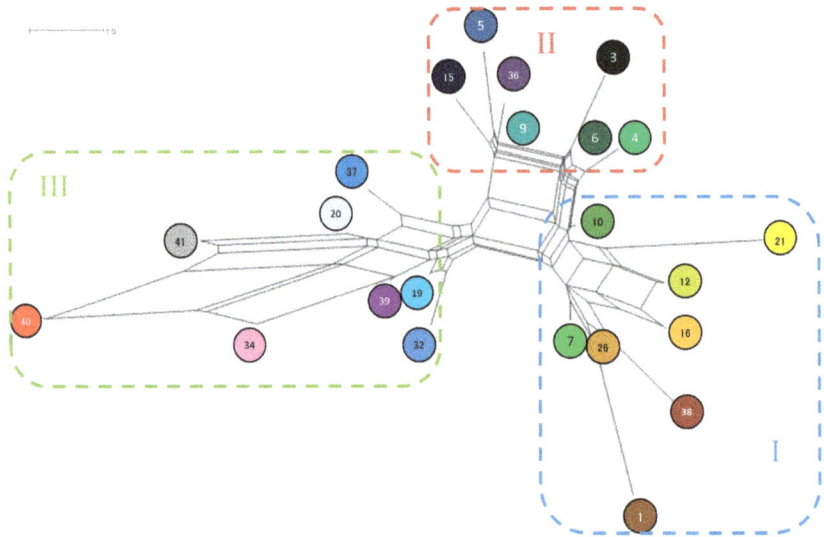

Figure 2. A Neighbor-net of 23 Chloroplast haplotypes for *Quercus acutissima*. Haplotypes of Group I were distributed mainly in Korea, haplotypes of Group II were in China and northern Korea and haplotypes of Group III were dominant in Japan.

Table 2. The 23 chloroplast haplotypes revealed by the six simple sequence repeat (SSR) markers and the fragment length of each marker, and frequency of each haplotype.

Haplotype	Maker						Haplotype Frequency *			
	μcd4 (VIC)	μcd5 (FAM)	μdt1 (PET)	μdt3 (FAM)	μdt4 (PET)	μkk4 (NED)	Japan		Continent	Total
							Natural	Artificial		
20	101	81	90	129	139	115	0.957	0.965	0.037	0.7217
7	101	81	90	128	136	115	0.028	0.018	0.399	0.1190
19	101	81	90	129	138	115	-	0.006	0.121	0.0353
3	101	81	90	126	137	115	-	0.002	0.112	0.0307
6	101	81	90	127	137	115	0.006	0.002	0.084	0.0237
9	101	81	90	127	138	115	0.009	0.001	0.061	0.0177
21	102	82	91	128	137	115	-	-	0.052	0.0135
26	101	81	89	128	136	115	-	-	0.027	0.0070
5	100	81	90	127	138	115	-	0.001	0.020	0.0056
16	102	81	90	129	136	115	-	0.001	0.018	0.0051
12	102	81	90	128	136	115	-	-	0.018	0.0046
15	102	81	90	127	138	115	-	0.002	0.012	0.0042
32	102	81	90	129	138	115	-	-	0.014	0.0037
36	101	81	89	127	138	115	-	-	0.012	0.0033
10	101	81	90	128	137	115	-	0.002	-	0.0009
39	101	81	89	129	138	115	-	-	0.004	0.0009
4	101	81	90	127	136	115	-	0.001	-	0.0005
38	101	81	89	129	135	115	-	0.001	-	0.0005
1	101	81	89	128	136	113	-	-	0.002	0.0005
34	101	82	88	129	138	115	-	-	0.002	0.0005
37	101	81	89	128	139	115	-	-	0.002	0.0005
40	101	82	88	130	139	116	-	-	0.002	0.0005
41	101	81	90	130	139	116	-	-	0.002	0.0005
Total number of allells	3	2	4	5	5	3				

VIC, FAM, PET, NED, Fluorescent Dye; *, The haplotype frequencies were rounding off four decimal place. The total of frequencies of raw data is 1.000 in each area.; -, no detection.

3.2. Geographical Distribution of Haplotypes of Q. acutissima

The relationship between geographical distribution and genetic variation is shown in Figure 1 and Tables 1 and 2. Haplotype 20 was found at high frequencies in all natural and artificial plantations in Japan and was also present at low frequencies in 11 of 13 continental populations, at most 10.7% occurrences in KW and 8.7% in SG in South Korea. On the other hand, 87.5% of the occurrences of Haplotype 7 were in 10 of 13 continental populations and only six of the 46 Japanese populations contained this haplotype. Ten of the 23 haplotypes were unique to a single population. On the continent, haplotypes 1, 39 and 41 were only found in Che; haplotypes 34, 36 and 40 were only found in DA; and haplotype 32 was only found in HA. In Japan, haplotype 38 was only found in MF, haplotype 4 only in IO and haplotype 10 only in the OK1 plantation. In Japan, 11 haplotypes, i.e., all except haplotype 20, were found at low frequencies and they were concentrated in eight of the 46 populations. Although no phylogeographic structure was detected because of the fixation of haplotype 20 in Japan, haplotypes of Group II were distributed mainly in China and northern Korea and haplotypes of Group III were dominant only in HA on the continent (Figure 3).

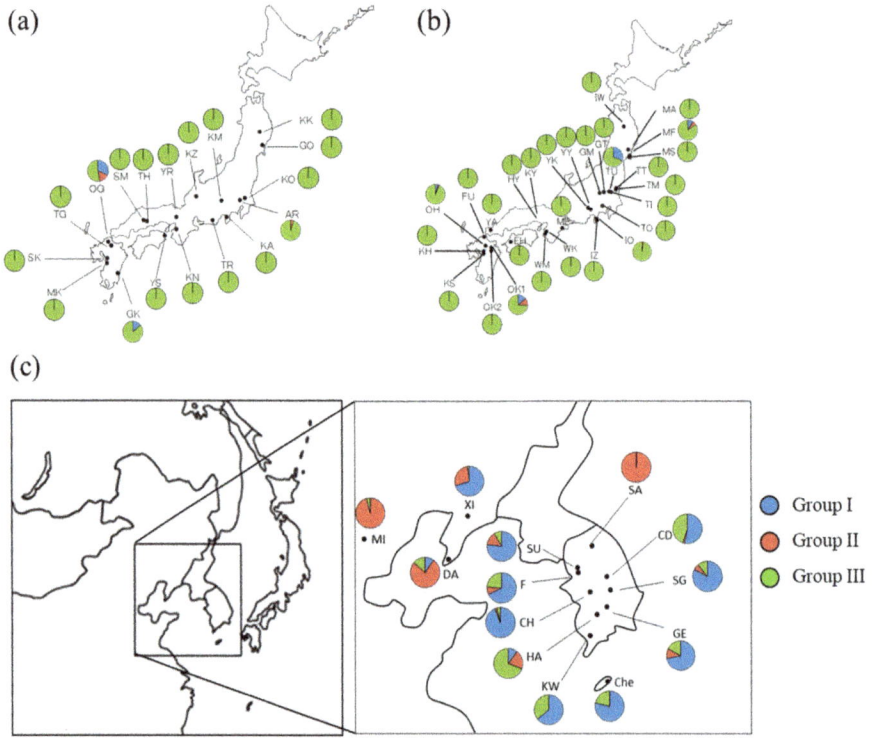

Figure 3. Three chloroplast haplotype group distribution of *Q. acutissima* for (**a**) 18 natural populations in Japan; (**b**) 28 artificial populations in Japan and (**c**) 13 continental populations.

3.3. Genetic Diversity within Populations and Differentiation between Populations of Q. acutissima

Most populations, both natural and artificial, in Japan were monomorphic, containing only haplotype 20; in contrast, on the continent, all populations except for SA were polymorphic. Within population variation is shown in Table 3. All calculated parameters were higher on the continent (*hs*: 0.488; *vs*: 0.482) than in Japan (*hs*: 0.055, 0.059; *vs*: 0.053, 0.057 in natural and artificial populations, respectively). In Japan, the values were almost the same in the natural and artificial populations. Although the number of haplotypes in artificial populations is more than in natural populations, most of the artificial populations were also dominated haplotype 20 (Table 1). Only five artificial populations had multiple haplotypes (Figure 1).The level of population subdivision was relatively low, for the natural and artificial populations in Japan, and the continental populations, respectively. These values were low for both types of population in Japan. R_{ST} was not significantly larger than G_{ST} for any dataset. Only in the continental populations was high population differentiation ($G'_{ST} = 0.856$) shown, and N_{ST} was significantly higher than G_{ST}, suggesting phylogeographic structure.

Table 3. Genetic diversity and genetic differentiation in the three population groups of *Quercus acutissima* Carruth.: Natural and artificial populations in Japan plus natural continental populations.

Region	Forest Type	No. of Populations	Sample Size	As	hs	vs	ht	vt	G_{ST}	G'_{ST}	R_{ST}	N_{ST}
Japan	Natural	18	322	1.111 (0.076)	0.055 (0.037)	0.053 (0.037)	0.073 (0.051)	0.073 (0.050)	0.246	0.261	0.238	0.238
Japan	Artificial	28	1269	1.122 (0.061)	0.059 (0.029)	0.057 (0.028)	0.072 (0.035)	0.072 (0.035)	0.180	0.191	0.201	0.179
Continent	Natural	13	561	2.088 (0.159)	0.488 (0.067)	0.482 (0.063)	0.843 (0.054)	0.844 (0.099)	0.421	0.856	0.429	0.480 *

As, haplotypic richness; *hs*, unorderd haplotypic diversity; *vs*, orderd haplotypic diversity; *ht*, unorderd total haplotypic diversity; *vt*, orderd total haplotypic diversity; G_{ST}, unorderd population differentiation index; G'_T, standardized value of G_{ST}; R_{ST}, orderd alleles based population differentiation index; N_{ST}, ordered loci based population differentiation index; standard errors in parentheses; * N_{ST} is significantly larger than G_{ST} ($p < 0.05$).

3.4. Genetic Relationship between Populations of Q. acutissima

A Neighbor-Joining tree for 59 populations was constructed based on the genetic distance $(\delta\mu)^2$ [44] (Appendix A). Two clades were recognized: one representing Japanese natural and artificial populations and the other continental populations. This suggests that Japanese populations and continental populations are genetically differentiated. Only OG, a Japanese natural population, belonged to the continental clade. No phylogeographic structure was apparent in the Japanese clade. However, the other clade had two subclades, one consisting of populations from northern Korea and China and the other of South Korean populations (Figure 4). Thus genetic structure was shown on the edge of the Eurasian continent.

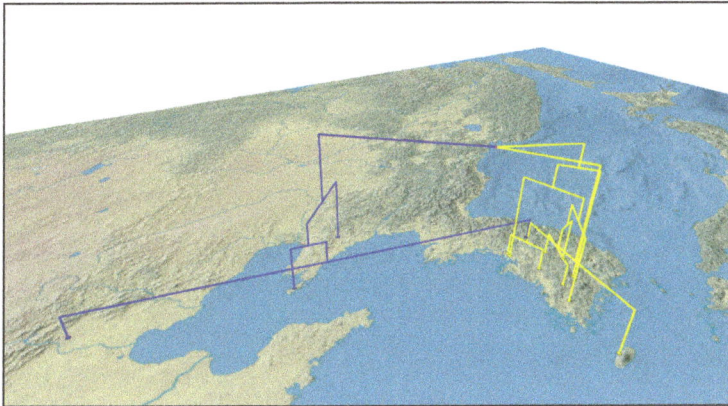

Figure 4. Neighbour-joining tree for *Quercus acutissima* Carruth. continental populations only, based on $(\delta\mu)^2$ genetic distance.

4. Discussion

In total, 59 *Q. acutissima* populations from Japan and Eurasia (South Korea and Northeast China), were analyzed using cpSSR markers and 23 haplotypes were detected in total and the markers used were highly polymorphic.

4.1. Genetic Characteristics of Q. acutissima Natural Populations in Northeast Asia

Genetic diversity was much lower in the Japanese populations than in the continental ones. Most populations were fixed for haplotype 20 alone, and genetic diversity with respect to chloroplast DNA was lacking in Japan despite the polymorphic markers. In tree species, the same haplotype or genetically related haplotypes are sometimes located close to each other geographically (e.g., *Betula maximowicziana* Regel [47], *Fraxinus angustifolia* Vahl. [48]); this is referred to as phylogeographic structure. Among the oak species present on the Japanese archipelago, *Q. crispra*, *Q. serrata*, *Quercus dentata* Thunb. and *Quercus aliena* Blume, all of them are autochthonous in Japan, do exhibit clear phylogeographic structure reflecting the presence of a refugium during the last glacial maximum and subsequent migration based on chloroplast DNA variation [31,32]. In this study, there was no phylogeographic structure identified for *Q. acutissima* in Japan. Thus the haplotype distribution of *Q. acutissima* is completely different from that found in these four native oak species. *Q. variabilis* in Japan, related species of *Q. acutissima*, also fixed for haplotype 20 as same as *Q. acutissima* [49]. *Q. variabilis* is inferred that the distribution was influenced by human activity as same as *Q. acutissima*. It must be chloroplast capture but it could not be determined which species incited the capture. Although it is not known when the introgression happened, it must have happened before they expanded over the Japanese archipelago. The haplotype distribution patterns of *C. sativa* [2] and *Prunus avium* L. [8] in Europe, which have both been influenced greatly by human activities, are similar to *Q. acutissima* in Japan, with one major haplotype covering almost all the distribution range. Alternatively, the absence of phylogeographic structure throughout Japan could be attributed to the monomorphism of cpDNA and long distance gene flow with seed among populations [5].

Japanese horse chestnut (*Aesculus turbinate* Blume), which is also a tree species used by human beings, has several haplotypes on the Japanese archipelago but one of these has a wide distribution extending circa 900 km [50]. The authors of that study discuss the discrepancy between a distribution range predicted based on migration capability and the actual distribution range, which is known as Reid's Paradox [51], and note that rare long-distance seed-dispersal events could account for more rapid colonization [50]. Similarly, in the case of *Q. acutissima*, haplotype 20 is present in all populations extending across a range of almost 1400 km. Acorns of *Q. acutissima* are transported by rodents and the maximum reported distance is 38.5 m [19]. The youngest reproductive age of the species is four years [52]. Based on these figures, if refugia of *Q. acutissima* during the last glacial maximum (LGM) existed around the southern part of Japan (e.g., Paleo-Yaku Peninsula) as suggested by the paleoecologial study [23], more than 300,000 years would be needed to cover the entire current distribution range. However, a recent study clearly suggested that *Q. acutissima* did not distribute in the main archipelago of Japan during the LGM [24]. Thus, it is difficult to explain the modern distribution of *Q. acutissima* only by natural population colonization although natural rare long-distance seed-dispersal events could account for rapid colonization. Indeed, haplotype 7, common on the continent, was also found in six Japanese populations. Although this may suggest the long distance dispersal, it would be more likely that this continental haplotype was introduced to Japan via past human activities, as discussed below.

4.2. Genetic Structure and Population Differentiation

Since most of the *Q. acutisima* populations in Japan were fixed for one haplotype, this tree species experienced a severe bottleneck when *Q. acutissima* expanded its distribution to the Japanese archipelago either naturally or artificially. On the other hand, some populations, mainly on Kyushu Island, have multiple haplotypes and these populations must have experienced several introductions of seeds from the continent. However, these seeds did not spread to other regions of Japan, as happened with haplotype 20. It is possible that these haplotypes arrived after haplotype 20 and after it had spread across the Japanese archipelago. There are plantations which were made by Korean seed sources in 1970s in Kyushu Island for the high demand of logs of *Q. acutissima* for Shiitake cultivation [53]. The haplotypes, except for haplotype 20, might have come over to Japan quite recently. Furthermore, there may not have been enough time for mutations to accumulate after haplotype 20 was spread over the archipelago.

The level of population differentiation was very low for *Q. acutissima*, particularly in Japan. Aguinagalde et al. [54] reviewed genetic structure associated with maternally inherited markers in European trees and shrubs, and the average for three species that are cached by animals (*Corylus avellana* L., *Fagus sylvatica* L., *Quercus robur* L.) was $G_{ST} = 0.83 \pm 0.05$ SD (standard deviation) and for 15 wind dispersed species was $G_{ST} = 0.56 \pm 0.07$ SD. In contrast, Duminil et al. [55] found that species with gravity-dispersed seeds had significantly higher G_{ST} values based on maternal DNA, although no significant relationships were identified between G_{ST} and the other modes of seed dispersal. The genetic differentiation revealed by maternal DNA markers for native tree species in Japan is summarized in Table 4, and there does not appear to be any link between seed dispersal mode and G_{ST} value, as Duminil et al. [55] described. Most species have values higher than 0.9 for G_{ST} and 0.97 for G'_{ST} and are highly differentiated because of restricted gene flow by seed. The values for Japanese natural populations of *Q. acutissima* ($G_{ST} = 0.251$, $G'_{ST} = 0.261$) are much lower than those for other Japanese species. Even *Q. gilva*, which was heavily influenced by human activity in Japan, because of its edible acorns and valuable timber, showed much higher differentiation ($G'_{ST} = 0.754$) [56] than this study. In general, seed or seedling transfer by humans reduces genetic differentiation and degrades genetic structure in the same way as high gene flow. *C. sativa* ($G_{ST} = 0.43$) [2] and *P. avium* ($G_{ST} = 0.29$) [5] also exhibit low genetic differentiation (Table 4). Both of these species are used by humans and their distributions have been influenced accordingly. The value of G_{ST} for *Q. robur* is much lower in roadside plantations than natural populations [4]. As described above, *Q. variabilis*, heavily influenced by human activity as same as *Q. acutissima*, has no cpDNA variation in Japan [49]. These support the suggestion that some non-natural factor influenced the current genetic structure of *Q. acutissima*, possibly human transport of seeds and plants over a long period. Moreover, a recent study by Zhang et al. [24] reconstructed the past distribution of *Q. acutissima* and they revealed that there was little distribution of the species during the LGM in the current land area of Japan. Thus, it is reasonable to consider that *Q. acutissima* started to distribute after the LGM in Japan. This also supports that this species was possibly transported through human impact.

Table 4. Genetic differentiation of organelle DNA of five Japanese tree species, which are little used by humans and two European tree species, which are intensively used.

Species	Dispersal Type	No. of Populations	No. of Individuals	DNA	G_{ST}	G'_{ST}	Reference
Quercus acutissima	G and A	18	322	C	0.246	0.261	This study
Quercus gilva	G and A	25	135	C	0.668	0.754	[55]
Quercus mongoloca	G and A	33	501	C	0.857	0.979 *	[32]
Fagus crenata	G and A	17	409	M	0.963	0.996 *	[57]
Fagus crenata	G and A	21	351	C	0.95	-	[58]
Betula maximowicziana	W	25	400	C	0.950	0.977	[47]
Picea jezoensis	W	33	264	M	0.901	0.974 *	[59]
Prunus avium	G and A	23	211	C	0.29	0.44 *	[5]
Castanea sativa	G and A	38	181	C	0.43	0.70 *	[2]

G, gravity; A, animal; W, wind; C, Chloroplast; M, Mitochondria; *, G'_{ST} is calculated in this paper.

Although the continental populations of *Q. acutissima* exhibited relatively low differentiation in G_{ST} (0.421), the standardized $G'ST$ value was high (0.856). Thus, the much lower value of G_{ST} on the continent was due to high polymorphism of the examined cpSSRs, as Hedrick [41] pointed out. Moreover, G_{ST}-N_{ST} tests suggested the existence of phylogeographic structure on the continent and indeed the NJ tree supports this finding, showing two groups. Thus, although human impact might affect the genetic structure of *Q. acutissima* on the continent, it is likely to have been much less than the influence on Japan populations.

5. Conclusions

Based on our results, it is probable that the genetic structure of *Q. acutissima* is heavily influenced by human activities in Japan and it is difficult to reconstruct the past introduction routes to Japan from the continent because of complicated multiple introductions including secondary transportation within Japan. The genetic characteristics of natural and non-natural populations in Japan are almost the same and it is likely that the natural populations originated from individuals that escaped from plantations. Thus, at least considering cpDNA variation, modern seed transfer may not affect the current genetic structure of this species, except for some populations on Kyusyu Island. A future study not only based on bi-parental nuclear DNA genetic variation but also on human geographic study of usage of this tree species, would provide more information on past demographic history of this species in Japan in order to propose a management program for these genetic resources.

Acknowledgments: This research was supported by a grant for Research on Genetic Guidelines for Restoration Programs using Genetic Diversity Information (FY2005–2009) from the Ministry of Environment, Japan. We thank Leanne Kay Faulks for English editing.

Author Contributions: Y. Saito conceived and designed the research program, conducted genetic experiments, analyzed the data and wrote the manuscript, Y. Tsuda analyzed the data and wrote the manuscript, K. Uchiyama contributed genetic experiments and data analysis, T. Fukuda contributed samplings and experiments of Japanese natural populations, Y. Seto contributed samplings and experiments of Japanese artificial populations, P. Kim contributed samplings of Korean populations, H. Shen contributed samplings of Chinese populations, Y. Ide conceived and designed the research program.

Conflicts of Interest: The authors declare no conflict of interest.

Appendix A

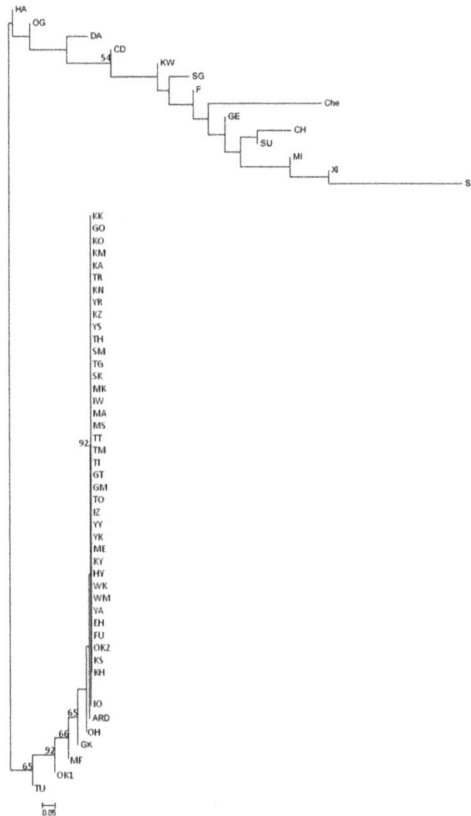

Figure A1. Neighbour-joining tree for *Quercus acutissima* Carruth. populations based on $(\delta\mu)^2$ genetic distance. Numbers indicate mean probabilities based on 1000 bootstraps and only those greater than 50 are shown.

References

1. Baldoni, L.; Tosti, N.; Ricciolini, C.; Belaj, A.; Arcioni, S.; Pannelli, G.; Germana, M.; Mulas, M.; Porceddu, A. Genetic structure of wild and cultivated olives in the central Mediterranean Basin. *Ann. Bot.* **2006**, *98*, 935–942. [CrossRef] [PubMed]

2. Fineschi, S.; Taurchini, D.; Villani, F.; Vendramin, G.G. Chloroplast DNA polymorphism reveals little geographical structure in *Castanea sativa* Mill. (*Fagaceae*) throughout southern European countries. *Mol. Ecol.* **2000**, *9*, 1495–1503. [CrossRef] [PubMed]

3. Gunn, B.; Baudouin, L.; Olsen, K. Independent origins of cultivated coconut (*Cocos nucifera* L.) in the old world tropics. *PLoS ONE* **2011**, *6*, e21143. [CrossRef] [PubMed]

4. König, A.O.; Ziegenhaben, B.; Dam, B.C.; Csaikl, U.M.; Coart, E.; Degen, B.; Burg, K.; Vries, S.M.G.; Petit, R.J. Chloroplast DNA variation of oaks in western central Europe and genetic consequences of human influences. *For. Ecol. Manag.* **2002**, *156*, 147–166. [CrossRef]

5. Mohanty, A.; Martin, J.P.; Aguinaglade, I. A population genetic analysis of chloroplast DNA in wild populations of *Prunus avium* L. in Europe. *Heredity* **2001**, *87*, 421–427. [CrossRef] [PubMed]

6. Besnard, G.; Khadari, B.; Baradat, P.; Bervill, A. *Olea europaea* (Oleaceae) phylogeography based on chloroplast DNA polymorphism. *Theor. Appl. Genet.* **2002**, *104*, 1353–1361. [PubMed]

7. Magri, D.; Fineschi, S.; Bellarosa, R.; Buonamici, A.; Sebastiani, F.; Schirone, B.; Simeone, M.C.; Vendramin, G.G. The distribution of *Quercus suber* chloroplast haplotypes matches the palaeogeographical history of the western Mediterranean. *Mol. Ecol.* **2007**, *16*, 5259–5266. [CrossRef] [PubMed]

8. Dumolin, S.; Demesure, B.; Petit, R.J. Inheritance of chloroplast and mitochondrial genomes in pedunculate oak investigated with an efficient PCR method. *Theor. Appl. Genet.* **1995**, *91*, 1253–1256. [CrossRef] [PubMed]

9. Petit, R.J.; Kremer, A.; Wagner, D. Geographic structure of chloroplast DNA polymorphisms in European oaks. *Theor. Appl. Genet. Int. J. Plant Breed. Res.* **1993**, *87*, 122–128. [CrossRef] [PubMed]

10. Rajora, O.P.; Dancik, B.P. Chloroplast DNA inheritance in Populus. *Theor. Appl. Genet.* **1992**, *84*, 280–285. [CrossRef] [PubMed]

11. Heuertz, M.; Fineschi, S.; Anzidei, M.; Pastorelli, R.; Salvini, D.; Paule, L.; Frascaria-Lacoste, N.; Hardy, O.J.; Vekemans, X.; Vendramin, G.G. Chloroplast DNA variation and postglacial recolonization of common ash (*Faxinus excelsio* L.) in Europe. *Mol. Ecol.* **2004**, *13*, 3437–3452. [CrossRef] [PubMed]

12. Palme, A.E.; Su, Q.; Rautenberg, A.; Manni, F.; Lascoux, M. Postglacial recolonization and cpDNA variation of silver birch, *Betula pendula*. *Mol. Ecol.* **2003**, *12*, 201–212. [CrossRef] [PubMed]

13. Ebert, D.; Peakall, R. Chloroplast simple sequence repeats (cpSSRs): Technical resources and recommendations for expanding cpSSR discovery and applications to a wide array of plant species. *Mol. Ecol. Res.* **2009**, *9*, 673–690. [CrossRef] [PubMed]

14. Bagnoli, F.; Tsuda, Y.; Fineschi, S.; Bruschi, P.; Magri, D.; Zhelev, P.; Paule, L.; Simeone, M.; González-Martínez, S.; Vendramin, G. Combining molecular and fossil data to infer demographic history of *Quercus cerris*: Insights on European eastern glacial refugia. *J. Biogeogr.* **2016**, *43*, 679–690. [CrossRef]

15. Provan, J.; Powell, W.; Hollingswoth, P.M. Chloroplast microsatellites: New tools for studies in plant ecology and evolution. *Trends Ecol. Evol.* **2001**, *16*, 142–147. [CrossRef]

16. Grivet, D.; Deguilloux, M.F.; Petet, R.J.; Sork, V.L. Contrasting patterns of historical colonization in white oaks (*Quercus* spp.) in California and Europe. *Mol. Ecol.* **2006**, *15*, 4085–4093. [CrossRef] [PubMed]

17. Marsico, T.D.; Hellmann, J.J.; Romero-Severson, J. Patterns of seed dispersal and pollen flow in *Quercus garryana* (Fagaceae) following post-glacial climatic changes. *J. Biogeogr.* **2009**, *36*, 929–941. [CrossRef]

18. Mohammad-Panah, N.; Shabanian, N.; Khadivi, A.; Rahmani, M.S.; Emami, A. Genetic structure of gall oak (*Quercus infectoria*) characterized by nuclear and chloroplast SSR markers. *Tree Genet. Genomes* **2017**, *13*, 70–81. [CrossRef]

19. Iida, S. Quantitative analysis of acorn transportation by rodents using magnetic locator. *Vegetatio* **1996**, *124*, 39–43. [CrossRef]

20. Satake, Y.; Hara, H.; Watari, S.; Tominari, T. *Wild Flowers of Japan—Woody Plants*; Heibonsha Ltd.: Tokyo, Japan, 1989. (In Japanese)

21. Kurata, S. *Trip for Plants and Literature*; Chikyu-Sha: Tokyo, Japan, 1976. (In Japanese)

22. Fukamachi, K.; Oku, H.; Rackham, O. A comparitive stydy on tees and hedgerows in Japan and England. In *Landscape Interfaces*; Palang, H., Fry, G., Eds.; Kluwer Academic Publishers: Dordrecht, The Netherlands, 2003; pp. 53–69.

23. Tsukada, M. Map of vegetation during the Last Glacial Maximum in Japan. *Quat. Res.* **1985**, *23*, 369–381. [CrossRef]

24. Zhang, X.; Li, Y.; Liu, C.; Xia, T.; Zhang, Q.; Fang, Y. Phylogeography of the temperate tree species *Quercus acutissima* in China: Inferences from chloroplast DNA variations. *Biochem. Syst. Ecol.* **2015**, *63*, 190–197. [CrossRef]

25. Yasuda, Y.; Miyoshi, N. *Illustration Vegetation History of Japan*; Asakura Pub.: Tokyo, Japan, 1998. (In Japanese)

26. Shimaji, K.; Itoh, T. *Comprehensive List of Unearthed Wood Product from the Remains in Japan*; Yuzankaku Pub.: Tokyo, Japan, 1988. (In Japanese)

27. Suzuki, M. *Japanese and Wood Culture*; Yasaka Shobo: Tokyo, Japan, 2002. (In Japanese)

28. Nakamura, K. *Historical Study on Shitake Mushroom Culture*; Tohsen Pub.: Tokyo, Japan, 1983. (In Japanese)

29. Abrams, M.D.; Copenheaver, C.A.; Terazawa, K.; Umeki, K.; Takiya, M.; Akashi, N. A 370-year dendroecological history of an old-growth Abies-Acer-Quercus forest in Hokkaido, northern Japan. *Can. J. For. Res.* **1999**, *29*, 1891–1899. [CrossRef]

30. Sakai, T.; Tanaka, H.; Shibata, M.; Suzuki, W.; Nomiya, H.; Kanazashi, T.; Iida, S.; Nakashizuka, T. Riparian disturbance and community structure of a *Quercus-Ulmus* forest in central Japan. *Plant Ecol.* **1999**, *140*, 99–109. [CrossRef]

31. Kanno, M.; Yokoyama, J.; Suyama, Y.; Ohyama, M.; Itoh, T.; Suzuki, M. Geographical distribution of two haplotypes of chloroplast DNA in four oak species (*Quercus*) in Japan. *J. Plant Res.* **2004**, *117*, 311–317. [CrossRef] [PubMed]

32. Okaura, T.; Quang, N.D.; Ubukata, M.; Harada, K. Phylogeographic structure and late Quaternary population history of the Japanese oak *Quercus mongolica* var. *crispula* and related species revealed by chloroplast DNA variation. *Genes Genet. Syst.* **2007**, *82*, 465–477. [PubMed]

33. Ohsawa, T.; Tsuda, Y.; Saito, Y.; Ide, Y. The genetic structure of *Quercus crispula* in northeastern Japan as revealed by nuclear simple sequence repeat loci. *J. Plant Res.* **2011**, *124*, 645–654. [CrossRef] [PubMed]

34. Lian, C.L.; Oishi, R.; Miyashita, N.; Nara, K.; Nakaya, H.; Wu, B.Y.; Zhou, Z.H.; Hogetsu, T. Genetic structure and reproduction dynamics of *Salix reinii* during primary succession on Mount Fuji, as revealed by nuclear and chloroplast microsatellite analysis. *Mol. Ecol.* **2003**, *12*, 609–618. [CrossRef] [PubMed]

35. Deguilloux, M.F.; Dumolin-Lapegue, S.; Gielly, L.; Grivet, D.; Petit, R.J. A set of primers for the amplification of chloroplast microsatellites in *Quercus*. *Mol. Ecol. Notes* **2003**, *3*, 24–27. [CrossRef]

36. Pons, O.; Petit, R.J. Mesuring and testing genetic differentiation with orderd versus unorderd allels. *Genetics* **1996**, *14*, 1237–1245.

37. Estoup, A.; Jarne, P.; Cornuet, J.M. Homoplasy and mutation model at microsatellite loci and their consequences for population genetics analysis. *Mol. Ecol.* **2002**, *11*, 1591–1604. [CrossRef] [PubMed]

38. Petit, R.J.; El Mousadik, A.; Pons, O. Identifying populations for conservation on the basis of genetic markers. *Conserv. Biol.* **1998**, *12*, 844–855. [CrossRef]

39. Nei, M. Analysis of gene diversity in subdivided populations. *Proc. Natl. Acad. Sci. USA* **1973**, *70*, 3321–3323. [CrossRef] [PubMed]

40. Slatkin, M. A measure of population subdivision based on microsatellite allele frequencies. *Genetics* **1995**, *139*, 457–462. [PubMed]

41. Hedric, P.W. A standardized genetic differentiation measure. *Evolution* **2005**, *59*, 1633–1638. [CrossRef]

42. Bryant, D.; Moulton, V. NeighborNet: An agglomerative method for the construction of planar phylogenetic networks. In *Algorithms in Bioinformatics, Proceedings of WABI, Rome, Italy, 17–21 September 2002*; Guigó, R., Gusfield, D., Eds.; Springer: Berlin, Germany, 2002; Volume 2452, pp. 375–391.

43. Huson, D.H.; Bryant, D. Application of Phylogenetic Networks in Evolutionary Studies. *Mol. Biol. Evol.* **2006**, *23*, 254–267. [CrossRef] [PubMed]

44. Goldstein, D.B.; Linares, A.R.; Feldman, M.W.; Cavallisforza, L.L. Genetic absolute dating based on microsatellites and origin of modern humans. *Proc. Natl. Acad. Sci. USA* **1995**, *92*, 6720–6727. [CrossRef]

45. Langella, O. Population 1.2.30. Logiciel de Génétique des Populations. Laboratoire Populations, Génétique et Évolution, CNRS UPR 9034, Gif-sur-Yvette. Available online: http://www.cnrs-gif.fr/pge/ (accessed on 2 October 2009).

46. Parks, D.H.; Mankowski, T.; Zangooei, S.; Porter, M.S.; Armanini, D.G.; Baird, D.J.; Langille, M.G.I.; Beiko, R.G. GenGIS 2: Geospatial Analysis of Traditional and Genetic Biodiversity, with New Gradient Algorithms and an Extensible Plugin Framework. *PLoS ONE* **2013**, *8*, e69885. [CrossRef] [PubMed]

47. Tsuda, Y.; Ide, Y. Chloroplast DNA phylogeography of *Betula maximowicziana*, a long-lived pioneer tree species and noble hardwood in Japan. *J. Plant Res.* **2010**, *123*, 343–353. [CrossRef] [PubMed]

48. Heuertz, M.; Carnevale, S.; Fineschi, S.; Sebastiani, F.; Hausman, J.F.; Paule, L.; Vendramin, G.G. Chloroplast DNA phylogeography of European ashes, *Fraxinus* sp. (Oleaceae): Roles of hybridization and life history traits. *Mol. Ecol.* **2006**, *15*, 2131–2140. [CrossRef] [PubMed]

49. Saito, Y.; Tsuda, Y.; Uchiyama, K.; Fukuda, T.; Ide, Y. Genetic structure of *Quercus variabilis* in Japan. *For. Genet. Tree Breed.* **2017**, in print. (In Japanese with English Summary).

50. Sugahara, K.; Kaneko, Y.; Ito, S.; Yamanaka, K.; Sakio, H.; Hoshizaki, K.; Suzuki, W.; Yamanaka, N.; Setoguchi, H. Phylogeography of Japanese horse chestnut (*Aesculus turbinata*) in the Japanese Archipelago based on chloroplast DNA haplotypes. *J. Plant Res.* **2011**, *124*, 75–83. [CrossRef] [PubMed]

51. Clark, J.S.; Fastie, C.; Hurtt, G.; Jackson, S.T.; Johnson, C.; King, G.A.; Lewis, M.; Lynch, J.; Pacala, S.; Prentice, C.; et al. Reid's Paradox of Rapid Plant Migration: Dispersal theory and interpretation of paleoecological records. *BioScience* **1998**, *48*, 13–24. [CrossRef]

52. Hashizume, H. Flowering habit in young trees of *Quercus acutissima* Carr. and *Quercus serrata* Thunb. *Hardwood Res.* **1983**, *2*, 49–54. (In Japanese with English Summary)

53. Toda, T.; Fujimoto, Y.; Nishimura, K.; Maeda, T. Reserch of 18 year old plantation *Quercus acutissima* established by imported seed. *Annu. Rep. Kyushu Reg. Breed.* **1986**, *13*, 100–106. (In Japanese)

54. Aguinagalde, I.; Hampe, A.; Mohanty, A.; Martin, J.P.; Duminil, J.; Petit, R.J. Effects of life-history traits and species distribution on genetic structure at maternally inherited markers in European trees and shrubs. *J. Biogeogr.* **2005**, *32*, 329–339. [CrossRef]

55. Duminil, J.; Fineschi, S.; Hampe, A.; Jordano, P.; Salvini, D.; Vendramin, G.G.; Petit, R.J. Can population genetic sturcture be predicted from life-history traits? *Am. Nat.* **2007**, *169*, 662–672. [PubMed]

56. Sugiura, N.; Tang, D.; Kurokochi, H.; Saito, Y.; Ide, Y. Genetic structure of *Quercus gilva* Blume in Japan as revealed by chloroplast DNA sequences. *Botany* **2015**, *93*, 873–880. [CrossRef]

57. Tomaru, N.; Takahashi, M.; Tsumura, Y.; Takahashi, M.; Ohba, K. Intraspecific variation and phylogeographic patterns of *Fagus crenata* (Fagaceae). *Am. J. Bot.* **1998**, *85*, 629–636. [CrossRef] [PubMed]

58. Okaura, T.; Harada, K. Phylogeographical structure revealed by chloroplast DNA variation in Japanese beech (*Fagus crenata* Blume). *Heredity* **2002**, *88*, 322–329. [CrossRef] [PubMed]

59. Aizawa, M.; Yoshimaru, H.; Saito, H.; Katsuki, T.; Kawahara, T.; Kitamura, K.; Shi, F.; Kaji, M. Phylogeography of a northeast Asian spruce, Picea jezoensis, inferred from genetic variation observed in organelle DNA markers. *Mol. Ecol.* **2007**, *16*, 3393–3405. [CrossRef] [PubMed]

© 2017 by the authors. Licensee MDPI, Basel, Switzerland. This article is an open access article distributed under the terms and conditions of the Creative Commons Attribution (CC BY) license (http://creativecommons.org/licenses/by/4.0/).

forests

MDPI

Article

Genetic Diversity and Structure of Natural *Quercus variabilis* Population in China as Revealed by Microsatellites Markers

Xiaomeng Shi [1,†], Qiang Wen [2,†], Mu Cao [1], Xin Guo [1] and Li-an Xu [1,*]

1   Co-Innovation Center for Sustainable Forestry in Southern China, Nanjing Forestry University, Nanjing 210037, China; shixiaomeng0603@163.com (X.S.); caomu1208@126.com (M.C.); guoxin0902@163.com (X.G.)
2   Plant Bio-tech Key Laboratory of Jiangxi Province, Jiangxi Academy of Forestry, Nanchang 330013, China; wenqiang1107@163.com
*   Correspondence: laxu@njfu.edu.cn; Tel.: +86-025-8542-7882
†   These two authors contribute equally to this work.

Received: 3 November 2017; Accepted: 7 December 2017; Published: 11 December 2017

Abstract: *Quercus variabilis* is a tree species of ecological and economic value that is widely distributed in China. To effectively evaluate, use, and conserve resources, we applied 25 pairs of simple sequence repeat (SSR) primers to study its genetic diversity and genetic structure in 19 natural forest or natural secondary forest populations of *Q. variabilis* (a total of 879 samples). A total of 277 alleles were detected. Overall, the average expected heterozygosity (H_e) was 0.707 and average allelic richness (AR) was 7.79. *Q. variabilis* manifested a loss of heterozygosity, and the mean of inbreeding coefficient (F_{IS}) was 0.044. Less differentiation among populations was observed, and the genetic differentiation coefficient (F_{ST}) was 0.063. Bayesian clustering analysis indicated that the 19 studied populations could be divided into three groups based on their genetic makeup, namely, the Southwest group, Central group, and Northeastern group. The Central group, compared to the populations of the Southwest and Northeast group, showed higher genetic diversities and lower genetic differentiations. As a widely distributed species, the historical migration of *Q. variabilis* contributed to its genetic differentiation.

Keywords: *Quercus variabilis*; SSR; genetic diversity; population structure

1. Introduction

Fagaceae is one of the most important components of northern sub-tropical forests and temperate forests. *Quercus* spp. consists of about 450 different species [1] and is the largest Fagaceae genus. Moreover, these have important ecological and economic values. Also known as Chinese cork oak, *Quercus variabilis* Bl. is a species of oak in the section *Quercus* Sect. *Cerris* [2] and native to a wide area of China [3], as well as the Korean peninsula and the southwestern Japanese archipelago. In China, the Qinling and Dabie mountain land areas are considered the geographical distribution centers of *Q. variabilis* [4]. Having one of the widest amplitudes of distribution and most complex climate types in worldwide geographic distribution [5], *Q. variabilis* forests have formed many stable forest ecosystems in warm temperate and northern subtropical regions in China and are playing a significant role in the conservation and improvement of water and soil [6]. It is a precious timber and economic tree species, and also considered as an important resource for natural raw materials in cork production due to its most unique feature, i.e., its bark [7,8]. In recent years, with the development of cork, tannin, and other industries in China, the phenomenon of over harvesting of *Q. variabilis* has become a prominent agricultural problem that has resulted in extensive destruction and waste of natural resources [9]. Currently, the main distribution areas of *Q. variabilis* in China are mostly natural secondary forests [4].

The original forest has become extremely rare; scattered forests are found in Guizhou, Yunnan, and Fujian (on-site investigations). The destruction of the natural populations has been strongly associated with climate change, soil erosion, and a series of ecological problems [10], hence a shortage in natural resources for *Q. variabilis* has caused a sharp decline in raw material production.

Genetic diversity in forests is determined by gene flow, genetic drift, selection, mutation, and other factors [11–13], and the levels of genetic variation and structures are the combined effects of evolutionary history, distribution area, life-styles, and breeding forms [14]. For a tree with a relatively long life cycle, genetic diversity determines its ability to adapt to changing environments, which in turn serves as the basis for maintaining long-term stability of forest ecosystems. The evaluation of genetic variability, especially for forests with wide distribution ranges and high ecological and economic values, is of great significance for forest ecosystem conservation and management of genetic resources [15]. Therefore, investigating the evolutionary history and distribution, even including the development of protection strategies and measures based on forest diversity levels and genetic structures, is imperative.

Studies on *Q. variabilis* have mainly concentrated on population structure and classification, biodiversity characteristics, population dynamics, and ecological function [16–18]. In addition, various features of reproduction, seedling cultivation, and planting techniques have also been studied [19–21]. Various reports on the population genetics of *Quercus* such as *Q. acutissima* [22], *Q. rubra* and *Q. ellipsoidalis* [23], *Q. mongolica* [24], and *Q. infectoria* [25] have been published in recent years. There are relatively few genetic investigations on *Q. variabilis*, but Xu [6] was the first to conduct a preliminary study on its genetic diversity of Chinese five natural populations by using simple sequence repeat (SSR) markers. Recent studies have focused on *Q. variabilis* pedigree geography such as Chen [26], who used chloroplast simple sequence repeat (cpSSR) to relatively and comprehensively analyze the historical evolution and its geographical distribution. However, this study was limited by the number of samples, and may therefore not fully represent the genetic variation of existing populations. Therefore, based on the natural resource distribution of all existing populations, the establishment of a suitable means of detection, a sufficient number of markers, and more comprehensive sample size is necessary to study the genetic structure and mating system.

China is considered as the global distribution center for *Q. variabilis* distribution. Furthermore, research studies on its genetic structure based on a broader sampling strategy involving its main distribution area, are of high significance to better understand the distribution pattern and formation of existing populations. Furthermore, such studies may provide important molecular bases for the protection and exploitation of the genetic resources.

2. Materials and Methods

2.1. Population Sample Information

The tender leaves of mature trees were collected from 19 different sub-locations in nine provinces (Figure 1), which is representative of the entire distribution area covered from the Henan township Neixiang, Henan (33°50′ N 111°18′ E) to Leye, Guangxi (24°47′ N 106°57′ E). Annual temperature varies from 8.3 °C to 18.4 °C, spanning two climatic zones: subtropical monsoon and Temperate monsoon. More details on the number of samples and geographical features of each group are listed in Table 1. The studied forests are typical natural secondary forests, and the plants of each population were separated by at least 50 m. The collected samples were brought back to the lab, dried by silica, and stored at room temperature until analysis.

Table 1. Geographic locations and mean values of climate parameter in 19 populations of *Q. variabilis*.

No.	Population	Code	Latitude (°N)	Longitude (°E)	N	Annual Temperature (°C)	Weather Patterns	Annual Precipitation (mm)
1	Ankang, Shaanxi	AK	32°40′	109°08′	45	15.3	S	803
2	Chuzhou, Anhui	CZ	32°17′	118°17′	39	15.3	S	1009
3	Macheng, Hubei	MC	31°17′	115°00′	45	16.4	S	1217
4	Xingshan, Hubei	XS	31°02′	110°07′	45	16.5	S	1498
5	Zhumadian, Henan	ZM	32°08′	114°01′	45	15.4	S	1098
6	Neixiang, Henan	NX	33°50′	111°18′	48	8.3	T	871
7	Leye, Guangxi	LY	24°47′	106°57′	45	18.4	S	1314
8	Nanjing, Jiangsu	NJ	32°03′	118°52′	45	15.6	S	1017
9	Xiaoxian, Anhui	XX	34°12′	116°56′	45	9.9	T	756
10	Pengze, Jiangxi	PZ	29°54′	116°34′	48	17.4	S	1460
11	Chengkou, Chongqing	CK	31°59′	108°40′	48	13.1	S	1165
12	Fengjie, Chongqing	FJ	31°04′	109°31′	48	13.6	S	1071
13	Pengshui, Chongqing	PS	29°12′	108°12′	48	15.6	S	1296
14	Wuxi, Chongqing	WX	31°25′	109°36′	48	12.9	S	1131
15	Jingshan, Hubei	JS	31°02′	113°07′	45	16.3	S	1035
16	Yingshan, Hubei	YS	30°45′	115°34′	48	17.1	S	1459
17	Yuanshan, Hube	YA	31°00′	111°36′	48	16.6	S	1018
18	Suizhou, Hubei	SZ	31°36′	113°18′	48	16.1	S	1007
19	Longquan, Zhejiang	LQ	28°02′	119°05′	48	17.6	S	1793
ALL					879			

Population: sub-location, province; N: sample size; S: Subtropical monsoon; T: Temperate monsoon.

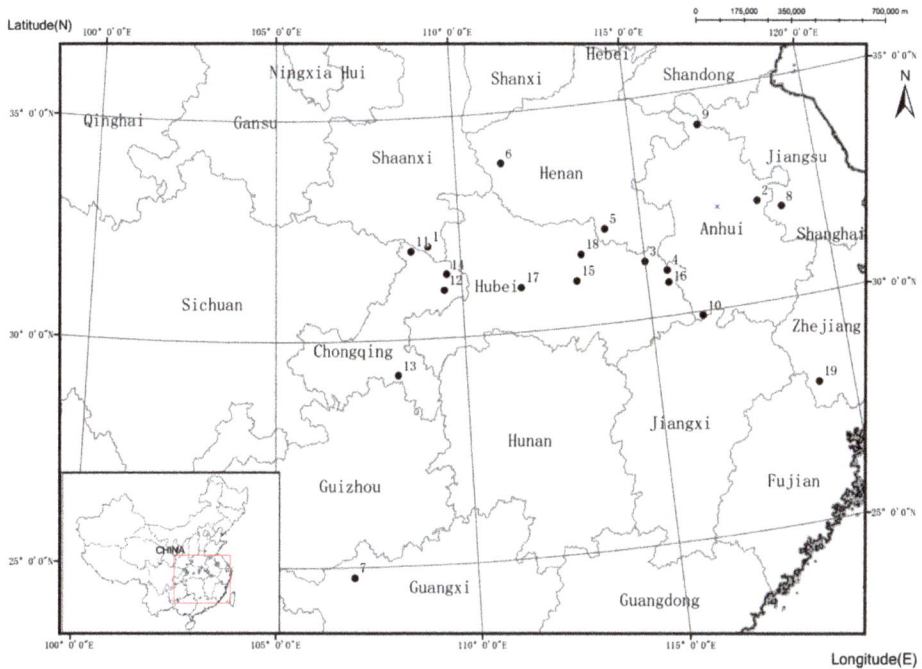

Figure 1. Locations of the 19 investigated populations of *Q. variabilis*. Numbers correspond to the population numbers in Table 1.

2.2. Experimental Methods

Total DNA was extracted from the samples by using a modified cetyl trimethylammonium bromide (CTAB) lysis-silica beads adsorption method [27]. The SSR primers used in population identification are shown in Table 2. The total volum e of each single locus polymerase chain reaction (PCR) was 20 µL, which consisted of 20 ng of DNA, 50 mM KCl, 20 mM Tris-HCl (pH 8.3), 2 mM MgCl$_2$, 0.2 mM dNTP, 0.5 µM of each primer (forward and reverse), and 1 U of Taq enzyme (Qiagen Inc., Hilden, Germany) PCR thermal cycling was performed on a GeneAmp PCR System 9700 (Applied Biosystems Inc., Carlsbad, CA, USA) using the following conditions: an initial denaturation step of 94 °C for 4 min; followed by 30 cycles of denaturation at 94 °C for 1 min, annealing at N °C for 2 min (N is the specific annealing temperature for each pair of primers), and extension at 72 °C for 1 min; and a final extension at 72 °C for 8 min and final storage at 4 °C. The PCR products were detected by using capillary electrophoresis (Bioptic100-automated nucleic acid analyzer, BiOptic Inc., Tucson, AZ, USA). Electrophoresis results were interpreted using Q-editor software [28].

Table 2. Source, repeat motif, and polymorphism information for the 25 microsatellite loci analyzed in the 879 *Q. variabilis* trees.

Locus	Repeat Motif	N_a	H_o	H_e	*PIC*	Source
2p24	$(CA)_{14}$	8	0.429	0.717	0.794	Alexis, R.S. et al. [29]
E71-72	$(GA)_{46}$	8	0.212	0.771	0.803	Qin, Y.Y. et al. [30]
PIE040	$(TTC)_8$	10	0.157	0.724	0.747	Alexis, R.S. et al. [29]
GOT040	$(GA)_{11}$	5	0.389	0.468	0.782	Durand, J. et al. [31]
G0T009	$(TC)_7$	6	0.208	0.590	0.673	Durand, J. et al. [31]
FIR053	$(GTG)_7$	8	0.294	0.751	0.786	Durand, J. et al. [31]
FIR039	$(CT)_7$	9	0.459	0.754	0.782	Durand, J. et al. [31]
FIR004	$(CT)_{18}$	8	0.472	0.760	0.778	Alexis, R.S. et al. [29]
G11	$(TC)_{22}$	4	0.324	0.551	0.615	Xu, X.L. et al. [6]
PL111-112	$(TC)_9$	6	0.534	0.694	0.720	Qin, Y.Y. et al. [30]
PL229-230	$(AG)_{15}$	9	0.387	0.669	0.689	Qin, Y.Y. et al. [30]
VIT107	$(TA)_{13}$	5	0.306	0.452	0.524	Durand, J. et al. [31]
DN949726	$(GAT)_6$	15	0.384	0.861	0.878	Saneyoshi, U. et al. [24]
E11-12	$(GA)_{32}$	14	0.578	0.851	0.887	Qin, Y.Y. et al. [30]
E79-80	$(TC)_{18}$	24	0.626	0.841	0.889	Qin, Y.Y. et al. [30]
EE812	$(AG)_7$	20	0.393	0.804	0.856	Zhang, Y.Y. et al. [22]
G7	$(TC)_{17}$	21	0.447	0.784	0.883	Xu, X.L. et al. [6]
G16	$(AG)_{21}$	20	0.606	0.829	0.860	Xu, X.L. et al. [6]
PL127-128	$(AG)_{12}$	18	0.408	0.846	0.874	Qin, Y.Y. et al. [30]
Q16	$(GA)_{18}$	26	0.704	0.875	0.911	Xu, X.L. et al. [6]
EE802	$(CT)_8$	7	0.427	0.748	0.790	Zhang, Y.Y. et al. [22]
EE856	$(GGT)_6$	4	0.308	0.418	0.423	Zhang, Y.Y. et al. [22]
FIR048	$(CT)_9$	8	0.438	0.758	0.784	Durand, J. et al. [31]
FIR110	$(TC)_{20}$	6	0.186	0.552	0.602	Alexis, R.S. et al. [29]
PIE125	$(GGAAGC)_3$	8	0.353	0.619	0.664	Durand, J. et al. [31]
mean		11	0.401	0.707	0.760	
min		4	0.157	0.418	0.423	
max		26	0.704	0.875	0.911	

N_a: number of alleles, H_o: observed, H_e: expected heterozygosities, *PIC*: polymorphism information content.

2.3. Data Analysis

Genetic parameters, calculated by the Fstat software, were: average number of alleles: N_a; observed heterozygosity: H_o; expected heterozygosity or gene diversity, H_e [32]; allelic richness, *AR* [33]; the fixation index (F_{IS}: the inbreeding coefficient) of every point and group level; the differentiation index between pairwise populations (F_{ST}, G_{ST}, R_{ST}) and the matrix of F_{ST} between various groups on which F-statistics were based [34,35]; Hardy-Weinberg equilibrium testing and estimation of allele frequency were calculated [36]. The polymorphic information index (*PIC*) and Linkage disequilibrium were calculated by using the PowerMarker v3.23 software (Bioinformatics Research Center; North Carolina State University, Raleigh, NC, USA) [37]. P values were adjusted by using a Bonferroni correction. Private alleles were calculated by using the GenAlEx software [38,39]. Gene flow (N_m) was calculated using the formula: $N_m = (1 - F_{ST})/4F_{ST}$ [40]. To investigate genetic diversity and geography-space-climate relationships [41], the average *AR* value was calculated using a total of 25 loci, whereas the relationships between H_e value and each geographic population (latitude and longitude), annual average temperature, and annual precipitation in climate parameters (obtained through DIVA-GIS software [42]) were calculated by using the Kendall rank-correlation test as provided in the SAS v9.1 software (SAS Institute 2001, Wallisellen, Switzerland). Detection of bottleneck effects was calculated by using BOTTLENECK version 1.2 software [43]. A two-phase mutation model (TPM) and a stepwise mutation model (SMM) were used for Wilcoxon signed-rank tests. The parameter settings included a 90% SMM and 10% TPM with a variance of 12%, and 1000 repeats [44].

For genetic relationships at the population level, a factorial correspondence analysis (FCA) was performed by using the Genetix 4.05 software (CNRSUMR 5000; Universite Montpellier II, Montpellier,

France) [45]. Population genetic structure was analyzed based on Bayesian clustering using the STRUCTURE software [46]. For clustering from $K = 1$ to $K = 20$ (populations + 1), an admixture ancestry model and correlated allele frequency model were used to perform a Markov chain Monte Carlo simulation algorithm (MCMC) [47]. The length of the burn-in period at start time was set as 100,000; MCMC after the length of the burn-in period was set as 100,000, and for each of K value, simulation calculation was repeated 10 times. The method from Evanno was used to determine the optimal K value [48]. The relative proportions of the geographical distribution diagram were plotted by using the Arc-GIS software (Environmental Systems Research Institute, Inc., Redlands, CA, USA). Molecular variance analysis (AMOVA) for population genetic structures [49] and a Mantel test to detect the presence of geographic segregation between populations by means of logarithmic normalization of geospatial distance and genetic distance [50] were calculated by using the Arlequin software (CMPG, Institute of Ecology and Evolution; University of Berne, Berne, Switzerland).

3. Results

3.1. Genetic Diversity

A total of 277 alleles were amplified from 25 SSR primers. On average, each microsatellite loci could be amplified 11 times, and the number of alleles ranged from a minimum of 4 (G11, EE856) to a maximum of 26 (Q16). The H_o of each microsatellite loci ranged from 0.157 (PIE040) to 0.704 (Q16), and the average observed heterozygosity was 0.401. For each microsatellite loci, H_e was higher than H_o, and ranged from 0.418 (EE856) to 0.875 (Q16), and the average H_e was 0.707. AR ranged from 3.50 (EE856) to 21.12 (Q16), and the average AR was 9.51. The PIC of SSR microsatellite loci ranged from 0.423 to 0.911, and the average PIC was 0.760 (Table 2). After detecting significance ($P < 0.05$), we did not find the existence of linkage disequilibrium between any two points.

Based on statistics of population genetic diversities of 25 microsatellite loci from 19 geographic populations (Table 3), the average number of alleles in each population was 8.01; the population with the highest number of alleles was PZ (8.76), and the lowest was XX (7.04). The average number of effective alleles was 4.52; the highest was LY, and the lowest was FJ. The overall average AR was 7.79; the highest one (8.56) was distributed in XS, whereas the lowest was in XX (6.90). *Q. variabilis* manifested a significant loss of heterozygosity, and the mean of inbreeding coefficient was 0.044. The average H_e was 0.707; the highest one was 0.745 with a distribution in CK, whereas the lowest one was 0.623 in FJ. BOTTLENECK analysis showed that in the 19 analyzed populations, three groups could be identified: CZ (SMM, $P < 0.05$), PS (SMM, $P < 0.05$), and FJ (TPM, SMM, $P < 0.05$), which may have recently undergone a severe decline in population size, i.e., the bottleneck effect had occurred.

Kendall rank-correlation analysis of genetic diversity of geographic populations with latitude/longitude and climatic parameters showed that in general, the AR of *Q. variabilis* geographic populations has a significant positive correlation with higher average annual temperatures ($r = 0.507$, $P < 0.01$), whereas no correlations among H_e, geographic latitude, and climate were observed.

Table 3. Mean values of genetic diversity statistics for 25 microsatellite loci in 19 *Q. variabilis* populations.

Code	N_a	AR	H_e	F_{IS}	TPM	SMM	F_{ST}	G_{ST}	R_{ST}
AK	8.20	8.03	0.710	0.038	*ns*	*ns*			
CZ	7.12	7.12	0.683	0.054	*ns*	0.037 *			
MC	8.32	8.16	0.720	0.033	*ns*	*ns*			
XS	8.68	8.56	0.723	0.045	*ns*	*ns*			
ZM	7.84	7.65	0.711	0.043	*ns*	*ns*			
NX	7.92	7.75	0.709	0.042	*ns*	*ns*			
LY	8.44	8.23	0.707	0.046	*ns*	*ns*			
NJ	7.96	7.81	0.725	0.049	*ns*	*ns*			
XX	7.04	6.90	0.690	0.041	*ns*	*ns*			
PZ	8.76	8.54	0.725	0.042	*ns*	*ns*			
CK	8.48	7.54	0.745	0.044	*ns*	*ns*			
FJ	7.12	6.95	0.623	0.045	0.045 *	0.002 *			
PS	8.68	7.73	0.699	0.050	*ns*	0.019 *			
WX	7.44	7.29	0.701	0.055	*ns*	*ns*			
JS	7.92	8.45	0.705	0.039	*ns*	*ns*			
YS	8.24	8.00	0.694	0.049	*ns*	*ns*			
YA	8.04	7.86	0.726	0.041	*ns*	*ns*			
SZ	8.16	7.93	0.728	0.043	*ns*	*ns*			
LQ	7.76	7.54	0.716	0.039	*ns*	*ns*			
mean	8.01	7.79	0.707	0.044			0.063	0.060	0.073
min	7.04	6.90	0.623	0.033					
max	8.76	8.56	0.745	0.055					

N_a: number of alleles; *AR*: allelic richness; H_e: expected heterozygosity; F_{IS}: fixation index. TPM: a two-phase mutation model, SMM: a stepwise mutation model; * Significant deviation from Wilcoxon signed-rank tests ($P < 0.05$). F_{ST}, G_{ST} differentiation among populations according to Weir and Cockerham [34]; R_{ST}, Slatkin [51].

3.2. Genetic Differentiation and Genetic Structure

The overall population differentiation degree among geographical populations of *Q. variabilis*: $F_{ST} = 0.063$, $G_{ST} = 0.060$ ($P < 0.001$) have been reported in Table 3. Gene flow: $N_m = 3.648$. AMOVA analysis [52] showed that 6.3% ($P < 0.001$) of the genetic variations was among populations. The greatest percentage of variation was contained within populations. The F_{ST} matrix between every two populations is shown in Table 4; the differentiation coefficient between populations CK and SZ was the smallest, ($F_{ST} = 0.028$); that between XX and FJ was the largest ($F_{ST} = 0.135$).

A FCA on genetic structure differences of pattern detection at the population level showed that the populations could be divided into three groups based on geographic location: (a) the Southwest populations: LY, PS, FJ, and WX; (b) the Central populations (the most resource-rich center of *Q. variabilis* distribution in China): AK, MC, XS, PZ, CK, JS, YS, YA, SZ, and LQ; and (c) the Northeast populations: NJ, CZ, ZM, NX, and XX. The Central populations were adjacent to the Southwest populations and Northeast populations in the FCA clustering groups, whereas the Southwest populations were distantly located from the Northeast populations in the FCA clustering groups (see Figure 2a,c). Small genetic differences between the Central populations and the other two populations were observed, whereas larger genetic differences were detected between the Southwest populations and the Northeast populations.

Table 4. The pairwise F_{ST} for all populations of *Q. variabilis*.

Code	AK	CZ	MC	XS	ZM	NX	LY	NJ	XX	PZ	CK	FJ	PS	WX	JS	YS	YA	SZ	LQ
AK	0.000																		
CZ	0.051	0.000																	
MC	0.067	0.079	0.000																
XS	0.040	0.042	0.049	0.000															
ZM	0.059	0.044	0.062	0.048	0.000														
NX	0.062	0.033	0.078	0.064	0.044	0.000													
LY	0.055	0.040	0.073	0.040	0.058	0.061	0.000												
NJ	0.063	0.048	0.051	0.035	0.050	0.053	0.050	0.000											
XX	0.076	0.068	0.111	0.072	0.077	0.061	0.082	0.075	0.000										
PZ	0.059	0.055	0.047	0.047	0.051	0.076	0.050	0.048	0.099	0.000									
CK	0.027	0.050	0.049	0.040	0.050	0.057	0.047	0.041	0.077	0.039	0.000								
FJ	0.077	0.081	0.105	0.078	0.108	0.112	0.059	0.091	0.135	0.082	0.079	0.000							
PS	0.039	0.056	0.069	0.040	0.064	0.074	0.047	0.057	0.081	0.057	0.050	0.068	0.000						
WX	0.078	0.082	0.101	0.066	0.091	0.096	0.053	0.078	0.101	0.076	0.079	0.082	0.053	0.000					
JS	0.058	0.076	0.067	0.071	0.086	0.093	0.084	0.065	0.113	0.049	0.045	0.093	0.079	0.111	0.000				
YS	0.040	0.040	0.049	0.039	0.041	0.058	0.043	0.035	0.088	0.029	0.033	0.065	0.036	0.070	0.060	0.000			
YA	0.048	0.047	0.068	0.054	0.063	0.061	0.054	0.050	0.082	0.057	0.036	0.082	0.066	0.099	0.064	0.054	0.000		
SZ	0.052	0.079	0.056	0.052	0.077	0.089	0.062	0.051	0.105	0.046	0.028	0.085	0.062	0.083	0.058	0.053	0.049	0.000	
LQ	0.052	0.050	0.050	0.031	0.052	0.069	0.049	0.043	0.082	0.039	0.037	0.085	0.054	0.084	0.060	0.038	0.045	0.053	0.000

See Table 1 for population information.

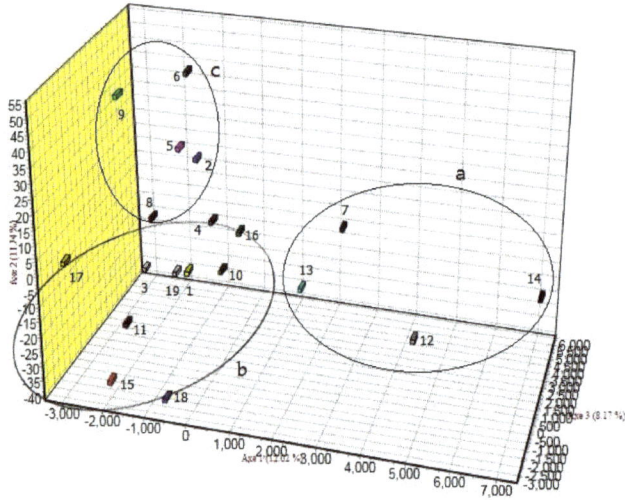

Figure 2. Results of factorial correspondence analysis at the population level based on simple sequence repeat markers. The population codes are as in Table 1.

For a more detailed understanding of the genetic structures, using a ΔK value that Evanno proposed to determine a reasonable K, the ΔK value reached a maximum when $K = 3$ (Figure 3). The 19 geographic populations of *Q. variabilis* could then be divided into three genetically distinct groups (Figure 4). In addition, we used the relative proportions of the geographical distribution diagram to visually compare the distribution ratio of each group within the population (Figure 5). The 19 populations as a whole were divided into three groups based on geographic distribution (Figures 4 and 5); the first group was the Southwest group (pop1–4), the second group was the Central group (pop5–14), and the third group was the Northeast group (pop15–19). Average distribution ratios [53] were 76%, 72% and 73%, respectively. The Central group had the lowest distribution ratio, i.e., the highest mixing degree among groups. XX (Northeast group, ratio: 87.9%), JS (Central group, ratio: 83.4%), and WX (Southwest group, ratio: 88.9%) were the source populations in each group, which had the highest distribution ratio.

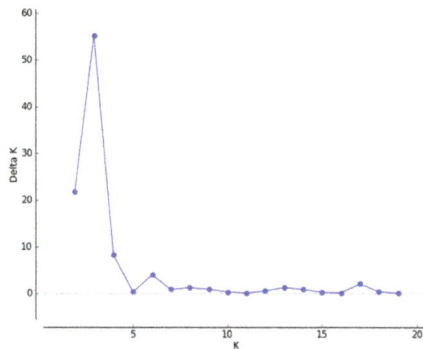

Figure 3. Relationships between the number of clusters (K) and the corresponding ΔK statistics calculated according to ΔK based on STRUCTURE analysis [46].

Forests **2017**, *8*, 495

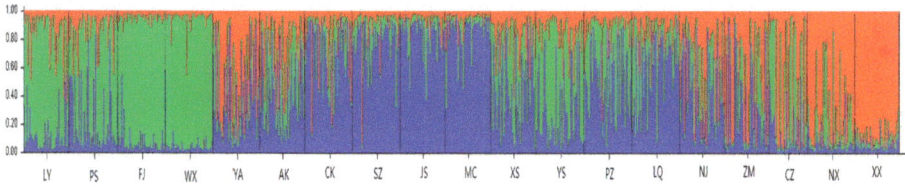

Figure 4. Results of the structure analysis of *Q. variabilis* populations when *K* = 3. Each individual is represented by a single vertical bar, which is partitioned into three different colors. Each color represents a genetic cluster and the colored segments shows the individual's estimated ancestry proportion to each of the genetic clusters. The population codes are as in Table 1.

The results of a FCA and STRUCTUE analysis indicated that the population as a whole could be divided into three genetic groups. Furthermore, AMOVA analysis showed 1.75% variation between groups (*P* < 0.001), 5.59% variation within groups (*P* < 0.001), and 92.66% variation within populations (*P* < 0.001). All genetic groups were ordered according to the calculated F_{ST} value: Northeast group (F_{ST} = 0.070) > Southwest group (F_{ST} = 0.039) > Central Group (F_{ST} = 0.027). The standard genetic distance and geographic distance of $F_{ST}/(1 - F_{ST})$ were calculated using the Mantel test, which revealed that the changes in the F_{ST} values did not follow the geographical distance separating mode (*R* = 0.0013, *P* = 0.371). The geographical distances between populations AK, CK and FJ, WX were very small, but FCA and Bayesian analyses did not cluster those into the same groups, and AMOVA analysis showed that most of the variations that existed within groups resulted in significant genetic structural differences in the populations AK, CK and FJ, WX.

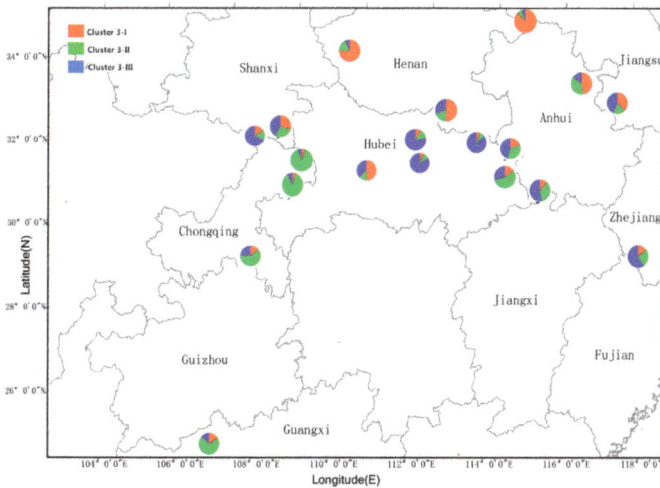

Figure 5. Mean proportions of cluster memberships of analyzed individuals in each of the 19 *Q. variabilis* populations, based on STRUCTURE analysis at *K* = 3.

4. Discussion

4.1. Genetic Diversity

The level of species genetic diversity is often associated with specific characteristics such as the length of life cycle, mating system and reproduction, the size of the geographic range, and genetic exchange [11,54–56]. The H_e (0.707) of *Q. variabilis* in our study was slightly lower than that of Xu's previous study (0.806) [6]. The main reason for this difference is that only five Chinese populations of *Q. variabilis* were studied in Xu's study, whereas we examined representative populations within the distribution range from the South to the North. Therefore, the population genetic variation of the species reflected in the present study was more comprehensive. Compared to the overall genetic variation level of *Q. variabilis* with other Fagaceae *Quercus* genera species such as the summer oak (*Q. robur*) (H_e = 0.764) [29] and Liaodong oak (*Q. liaotungensis*) (H_e = 0.754) [30], the levels of diversity were about the same yet slightly higher than that of the Sawtooth oak (*Q. acutissima*) (H_e = 0.660) [22] and Mongolian oak (*Q. mongolica*) (H_e = 0.630) [24]. Compared to the other endangered species or narrow domains that were left over from the Quaternary Ice Age in China such as *Ginkgo biloba* (H_e = 0.241) [57], the overall genetic diversity level of *Q. variabilis* was high, which was mainly due to the extensive distribution of the species distributed from 19° N to 42° N across the temperate forests. Because of huge differences in climate and habitat conditions among populations in the distribution area, *Q. variabilis* may have differentiated adapted ecological and genetic types, thereby resulting in a wide range of genetic variations. In addition, affected by the fact that the populations were sampled from natural secondary forest, heterozygote deficits were shown (F_{IS} = 0.044).

H_e and *AR* of each population by Kendall rank-correlation analysis revealed that neither followed the geographical space separation mode. However, changes in *AR* clearly showed significant correlations with changes in the average annual temperature, indicating that temperature has a significant impact on *Q. variabilis*. Among the 19 populations studied, the genetic diversity of the central populations (*AR* = 8.06, H_e = 0.718) was higher than that of the Southwest populations (*AR* = 7.55, H_e = 0.683) and the Northeast populations (*AR* = 7.45, H_e = 0.704). The two populations with the highest genetic diversity—XS (*AR* = 8.56) and CK (H_e = 0.745)—both belonged to the Central group; these results supported the conclusions of Chen [26] in the populations of *Q. variabilis* in the Central region, which experiences an average annual temperature of 15 °C or above and has a higher level of genetic diversities than other areas. Bottleneck detection revealed that the Southwest populations FJ and PS, and Northeast population CZ experienced a genetic bottleneck, which may be the main reason for the low level of genetic diversity in all three populations [14,43,58].

Mayr proposed a "core–periphery" hypothesis that states that compared to the core groups in general [59], the genetic diversity of peripheral groups decreased and genetic differentiation increased under the pressures of the bottleneck effect, genetic drift, and selection pressures [60]. Based on the genetic diversity of five *Q. variabilis* populations, Xu concluded that the genetic diversity of the central groups was higher than that of the peripheral groups [6], but because of the limitations of the sample population scales, these findings could only be used as a reference. In our study, the samples of 19 populations were collected from a representative distribution in the main distribution areas. For the populations located in the geographical centers of existing distribution areas (the Central group), overall genetic diversities were higher than those in the peripheral regions (the Southwest group and the Northeast group). In addition, private alleles were only found in the southwestern population, such as LY; this indicated that there was genetic variation in the peripheral population in adaptation to the climate and environment. Accordingly, the diversity distribution of *Q. variabilis* was in line with the "core–periphery" hypothesis.

4.2. Genetic Differentiation

The genetic differentiation level of *Q. variabilis* was moderately low (F_{ST} = 0.063), which was in agreement with the results of AMOVA analysis (among populations variation = 6.3%). The genetic differentiation level of *Q. variabilis* is similar to that of the summer oak (*Q. robur*) (F_{ST} = 0.080) [29] and Mongolian oak (*Q. mongolica*) (F_{ST} = 0.077) [24]. Its genetic differentiation level is close to that reported by Hamrick [61], who investigated diversity in widespread species, including long-lived woody perennials and wind-pollinated outcrossing species. As a perennial tree, *Q. variabilis* is a typical wind-pollinated and outcrossing species, which undoubtedly increases gene flow between populations and reduces differentiation among populations.

Gene flow determines the genetic structure and survival potential of future populations of a species [12]. The present study showed that *Q. variabilis* populations are characterized by a relatively large gene flow (N_m = 3.648) that may inhibit genetic drifting and prevent population genetic differentiation [11]. The large gene flow is mainly determined by its biological means of spreading via pollen and seeds. Xu [6] reported a higher gene flow (N_m = 5.239) than this study, which objectively reflected a decreasing gene flow for *Q. variabilis* in the last decade; the original natural forest of *Q. variabilis* has degenerated into a natural secondary forest, which may have caused a reduction in gene flow.

The clustering results of STRUCTURE Bayesian clustering showed that the 19 populations could be divided into three groups: the Southwest group, the Central group, and the Northeast group. These findings were in agreement with the results of a FCA, wherein the mix in the population genetic structure of the Southwest group and the Northeast group was relatively low, and there was a large difference in the genetic composition between the two groups. On the other hand, the Central populations were the largest; they showed a high mixing degree and the lowest differentiation. In general, the population that is capable of maintaining its level of genetic diversity is often the largest population group [62]. Therefore, as expected, the Central group showed the highest capacity to maintain its level of genetic diversity. The Mantel test showed that correlations of genetic and geographic distances between the groups were not significant (R = 0.0013, P = 0.371), which showed that geographical distance is not the main reason for genetic differentiation among populations.

As a widely distributed species and perennial tree, the genetic structure formation of *Q. variabilis* should not be analyzed from a single geographic distribution and distance. During the Quaternary Ice Age, which was affected by the global glacial climate, *Q. variabilis* in northern China was unable to adapt and eventually disappeared. Furthermore, the South of Qinling and Dabie Mountains had become barriers that blocked the Quaternary glaciers, with their complex landforms and diverse ecological environments rendering them sources of plant heterogeneous differentiations [63]. The area thus served as a sanctuary for *Q. variabilis* during Quaternary glaciation [64] and eventually became the geographical distribution center for *Q. variabilis* [4]. At the end of the Ice Age and with the recovery of temperature as well as other natural conditions, it is assumed that some populations in the distribution center differentiated into varying types which adapted to a cold, dry climate and began to migrate to high latitudes in the northeast, while some others adapted to warm and moist types and began to spread southwestward, and gradually formed the geographical distribution pattern of *Q. variabilis* (in the present study, there were significant correlations between *AR* and the average annual temperature). This evolutionary geographic history has been verified by simple haplotypes in peripheral populations, and populations in the central region represented almost all haplotypes of *Q. variabilis* [26]. It is in this complex historical dynamic migration process that the genetic structure of the existing distribution of *Q. variabilis* was formed.

In addition, human activities may also have an impact on the genetic structure of various species [65–67]. However, the results of the present study showed that among the 19 *Q. variabilis* populations, the population in the center of the distribution did not experience a bottleneck effect, whereas three populations at the periphery (FJ, PS, CZ) did, and three genetic lineages (cluster I,II and III) obtained from STRUCTURE showed distributions in every population (Figures 4 and 5). Based on these results, we inferred that *Q. variabilis* has been a widely distributed species for several historical periods, and the history of its population dynamics and genetic variations stretches back further than the time of human activities [68], so human activity is not the major factor that has affected its current genetic structure.

5. Conclusions

This represents the first research carried out on the genetic diversity and structure of *Q. variabilis* based on a broader sampling strategy involving its main distribution area. *Q. variabilis*, in general, was in line with the "core–periphery" hypothesis. The overall genetic diversity level of *Q. variabilis* was relatively high; the genetic diversity level of populations in the geographical distribution center was higher than that of peripheral populations. The accessions from southwest, center and northeast areas clustered into three separate groups, and the genetic structure of *Q. variabilis* was mainly affected by the preferable adaptability to the climate and environment in the complex historical dynamic migration process. Due to the development of cork, tannin, and other industries, the distribution scale of *Q. variabilis* has gradually reduced; we should establish conservation measures to prevent population decline; peripheral populations FJ, PS and CZ, in particular, suffered the bottleneck effect. Additionally central populations with abundant genetic variation could be used as the preferred germplasm resource in plantations. Furthermore, private alleles were found only in peripheral population LY; they are of great research value to the maintenance and evolution of species in disadvantageous habitats.

Acknowledgments: This research was supported by the Priority Academic Program Development of Jiangsu Higher Education Institutions (PAPD).

Author Contributions: This study was carried out with collaboration among all authors. Li-an Xu conceived and designed the experiments; Xiaomeng Shi and Xin Guo performed the experiments; Qiang Wen carried out data correction and manuscript revision; Xiaomeng Shi and Mu Cao provided the experimental data; Xiaomeng Shi wrote the paper.

Conflicts of Interest: The authors declare no conflict of interest.

References

1. Johnson, P.S.; Shifley, S.R. *The Ecology and Silviculture of Oaks*; CABI Publishing: New York, NY, USA, 2001; pp. 9–10, ISBN 9780851995700.
2. Kremer, A.; Petit, R.J. Gene diversity in natural Population of oak species. *Ann. For. Sci.* **1993**, *50*, 186–202. [CrossRef]
3. eFlora. Org., Flora of China, *Quercus variabilis* Blume. Available online: http://www.efloras.org/ (accessed on 10 October 2017).
4. Wang, J.; Wang, S.B.; Kang, H.Z.; Xin, Z.J.; Qian, Z.H.; Liu, C.J. Distribution Pattern of Oriental Oak (*Quercus variabilis* Blume) and the Characteristics of Climate of Distribution Area in Eastern Asia. *J. Shanghai Jiaotong Univ.* **2009**, *27*, 235–241.
5. Wei, L. Preliminary investigation on the distribution of the cork oak. *Sci. Silvae Sin.* **1960**, *6*, 70–71.
6. Xu, X.L.; Xu, L.A.; Huan, M.R.; Wang, Z.R. Genetic Diversity of Microsatellites (SSRs) of Natural Populations of *Quercus variabilis*. *Hereditas* **2004**, *26*, 683–688. [PubMed]
7. Zhao, G.; Duan, X.F.; Guan, T.; Huang, L.H. Situation of Cork Utilization in the World and the Development Countermeasure of the China's Cork Industry. *World For. Res.* **2004**, *17*, 25–28.
8. Flores, M.; Rosa, M.E. Properties and uses of consolidated cork dust. *J. Mater. Sci.* **1992**, *27*, 5629–5634. [CrossRef]
9. Zhang, W.H.; Lu, Z.J. A study on the biological and ecological property and geographical distribution of *Quercus variabilis* population. *Acta Bot. Boreali-occident. Sin.* **2002**, *22*, 1093–1101.
10. Lei, J.P.; Xiao, W.F.; Liu, J.F. Distribution of *Quercus variabilis* Blume and Its Ecological Research in China. *World For. Res.* **2013**, *26*, 57–62.
11. Hamrick, J.L.; Godt, M.J.W. Allozyme diversity in plant species. In *Plant Population Genetics, Breeding, and Genetic Resources*; Brown, A.D.H., Clegg, M.T., Kahler, A.L., Weit, B.S., Eds.; Sinauer Associates: Sunderland, MA, USA, 1990; pp. 43–63, ISBN 9780878931170.
12. Hamrick, J.L.; Godt, M.J.W.; Sherman-Broyles, S.L. Factors influencing levels of genetic diversity in woody plant species. *New For.* **1992**, *6*, 95–124. [CrossRef]
13. Heuertz, M.; Hausman, J.F.; Hardy, O.J.; Vendramin, G.G.; Frascarialacoste, N. Nuclear microsatellites reveal contrasting patterns of genetic structure between western and southeastern European populations of the common ash (*Fraxinus excelsior* L.). *Evolution* **2004**, *58*, 976–988. [PubMed]
14. Frankham, R.; Ballou, J.D.; Briscoe, D.A. *Introduction to Conservation Genetics*; Cambridge University Press: Cambridge, UK, 2004; Volume 190, pp. 385–386, ISBN 9780521702713.

15. Sutherland, W.J.; Albon, S.D.; Allison, H.S.; Armstrong, B.; Bailey, M.J.; Brereton, T.; Boyd, I.L.; Carey, P.; Edwards, J.; Gill, M.; et al. The identification of priority policy options for UK nature conservation. *J. Appl. Ecol.* **2010**, *47*, 955–965. [CrossRef]

16. Li, Y.; Zhang, X.W.; Fang, Y.M. Predicting the impact of global warming on the geographical distribution pattern of *Quercus variabilis* in China. *Chin. J. Appl. Ecol.* **2014**, *25*, 3381–3389.

17. Zhou, J.Y.; Lin, J.; He, J.F.; Zhang, W.H. Review and Perspective on *Quercus variabilis* Research. *J. Northwest For. Univ.* **2010**, *25*, 43–49.

18. Yuan, Z.L.; Wang, T.; Zhu, X.L.; Sha, Y.Y.; Ye, Y.Z. Patterns of spatial distribution of *Quercus variabilis* in deciduous broadleaf forests in Baotianman Nature Reserve. *Biodivers. Sci.* **2011**, *19*, 224–231.

19. Hu, X.J.; Zhang, W.H.; He, J.F.; Yin, Y.N. Architectural analysis of crown geometry of *Quercus variablis* BL. natural regenerative seedlings in different habitats. *Acta Ecol. Sin.* **2014**, *35*, 789–795.

20. Li, Z.P.; Zhang, W.H.; Cui, Y.C. Effects of NaCl and Na_2CO_3 stresses on seed germination and seedling growth of *Quercus variabilis*. *Acta Ecol. Sin.* **2015**, *35*, 742–751.

21. Ran, R.; Zhang, W.H.; He, J.F.; Zhou, J.Y. Effects of thinning intensities on population regeneration of natural *Quercus variabilis* forest on the south slope of Qinling Mountains. *Chin. J. Appl. Ecol.* **2014**, *25*, 695–701.

22. Zhang, Y.Y.; Fang, Y.M.; Yu, M.K.; Li, X.X.; Xia, T. Molecular characterization and genetic structure of *Quercus acutissima* germplasm in China using microsatellites. *Mol. Biol. Rep.* **2013**, *40*, 4083–4090. [CrossRef] [PubMed]

23. Jennifer, F.L.; Oliver, G. Genetic structure of *Quercus rubra* L. and *Quercus ellipsoidalis* E.J. Hill populations at gene-based EST-SSR and nuclear SSR markers. *Tree Genet. Genomes* **2013**, *9*, 707–722.

24. Saneyoshi, U.; Yoshihiko, T. Development of ten microsatellite markers for *Quercus mongolica* var. crispula by database mining. *Conserv. Genet.* **2008**, *9*, 1083–1085.

25. Neophytou, C.H.; Dounavi, A.; Aravanopoulos, F.A. Conservation of Nuclear SSR Loci Reveals High Affinity of *Quercus infectoria* ssp. *veneris* A. Kern (Fagaceae) to Section *Robur*. *Plant Mol. Biol. Rep.* **2008**, *26*, 133–141. [CrossRef]

26. Chen, D.M.; Zhang, X.X.; Kang, H.Z.; Sun, X.; Yin, S.; Du, H.M.; Yamanaka, N.; Gapare, W.; Wu, H.X.; Liu, C.J. Phylogeography of *Quercus variabilis* Based on Chloroplast DNA Sequence in East Asia: Multiple Glacial Refugia and Mainland-Migrated Island Populations. *PLoS ONE* **2013**, *7*, e47268. [CrossRef] [PubMed]

27. Zhu, Q.H.; Pan, H.X.; Zhu, G.Q.; Yin, T.M.; Zou, H.Y.; Huang, M.R. Analysis of Genetic Structure of Natural Populations of *Castanopsis fargesii* by RAPDs. *Acta Bot. Sin.* **2002**, *44*, 1321–1326.

28. BiOptic Inc. Available online: http://www.bioptic.com.tw (accessed on 5 July 2016).

29. Alexis, R.S.; Jennifer, F.L.; Tim, S.M.; Jeanne, R.S.; Oliver, G. Development and Characterization of Genomic and Gene-Based Microsatellite Markers in North American Red Oak Species. *Plant Mol. Biol. Rep.* **2012**, *31*, 231–239.

30. Qin, Y.Y.; Han, H.R.; Kang, F.F.; Zhao, Q. Genetic diversity in natural populations of *Quercus liaotungensis* in Shanxi Province based on nuclear SSR markers. *J. Beijing For. Univ.* **2012**, *34*, 61–65.

31. Durand, J.; Bodenes, C.; Chancerel, E.; Frigerio, J.M.; Vendramin, G.; Sebastiani, F.; Buonamici, A.; Gailing, O.; Koelewijn, H.P.; Villani, F.; et al. A fast and cost-effective approach to develop and map EST-SSR markers: Oak as a case study. *BMC Genom.* **2010**, *11*, 570. [CrossRef] [PubMed]

32. Nei, M. *Molecular Evolutionary Genetics*; Columbia University Press: New York, NY, USA, 1987; pp. 73–95, ISBN 9780231063210.

33. Mousadik, E.L.; Petit, R.J. High level of genetic differentiation for allelic richness among populations of the argan tree [*Argania spinosa* (L.) Skeels endemic to Morocco. *Theor. Appl. Genet.* **1996**, *92*, 832–839. [CrossRef] [PubMed]

34. Weir, B.S.; Cockerham, C.C. Estimating F-statistics for the analysis of population structure. *Evolution* **1984**, *38*, 1358–1370. [PubMed]

35. Nei, M. Analysis of gene diversity in subdivided populations. *Proc. Natl. Acad. Sci. USA* **1973**, *70*, 3321–3323. [CrossRef] [PubMed]

36. Goudet, J. FSTAT Version 2.9.3, A Program to Estimate and Test Gene Diversities and Fixation Indices. 2002. Available online: http://www.unil.ch/izea/softwares/fstat.html (accessed on 5 July 2016).

37. Liu, J. *POWERMARKER: A Powerful Software For Marker Data Analysis*; North Carolina State University, Bioinformatics Research Center: Raleigh, NC, USA, 2002.

38. Peakall, R.; Smouse, P.E. GENEALEX: Genetic analysis in excel. Population genetic software for teaching and research. *Mol. Ecol. Notes* **2006**, *6*, 288–295. [CrossRef]

39. Curtu, A.L.; Gailing, O.; Leinemann, L.; Finkeldey, R. Genetic variation and differentiation within a natural community of five oak species (*Quercus* spp.). *Plant Biol.* **2007**, *9*, 116–126. [CrossRef] [PubMed]

40. Slatkin, M.; Barton, N.H. Methods for estimating geneflow. *Evolution* **1989**, *43*, 1349–1368. [CrossRef] [PubMed]

41. Iwaizumi, M.G.; Tsuda, Y.; Ohtani, M.; Ohtani, M.; Tsumura, Y.; Takahashi, M. Recent distribution changes affect geographic clines in genetic diversity and structure of *Pinus densiflora* natural populations in Japan. *For. Ecol. Manag.* **2013**, *304*, 407–416. [CrossRef]

42. Tian, C.; Mu, N.; Zhu, Z.Y.; Wang, C.J. Method for the Rapid Obtaining of Climate Data Based on DIVA-GIS. *J. Agric.* **2015**, *5*, 109–113.

43. Piry, S.; Luikart, G.; Cornuet, J.M. BOTTLENECK: A computer program for detecting recent reductions in the effective population size using allele frequency data. *J. Hered.* **1999**, *90*, 502–503. [CrossRef]

44. Luikart, G.; Cornuet, J.M. Empirical evaluation of a test for identifying recently bottlenecked populations from allele frequency data. *Conserv. Biol.* **1998**, *12*, 228–237. [CrossRef]

45. Belkhir, K.; Borsa, P.; Chikhi, L.; Raufaste, N.; Bonhomme, F. *GENETIX (ver.4.02): Logiciel sous Windows TM pour la Genetique des Populations, Laboratoire Genome, Populations, Interactions*; CNRSUMR 5000; Universite Montpellier II: Montpellier, France, 2004.

46. Pritchard, J.K.; Wen, X.; Falush, D. STRUCTURE Version 2.3.1. 2009. Available online: http://pritch.bsd. uchicago.edu/structure (accessed on 5 July 2016).

47. Falush, D.; Stephens, M.; Pritchard, J.K. Inference of population structure using multilocus genotype data: Linked loci and correlated allele frequencies. *Genetics* **2003**, *164*, 1567–1587. [PubMed]

48. Evanno, G.; Regnaut, S.; Goudet, J. Detecting the number of clusters of individuals using the software STRUCTURE: A simulation study. *Mol. Ecol.* **2005**, *14*, 2611–2620. [CrossRef] [PubMed]

49. Excoffier, L.; Laval, G.; Schneider, S. ARLEQIN (ver.3.0): An integrated software package for population genetics data analysis. *Evol. Bioinform. Online* **2007**, *1*, 47–50. [PubMed]

50. Excoffier, L.; Lischer, H. Arlequin suite ver 3.5: A new series of programs to perform population genetics analyses under Linux and Windows. *Mol. Ecol. Resour.* **2010**, *10*, 564–567. [CrossRef] [PubMed]

51. Slatkin, M. A measure of population subdivision based on microsatellite allele frequencies. *Genetics* **1995**, *139*, 1463.

52. Meirmans, P.G. Using the AMOVA framework to estimate a standardized genetic differentiation measure. *Evolution* **2010**, *60*, 2399–2402. [CrossRef]

53. Yan, B.Q.; Wang, T.; Hu, L.L. Population genetic diversity and structure of *Schisandra sphenanthera*, a medicinal plant in China. *Chin. J. Ecol.* **2009**, *28*, 811–819.

54. Gaudeul, M.; Taberlet, P.; Till, B.I. Genetic diversity in an endangered alpine plant, *Eryngium alpinum* L. (Apiaceae), inferred from a mplified fragment length polymorphism markers. *Mol. Ecol.* **2000**, *9*, 1625–1637. [CrossRef] [PubMed]

55. Li, J.H.; Jin, Z.X.; Lou, W.Y.; LI, J.M. Genetic diversity of *Lithocarpus harlandii* populations in three forest communities with different succession stage. *Chin. J. Ecol.* **2007**, *26*, 509–514. [CrossRef]

56. Hellmanna, J.J.; Pineda, K.M. Constraints and reinforcement on adaptation under climate change: Selection of genetically correlated traits. *Biol. Conserv.* **2007**, *13*, 599–609. [CrossRef]

57. Ge, Y.Q.; Qiu, Y.X.; Ding, B.Y.; Fu, C.X. An ISSR analysis on population genetic diversity of the relict plant *Ginkgo biloba*. *Chin. Biodivers.* **2003**, *4*, 276–287.

58. Cornuet, J.M.; Luikart, G. Description and power analysis of two tests for detecting recent population bottlenecks from allele frequency data. *Genetics* **1996**, *144*, 2001–2014. [PubMed]

59. Mayr, E. *Animal Species and Evolution*; Harvard University Press: Cambridge, UK, 1963; pp. 425–457, ISBN 9780674037502.

60. Hampe, A.; Petit, R.J. Conserving biodiversity under climate change: The rear edge matters. *Ecol. Lett.* **2005**, *8*, 461–467. [CrossRef] [PubMed]

61. Hamrick, J.L. IsoZymes ananysis of genetic structure of plant population. In *Isozymes in Plant Biology*; Soltis, D., Soltis, P., Eds.; Dioscorids Press: Washington, DC, USA, 1989; pp. 87–105, ISBN 9780412365003.

62. Ohsawa, T.; Tsuda, Y.; Saito, Y.; Ide, Y. The genetic structure of *Quercus crispula* in northeastern Japan as revealed by nuclear simple sequence repeat loci. *J. Plant Res.* **2011**, *124*, 645–654. [CrossRef] [PubMed]

63. He, C.R.; Chen, F.Q. The endemic genera of Chinese seed plants distributed in Three Gorges Reservoir Area of Changjiang River. *Guihaia* **1999**, *19*, 43–46.

64. Jin, J.H.; Liao, W.B.; Wang, B.S.; Peng, S.L. Global change in Cenozoic and evolution of flora in China. *Guihaia* **2003**, *23*, 217–225.

65. Knowles, P.; Perry, D.J.; Foster, H.A. Spatial genetic structure in two tamarack (*Larix laricina*) populations with differing establish menthi stories. *Evolution* **1992**, *46*, 572–576. [CrossRef] [PubMed]

66. Young, A.C.; Merriam, H.G. Effect of forest fragmentation on the spatial genetic structure of *Acer saccharum* Marsh. (sugar maple) populations. *Heredity* **1994**, *72*, 201–208. [CrossRef]

67. Aldrich, P.R.; Hamrick, J.L.; Chavarriaga, P. Microsatellite analysis of de mographic genetic structure in fragmented populations of the tropical tree *Symphonia globulifera*. *Mol. Ecol.* **1998**, *7*, 933–944. [CrossRef] [PubMed]

68. Zhou, Z.K. Origian, Phylogeny and Disperal of *Quercus* from China. *Acta Bot. Yunnanica* **1992**, *14*, 227–236.

© 2017 by the authors. Licensee MDPI, Basel, Switzerland. This article is an open access article distributed under the terms and conditions of the Creative Commons Attribution (CC BY) license (http://creativecommons.org/licenses/by/4.0/).

forests

MDPI

Article

Expression Profiling in *Pinus pinaster* in Response to Infection with the Pine Wood Nematode *Bursaphelenchus xylophilus*

Daniel Gaspar [1,2], **Cândida Trindade** [3], **Ana Usié** [1,2], **Brígida Meireles** [1,2], **Pedro Barbosa** [1,2], **Ana M. Fortes** [4], **Cátia Pesquita** [5], **Rita L. Costa** [3,*] and **António M. Ramos** [1,2]

1   Centro de Biotecnologia Agrícola e Agro-alimentar do Alentejo (CEBAL)/Instituto Politécnico de Beja (IPBeja), Rua Pedro Soares, s.n.—Campus IPBeja/ESAB, Apartado 6158, 7801-908 Beja, Portugal; daniel.gaspar@cebal.pt (D.G.); ana.usie@cebal.pt (A.U.); brigida.meireles@cebal.pt (B.M.); pedro.barbosa@cebal.pt (P.B.); marcos.ramos@cebal.pt (A.M.R.)
2   Instituto de Ciências Agrárias e Ambientais Mediterrânicas (ICAAM), Universidade de Évora, Núcleo da Mitra, Apartado 94, 7006-554 Évora, Portugal
3   Instituto Nacional de Investigação Agrária e Veterinária, I.P. (INIAV), Avenida da República, Quinta do Marquês (edifício sede), 2780-157 Oeiras, Portugal; candida.smpp.trindade@gmail.com
4   Faculdade de Ciências da Universidade de Lisboa (FCUL), BioSystems & Integrative Sciences Institute (BIOISI), Campo Grande, 1749-016 Lisboa, Portugal; margafortes@yahoo.com
5   Faculdade de Ciências da Universidade de Lisboa (FCUL), LaSIGE, Campo Grande, 1749-016 Lisboa, Portugal; cpesquita@di.fc.ul.pt
*   Correspondence: rita.lcosta@iniav.pt; Tel.: +35-191-907-3379

Received: 22 June 2017; Accepted: 31 July 2017; Published: 3 August 2017

Abstract: Forests are essential resources on a global scale, not only for the ecological benefits, but also for economical and landscape purposes. However, in recent years, a large number of forest species have suffered a serious decline, with maritime pine being one of the most affected. In Portugal, the maritime pine forest has been devastated by the pine wood nematode (PWN), the causal agent of pine wilt disease. In this study, RNA-Seq data was used to characterize the maritime pine response to infection with PWN, by determining the differentially expressed genes and identifying the regulatory networks and pathways associated. The analyses showed clear differences between an early response that occurs immediately after inoculation and a late response that is observed seven days after inoculation. Moreover, differentially expressed genes related to secondary metabolism, oxidative stress and defense against pathogen infection were identified over different time points. These results provide new insights about the molecular mechanisms and metabolic pathways involved in the response of *Pinus pinaster* against PWN infection, which will be a useful resource in follow-up studies and for future breeding programs to select plants with lower susceptibility to this disease.

Keywords: RNA-sequencing; transcriptome analysis; gene expression; pine wilt disease; *Pinus pinaster*; *Bursaphelenchus xylophilus*

1. Introduction

Forests are much more than a large area of land covered with trees. They represent one of life's support systems on Earth, providing essential resources for a range of ecosystems. Furthermore, forests supply various products and services, generating a huge number of economic and social benefits. Due to the high commercial value of wood products, maritime pine (*Pinus pinaster* Ait.) is one of the main conifer species in southwestern Europe, covering approximately four million hectares in this region [1]. In Portugal, maritime pine is one of the predominant tree species, and by far the most widespread, mainly in the regions of Atlantic influence, covering more than 700 thousand hectares, which corresponds to 23% of the total forest surface [2].

In recent years there has been a worrying decline of a large number of forest species around the world, with maritime pine being one of the most affected [3]. This alarming decrease is caused by abiotic and biotic factors, of which the pine wood nematode (PWN), *Bursaphelenchus xylophilus* Steiner & Buhrer, 1934 (Nickle, 1970) is one of the main biotic factors [4].

PWN is a quarantine organism in the European Union (Directive 77/93 EEC), being the causal agent of the pine wilt disease (PWD) that may kill a host tree within a short period of time after infection [5]. Mostly due to this pathogen, the total area occupied by *P. pinaster* suffered an abrupt decline in Portugal, with losses of 263,000 hectares between 1995 and 2010 [2]. As a result, *P. pinaster* went from being the main forest species, in terms of distribution and area, to the third, behind eucalyptus and cork oak. Recently, it was classified as an endangered species by the IUCN red list of threatened species [6].

PWN was reported for the first time in Portugal in 1999 [5], and in less than 10 years the whole *P. pinaster* area had been affected. PWN is transported between host trees by an insect vector, a longhorn cerambycid beetle (*Monochamus galloprovincialis* Oliv.) [7]. The transmission may occur in two forms: (i) by oviposition, where the female beetles lay their eggs under the bark of stressed or recently killed trees by the PWN, and the nematodes migrate to pupae just before adult beetles emerge, ensuring successful survival of the parasite. Note that due to the low frequency and efficiency in susceptible trees, such as *P. pinaster*, transmission by oviposition represents a secondary inoculation way [8]; (ii) by feeding, considered the most common pathway of transmission on susceptible trees, that occurs through beetle feeding wounds (primary transmission). Nematodes carried by beetles move into wounds and breed in the xylem, but the survival of nematodes is not guaranteed [9,10]. This is a close relationship between PWN and its vector beetle, resulting in the epidemiological cycle of PWD [11].

PWD expression depends not only on the pathogenicity of PWN and susceptibility of host trees but also on environmental conditions such as high temperature and large soil moisture, the optimal conditions for PWN proliferation [9]. The symptoms caused by PWD are common to other diseases, and therefore can easily be confused. A typical early symptom is needle discoloration. Needles turn grayish green, then tan, and finally brown. Then, resin flow ceases and the wood is dry when cut [4].

The defensive mechanisms of host trees can be divided into early and advanced stages [12]. In the first stage, defensive response occurs in both susceptible and resistant trees, while late response is found only in susceptible trees [12]. In the same species, it has been verified the existence of trees with different levels of susceptibility, some of which survive the infection, thus, constituting an opportunity for selective breeding. This has been the approach in breeding programs developed in China and Japan in the early sixties [13].

Transcriptome analysis based on next-generation sequencing data provides information about all transcriptional activity in a cell or organism. It is now the most commonly used approach, and has been applied to disease pathogenesis studies and identification of biomarkers [14]. For non-model organisms like *P. pinaster*, for which there is no reference genome sequence available, RNA-Seq is an efficient means to generate functional genomic data [15].

In order to understand the pathogenic mechanisms and reduce the damage caused by the PWD in Portuguese forests and respective ecosystems, several studies were performed [16–19] However, to our knowledge, the analysis of maritime pine molecular response based on RNA-Seq data was reported in only one approach [20]. In this study, was pointed out a set of candidate genes potentially involved in the response to PWN, mainly related with terpenoid metabolism, defense against pathogen attack and oxidative stress.

This work is an approach to PWD, using RNA-Seq data to characterize the maritime pine transcriptomic profile in the response to infection with *Bursaphelenchus xylophilus*, over three different time points after inoculation, by determining the differentially expressed (DE) genes, regulatory networks and pathways, with the purpose of identifying potential candidate genes that may later on be used in the selection of *P. pinaster* trees displaying resistance against PWD.

2. Materials and Methods

2.1. Biological Material, Pine Wood Nematode Inoculation and Sampling

A total of fourteen potted three-year-old *Pinus pinaster* trees were used in this study. These plants were derived from seeds and maintained in natural environmental conditions during the assay. *Bursaphelenchus xylophilus* culture was grown in PDA (Potato Dextrose Medium) with Botrytis cinerea. After a significant growth, a suspension of nematodes was transferred to test tubes with 5 mL of water and barley grains previously autoclaved. Later they were incubated for a week at 25 °C and relative humidity of 70%, which represent optimal conditions for nematodes growth. Before inoculation, nematodes were extracted from test tubes using the Baermann funnel technique [21]. Then, the culture was placed at 4 °C to stop multiplication and passing from juvenile stage to adult stage.

Inoculation with PWN was conducted following the method of Futai and Furuno [22]. Shortly, a suspension with 2000 nematodes was pipetted into a small vertical wound (1 cm) made on the upper part of the main pine stem with a sterile scalpel. A sterilized piece of gauze was placed around the wound site and fixed with parafilm to maintain the optimal humidity level. This procedure was done in twelve *P. pinaster* plants, while the two remaining plants were used as control (inoculation with water).

Four sampling time points were established, including 6 h, 24 h, 48 h and 7 days after inoculation. For each time point, a set of three *P. pinaster* plants was collected. Briefly, a small piece of stem tree above inoculation point was cut and flash frozen at −80 °C for further RNA extraction.

2.2. RNA Extraction, cDNA Synthesis, Library Preparation and Sequencing

All collected samples were ground in liquid nitrogen and a total RNA extraction was performed from 2 g of plant material, according to an optimized method from Provost and colleagues [23]. Then, a DNase treatment was carried out following the instructions of the manufacturer (Kit TURBO DNA-free by Life Technologies, Hong Kong, China).

Approximately 1 microgram of total RNA was used for cDNA synthesis, following the ImProm-IITM Reverse Transcription System protocol kit (Promega, Madison, WI, USA). Before sequencing, four pools of cDNA were constructed (pool 1—control; pool 2—6 + 24 h; pool 3—48 h; pool 4—7 days).

cDNA libraries were constructed with the Ion Total RNA-Seq Kit v2 (Life Technologies, Hong Kong, China). Briefly, mRNA was fragmented with RNAse III. After short fragment removal, RNA adapters were ligated and the cDNA first and second strands synthesized. cDNA was then amplified with specific barcoded primers by PCR amplification and the resulting fragments selected for the correct size with magnetic beads.

Finally, the positive spheres from the four libraries were loaded into an Ion PI chip v2 and the transcriptomes were sequenced as single-end reads in the Ion Proton System (Thermo Fisher Scientific, Waltham, MA, USA) at Biocant (Cantanhede, Portugal). All procedures were carried out according to manufacturer's instructions.

2.3. Pre-Processing RNA-Sequencing Data and Transcriptome Assembly

The quality of the RNA-Seq reads from the four sequenced libraries was checked using FastQC software Version 0.11.5 [24], a quality control tool for high throughput sequence data. Based on the FastQC results, the thresholds for minimum average read quality and read length were established as 12 and 80 bp, respectively. These parameters were used to run Sickle tool Version 1.33 [25], which trimmed poor quality bases and adapters sequences from the raw reads, and produced a set of processed reads that were then used in the downstream analyses.

Since no reference genome sequence is available for *P. pinaster*, it was necessary to perform a de novo transcriptome assembly. The processed reads from all libraries were assembled into contigs using Trinity 2.1.1 [26]. The contigs generated with the Trinity assembly were used as input for a run with CAP3 [27]. The resulting assembly was the basis for the next procedures, being used as the reference transcriptome assembly.

2.4. Prediction of Candidate Coding Regions

The sequences from the reference transcriptome were analyzed with TransDecoder-2.0.1 [28] to identify the open reading frames (ORF). The ORF transcripts identified were further scanned for homology to known proteins against the Swiss-Prot [29] and Pfam [30] databases by running BlastP [31] and HmmScan [32], respectively. In the end, TransDecoder provided the final set of candidate coding regions, namely the predicted genes representing the basis for their annotation.

2.5. Mapping and Differential Expression Analysis

Mapping the reads against the transcriptome assembly was performed using RapMap [33]. Before performing a differential gene expression analysis, it is common to determine the number of unique mapped reads, which was accomplished with SAMtools-1.3 [34]. Only the reads that mapped to a unique location in the reference transcriptome were used for downstream analyses.

The EdgeR package [35] of Bioconductor was used to identify transcripts that were differentially expressed between the conditions. To adjust for library sizes and skewed expression of transcripts, the estimated abundance values were normalized using the Trimmed Mean of M-values normalization method [36] included in the EdgeR package. As our experiment did not have replicates, it was necessary to determine the biological variability. Thus, in accordance with the EdgeR guidelines, a BCV (biological coefficient variation) of 0.1 was assigned [37]. This procedure has been successfully used previously in other studies, for which biological replicates were also not available [38]. After the identification of the differentially expressed (DE) genes, correction for multiple testing was performed by applying the Benjamini-Hochberg method [39] on the *p*-values, to control the false discovery rate (FDR). The final list of differentially expressed genes was generated after employing a threshold of 0.01 for the FDR.

2.6. qPCR Validation

To perform the validation of the data from RNA-Seq, five DE genes were selected (water deficit inducible (Wdip), WRKY transcription factor 1 (WRKY), PR10 protein (Pr10), MYB-like transcriptional factor (Myb), TIR/NBS/LRR disease resistance protein (LRR) and primers designed using Primer3 software [40]. cDNA synthesis was performed using the ImProm-II TM Reverse Transcription System Kit (Promega) with 1 µg of total RNA following manufacturer instructions. Relative expression quantification was performed with Rotor-Gene Q software 1.7 (Qiagen, Venlo, The Netherlands) using the SsoFast™ Eva Green SuperMix 1x (SYBR based system, Bio-Rad, Hercules, CA, USA); 250 nM of each primer and 1 µL of cDNA in a final volume of 20 µL. All samples were run in triplicate, and a no template control (NTC) and a housekeeping gene were used for every primer pair. PCR cycling conditions were 95 °C for 3 min, followed by 40 cycles at 95 °C 10 s, 60 °C 60 s and 72 °C 30 s. A melting curve was generated for each reaction to assure specificity of the primers and the presence of primer-dimer. Primers efficiencies were assessed using a serial dilution of cDNA stock. The Elongation factor-1 alpha was used as housekeeping gene and for normalization of expression of each gene.

To compare the RNA-Seq and qPCR results, a Pearson correlation was calculated using the Log2 of the normalized expression values.

2.7. Transcriptome Annotation

The ORFs transcripts identified by TransDecoder were used for transcriptome annotation. This procedure was performed using InterProScan [41,42]. The protein domains, gene ontology (GO) terms [43] and Kyoto Encyclopedia of Genes and Genomes (KEGG) pathways [44] associated with the genes annotated that are encoding enzymes were identified. A custom python script was run to filter GOs and KEGGs from the InterProScan output. Categorizer [45] was used for the analysis of the GOs. The list of GO IDs belonging to one of the GO categories, which includes Biological process (BP), Cellular component (CC) and Molecular function (MF), was used and classified by its corresponding subcategories against the GO Slim plant database [45]. The number of GOs was counted within each subcategory, and its percentage over the total set of GO IDs provided was reported.

Regarding the functional annotation for differential expressed genes, the contigs were annotated against the non-redundant National Center for Biotechnology Information (NCBI) plants database (version of August 2015) using BlastP (e-value 1×10^{-5}).

2.8. Biological Networks Analysis

In this study, Cytoscape [46] was used for visualization of molecular interaction networks. This type of analysis provided a deep knowledge about resistance mechanisms at the molecular level. Cytoscape analysis procedures began by establishing interactions between DE genes associated with KEGG pathways and GO terms. From the large amount of plugins and features available in Cytoscape to perform different types of studies, BiNGO [47] was selected to identify which GOs were statistically overrepresented in the sets of DE genes. Moreover, the Enrichment Map plugin [48] was also used with the results from BiNGO to visualize enrichment of specific functions. The statistical analysis was carried out with customized default values recommended by user's guidelines.

2.9. SNP Calling

Variant calling was performed with the GATK toolkit [49], which offers a variety of tools for variant discovery. Similarly to the differential expression analysis, only the unique mapped reads were used for SNP calling. A first set of variants was identified using the UnifiedGenotyper tool available in the GATK toolkit. This initial set of variants was then filtered, using the SelectVariants option with the parameters SNP quality (QUAL $\geq$ 60), individual coverage (DP $\geq$ 15) and genotype quality (GQ-phred quality $\geq$ 40), in order to produce the final set of high-confidence SNPs. Finally, SnpEff was used to annotate and predict the effects of the filtered SNPs.

2.10. Data Archiving Statement

The raw sequences used in this work were submitted to the Sequence Read Archive (SRA) with the BioProject accession number PRJNA378402 and accession name "*Pinus pinaster* Transcriptome sequencing".

3. Results

3.1. Pre-Processing of RNA-Sequencing Data and Transcriptome Assembly

A total of 176,282,168 raw reads were generated for all libraries. After quality control using FastQC, low-quality bases were trimmed by Sickle and 144,422,207 high quality reads were obtained, with an average range length between 119 bp and 122 bp (Table 1). A total of 81.9% of the original number of reads were retained after applying the quality control procedures.

Table 1. Pre-processing statistics for each library.

Sample	Number of Sequenced Reads	Average Read Length (bp)	Number of Reads after QC	% Reads after QC
Pp01—Control (0 h)	47,903,109	122	39,091,399	81.6
Pp02—6 h + 24 h	38,483,969	119	30,863,177	80.2
Pp03—48 h	44,943,925	122	37,186,370	82.7
Pp04—7 days	44,951,165	121	37,281,261	82.9
Total	176,282,168	121	144,422,207	81.9

The de novo assembly performed with Trinity produced 483,428 contigs. The additional clustering of these contigs performed with CAP3 resulted in an improved assembly comprising 355,287 contigs with a total length of 147,022,102 base pairs.

Regarding ORF prediction for the transcriptome assembly, 83,468 genes were predicted from the 355,287 assembled contigs.

3.2. Mapping and Differential Expression Analysis

The results regarding the mapped reads (MR) for each library are presented in Table 2. A total of 103,269,178 pre-processed reads were mapped by RapMap against the transcriptome assembly for all libraries, which corresponded to an average of 71.6% of the total number of pre-processed reads. The lowest percentage of MR (70.8%) was obtained for the control library (Pp01). On the other hand, for the Pp02 library, which corresponds to the 6 h + 24 h sampling time points after inoculation, the highest percentage of MR (73.3%) was obtained (Table 2).

Table 2. Number of mapped reads, unique mapped reads and their percentages for each library.

Sample	Number of Reads Mapped	Number of Unique Mapped Reads	% of Mapped Reads	% of Unique Mapped Reads
Pp01—Control	27,683,922	14,432,190	70.8	36.9
Pp02—6 h + 24 h	22,608,382	12,162,427	73.3	39.4
Pp03—48 h	26,553,659	14,077,794	71.4	37.9
Pp04—7 days	26,423,215	13,828,813	70.9	37.1
Total	103,269,178	54,501,224	71.6	37.7

For all downstream analyses it was essential to filter the unique mapped reads (UMR) from this set of MR. A total of 54,501,224 UMR were retained, which corresponded to approximately 37.7% of total processed reads (Table 2). Similarly to MR, the lower percentage of UMR was detected in Pp01 (36.9%), while the Pp02 library had the highest percentage (39.4%) (Table 2).

Statistical analysis with EdgeR identified a total of 17,533 DE genes (adjusted p-value ≤ 0.01 and FDR value = 0.01) within the 42,594 significant tests. The number of significant tests (up and down regulated) for each comparison between two different stages is summarized in Figure S1. The highest number of significant tests was identified between the control sample (Pp01) and Pp02, where 4968 genes were up regulated and 5103 genes were down regulated. This result highlights the importance of studying early responses to pathogen attack. Moreover, 85 genes were always differentially expressed (up or down) in all comparisons.

Figure 1 shows the number of genes differentially expressed (up and down) uniquely for each comparison. These results were in agreement with the total number of significant tests, since the highest number of DE genes was present between Pp01 and Pp02.

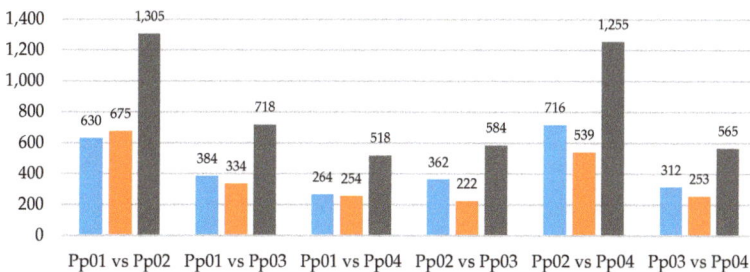

Figure 1. Number of differentially expressed genes (up and down) uniquely for each comparison. Tests up-regulated are shown in blue, tests down regulated are shown in orange and the total number of tests is shown in gray.

3.3. qPCR Validation

In order to validate the results from RNA-Seq data, five DE genes were selected to assess their expression by qPCR. The qPCR assay for the selected transcripts shows an expression pattern similar to the one obtained by RNA-Seq data analysis. Differential expression detected by RT-PCR was consistent with the RNA-Seq expression profiling of each of the selected transcripts. A correlation between the Log2 of the expression values of both RNA-Seq and qPCR data was performed and showed a positive correlation between the values obtained by these two different approaches (Figure 2 and Figure S2), which supports the accuracy of the RNA-Seq analyses.

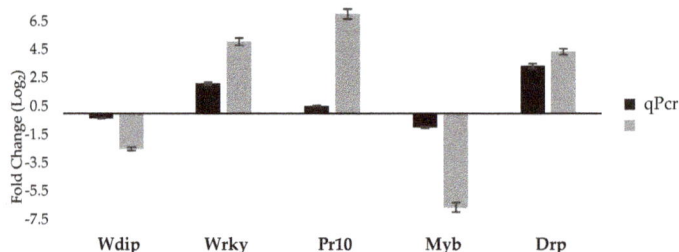

Figure 2. Fold Change for each of the transcripts selected for validation. The values are represented as the Log2 of the normalized expression value of both RNA-Seq and qPCR data, at time point Pp04.

3.4. Transcriptome Annotation

Functional annotation over the 83,468 predicted genes by TransDecoder was performed using BlastP against the NCBI NR-plants database, with results showing a total of 70,646 (84.6%) annotated genes. However, 30.6% (25,545) of annotated genes had "Unknown" description, predominantly being associated to *Picea sitchensis*, a conifer of the *Pinaceae* family. From this set of annotated genes, the subset containing only the DE genes also contained 8996 with an "Unknown" description or no description available. Likewise for the DE genes, most of the "Unknown" descriptions were related to *Picea sitchensis*. A list of selected DE genes for analysis is showed in Table 3.

Table 3. Summary of selected DE genes (Up and Down) between different comparisons.

Gene Annotation	Condition	Comparison	Log Fold-Change
GDSL esterase/lipase	Up	Control vs. Pp02	12.4
Translationally-controlled tumor protein homolog	Up	Control vs. Pp02	11.8
Jacalin-related lectin 3 protein	Up	Control vs. Pp02	6.9
Cytokinin dehydrogenase 6-like	Up	Control vs. Pp02	6.9
Endoglucanase	Up	Control vs. Pp02	9.4
Acyl-CoA oxidase	Up	Control vs. Pp02	4.4
Thaumatin-like protein	Up	Control vs. Pp02	4.3
Nucleotide-binding site leucine-rich repeat (NBS-LRR)	Down	Control vs. Pp02	−3.1
12-oxophytodienoate reductase 3-like	Up	Control vs. Pp02	2.9
Iron superoxide dismutase	Up	Control vs. Pp02	1.9
Auxin-induced cell wall protein	Up	Control vs. Pp02	7.7
Multifunctional protein (MFP)	Up	Control vs. Pp02	1.3
Mildew resistance locus 6 calmodulin binding protein	Up	Control vs. Pp02	2.7
Sucrose synthase	Up	Control vs. Pp02	2.5
TMV resistance protein N-like	Down	Control vs. Pp02	−7.1
Phenylalanine ammonia-lyase	Up	Control vs. Pp02	2.1
Peroxidase	Up	Control vs. Pp02	7.7
GDSL esterase/lipase	Up	Control vs. Pp03	9.9
Translationally-controlled tumor protein homolog	Up	Control vs. Pp03	10.4
Endoglucanase	Up	Control vs. Pp03	9.7
Thaumatin-like protein	Up	Control vs. Pp03	2.5
(E)-4-hydroxy-3-methylbut-2-enyl diphosphate synthase	Up	Control vs. Pp03	4.3
Nucleotide-binding site leucine-rich repeat (NBS-LRR)	Down	Control vs. Pp03	−12.5
GDSL esterase/lipase	Up	Control vs. Pp04	8.1

Table 3. *Cont.*

Gene Annotation	Condition	Comparison	Log Fold-Change
Translationally-controlled tumor protein homolog	Up	Control vs. Pp04	9.6
Jacalin-related lectin 3 protein	Up	Control vs. Pp04	8.6
Cytokinin dehydrogenase 6-like	Up	Control vs. Pp04	6.9
Endoglucanase	Up	Control vs. Pp04	9.1
Acyl-CoA oxidase	Up	Control vs. Pp04	4.9
Thaumatin-like protein	Up	Control vs. Pp04	3.1
(E)-4-hydroxy-3-methylbut-2-enyl diphosphate synthase	Up	Control vs. Pp04	3.7
Pinosylvin synthase	Up	Control vs. Pp04	2.8
Nucleotide-binding site leucine-rich repeat (NBS-LRR)	Down	Control vs. Pp04	−8.5
Auxin-induced protein 1	Down	Pp02 vs. Pp03	−1.9
Laccase	Up	Pp02 vs. Pp03	6.7
Dehydrin	Down	Pp02 vs. Pp04	−11.0
Pathogenesis related 10	Down	Pp02 vs. Pp04	−3.0
Pinosylvin synthase	Up	Pp02 vs. Pp04	2.6
Heat shock protein	Up	Pp02 vs. Pp04	7.6
Light harvesting complex protein	Up	Pp02 vs. Pp04	7.9
Nucleotide-binding site leucine-rich repeat (NBS-LRR)	Up	Pp02 vs. Pp04	8.8
Phospholipase D alpha 1-like	Up	Pp03 vs. Pp04	2.7
Tau class glutathione S-transferase	Up	Pp03 vs. Pp04	3.6
Pinosylvin synthase	Up	Pp03 vs. Pp04	3.5

The protein domains were also identified for the set of predicted genes using InterProScan, providing information about the functional classes of GO terms and KEGG pathways. Over the set of predicted genes, 77.7% (64,853) had at least one protein domain identified, 46.4% (38,762) were associated with at least one GO term (BP: 29.6%, CC: 11.3% and MF: 41.4%) and 5.9% (4904) were associated with at least one KEGG pathway. The total number of different GO terms and KEGG pathways identified over the whole gene set was 1810 and 111, respectively. Focusing just over the set of 17,553 DE genes, 9119 (52.0%) were associated with at least one GO term and a total of 1292 different GO terms were found. Moreover, 1154 DE genes were associated with at least one KEGG pathway and a total of 102 different KEGG pathways were found. The most representative pathways for predicted genes and for DE genes, in terms of enzymes associated, are shown in Table 4.

Table 4. Summary of KEGG pathways ranked by the number of associated enzymes, detected in predicted genes and in DE genes.

Pathways	Enzymes in All Set of Genes	Enzymes in DE Genes
Purine metabolism	37	24
Pyrimidine metabolism	26	18
Aminoacyl-tRNA biosynthesis	21	15
Cysteine and methionine metabolism	20	16
Starch and sucrose metabolism	19	16
Porphyrin and chlorophyll metabolism	20	12
Phenylalanine, tyrosine and tryptophan biosynthesis	18	14
Pyruvate metabolism	17	14
Glycolysis/Gluconeogenesis	17	15
Carbon fixation in photosynthetic organisms	13	12
Terpenoid backbone biosynthesis	17	7

Additionally, CateGOrizer was used for further analysis of the GO terms, for both the whole set of predicted genes and the set of DE genes, identifying subcategories of each of the three GO categories (BP, MF and CC), against the GOSlim plant database.

With respect to the biological process branch, a total of 30 subcategories of GO terms were found. The most significant were cellular process (GO: 0009987) (32.8%), metabolic process (GO: 0008152) (26.9%) and biosynthetic process (GO: 0009058) (11.1%) (Figure S3). In the case of cellular component, 26 subcategories of GO terms were identified. The largest proportion GOs were assigned to cell (GO: 0005623) (28.8%), intracellular (GO: 0005622) (27%) and cytoplasm (GO: 0005737) (11.4%) (Figure S4). In the molecular function category, a total of 24 subcategories of GO terms were detected.

In this category, the most representative terms were catalytic activity (GO: 0003824) (44.2%), transferase activity (GO: 0016740) (13.8%) and hydrolase activity (GO: 0016787) (11.7%) (Figure S5).

Subsequently, to further investigate the biological response associated to PWN infection, a GOs analysis for DE genes among all conditions was performed. In this analysis, a total of 9119 DE genes (52.0%) associated with at least one GO term were identified and a total of 1292 different GO terms were found. For the biological process term 36 GO subcategories were identified, with a total of 1477 hits. The most representative subcategories were cellular process (GO: 0009987) (27.4%), metabolic process (GO: 0008152) (22.6%) and biosynthetic process (GO: 0009058) (9.1%) (Figure S3). Regarding cellular component terms, 24 subcategories were found with a total of 486 hits. The subcategories with more hits were cell (GO: 0005623) (28.6%), intracellular (GO: 0005622) (27.4%) and cytoplasm (GO: 0005737) (12.4%) (Figure S4). Lastly, for the molecular function term 24 subcategories were identified, with a total of 1039 hits. The most relevant subcategories were catalytic activity (GO: 0003824) (42.9%), transferase activity (GO: 0016740) (12.9%) and hydrolase activity (GO: 0016787) (11.6%) (Figure S5).

3.5. Biological Networks Analysis

Understanding the molecular mechanisms involved in pine response to PWN infection depends on clarifying the biological activity of proteins. In this sense, a molecular interaction network analysis was performed using Cytoscape. Within the total of 102 KEGG pathways significantly regulated over the DE genes, phenylpropanoid biosynthesis, terpenoid backbone biosynthesis and flavonoid biosynthesis were prioritized for further discussion, because of the important role they play in the mechanisms of defense against PWN infection.

A total of 72 DE genes were associated with the phenylpropanoid biosynthesis pathway (Figure 3). This pathway, involved in the synthesis of secondary metabolites, plays a crucial role in the signaling mechanism, being a key mediator of the higher plants resistance towards pests [50]. The first reaction in the pathway is catalyzed by phenylalanine ammonia-lyase (PAL) (EC: 4.3.1.24). PAL links phenylpropanoid biosynthesis with the phenylalanine metabolism and can be induced by wounding or under other stress conditions. Its synthesis has been strongly associated to defense responses [51]. One of the main end products of this pathway is lignin. This polymer is determinant for the trees stability and robustness, acting mainly as physical barrier to pathogens [52]. Regulation of lignin components production is controlled by peroxidase (EC: 1.11.1.7). Both, PAL and peroxidase enzymes are found over-expressed in all the stages after inoculation with PWN.

For the pathways indicated, several proteins without functional annotation were named as unknown. Nevertheless, a metabolic network can still be associated to these proteins, by similarity, contextual methods (for example, homology-based function predictions) or both. In our work, the unknown proteins that are presented in the metabolic pathways were annotated in the UniProt database, which provides the link to the pathway information generated with Cytoscape (Table S1).

Terpenoid backbone biosynthesis pathway (Figure 4) has been associated to defensive mechanisms against biotic and abiotic stresses, which may act as insect repellents in a chemical response. This main pathway can be subdivided in two biosynthetic pathways, the mevalonate and the non-mevalonate (MEP/DOXP) pathway. A total of 31 DE genes were associated to terpenoid backbone biosynthesis. Most of them codified two enzymes: hydroxymethylglutaryl-CoA reductase (HMGR) (EC: 1.1.1.34) and 1-deoxy-D-xylulose-5-phosphate (DXP) synthase (EC: 2.2.1.7). HMGR is involved in the synthesis of mevalonate, being an important control point in the mevalonate pathway [53]. HMGR induction is especially high in the first stages after PWN inoculation. Regarding 1-deoxy-D-xylulose-5-phosphate synthase, this enzyme is the first of the MEP/DOXP pathway, being considered important in the regulation of plastidial isoprenoid production [54].

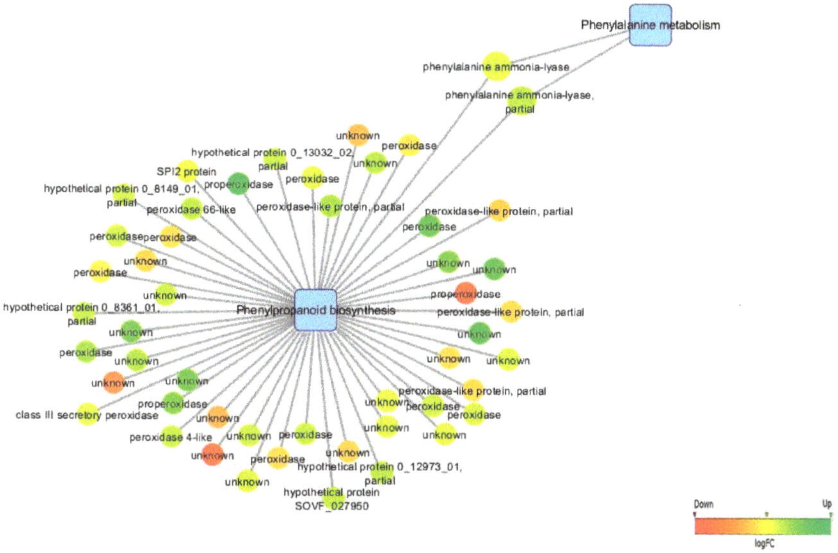

Figure 3. Phenylpropanoid biosynthesis pathway with interactions related to DE genes in the comparison Pp01–Pp02. The blue square nodes represent the pathways. The rest of the nodes represent all DE genes associated to the pathways. Note that due to the complexity of representing the expression values of those genes, the RGB color scale goes from the most down regulated in red to the most up regulated in green.

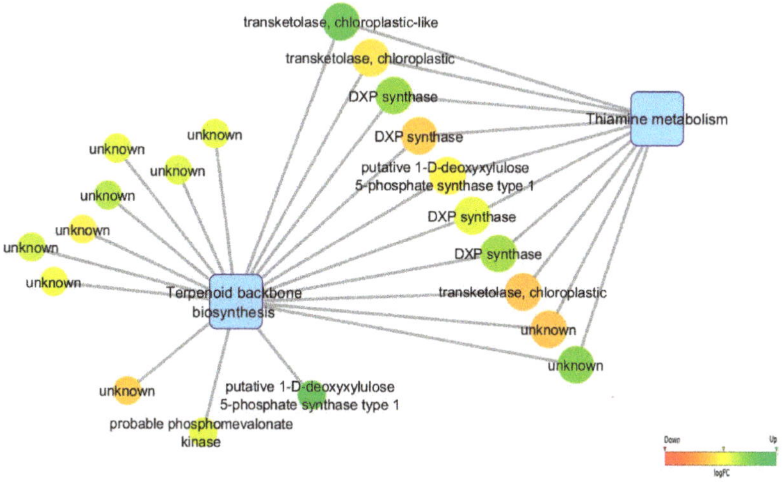

Figure 4. Terpenoid backbone biosynthesis pathway with interactions related to DE genes in the comparison Pp01–Pp02. The blue square nodes represent the pathways. The rest of the nodes represent all DE genes associated to the pathways. Note that due to the complexity of representing the expression values of those genes, the RGB color scale goes from the most down regulated in red to the most up regulated in green.

In the flavonoid biosynthesis pathway a total of 9 DE genes were identified (Figure 5), all being highly expressed in early stages after PWN inoculation. Flavonoids are one of the largest classes of phenolics, a group of secondary metabolites produced by plants to defend themselves against pathogens. A key enzyme in this pathway is chalcone synthase (EC: 2.3.1.74), which is induced in plants under stress conditions. This enzyme plays a decisive role in the synthesis of a large set of flavonoid metabolites, being involved in the salicylic acid defense pathway [55]. Another important enzyme is chalcone isomerase (EC: 5.5.1.6), which catalyzes the cyclization of a chalcone into a flavone [56]. While chalcone synthase is highly induced in the first stage after PWN inoculation, chalcone isomerase is highly induced in the late stages after inoculation, both causing the accumulation of flavonoid and isoflavonoid phytoalexins in plant tissues.

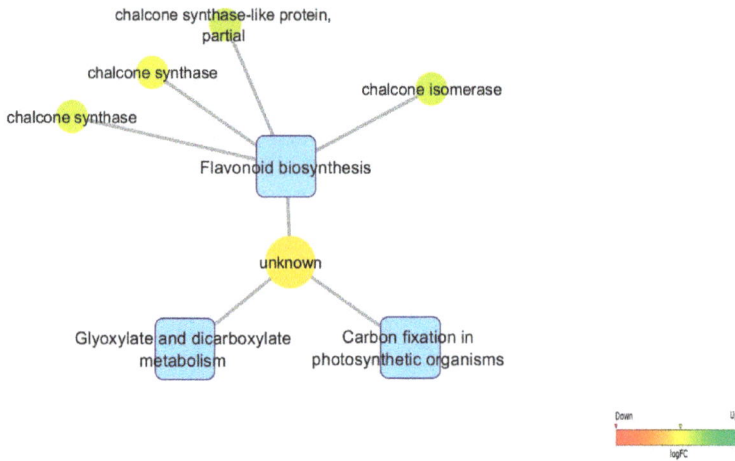

Figure 5. Flavonoid biosynthesis pathway with interactions related to DE genes in the comparison: Pp01–Pp02. The blue square nodes represent the pathways. The rest of the nodes represent all DE genes associated to the pathways. Note that due to the complexity of representing the expression values of those genes, the RGB color scale goes from the most down regulated in red to the most up regulated in green.

Regarding the GO networks, the analysis was focused in subcategories related to signaling system and to plant survival mechanisms under stress conditions. Thus, with respect to signal transduction (GO: 0007165), which mediates the sensing and processing of stimuli, 120 interactions over DE genes were detected, identifying a high level of some capital metabolites in all stages after inoculation. Molecular switches in plant signal, such as Rab7/RabG-famlily small GTPase and a small tab-related GTPase were induced immediately after PWN inoculation and remain highly induced over time. This suggests that these two proteins may act as a trigger in the maritime pine defensive response against PWN infection. A study in other plant species [57] reported that GTPases family is induced by external signals and plays a crucial role in the regulation of multiple responses in plants against pathogens.

A total of 116 DE genes were associated with response to oxidative stress (GO: 0006979). This high number of associated genes makes sense, since oxidative stress is one of the typical symptoms caused by PWN infection. Within this term, glutathione peroxidase 1 (Gpx) was identified as highly induced in all the stages after inoculation. Gpx is an enzyme whose main biological role is to protect the organism from oxidative damage, acting as antioxidant, by the inactivation of hydrogen and lipids peroxides [58].

Concerning response to stress (GO: 0006950), a total of 75 DE genes related with this term were identified. Over this term, dehydrin1 is hugely expressed in all stages after inoculation. Dehydrins are involved in plant response and adaptation to abiotic and biotic stress [59]. Moreover, abscisic acid stress ripening protein 1-like is also over-expressed after inoculation; however, its biological role in stress conditions remains understudied.

The results of the over-expression analysis for the gene ontology terms, performed by BiNGO, related with the DE genes identified between the control sample and the first stage after inoculation, are presented in Figures S6 and S7. In total, two cellular components and three molecular functions were over represented in the control sample. Regarding the biological process, there was no statistical evidence of overrepresentation (Figure S6). On the other hand, five biological processes, four cellular components and one molecular function were over represented in the first stage after inoculation (Figure S7).

3.6. SNP Calling Analysis

For SNP discovery and filtering, GATK was used with stringent parameters. Variants were called using the UnifiedGenotyper and further filtering was performed using the SelectVariants option. In total, 36,295 different SNPs were detected. Among these SNPs, 31.9% were found in exons, while 30.6% were detected in an intergenic region, a portion of a contig without gene prediction (Table 5). Moreover, with respect to the SNPs found in each functional class, 48.5% were associated to missense mutations, 50.7% to silent mutations and less than 1% to nonsense mutations (Table 6).

Table 5. SNP calling analysis. Number and percentage of effects by region.

Region	Count	Percentage
Exon	15,232	31.9%
Intergenic	14,600	30.6%
Splice site region	1	<0.1%
Transcript	31	0.1%
UTR 3 Prime	9072	19.0%
UTR 5 Prime	8718	18.3%

Table 6. SNP calling analysis. Number and percentage of effects by functional class.

Type	Count	Percentage
MISSENSE	7410	48.5%
NONSENSE	121	0.8%
SILENT	7732	50.7%

4. Discussion

In this study, an RNA-Seq based approach was used to determine the transcriptomic profile of maritime pine in different stages after inoculation with PWN, to identify candidate genes associated with the response mechanisms to the infection.

One of the main challenges in RNA-Seq studies for non-model organisms like maritime pine is to produce de novo transcriptome assembly. This is a crucial step, which can yield some erroneous assembled contigs, due to the nature of high-throughput sequencing reads, which can contain sequencing errors, and the algorithms used for de novo assembly, which can also generate assembly artifacts, i.e., contigs that do not represent true regions of the transcriptome. These factors may have had an impact in this study since the rate of predicted genes from the set of assembled contigs was low (83,468 genes were predicted from 355,287 assembled contigs). Even though genomic information for several conifer species has been generated, and compiled in the TreeGenes database [60], the number of genomic resources for maritime pine remains limited, which is another relevant factor that may have contributed to the low rate of predicted genes. When a reference genome is not available, the genetic description contained in the assembled transcripts can only be successfully identified by homology if the protein products have homologies in different protein databases, giving a set of predicted genes.

Hence, in the case this type of analysis is applied in species that yield a distinct set of expressed genes, with little or no homology when compared with all gene sequences deposited in the databases commonly used in these studies, it is possible to end up with a significant percentage of contigs for which no gene could be predicted. From the total genes predicted in this study, 70,646 (84.6%) of them were annotated, providing a genomic resource to further deepen the study of genes involved in the transcriptomic response of pine wood to infection with PWN. However, 25,545 annotated genes had "Unknown" description, mainly associated to *Picea sitchensis*, which reinforces the need to strengthen the genomic resources for maritime pine, ideally with the availability of a fully sequenced and annotated reference genome.

The comparison of sequence data from all libraries revealed a total of 17,533 DE genes, a number that was obtained using a FDR value of 0.01, which is a more stringent correction for multiple testing when compared with the traditionally used FDR value of 0.05. Despite the stringency applied in the statistical methodologies used to generate the final list of differentially expressed genes, the unavailability of replicates may be responsible for increasing the number of false-positive results. Functional annotation with GO terms for predicted genes resulted in 38,762 (46.4%) unigenes with at least one assignment into one of the three categories of GO terms (BP, MF and CC). In each of the GO categories, the GO terms fell mainly into two or three subcategories. The GO subcategories identified with more evidences are in accordance with other reports [20], and may represent a typical gene expression profile for *P. pinaster* after infection with PWN.

Most plant defensive responses to pathogens have evolved into a complex system, simultaneously combining several mechanisms and pathways. To identify possible pathways involved in defense against PWN, a KEGG analysis for our set of predicted genes was performed. The different KEGG pathways associated with the predicted genes are in agreement with the Physiome Project Models [61] for *P. pinaster*. The most prevalent pathways were purine and pyrimidine metabolism. These subunits of nucleic acids are major energy carriers and precursors for the synthesis of nucleotide cofactors such and NAD and SAM [62].

In this study, the identification of several DE genes related to biotic and abiotic stresses, hormonal regulation and cell wall defense further validates the hypothesis that these mechanisms play a crucial role in the plant defense system involved in the response to PWD.

4.1. Infection Leads to de novo Transcription of Genes Involved in Biotic Stress Response, Phenylpropanoid/Terpenoid Metabolisms and Hormonal Regulation

The genes induced only after inoculation with PWN are important to understand the mechanisms activated when the plant is infected by this pathogen. Indeed, a set of genes over-expressed in all conditions after inoculation (Pp02—6 h + 24 h, Pp03—48 h and Pp04—7 days) were identified, including the GDSL esterase/lipase that is potentially involved in defensive reactions [63,64], and a translationally-controlled tumor protein homolog, which participates in important cellular processes like the protection of cells against various stresses and apoptosis [65]. Moreover, the pattern of over-expression after inoculation was also detected for the gene that codifies a jacalin-related lectin 3 protein (JRL), which is often associated with biotic and abiotic stimuli. JRLs proteins have been referenced as a component of the plant defense system, although their role is not well understood yet, due to their structural diversity [66]. In general, the results obtained in the present study are in agreement with the set of candidate genes, related with the response to PWN infection, reported in different studies [20,67]. For instance, the relevance of terpenoid metabolism in maritime pine defense against PWN, important in the resin production process, was pinpointed by Santos and colleagues [20]. In that study, the (E)-4-hydroxy-3-methylbut-2-enyl diphosphate (HMB-PP) reductase gene was found highly expressed after inoculation, an expression pattern that was also identified in Pp03 and Pp04 for HMB-PP synthase. Likewise, the thaumatin-like protein (TLP) was found to be deeply involved in plant defense system as a response to biotic and abiotic stress. Various studies imply its importance in plant resistance [68,69], even though its role remains unclear. For example, towards a pathogen attack and under stress, these proteins confer tolerance and induce stress resistance [68]. In our work the gene codifying this protein is highly induced immediately after inoculation, decreasing its expression levels slightly over time. The difference observed for the expression pattern of this gene, relative to the study carried by Xu and colleagues [67] in *Pinus massoniana*, may indicate a specific response for maritime pine.

Genes involved in hormonal regulation, such as cytokinin dehydrogenase 6-like and auxin-induced cell wall protein, were identified as highly expressed after infection with PWN. In our study, the expression of cytokinins supports the possibility that they are involved in signaling defense responses after a pathogen attack, improving the resistance to pathogens [70].

Phytohormones are responsible for various important physiological processes, from plant growth to plant defense [71]. The response to an insect attack is mediated by plant hormones, as primary signal in the regulation of plant defense. The salicylic acid (SA), hormone ethylene (ET), jasmonic acid (JA) and its derivates are the major defense hormones. Differentially expressed genes codifying important enzymes in jasmonate biosynthesis, such as 12-oxophytodienoate reductase 3-like (OPR3), which is fundamental in this biosynthesis, acyl-CoA oxidase and a multifunctional protein (MFP) found to be associated with wound-induced, were identified through the different comparisons. Genes codifying transcription factors involved in the hormone signaling at local and distant tissues were also identified differentially expressed after inoculation. Most of the genes identified have higher levels of expression in the last stages post inoculation. A faster recognition of the pathogen is important for an efficient response, since the timing of the hormonal production can determine if the plant becomes more resistant to the nematode.

As already mentioned, the highest number of DE genes was identified between the control sample (Pp01) and Pp02, which clearly indicates an immediate response to PWN after inoculation. This observation is in accordance with previous results obtained in *Pinus thunbergii* Parl., that suggested an early response to PWN in susceptible and in resistant trees [72]. Within this early stage of response, and when compared with the control sample, several genes potentially involved in the defensive response were detected, evidencing the activation of the defense mechanisms in the infected plant. These genes included the mildew resistance locus 6 calmodulin binding protein gene, which triggers a defensive response in the occurrence of an infection caused by a foreign body [73]. The processes used by the PWN to invade the *Pinus pinaster* tissues are likely to represent a very similar mechanism that is used by the powdery mildew, hence these results provide further support for the involvement of the mildew resistance locus 6 calmodulin binding protein gene in the initial response of plants to infections. The sucrose synthase gene also displayed over-expression in the Pp02 time point. This gene codes for an enzyme that provides metabolites for the synthesis of cellulose and callose, and plays an important role in secondary cell wall synthesis [74,75]. Genes codifying TMV resistance protein N-like and nucleotide-binding site leucine-rich repeat (NBS-LRR) disease resistance proteins were identified in the control sample. The high expression of these proteins, which are relevant in the plant immune response, suggests that the plant recognizes and triggers the defense system [76]. Comparing the control with all stages after inoculation, the NBS-LRR proteins showed a lower expression in the inoculation stages. These proteins are involved in the pathogen recognition and should act at an early stage for an effective response. In our work, these proteins only increased their values of expression in the last stage of inoculation, which may be too late for the plants to recover and combat the nematode.

Cytoscape analysis showed relevant interactions between the DE genes and KEGG pathways. Some enzymes appear to be directly associated with plants defense, being an important connection between relevant pathways, such as terpenoid backbone biosynthesis and thiamine metabolism, against biotic and abiotic stresses. In our work, the phenylpropanoid biosynthesis can be a relevant pathway, not only due to its connection with the phenylalanine metabolism by PAL, but also for the high number of peroxidase genes over-expressed in the Pp02. The higher expression of PAL genes, responsible for the first step in this pathway, can be a response to the wounds caused by the vector.

4.2. Infection Leads to a Reprogramming of Cell Wall Metabolism Putatively Involved in Cell Wall Reinforcement

The over-expression of peroxidase genes during PWD may be related with the oxidation of phenolic residues, likely the substrates of lignin and suberin, into cell wall polymers in the infected tissues [51].

Enzymes involved in cell wall modifications were found differentially expressed, endoglucanase being the most highly expressed in the second and third time points. The endoglucanase is responsible for the catalysis of cellodextrin in cellobiose, the smallest subunit of cellulose, being also related with cell wall modifications in infected plants [77]. Thus, the over-expression of this enzyme as a response to infection, suggests that not just proteins related to defensive mechanisms are used to fight the infection, since some mechanisms are activated to reconstruct the cell damage originated by the PWN.

Additionally, when comparing the Pp02 with Pp03 a gene encoding a laccase was found highly expressed in the Pp03. These kinds of proteins are involved in lignin biosynthesis and plant pathogenesis [78]. Lignin forms important structural materials in the support tissues of vascular plants. Hence, it makes sense that one of the mechanisms activated is to reinforce the cell walls, especially in wood and bark.

4.3. Late Responses to Infection Seem to Be Involved in the Mitigation of Stress Caused by an Inefficient Early Response

A previous study reported a late response to infection with PWN in susceptible trees [72]. This late response is observed in our study, and may occur approximately one week after inoculation, due to the large amount of DE genes that were identified between Pp02 and Pp04. Measuring differences between early and late responses can elucidate the different mechanisms activated. When comparing the two stages, the higher expression of the dehydrin gene in Pp02 is relevant since it has been associated to plant response and adaptation to abiotic stress such as water stress, being involved in a mechanism commonly developed in these stages [59]. Thus, considering that the PWD results in destruction of parenchyma cells surrounding xylem resin ducts, causing a dysfunction of the water-conducting system, the involvement of this gene to prevent water loss makes sense. The pathogenesis related type 10 gene was also highly expressed at this stage. This gene was already found in conifers, displaying a transient accumulation in needles of drought-stressed trees [79]. As a consequence of the water stress, the needles become drought stressed, which is one of the most characteristic symptoms of PWD. Stress proteins, such as heat shock, were encoded by over-expressed genes in Pp04. Under stressful conditions they protect cells by stabilizing unfolded proteins, giving the cell time to repair damaged proteins. Heat shock proteins are highly conserved among different organisms [63]. Although it is unclear its precise role in *Pinus pinaster*, it seems that this protein is involved in the plant's effort to combat the PWN. The gene coding the light harvesting complex protein was highly expressed in Pp04. It is involved in light energy transfer to one chlorophyll-a molecule at the reaction center of a photosystem. Although this protein is not directly related with the defensive mechanism, it plays an important role trying to maximize the production of energy, which could be essential in helping the cellular systems triggered within the response. Genes encoding NBS-LRR proteins were identified highly expressed in this time point. NBS-LRR proteins are capable of recognizing a wide variety of pathogens and initiate a hypersensitive response (HR), resulting in cell death. The NBS-LRR proteins are divided in two major groups, involved in downstream specificity and signaling regulation [80], both of them showing high levels of expression in the Pp04. The behavior of NBS-LRR proteins through the different time point comparisons pointed out one more difference between the early and the late phase of response and a possible reason why the early response is not effective.

When comparing the Pp02 and Pp03 time points, the auxin-induced protein 1 was found over-expressed in Pp02. Auxins regulate and control vital mechanisms, being involved in growth, development and in defense via signaling, involving different interactions of molecules [81]. This protein seems to have an important role in the first stage of the response against the infection.

Lastly, regarding the comparison between Pp03 and Pp04, a phospholipase D alpha 1-like and a tau class glutathione S-transferase were over-expressed in Pp04. The former plays an important role in various cellular processes, including response to stress [82], while the latter has been associated to the oxidative stress response mechanism [83].

Recent studies related to PWD reported a set of genes associated to response to PWN infection. Several biotic-stress resistance genes were identified after PWN inoculation by Shin and colleagues in Japanese red pine [72], including the pinosylvin synthase and iron superoxide dismutase genes. The pinosylvin synthase was found over-expressed only in Pp04, while the iron superoxide dismutase genes were over expressed immediately after inoculation. Pinosylvin, which belongs to stilbenoids family, has been associated to phytoalexin induction, mainly in young pine trees exposed to biotic stress [84]. Moreover, pinosylvin is a key metabolite that can kill nematodes [85]. The presence of pinosylvin observed only in the late response against PWD, is an important feature that must be further analyzed, namely in tolerant and resistant *Pinus* species to PWN, since it can be one of the reasons why the first response to the infection is inefficient. The iron superoxide dismutase gene plays an important role in cell protection against oxidative stress [86].

The response displayed by maritime pine against PWN infection is clearly very complex and dynamic, however, in the end it still fails to prevent plant death as a result of the infection. This study provides an explanation for the possible reasons why the response of maritime pine to PWN infection is so inefficient. The comparison with the response of a tolerant *Pinus* species, for the same time points after infection (study under progress) will help unveiling what are the preponderant genes and pathways associated with resistance to PWD.

The SNP calling analysis performed in this study yielded a total of 36,295 SNPs, of which 69.2% were identified in exons and 30.6% were located in intergenic regions. The analysis of the SNP effects by functional class, performed only for the 15,263 SNPs located in exons and transcripts, revealed that over than 50% had a silent effect, which means that the SNP does not change the protein sequence. However, about 48.5% displayed a missense effect. In these situations, these changes are responsible for coding a different amino acid. When a new amino acid is coded, the sequence of the protein coded by a particular gene is also changed. These changes may occur between amino acids with markedly different properties, which in turn can affect the enzyme catalytic activity, or affect the secondary and tertiary structure of the protein, among others. Additionally, from a total of 4061 SNPs identified within the 17,533 DE genes, 15 were found in the sequences of genes discussed in more detail in the present work. There were 11 SNPs displaying a missense effect in the exon regions of the genes that codify the GDSL esterase/lipase, auxin induced cell wall and dehydrin proteins. Moreover, the two genes codifying the phenylalanine ammonia-lyase and auxin induced cell wall proteins respectively present a SNP in their 3′ UTR region. This type of variation might affect transcription and translation and could be responsible for differences in gene expression. The two remaining SNPs were identified in the genes that encode the pathogenesis related 10 and jacalin-related lectin 3 proteins, generating a new stop codon, known as nonsense SNP. This type of SNP is responsible for the change of a coding codon to a stop codon, resulting in the inactivation of the respective gene [87]. Hence, these are very important SNPs whose effect on potential resistance to PWD can be tested in larger pine tree populations where resistance phenotypic data might be available for genome wide association studies (GWAS).

5. Conclusions

Currently, PWD, caused by *Bursaphelenchus xylophilus*, is the most deadly maritime pine disease. This study establishes a new approach for the understanding of the molecular response of maritime pine, which is susceptible to PWN, over different time points after inoculation with PWN. Clear insights related with the defense mechanisms of *Pinus pinaster* against PWN were identified. The functional annotation of the predicted genes revealed the complexity of the system involved in the response against PWN, combining a number of mechanisms and pathways, simultaneously. As pointed out in previous studies, the occurrence of two phases of response against PWN was identified from the results of the differential expression analysis: an early response that occurs immediately after infection, and a late response that is developed approximately seven days after infection. Future studies will focus the analysis on the comparisons between for *P. pinaster* and a tolerant *Pinus* species, for the same time points, in order to try to understand which response is more effective to prevent the pathogen progression after infection. Moreover, the high number of DE genes found between the early and the late responses suggests that these two phases may have significant differences at the molecular level. The set of candidate genes identified over the different time points after inoculation will be a useful resource in future studies and breeding programs to select plants with lower susceptibility to PWD. Moreover, the SNPs identified in this study will be available to be tested in larger populations with available phenotypic records for resistance to PWD, thereby enabling the possibility of identification of molecular markers linked with this very important economic and biological trait in maritime pine.

Supplementary Materials: The following are available online at www.mdpi.com/1999-4907/8/8/279/s1, Figure S1: Total number of significant tests (up and down) between each comparison. Tests up-regulated are shown in blue, tests down regulated are shown in orange and the total number of tests is shown in gray, Figure S2: Correlation between the results from RNA-Seq and qPCR for the transcripts selected to perform validation, at time point Pp04, Figure S3: Distribution of the most representative biological process subcategories. The results for predicted genes are shown in blue and for DE genes in orange, Figure S4: Distribution of the most representative cellular component subcategories. The results for predicted genes are shown in blue and for DE genes in orange, Figure S5: Distribution of the most representative molecular function subcategories. The results for predicted genes are shown in blue and for DE genes in orange, Figure S6: Overrepresentation of cellular component and molecular function terms in control, obtained by BINGO. Color bar in the right lower quadrant indicates level of significance from low (yellow) to high (orange). The size of the nodes is proportional to the number of genes in GO category. Statistical analysis was performed with a hypergeometrical test and a *p*-value of <0.05, Figure S7: Overrepresentation of biological processes, cellular component and molecular function terms in first stage (Pp01), obtained by BINGO. Color bar in the right lower quadrant indicates level of significance from low (yellow) to high (orange). The size of the nodes is proportional to the number of genes in GO category. Statistical analysis was performed with a hypergeometrical test and a *p*-value of <0.05, Table S1: List of proteins with "unknown" annotation associated to flavonoid biosynthesis, terpenoid backbone biosynthesis and phenylpropanoid biosynthesis pathways. GI: Gene Identifier; GB: Gene Bank identifier.

Acknowledgments: Authors acknowledge the funding provided by Project REPHRAME of 7th Framework programme: "Development of improved methods for detection, control and eradication of pine wood nematode in support of EU Plant Health policy". Financial support for D. Gaspar, A. Usié, B. Meireles, P. Barbosa and A.M. Ramos was provided by Investigador FCT project IF/00574/2012/CP1209/CT0001: "Genetic characterization of national animal and plant resources using next-generation sequencing". A.M. Fortes was funded by FCT Investigator FCT050 (IF/00169/2015) and PEst-OE/BIA/UI4046/2014.

Author Contributions: Rita L. Costa conceived and designed the study and supervised the laboratorial experiments; Cândida Trindade performed the laboratorial experiments; Daniel Gaspar performed bioinformatics analyses of the data; Ana Usié, Brígida Meireles and Pedro Barbosa contributed in the bioinformatics analyses; António M. Ramos coordinated the bioinformatics analyses; Rita L. Costa, Daniel Gaspar, Ana Usié, Brígida Meireles, Ana M. Fortes, Cátia Pesquita and António M. Ramos interpreted the results; Daniel Gaspar and António M. Ramos wrote the manuscript; Ana Usié, Ana M. Fortes, Cátia Pesquita and Rita L. Costa revised the manuscript; All authors have read and approved this version of the manuscript.

Conflicts of Interest: The authors declare no conflict of interest.

References

1. Plomion, C.; Pionneau, C.; Brach, J.; Costa, P.; Baillères, H. Compression wood-responsive proteins in developing xylem of maritime pine (*Pinus pinaster* Ait.). *Plant Physiol.* **2000**, *123*, 959–969. [CrossRef] [PubMed]

2. Uva, J.S. IFN6—Áreas dos usos do solo e das espécies florestais de Portugal continental. *Inst. Conserv. Nat. Florestas I.P.* **2013**, *1*, 1–35.

3. Mota, M.M.; Futai, K.; Vieira, P. Pine Wilt Disease And The Pinewood Nematode, Bursaphelenchus Xylophilus. In *Integrated Management of Fruit Crops Nematodes*; Springer: Dordrecht, The Netherlands, 2009; pp. 253–274.

4. Futai, K.; Sutherland, J.R.; Takeuchi, Y. *Pine Wilt Disease*; Springer: Tokyo, Japan, 2008; ISBN 978-4-431-75655-2.
5. Mota, M.; Braasch, H.; Bravo, M.A.; Penas, A.C.; Burgermeister, W.; Metge, K.; Sousa, E. First report of Bursaphelenchus xylophilus in Portugal and in Europe. *Nematology* **1999**, *1*, 727–734. [CrossRef]
6. Farjon, A. *A Handbook of the World's Conifers*; Brill: Leiden, The Netherlands, 2010; ISBN 9789004177185.
7. Sousa, E.; Bravo, M.A.; Pires, J.; Naves, P.; Penas, A.C.; Bonifácio, L.M.M. Bursaphelenchus xylophilus (Nematoda: Aphelenchoididae) associated with Monochamus galloprovincialis (Coleoptera: Cerambycidae) in Portugal. *Nematology* **2001**, *3*, 89–91. [CrossRef]
8. Naves, P.M.; Camacho, S.; De Sousa, E.; Quartau, J.A. Transmission of the pine wood nematode Bursaphelenchus xylophilus through oviposition activity of Monochamus galloprovincialis (Coleoptera: Cerambycidae). *Entomol. Fenn.* **2007**, *18*, 193–198.
9. Fielding, N.J.; Evans, H.F. The pine wood nematode Bursaphelenchus xylophilus (Steiner and Buhrer) Nickle (= B. lignicolus Mamiya and Kiyohara): An assessment of the current position. *Forestry* **1996**, *69*, 35–46. [CrossRef]
10. Edwards, O.R.; Linit, M.J. Transmission of Bursaphelenchus xylophilus through Oviposition Wounds of Monochamm carolinensis (Coleoptera: Cerambycidae). *J. Nematol.* **1992**, *24*, 133–139. [PubMed]
11. Naves, P.M.; Camacho, S.; de Sousa, E.M.; Quartau, J.A. Transmission of the pine wood nematode Bursaphelenchus xylophilus through feeding activity of Monochamus galloprovincialis (Col., Cerambycidae). *J. Appl. Entomol.* **2007**, *131*, 21–25. [CrossRef]
12. Fukuda, K. Physiological process of the symptom development and resistance mechanism in pine wilt disease. *J. For. Res.* **1997**, *2*, 171–181. [CrossRef]
13. Jusheng, H. A brief account of forest tree improvment in China. *Genet. Resour. Inf.* **1985**, *14*, 2–6.
14. Wang, Z.; Gerstein, M.; Snyder, M. RNA-Seq: A revolutionary tool for transcriptomics. *Nat. Rev. Genet.* **2009**, *10*, 57–63. [CrossRef] [PubMed]
15. Parchman, T.L.; Geist, K.S.; Grahnen, J.A.; Benkman, C.W.; Buerkle, C.A. Transcriptome sequencing in an ecologically important tree species: Assembly, annotation, and marker discovery. *BMC Genom.* **2010**, *11*, 180. [CrossRef] [PubMed]
16. Mota, M.M.; Takemoto, S.; Takeuchi, Y.; Hara, N.; Futai, K. Comparative Studies between Portuguese and Japanese Isolates of the Pinewood Nematode, Bursaphelenchus xylophilus. *J. Nematol.* **2006**, *38*, 429–433. [PubMed]
17. Naves, P.M.; Sousa, E.; Rodrigues, J.M. Biology of Monochamus galloprovincialis (Coleoptera, Cerambycidae) in the Pine Wilt Disease Affected Zone, Southern Portugal. *Silva Lusit.* **2008**, *16*, 133–148.
18. Valadas, V.; Oliveira, S.; Espada, M.; Laranjo, M.; Mota, M.; Barbosa, P. The pine wood nematode, Bursaphelenchus xylophilus, in Portugal: Possible introductions and spread routes of a serious biological invasion revealed by molecular methods. *Nematology* **2012**, *14*, 899–911. [CrossRef]
19. Vicente, C.S.L.; Nascimento, F.; Espada, M.; Barbosa, P.; Mota, M.; Glick, B.R.; Oliveira, S. Characterization of Bacteria Associated with Pinewood Nematode Bursaphelenchus xylophilus. *PLoS ONE* **2012**, *7*, e46661. [CrossRef] [PubMed]
20. Santos, C.S.; Pinheiro, M.; Silva, A.I.; Egas, C.; Vasconcelos, M.W. Searching for resistance genes to Bursaphelenchus xylophilus using high throughput screening. *BMC Genom.* **2012**, *13*, 599. [CrossRef] [PubMed]
21. Baermann, G. Ein einfache Methode zur Auffindung von Anklyostomum (Nematoden) Larven in Erdproben. *Ned. Tijdschr. Geneeskd.* **1917**, *57*, 131–137.
22. Futai, K.; Furuno, T. The variety of resistances among pine species to pine wood nematode, Bursaphelenchus lignicolus. *Bull. Kyoto Univ. For.* **1979**, *51*, 23–36.
23. Le Provost, G.; Herrera, R.; Paiva, J.A.; Chaumeil, P.; Salin, F.; Plomion, C. A micromethod for high throughput RNA extraction in forest trees. *Biol. Res.* **2007**, *40*, 291–297. [CrossRef] [PubMed]
24. FastQC—A Quality Control Tool for High Throughput Sequence Data. Available online: http://www.bioinformatics.babraham.ac.uk/projects/fastqc (accessed on 20 September 2016).
25. Sickle: A Windowed Adaptive Trimming Tool for FASTQ Files Using Quality. Available online: https://github.com/najoshi/sickle (accessed on 20 September 2016).

26. Grabherr, M.G.; Haas, B.J.; Yassour, M.; Levin, J.Z.; Thompson, D.A.; Amit, I.; Adiconis, X.; Fan, L.; Raychowdhury, R.; Zeng, Q.; et al. Full-length transcriptome assembly from RNA-Seq data without a reference genome. *Nat. Biotechnol.* **2011**, *29*, 644–652. [CrossRef] [PubMed]

27. Huang, X.; Madan, A. CAP3: A DNA sequence assembly program. *Genome Res.* **1999**, *9*, 868–877. [CrossRef] [PubMed]

28. Haas, B. TransDecoder (Find Coding Regions Within Transcripts). Available online: http://transdecoder. github.io (accessed on 16 May 2016).

29. Boeckmann, B.; Bairoch, A.; Apweiler, R.; Blatter, M.-C.; Estreicher, A.; Gasteiger, E.; Martin, M.J.; Michoud, K.; O'Donovan, C.; Phan, I.; et al. The SWISS-PROT protein knowledgebase and its supplement TrEMBL in 2003. *Nucleic Acids Res.* **2003**, *31*, 365–370. [CrossRef] [PubMed]

30. Finn, R.D.; Coggill, P.; Eberhardt, R.Y.; Eddy, S.R.; Mistry, J.; Mitchell, A.L.; Potter, S.C.; Punta, M.; Qureshi, M.; Sangrador-Vegas, A.; et al. The Pfam protein families database: Towards a more sustainable future. *Nucleic Acids Res.* **2015**, *44*, D279–D285. [CrossRef] [PubMed]

31. Altschul, S.F.; Gish, W.; Miller, W.; Myers, E.W.; Lipman, D.J. Basic local alignment search tool. *J. Mol. Biol.* **1990**, *215*, 403–410. [CrossRef]

32. Eddy, S.R. Multiple alignment using hidden Markov models. *Proc. Int. Conf. Intell. Syst. Mol. Biol.* **1995**, *3*, 114–120. [PubMed]

33. Srivastava, A.; Sarkar, H.; Gupta, N.; Patro, R. RapMap: A Rapid, Sensitive and Accurate Tool for Mapping RNA-seq Reads to Transcriptomes. *Bioinformatics* **2016**, *32*, i192–i200. [CrossRef] [PubMed]

34. Li, H.; Handsaker, B.; Wysoker, A.; Fennell, T.; Ruan, J.; Homer, N.; Marth, G.; Abecasis, G.; Durbin, R. The Sequence Alignment/Map format and SAMtools. *Bioinformatics* **2009**, *25*, 2078–2079. [CrossRef] [PubMed]

35. Robinson, M.D.; McCarthy, D.J.; Smyth, G.K. edgeR: A Bioconductor package for differential expression analysis of digital gene expression data. *Bioinformatics* **2009**, *26*, 139–140. [CrossRef] [PubMed]

36. Robinson, M.D.; Oshlack, A. A scaling normalization method for differential expression analysis of RNA-seq data. *Genome Biol.* **2010**, *11*, R25. [CrossRef] [PubMed]

37. McCarthy, D.J.; Chen, Y.; Smyth, G.K. Differential expression analysis of multifactor RNA-Seq experiments with respect to biological variation. *Nucleic Acids Res.* **2012**, *40*, 4288–4297. [CrossRef] [PubMed]

38. Sebastiana, M.; Vieira, B.; Lino-Neto, T.; Monteiro, F.; Figueiredo, A.; Sousa, L.; Pais, M.S.; Tavares, R.; Paulo, O.S. Oak Root Response to Ectomycorrhizal Symbiosis Establishment: RNA-Seq Derived Transcript Identification and Expression Profiling. *PLoS ONE* **2014**, *9*, e98376. [CrossRef] [PubMed]

39. Yoav, B.; Yosef, H. Controlling the False Discovery Rate: A Practical and Powerful Approach to Multiple Testing. *J. R. Stat. Soc.* **1995**, *57*, 289–300.

40. Untergasser, A.; Nijveen, H.; Rao, X.; Bisseling, T.; Geurts, R.; Leunissen, J.A.M. Primer3Plus, an enhanced web interface to Primer3. *Nucleic Acids Res.* **2007**, *35*, W71–W74. [CrossRef] [PubMed]

41. Quevillon, E.; Silventoinen, V.; Pillai, S.; Harte, N.; Mulder, N.; Apweiler, R.; Lopez, R. InterProScan: Protein domains identifier. *Nucleic Acids Res.* **2005**, *33*, W116–W120. [CrossRef] [PubMed]

42. Jones, P.; Binns, D.; Chang, H.-Y.; Fraser, M.; Li, W.; McAnulla, C.; McWilliam, H.; Maslen, J.; Mitchell, A.; Nuka, G.; et al. InterProScan 5: Genome-scale protein function classification. *Bioinformatics* **2014**, *30*, 1236–1240. [CrossRef] [PubMed]

43. Ashburner, M.; Ball, C.A.; Blake, J.A.; Botstein, D.; Butler, H.; Cherry, J.M.; Davis, A.P.; Dolinski, K.; Dwight, S.S.; Eppig, J.T.; et al. Gene ontology: Tool for the unification of biology. The Gene Ontology Consortium. *Nat. Genet.* **2000**, *25*, 25–29. [CrossRef] [PubMed]

44. Kanehisa, M.; Goto, S. KEGG: kyoto encyclopedia of genes and genomes. *Nucleic Acids Res.* **2000**, *28*, 27–30. [CrossRef] [PubMed]

45. Zhi-Liang, H.; Jie, B.; James, M.R. CateGOrizer: A Web-Based Program to Batch Analyze Gene Ontology Classification Categories. *Online J. Bioinform.* **2008**, *9*, 108–112.

46. Shannon, P.; Markiel, A.; Ozier, O.; Baliga, N.S.; Wang, J.T.; Ramage, D.; Amin, N.; Schwikowski, B.; Ideker, T. Cytoscape: A software environment for integrated models of biomolecular interaction networks. *Genome Res.* **2003**, *13*, 2498–2504. [CrossRef] [PubMed]

47. Maere, S.; Heymans, K.; Kuiper, M. BiNGO: A Cytoscape plugin to assess overrepresentation of gene ontology categories in biological networks. *Bioinformatics* **2005**, *21*, 3448–3449. [CrossRef] [PubMed]

48. Merico, D.; Isserlin, R.; Stueker, O.; Emili, A.; Bader, G.D. Enrichment map: A network-based method for gene-set enrichment visualization and interpretation. *PLoS ONE* **2010**, *5*, e13984. [CrossRef] [PubMed]

49. McKenna, A.; Hanna, M.; Banks, E.; Sivachenko, A.; Cibulskis, K.; Kernytsky, A.; Garimella, K.; Altshuler, D.; Gabriel, S.; Daly, M.; et al. The Genome Analysis Toolkit: A MapReduce framework for analyzing next-generation DNA sequencing data. *Genome Res.* **2010**, *20*, 1297–1303. [CrossRef] [PubMed]

50. Vogt, T. Phenylpropanoid Biosynthesis. *Mol. Plant* **2010**, *3*, 2–20. [CrossRef] [PubMed]

51. Gómez-Vásquez, R.; Day, R.; Buschmann, H.; Randles, S.; Beeching, J.R.; Cooper, R.M. Phenylpropanoids, phenylalanine ammonia lyase and peroxidases in elicitor-challenged cassava (Manihot esculenta) suspension cells and leaves. *Ann. Bot.* **2004**, *94*, 87–97. [CrossRef] [PubMed]

52. Sattler, S.E.; Funnell-Harris, D.L. Modifying lignin to improve bioenergy feedstocks: Strengthening the barrier against pathogens? *Front. Plant Sci.* **2013**, *4*, 70. [CrossRef] [PubMed]

53. Stermer, B.A.; Bianchini, G.M.; Korth, K.L. Regulation of HMG-CoA reductase activity in plants. *J. Lipid Res.* **1994**, *35*, 1133–1140. [PubMed]

54. Wright, L.P.; Phillips, M.A. Measuring the activity of 1-deoxy-D-xylulose 5-phosphate synthase, the first enzyme in the MEP pathway, in plant extracts. *Methods Mol. Biol.* **2014**, *1153*, 9–20. [CrossRef] [PubMed]

55. Dao, T.T.H.; Linthorst, H.J.M.; Verpoorte, R. Chalcone synthase and its functions in plant resistance. *Phytochem. Rev.* **2011**, *10*, 397–412. [CrossRef] [PubMed]

56. Gensheimer, M.; Mushegian, A. Chalcone isomerase family and fold: No longer unique to plants. *Protein Sci.* **2004**, *13*, 540–544. [CrossRef] [PubMed]

57. Liu, F.; Guo, J.; Bai, P.; Duan, Y.; Wang, X.; Cheng, Y.; Feng, H.; Huang, L.; Kang, Z. Wheat TaRab7 GTPase Is Part of the Signaling Pathway in Responses to Stripe Rust and Abiotic Stimuli. *PLoS ONE* **2012**, *7*, e37146. [CrossRef] [PubMed]

58. Navrot, N.; Collin, V.; Gualberto, J.; Gelhaye, E.; Hirasawa, M.; Rey, P.; Knaff, D.B.; Issakidis, E.; Jacquot, J.-P.; Rouhier, N. Plant Glutathione Peroxidases Are Functional Peroxiredoxins Distributed in Several Subcellular Compartments and Regulated during Biotic and Abiotic Stresses. *Plant Physiol.* **2006**, *142*, 1364–1379. [CrossRef] [PubMed]

59. Hanin, M.; Brini, F.; Ebel, C.; Toda, Y.; Takeda, S.; Masmoudi, K. Plant dehydrins and stress tolerance: Versatile proteins for complex mechanisms. *Plant Signal. Behav.* **2011**, *6*, 1503–1509. [CrossRef] [PubMed]

60. Wegrzyn, J.L.; Lee, J.M.; Tearse, B.R.; Neale, D.B. TreeGenes: A forest tree genome database. *Int. J. Plant Genom.* **2008**, 412875. [CrossRef] [PubMed]

61. Bassingthwaighte, J.B. Strategies for the Physiome Project. *Ann. Biomed. Eng.* **2000**, *28*, 1043–1058. [CrossRef] [PubMed]

62. Moffatt, B.A.; Ashihara, H. Purine and pyrimidine nucleotide synthesis and metabolism. *Arabidopsis Book* **2002**, *1*, e0018. [CrossRef] [PubMed]

63. Ling, H. Sequence analysis of GDSL lipase gene family in Arabidopsis thaliana. *Pak. J. Biol. Sci.* **2008**, *11*, 763–767. [CrossRef]

64. Oh, S., II; Park, A.R.; Bae, M.S.; Kwon, S.J.; Kim, Y.S.; Lee, J.E.; Kang, N.Y.; Lee, S.; Cheong, H.; Park, O.K. Secretome Analysis Reveals an Arabidopsis Lipase Involved in Defense against Alternaria brassicicola. *Plant Cell* **2005**, *17*, 2832–2847. [CrossRef] [PubMed]

65. Bommer, U.A.; Thiele, B.J. The translationally controlled tumour protein (TCTP). *Int. J. Biochem. Cell Biol.* **2004**, *36*, 379–385. [CrossRef]

66. Xiang, Y.; Song, M.; Wei, Z.; Tong, J.; Zhang, L.; Xiao, L.; Ma, Z.; Wang, Y. A jacalin-related lectin-like gene in wheat is a component of the plant defence system. *J. Exp. Bot.* **2011**, *62*, 5471–5483. [CrossRef] [PubMed]

67. Xu, L.; Liu, Z.-Y.; Zhang, K.; Lu, Q.; Liang, J.; Zhang, X.-Y. Characterization of the Pinus massoniana transcriptional response to Bursaphelenchus xylophilus infection using suppression subtractive hybridization. *Int. J. Mol. Sci.* **2013**, *14*, 11356–11375. [CrossRef] [PubMed]

68. Cao, J.; Lv, Y.; Hou, Z.; Li, X.; Ding, L. Expansion and evolution of thaumatin-like protein (TLP) gene family in six plants. *Plant Growth Regul.* **2016**, *79*, 299–307. [CrossRef]

69. Liu, J.-J.; Sturrock, R.; Ekramoddoullah, A.K.M. The superfamily of thaumatin-like proteins: Its origin, evolution, and expression towards biological function. *Plant Cell Rep.* **2010**, *29*, 419–436. [CrossRef] [PubMed]

70. Zalabák, D.; Pospíšilová, H.; Šmehilová, M.; Mrízová, K.; Frébort, I.; Galuszka, P. Genetic engineering of cytokinin metabolism: Prospective way to improve agricultural traits of crop plants. *Biotechnol. Adv.* **2013**, *31*, 97–117. [CrossRef] [PubMed]

71. Wani, S.H.; Kumar, V.; Shriram, V.; Sah, S.K. Phytohormones and their metabolic engineering for abiotic stress tolerance in crop plants. *Crop J.* **2016**, *4*, 162–176. [CrossRef]

72. Shin, H.; Lee, H.; Woo, K.S.; Noh, E.W.; Koo, Y.B.; Lee, K.J. Identification of genes upregulated by pinewood nematode inoculation in Japanese red pine. *Tree Physiol.* **2009**, *29*, 411–421. [CrossRef] [PubMed]

73. Bouché, N.; Yellin, A.; Snedden, W.A.; Fromm, H. Plant-specific calmodulin-binding proteins. *Annu. Rev. Plant Biol.* **2005**, *56*, 435–466. [CrossRef] [PubMed]

74. Nairn, C.J.; Lennon, D.M.; Wood-Jones, A.; Nairn, A.V.; Dean, J.F.D. Carbohydrate-related genes and cell wall biosynthesis in vascular tissues of loblolly pine (Pinus taeda). *Tree Physiol.* **2008**, *28*, 1099–1110. [CrossRef] [PubMed]

75. Parrotta, L.; Faleri, C.; Cresti, M.; Cai, G. Heat stress affects the cytoskeleton and the delivery of sucrose synthase in tobacco pollen tubes. *Planta* **2016**, *243*, 43–63. [CrossRef] [PubMed]

76. Belkhadir, Y.; Subramaniam, R.; Dangl, J.L. Plant disease resistance protein signaling: NBS–LRR proteins and their partners. *Curr. Opin. Plant Biol.* **2004**, *7*, 391–399. [CrossRef] [PubMed]

77. Doi, R.H.; Kosugi, A. Cellulosomes: Plant-cell-wall-degrading enzyme complexes. *Nat. Rev. Microbiol.* **2004**, *2*, 541–551. [CrossRef] [PubMed]

78. Christopher, L.P.; Yao, B.; Ji, Y. Lignin Biodegradation with Laccase-Mediator Systems. *Front. Energy Res.* **2014**, *2*, 12. [CrossRef]

79. Dubos, C. Drought differentially affects expression of a PR-10 protein, in needles of maritime pine (*Pinus pinaster* Ait.) seedlings. *J. Exp. Bot.* **2001**, *52*, 1143–1144. [CrossRef] [PubMed]

80. McHale, L.; Tan, X.; Koehl, P.; Michelmore, R.W. Plant NBS-LRR proteins: Adaptable guards. *Genome Biol.* **2006**, *7*, 212. [CrossRef] [PubMed]

81. Carna, M.; Repka, V.; Skupa, P.; Sturdik, E. Auxins in defense strategies. *Biologia (Bratisl)* **2014**, *69*, 1255–1263. [CrossRef]

82. Canonne, J.; Froidure-Nicolas, S.; Rivas, S. Phospholipases in action during plant defense signaling. *Plant Signal. Behav.* **2011**, *6*, 13–18. [CrossRef] [PubMed]

83. Kilili, K.G.; Atanassova, N.; Vardanyan, A.; Clatot, N.; Al-Sabarna, K.; Kanellopoulos, P.N.; Makris, A.M.; Kampranis, S.C. Differential roles of tau class glutathione S-transferases in oxidative stress. *J. Biol. Chem.* **2004**, *279*, 24540–24551. [CrossRef] [PubMed]

84. Preisig-müller, R.; Schwekendiek, A.; Brehm, I.; Reif, H.-J.; Kindl, H. Characterization of a pine multigene family containing elicitor-responsive stilbene synthase genes. *Plant Mol. Biol.* **1999**, *39*, 221–229. [CrossRef] [PubMed]

85. Kodan, A.; Kuroda, H.; Sakai, F. A stilbene synthase from Japanese red pine (Pinus densiflora): Implications for phytoalexin accumulation and down-regulation of flavonoid biosynthesis. *Proc. Natl. Acad. Sci. USA* **2002**, *99*, 3335–3339. [CrossRef] [PubMed]

86. Karlsson, M.; Stenlid, J.; Olson, Å. Identification of a superoxide dismutase gene from the conifer pathogen Heterobasidion annosum. *Physiol. Mol. Plant Pathol.* **2005**, *66*, 99–107. [CrossRef]

87. Cingolani, P.; Platts, A.; Wang, L.L.; Coon, M.; Nguyen, T.; Wang, L.; Land, S.J.; Lu, X.; Ruden, D.M. A program for annotating and predicting the effects of single nucleotide polymorphisms, SnpEff: SNPs in the genome of Drosophila melanogaster strain w1118; iso-2; iso-3. *Fly (Austin)* **2012**, *6*, 80–92. [CrossRef] [PubMed]

© 2017 by the authors. Licensee MDPI, Basel, Switzerland. This article is an open access article distributed under the terms and conditions of the Creative Commons Attribution (CC BY) license (http://creativecommons.org/licenses/by/4.0/).

![forests logo] *forests*

MDPI

Article

Long Non-Coding RNAs Responsive to Witches' Broom Disease in *Paulownia tomentosa*

Zhe Wang [1], Xiaoqiao Zhai [2], Yabing Cao [1], Yanpeng Dong [1,3] and Guoqiang Fan [1,3,*]

[1]   Institute of Paulownia, Henan Agricultural University, 95 Wenhua Road, Jinshui District, Zhengzhou 450002, China; wangzhe6636@foxmail.com (Z.W.); cyb201406@163.com (Y.C.); dongdyp@163.com (Y.D.)
[2]   Forestry Academy of Henan, Zhengzhou 450002, China; user7117@163.com
[3]   College of Forestry, Henan Agricultural University, 95 Wenhua Road, Jinshui District, Zhengzhou 450002, China
*   Correspondence: fanguoqiangdr@163.com; Tel.: +86-0371-6355-8606

Received: 16 August 2017; Accepted: 12 September 2017; Published: 15 September 2017

Abstract: Paulownia witches' broom (PaWB) disease caused by phytoplasmas is a fatal disease that leads to considerable economic losses. Long non-coding RNAs (lncRNAs) have been demonstrated to play critical regulatory roles in posttranscriptional and transcriptional regulation. However, lncRNAs and their functional roles remain poorly characterized in Paulownia. To identify lncRNAs and investigate their roles in the response to PaWB phytoplasmas, RNA sequencing was performed for healthy *Paulownia tomentosa*, PaWB-infected *P. tomentosa*, and for healthy and PaWB-infected *P. tomentosa* treated with 100 mg L^{-1} rifampicin. A total of 28,614 unique mRNAs and 3693 potential lncRNAs were identified. Comparisons between lncRNAs and coding genes indicated that lncRNAs tended to have shorter transcripts and fewer exon numbers, and displayed significant expression specificity. Based on our comparison scheme, 1063 PaWB-related mRNAs and 110 PaWB-related lncRNAs were identified; among them, 12 PaWB-related candidate target genes that were regulated by nine PaWB-related lncRNAs were characterized. This study provides the first catalog of lncRNAs expressed in Paulownia and gives a revealing insight into the molecular mechanism responsible for PaWB.

Keywords: PaWB; phytoplasma; lncRNAs; candidate target genes

1. Introduction

Phytoplasmas are minute bacteria without cell walls and are the causal agents of witches' broom disease in Paulownia. Witches' broom disease occurs in many countries and is the biggest threat to Paulownia production, causing serious economic losses [1,2]. In Paulownia, witches' broom disease has been found in saplings and big trees, and leads to malformations of branch, leaf, stem, and flower, such as witches' broom, yellowing, phloem necrosis, and phyllody, respectively [3,4].

Paulownia trees are native to China and have been introduced into many countries [5]. The superior traits of Paulownia, such as distinguished adaptive capacity to poor habitat, extremely fast growth, and high-quality timber, make it a popular tree species with important economic and ecological values [6]. Accordingly, Paulownia witches' broom (PaWB) disease has been studied by biologists for many years. Vitamin C plays an important role in the resistance to PaWB [7], and plant hormones are known to be related to the morphogenesis of PaWB [8]. The Sec protein translocation system [9], elongation factor EF-Tu [10], and two plasmids [11] of the PaWB phytoplasma have been analyzed. Transgenic technology has been used in the breeding of PaWB-resistant Paulownia [12]. Rifampicin treatment can make PaWB-infected Paulownia recover to normal morphology [13]. Furthermore, the rapid development of 'omics' has allowed for the investigation of the transcriptomes [3,14–18], microRNAs (miRNAs), degradomes [19–22], and proteomes [4,23] of Paulownia species, and the results have revealed changes after PaWB infection at transcriptional, post-transcriptional, and translational levels.

Long non-coding RNAs (lncRNAs) are generally defined as non-coding transcripts that are more than 200 nucleotides in length [24]. Large numbers of lncRNAs have been identified in *Arabidopsis thaliana* [25–27], rice [28,29], wheat [30,31], maize [32], cotton [33,34], *Brassica napus* [35], and *Populus* [36,37], and have been found to play important roles in growth, development, differentiation, and stress responses. LncRNAs involved in the responses of plants to biotic stresses have also been identified. For example, in wheat, four lncRNAs involved in the response to stripe rust pathogen infection have been identified [31]; in *Arabidopsis*, lncRNAs responsive to *Fusarium oxysporum* infection were revealed and played an important role in antifungal immunity [38]; and in tomato, lncRNAs have been identified as endogenous target mimics for miRNAs in response to Tomato yellow leaf curl virus infection [39], and a lncRNA (lncRNA16397) that conveys resistance to *Phytophthora infestans* by co-expressing glutaredoxin has also been detected [40]. However, in studies of PaWB, lncRNAs that play important roles in the Paulownia–phytoplasma interaction have not yet been identified.

In this study, we used next-generation deep-sequencing technologies to identify PaWB-related genes and lncRNAs among healthy *Paulownia tomentosa* (PT), PaWB-infected *P. tomentosa* (PTI), and PT and PTI treated with 100 mg L^{-1} rifampicin (PT-100 and PTI-100, respectively). Our results may help to clarify the molecular mechanisms of PaWB, and accelerate the process of preventing the disastrous effects of PaWB.

2. Materials and Methods

2.1. Plant Materials

All the biological materials used in this study were obtained from the Institute of Paulownia, Henan Agricultural University, China. The following four groups of *P. tomentosa* seedlings were set up: PT, PTI, PT-100, and PTI-100. The cultivation and rifampin treatment procedures were as described by Fan et al. [13]. The terminal buds from three individual plants were combined to form one biological replicate, and at least three biological replicates were used for each treatment.

2.2. Library Construction and Sequencing

Total RNA was extracted from the terminal buds of *P. tomentosa* using TRIzol reagent (Invitrogen, Carlsbad, CA, USA) and treated with RNase-free DNase (Promega, Madison, WI, USA) according to the manufacturer's instructions. The quality and quantity of total RNA were measured using an Agilent 2100 Bioanalyzer (Agilent Technologies, Santa Clara, CA, USA). Total RNA was used to construct RNA sequencing (RNA-seq) libraries using a TruSeq Stranded Total RNA preparation kit with RiboZero Plant (Illumina Inc., San Diego, CA, USA) according to the manufacturer's instructions. The libraries were sequenced on an Illumina HiSeq 2000 platform by the Beijing Genomics Institute, Shenzhen, Guang dong, China.

2.3. Genome Mapping and Transcript Assembly

Reads with more than 10% N (unable to determine base information), with adapter sequence, or of low quality were removed from the raw reads to obtain clean reads. To remove ribosomal RNA reads, the clean reads were mapped to the SILVA rRNA database (http://www.arb-silva.de/) using the short read alignment software SOAP2 (BGI, Shenzhen, China). Finally, the clean reads were mapped to the Paulownia genome sequence (http://paulownia.genomics.cn) using TopHat2 [41]. Cufflinks 2.0.0 was used to assemble the mapped reads [42]. Transcripts that were no less than 200 nucleotides in length were selected for further analysis.

2.4. LncRNAs Identification

Based on the assembled transcripts annotated to the Paulownia genome sequence, transcripts encoding known proteins were identified and the unknown transcripts (<200 bp) were excluded. Non-coding transcripts were identified based on a coding potential score of less than 0, as calculated by the Coding Potential Calculator (CPC; http://cpc.cbi.pku.edu.cn). The candidate lncRNAs were classified into several categories, according to their genomic locations and previous descriptions of Sun et al. [43].

2.5. Genomic Characterization and Function Prediction of lncRNAs

We determined the distribution of the identified lncRNA on the Paulownia chromosomes according to the description of Li et al. [44]. To explore lncRNA conservation, all the lncRNA sequences identified in this study were aligned against the CANTATAdb (http://yeti.amu.edu.pl/CANTATA) using BLASTN (*E*-value < 1×10^{-5}). To identify lncRNAs that may act as miRNA precursors, we aligned the lncRNA sequences to the miRNA precursor sequences in miRBase (Release 21, http://www.mirbase.org/). The secondary structures of the lncRNA and miRNA precursor sequences were predicted using RNAfold in the Vienna RNA package (http://rna.tbi.univie.ac.at/). To explore the homology of the lncRNAs based on their consensus RNA secondary structures, we classified the predicted lncRNAs into different non-coding RNA families using INFERNAL [45].

We used an algorithm to search for potential *cis* target genes within a 10-kb region upstream and downstream of the identified lncRNAs. Potential trans target genes were detected based on sequence complementarity and RNA duplex energy prediction to assess the impact of lncRNA binding on complete mRNAs. BLAST was used to determine sequence complementarity (identify $\geq$ 95%, *E*-value < 0.05), and RNAplex was used to calculated the energy of potential mRNA–lncRNA interactions. The RNAplex parameters were set as e < −30. The criteria used for the prediction of potential trans targets are described in previous studies [36,46,47].

2.6. Identification of PaWB-Related Genes and lncRNAs

All the sequencing data were uploaded to SRA database with accession number SRP117715. Expression levels of genes and lncRNAs were measured as fragments per kilobase of exon model per million mapped reads (FPKM). Differentially expressed genes (DEGs) and differentially expressed lncRNAs (DELs) were identified according to the methods described in previous studies [36,40].

To identify genes and lncRNAs related to PaWB, we made comparisons among the four libraries (Figure 1). The following comparison schemes were considered: (1) DEGs and DELs in PT/PT-100 may be related to the influence of rifampicin; (2) DEGs and DELs in PTI/PTI-100 may be related to the influence of rifampicin and PaWB; (3) Differences between comparisons 1 and 2 may exclude DEGs and DELs related to the influence of rifampicin; (4) DEGs and DELs in PT/PTI may be related to PaWB; (5) The common DEGs and DELs between comparisons 3 and 4 may be related directly to PaWB.

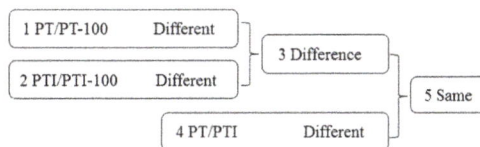

Figure 1. Comparison schemes of the four samples. PT represents the healthy wild-type sample of *P. tomentosa*, PTI represents the sample of phytoplasma infected PT. PT-100 represents the sample of 100 mg L^{-1} rifampicin treated PT, PTI-100 represents the sample of 100 mg L^{-1} rifampicin treated PTI. Different represents differentially expressed genes/long non-coding RNAs (lncRNAs). Difference represents the union of two groups, except their intersection. Same represents the intersection of two groups.

2.7. Nested-PCR and Quantitative RT-PCR

We applied nested-PCR to detect phytoplasma in seedlings according to the method described by Fan et al. [13]. Total RNA was reverse transcribed into cDNA and used to measure the expression of lncRNAs by quantitative real-time PCR (qRT-PCR). The qRT-PCRs were performed on a DNA Engine Opticon machine (MJ Research, Waltham, MA, USA) using a LightCycler FastStar DNA master SYBR Green I kit (Roche Diagnostics, Mannheim, Germany). The primers were designed using Primer Express 3.0 (Applied Biosystems, Stockholm, Sweden) and the specificity of primer pairs was checked by sequencing the PCR products. All qRT-PCR amplifications were carried out in triplicate, with the standard reaction program. The specificity of the amplified fragments was checked using the generated melting curve. The generated real-time data were analyzed using the Opticon Monitor Analysis Software 3.1 (Bio-Rad, Hercules, CA, USA) tool and standardized to the levels of 18S RNA using the $2^{-\Delta\Delta Ct}$ method [48]. The primers used for the qRT-PCRs are listed in Table S1.

3. Results

3.1. Changes in Morphology of the Four Samples

Compared with PT (Figure 2a), PTI (Figure 2b) showed typical PaWB symptom, such as witches' broom, and yellowing and relatively small leaves. In the rifampicin treated-samples (PT-100 and PTI-100), PT-100 (Figure 2c) showed no significant differences compared with PT. PTI-100 (Figure 2d) showed asymptomatic morphology. We applied nest-PCR to detect phytoplasma DNA in the four samples, and found that phytoplasma DNA was detected only in PTI (Figure S1).

Figure 2. Rifampin treated plantlets morphology. (**a**) PT; (**b**) PTI; (**c**) PT-100; (**d**) PTI-100.

3.2. High-Throughput Sequencing and Differentially Expressed Genes Analysis

Four cDNA libraries were constructed using the samples, and sequenced on an Illumina HiSeq 2000 platform. Among the clean reads, we identified 26,155, 27,017, 26,334, and 26,826 unique mRNAs from the PT, PTI, PT-100, and PTI-100 libraries, respectively (Table 1 and Table S2). A comparative summary of the mRNAs in the four libraries is provided as a Venn diagram (Figure S2a); 24,282 unique mRNAs were common among the four libraries, whereas 367, 337, 229, and 320 unique mRNAs were only presented in PT, PTI, PT-100, and PTI-100 libraries, respectively. In the PT/PT-100, PTI/PTI-100, and PT/PTI comparisons, we detected 1491 DEGs (706 upregulated and 785 downregulated), 1802 DEGs (725 upregulated and 1077 downregulated), and 3744 DEGs (2208 upregulated and 1536 downregulated), respectively (Table S2). Based on our comparison scheme (Figure 1), 2481 specific DEGs from comparison 1 (PT/PT-100) and comparison 2 (PTI/PTI-100) were identified (comparison 3), and 1063 common DEGs from comparison 3 and comparison 4 (PT/PTI) were PaWB-related genes. (Figure S3a and Table S2).

Table 1. Statistical data of the RNA-Seq reads.

	PT	PTI	PT-100	PTI-100
Raw reads	53,370,668	53,369,943	41,161,819	47,621,768
Clean reads	52,474,513	52,726,949	40,258,456	46,657,407
mRNAs	26,155	27,017	26,334	26,826
lncRNAs	3434	3504	3520	3479

3.3. Genome-Wide Identification and Characterization of lncRNAs in P. tomentosa

Based on our analysis, we identified 3434, 3504, 3520, and 3479 lncRNAs in the PT, PTI, PT-100, and PTI-100 libraries, respectively (Table S3). A comparative summary of the mRNAs in the four libraries is provided as a Venn diagram (Figure S2b); 3102 unique lncRNAs were common among the four libraries, whereas 10, 2, 1 and 3 unique lncRNAs were only presented in PT, PTI, PT-100, and PTI-100 libraries, respectively. A total of 3693 lncRNAs were revealed from the four libraries and classified into four categories: 41 into classcode "j", novel long noncoding isoforms with at least one splice junction shared with the reference genes; 73 into classcode "o", exonic overlap with a known transcript; 578 into classcode "x", exonic overlap with reference on the opposite strand; and 3001 into classcode "u", intergenic lncRNAs (Figure S4a). The INFERNAL analysis results showed that 493 of the unique lncRNAs belonged to 155 conserved lncRNAs families (Table S4). About half of the lncRNAs belonged to miRNA families, such as MIR159, MIR169, and MIR171, and five lncRNAs belonged to two small nucleolar RNA families (SNORD25 and SNORD27). Some non-plant-specific lncRNA families (X inactive specific transcript, small cajal body-specific RNA 8, fungal small nucleolar RNA, bacterial small transcript non-coding 150) were also found. Almost all the lncRNAs were distributed among the chromosomes and did not show significant chromosome location preferences (Figure 3a). The lengths of the lncRNAs were usually shorter than the lengths of the mRNAs, and more than 10% of all lncRNAs were ≤300, 300–400, or 400–500 nucleotides long, whereas about 10% of the mRNA were ≥3000 nucleotides in length (Figure 3b). Most lncRNAs have only exons, while most mRNAs have more than one exon (Figure 3c). All the unique lncRNAs were searched against the genomes of *Arabidopsis thaliana*, *Oryza sativa*, *Glycine max*, *Selaginella*, *Chlamydomonas*, *Physcomitrella*, *Amborella*, potato, *Vitis vinifera*, and *Zea mays* using BLAST. The results showed that only a small number of the *P. tomentosa* lncRNAs were conserved across the ten species; the highest number on lncRNAs matched *V. vinifera* lncRNAs, likely because Paulownia and *V. vinifera* are the most closely related in evolution (Table S5).

(a)

(b)

(c)

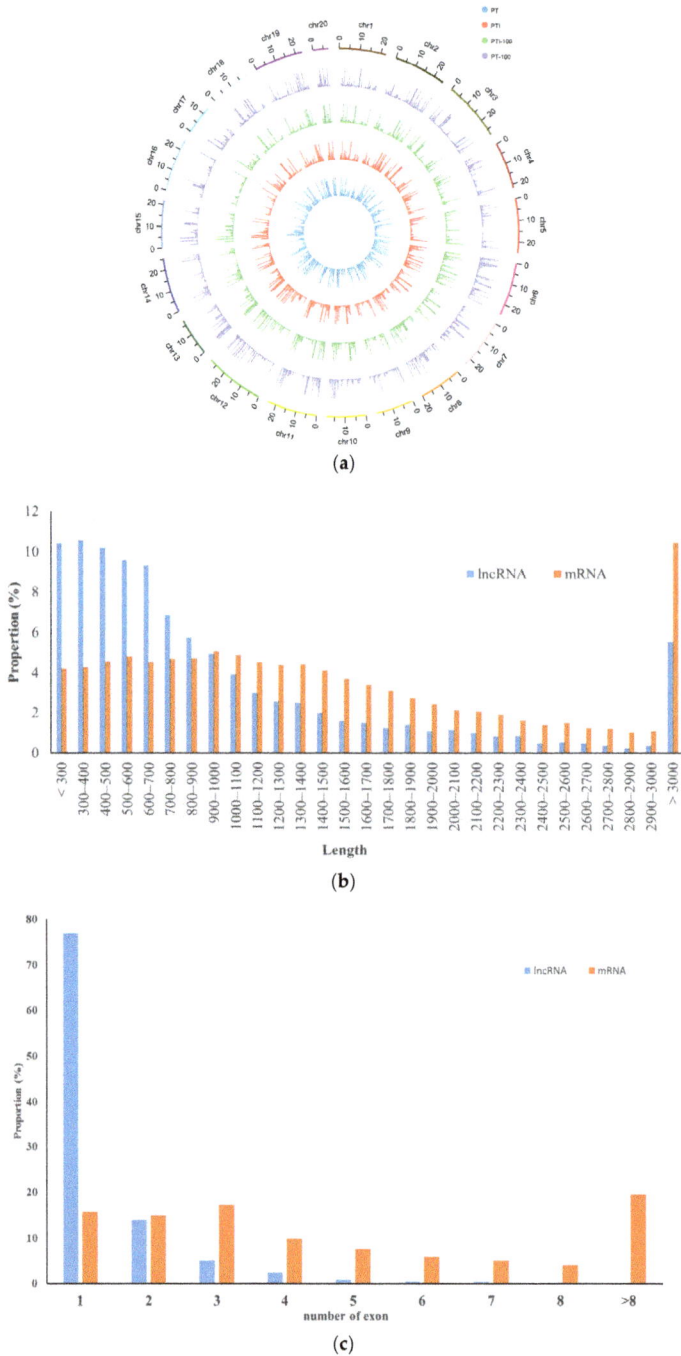

Figure 3. Characteristics of Paulownia lncRNAs. (a) Distribution of lncRNAs along each chromosome; (b) Transcript length distribution of lncRNAs and mRNAs; (c) Number of exons per transcripts for lncRNAs and mRNAs.

3.4. Differential Expression of lncRNAs

The expression levels of the 3693 lncRNAs in the four libraries were measured by FPKM. Based on log2 fold change >1 and $p < 0.05$, we identified 750 (200 upregulated and 550 downregulated), 123 (63 upregulated and 60 downregulated), and 173 (98 upregulated and 75 downregulated) differentially expressed lncRNAs (DELs) in PT/PTI, PT/PT-100, and PTI/PTI-100, respectively (Table S3).

3.5. Prediction of lncRNA Candidate Target Genes

Because lncRNAs play important roles in regulating gene expression, identification and analysis of their candidate target genes could help understand their functions. A total of 6746 lncRNA–mRNA pairs were identified (4175 were *cis*-regulated and 2571 were *trans*-regulated, Table S6). In total, we detected 5333 candidate target genes for 3222 lncRNAs; among them, 3218 were *cis* target genes for 2290 lncRNAs, and 2401 were *trans* target genes for 2571 lncRNAs. We found that one lncRNA could have more than one target gene, and one target gene could be targeted by one or more lncRNAs. Among the 3222 lncRNAs, 1246 had a single candidate target gene, while nine candidate target genes were predicted for each of two lncRNAs (TCONS_00019479 and TCONS_00030009). Furthermore, among the 5333 candidate target genes, 4313 were targeted by one lncRNA, while the other two target genes (TCONS_00013387 and TCONS_00019818) were targeted by more than one lncRNA (each of the target genes were targeted by 11 lncRNAs) (Table S6).

3.6. Analysis of PaWB-Related lncRNAs

Based on our comparisons (Figure 1), 234 specific DELs from comparison 1 (PT/PT-100) and comparison 2 (PTI/PTI-100) were identified (comparison 3), 110 common DELs from comparison 3 and comparison 4 (PT/PTI) were PaWB-related lncRNAs (Figure S3b and Table S3). Among the 110 PaWB-related DELs, 5, 30, and 75 lncRNAs were assigned classcode "o", "x", and "u", respectively (Figure S4b). *Cis*- or *trans*-regulated target genes were predicted for 80 of the 110 PaWB-related lncRNAs, and 3, 23, and 54 of them were assigned classcode "o", "x", and "u", respectively. A total of 147 lncRNA–mRNA pairs were identified for the 80 PaWB-related lncRNAs (91 pairs were *cis*-regulated and 56 were *trans*-regulated). In total, 138 candidate target genes were predicted for the 80 lncRNAs: 82 *cis* target genes for 58 lncRNAs, and 56 *trans* target genes for 56 lncRNAs.

We compared the 138 candidate targets of the 80 PaWB-related lncRNAs with the 1063 PaWB-related DEGs, and identified 12 PaWB-related candidate target genes among the DEGs that were regulated by nine PaWB-related lncRNAs (Table 2 and Figure 4). The KEGG (Kyoto Encyclopedia of Genes and Genomes) analysis results indicated that the 12 candidate target genes were mapped to eight pathways, namely "Phenazine biosynthesis", "RNA transport", "Cysteine and methionine metabolism", "Amino sugar and nucleotide sugar metabolism", "Glycerolipid metabolism", "Monoterpenoid biosynthesis".

Table 2. Paulownia witches' broom (PaWB)-related lncRNAs targeting PaWB-related mRNAs.

mRNA	Description	lncRNA
PAU008069.1	indol-3-yl-methylglucosinolate hydroxylase	TCONS_00007994
PAU008371.1	*trans*-2,3-dihydro-3-hydroxyanthranilate isomerase	TCONS_00009450
PAU017744.2	hypothetical protein	TCONS_00019908
TCONS_00022138	nucleoporin-like protein 2	
TCONS_00022139	xyloglucan:xyloglucosyl transferase	TCONS_00022242
TCONS_00022137	nucleoporin-like protein 2	

Table 2. *Cont.*

mRNA	Description	lncRNA
PAU012048.1	neomenthol dehydrogenase	TCONS_00026472
TCONS_00030164	uncharacterized protein	TCONS_00030312
PAU029256.2	DNA (cytosine-5)-methyltransferase	TCONS_00032704
PAU029162.1	hypothetical protein	TCONS_00032729
PAU029158.1	UDP-sulfoquinovose synthase	
PAU029184.1	hypothetical protein	TCONS_00032741

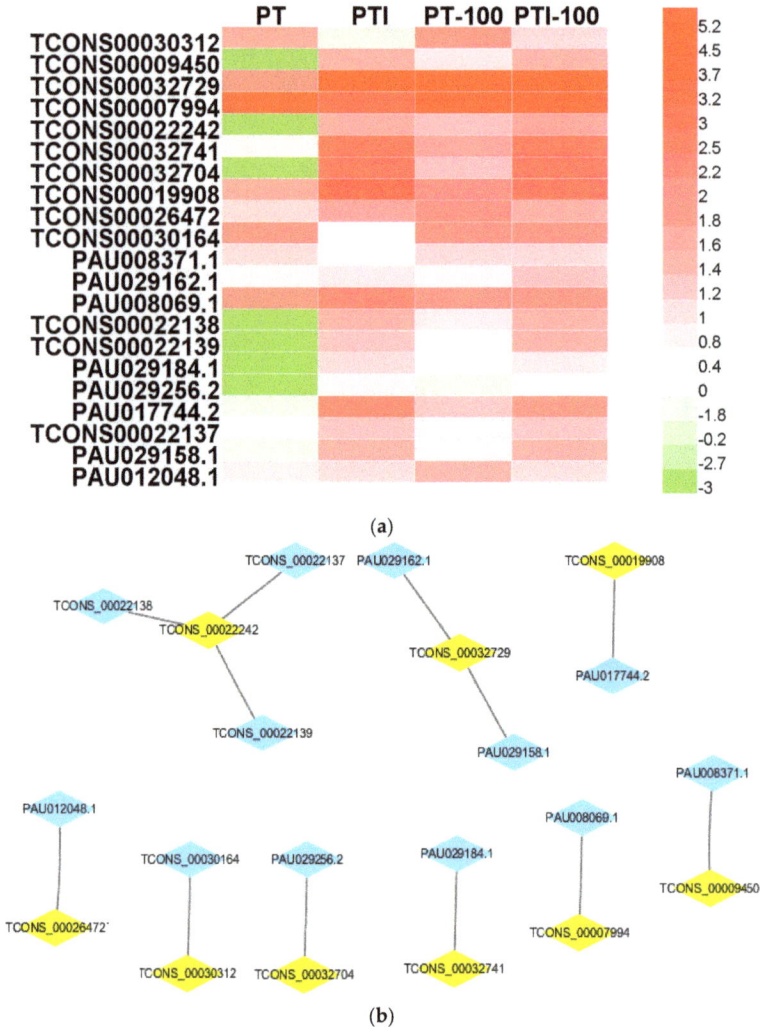

(a)

(b)

Figure 4. Characteristics of PaWB-related lncRNAs and mRNAs. (**a**) Expression profiling in PT, PTI PT-100, and PTI-100. The data were based on lg(FPKM); (**b**) LncRNA-mRNA networks. The blue polygons represent mRNAs, the yellow polygons represent lncRNAs. FPKM, fragments per kilobase of exon model per million mapped reads.

3.7. Quantitative RT-PCR Analysis of DEGs and DELs

To confirm the expression of the *P. tomentosa* lncRNAs and their response to PaWB, qRT-PCR analysis was performed to verify the results of the high-throughput RNA-seq. We randomly selected seven lncRNAs and six mRNAs for qRT-PCR validation. The results demonstrated that, except TCONS_00029942, the qRT-PCR results were consistent with those from the RNA-seq analysis (Figure 5). This correlation confirmed that the RNA-seq data were reliable.

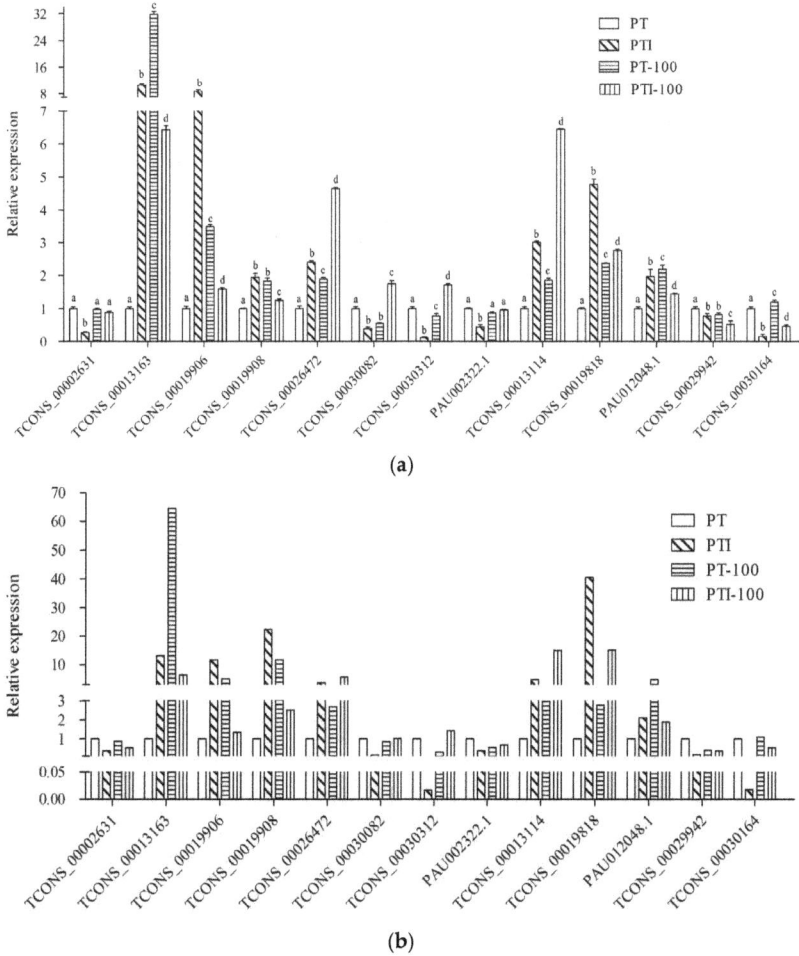

(a)

(b)

Figure 5. Expression pattern confirmation by qRT-PCR. (**a**) Changes in the relative expression levels as determined by qRT-PCR. Standard error of the mean for three technical replicates is represented by the error bars. Samples marked with various letters show a significant difference at $p < 0.05$; (**b**) Changes in the relative expression levels as determined by RNA-seq. The expression level of PT was set to 1.

4. Discussion

LncRNAs play key roles in various biological pathways in animals and plants. There have been recent advances in RNA-seq which, when combined with genome-wide mapping, have resulted in the identification of lncRNAs in plants. The identification of plant lncRNAs has opened up the investigation of novel regulatory pathways in plants. For example, 7655 lncRNAs from control and gibberellin-treated *Populus tomentosa* have been identified, indicating that lncRNAs may participate in auxin signal transduction and synthesis of cellulose and pectin, and may influence growth and wood properties. In the present study, we investigated transcriptomic changes in PT, PTI, PT-100, and PTI-100 seedlings, and systematically identified genes and lncRNAs associated with PaWB. Our results represent the first comprehensive analysis of lncRNAs in Paulownia. The identified lncRNAs will be useful for other woody plants researchers and provide an important resource for future functional genomics and regulatory expression studies.

The Paulownia lncRNAs had fewer exons and were shorter than mRNAs in length, which is consistent with other studies [35,44]. BLAST searches of the Paulownia lncRNAs against the genomes of species such as *Arabidopsis thaliana*, *Oryza sativa*, and *Glycine max* revealed that the majority of Paulownia lncRNAs were not conserved with lncRNAs in other species. The low conservation suggested that lncRNAs may undergo rapid evolution. It is not surprising that lncRNAs are not well conserved for various reasons. First, lncRNAs are not constrained by codon usage. Second, although lncRNAs may possess short conserved motifs, these short motifs are not easily identifiable by local alignment software such as BLAST. Third, some lncRNAs may interact directly with RNA-binding proteins through conserved secondary structures.

Xyloglucan (XG) is the main hemicellulose of the primary cell wall in dicotyledons. Xylosyl transferases (XT, EC:2.4.1.207) are a family of enzymes that catalyze xyloglucan endotransglucosylase and/or xyloglucan endohydrolase [49]. XTs can cut xyloglucan chains and transfer the fragment with the new reducing end either to another XG or to water, and play an important role in cell wall remodeling, affecting plant growth and development [50]. Transglycosylation can contribute to both cell wall assembly and cell wall loosening [51]. In this study, *XT* was the predicted target gene of lncRNA (TCONS_00022242), and it was upregulated in PT/PTI. XT may be related to witches' broom and hyperplasia because of its roles in cell wall assembly and loosening.

UDP-sulfoquinovose synthase (SQD1) catalyzes the transfer of sulfite to UDP-glucose to produce UDP-sulfoquinovose, which is the head group donor for the biosynthesis of the plant sulfolipid sulfoquinovosyldiacylglyerol (SQDG), a unique nonphosphorous lipid found in the photosynthetic membranes of plants [52]. *SQD1* (PAU029158.1) was the candidate target of PaWB-related lncRNA (TCONS_00032729), and it was upregulated in PT/PTI and downregulated in PTI/PTI-100. The changes in *SQD1* expression may influence SQDG and then photosynthesis. In addition, UDP-glucose is involved in many pathways (Figure 6), including Pentose and glucuronate interconversions (map00040), Galactose metabolism (map00052), Ascorbate and aldarate metabolism (map00053), Pyrimidine metabolism (map00240), Starch and sucrose metabolism (map00500), Amino sugar and nucleotide sugar metabolism (map00520), Neomycin, kanamycin and gentamicin biosynthesis (map00524), Glycerolipid metabolism (map00561), Zeatin biosynthesis (map00908), Biosynthesis of plant secondary metabolites (map01060), and Metabolic pathways (map01100). Therefore, changes in *SQD1* expression may influence UDP-glucose content and the pathways in which UDP-glucose is involved.

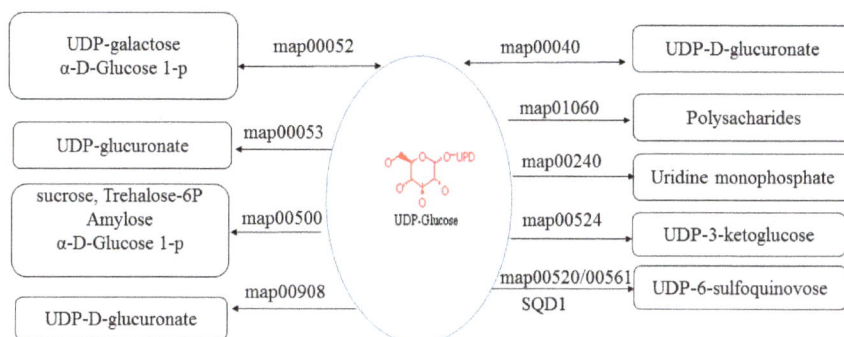

Figure 6. KEGG (Kyoto Encyclopedia of Genes and Genomes) pathways which UDP-glucose were involved in. map00052: Galactose metabolism; map00040: Pentose and glucuronate interconversions; map00053: Ascorbate and aldarate metabolism; map01060: Biosynthesis of plant secondary metabolites; map00240: Pyrimidine metabolism; map00500: Starch and sucrose metabolism; map00524: Neomycin, kanamycin and gentamicin biosynthesis; map00520: Amino sugar and nucleotide sugar metabolism; map00561: Glycerolipid metabolism; SQD1: UDP-sulfoquinovose synthase.

Plants constitutively synthesize a wide variety of secondary metabolites to aid fitness by preventing pathogen invasion [53]. Among these metabolites, terpenoids are frequently described as natural products that are active against pathogens [54]. Neomenthol dehydrogenase (MNR, PAU012048.1), which participates in the monoterpene synthesis of neomenthol and isomenthol, is a monoterpenoid dehydrogenase [55]. In *Capsicum annuum*, the menthone reductase gene *CaMNR1* regulates plant defenses against a broad spectrum of pathogens [56]. In our study, *MNR* was the target of a PaWB-related lncRNA (TCONS_00026472), and was upregulated in PT/PTI. The change in *MNR* expression may be in response to PaWB-phytoplasmas infection.

DNA methylation plays an essential role in regulating plant development [57]. DNA methyltransferase (Dnmt), an important enzyme for DNA methylation, is not only associated with DNA methylation, but is also linked to many important biological activities, including cell proliferation and senescence [58]. In previous studies, we found that PaWB may be related to changes in DNA methylation [59,60]. In this study, a PaWB-related gene, DNA (cytosine-5-)-methyltransferase (PAU029256.2, *Dnmt1*), was identified as the target gene of a PaWB-related lncRNA (TCONS_00032704), and was upregulated in PT/PTI. After infection, the change in *Dnmt1* expression might influence DNA methylation and result in the phenotype of infected Paulownia.

5. Conclusions

Although the functions of most lncRNAs are still unknown, the evidence suggests that lncRNAs may play regulatory roles by interacting with RNA, DNA, and protein-coding genes. For example, it has been reported that lncRNAs can act as *cis*- or *trans*-regulators of protein-coding gene expression in animals and plants. The lncRNAs detected in this study will lay the groundwork for future functional studies of lncRNAs in Paulownia. We revealed 26,155 and 27,017 genes, and 3434 and 3504 lncRNAs in healthy and PaWB-infected *P. tomentosa*, respectively. A total of 6746 lncRNA–mRNA pairs were identified (4175 were *cis*-regulated and 2571 were *trans*-regulated). In PT/PTI, 3744 DEGs (2208 upregulated and 1536 downregulated) and 750 (200 upregulated and 550 downregulated) were identified. According to our analysis, 1063 mRNAs and 110 lncRNAs were related to PaWB, in which 12 mRNAs were candidate target genes of nine lncRNAs. To understand the specific biological role of lncRNAs and their regulatory mechanisms in Paulownia, future research should include functional analyses of the lncRNA candidate target genes using overexpression, CRISPR-Cas9, or RNA interference gene silencing strategies.

Supplementary Materials: The following are available online at www.mdpi.com/1999-4907/8/9/348/s1, Figure S1: Detection of phytoplasma in Paulownia seedings. 1: PTI, 2: PTI -100, 3: PT, 4: PT-100, 5: ddH$_2$O, D: DNA maker. Figure S2: A Venn diagram showing mRNAs (a) and lncRNAs (b) that are commonly expressed in PT, PTI, PT-100, and PTI-100 samples as well as those specifically expressed under one but not the other. Figure S3: Details of the comparison schemes. (a) PaWB-related mRNAs; (b) PaWB-related lncRNAs. Figure S4: Classification of lncRNAs, (a) all-lncRNAs; (b) PaWB-related lncRNAs. class code 'u': intergenic lncRNAs, class code 'o': lncRNAs that had exonic overlap with a known transcript, class code 'x': lncRNAs that had exonic overlap with reference on the opposite strand, class code 'j': novel long noncoding isoforms with at least one splice junction shared with the reference genes; Table S1: Primer sequence used in this study. Table S2: The revealed mRNAs in this study. Table S3: The revealed lncRNAs in this study. Table S4: Classification of predicted lncRNAs into different ncRNA families. Table S5: The conservation of Paulownia lncRNAs. Table S6: The identified target genes in this study.

Acknowledgments: The authors wish to acknowledge the support received from Distinguished Talents Foundation of Henan Province of China (174200510001) and the Natural Science Foundation of Henan Province of China (162300410158).

Author Contributions: Z.W. and G.F. conceived and designed the experiments; Z.W. performed the experiments; Y.C. and Y.D. analyzed the data; X.Z. contributed reagents/materials/analysis tools; Z.W. wrote the paper.

Conflicts of Interest: The authors declare no conflict of interest.

References

1. Fan, G.; Zhai, X.; Qin, H.; Yang, X.; Dong, Z.; Jiang, J. Study on In Vitro plantlet regeneration of paulownia plant with witches' broom. *J. Henan Agric. Univ.* **2005**, *39*, 254–258.

2. Hogenhout, S.A.; Oshima, K.; Ammar el, D.; Kakizawa, S.; Kingdom, H.N.; Namba, S. Phytoplasmas: Bacteria that manipulate plants and insects. *Mol. Plant Pathol.* **2008**, *9*, 403–423. [CrossRef] [PubMed]

3. Fan, G.; Dong, Y.; Deng, M.; Zhao, Z.; Niu, S.; Xu, E. Plant-pathogen interaction, circadian rhythm, and hormone-related gene expression provide indicators of phytoplasma infection in *Paulownia fortunei*. *Int. J. Mol. Sci.* **2014**, *15*, 23141–23162. [CrossRef] [PubMed]

4. Wang, Z.; Liu, W.; Fan, G.; Zhai, X.; Zhao, Z.; Dong, Y.; Deng, M.; Cao, Y. Quantitative proteome-level analysis of paulownia witches' broom disease with methyl methane sulfonate assistance reveals diverse metabolic changes during the infection and recovery processes. *PeerJ* **2017**, *5*, e3495. [CrossRef] [PubMed]

5. Ipekci, Z.; Gozukirmizi, N. Direct somatic embryogenesis and synthetic seed production from paulownia elongata. *Plant Cell Rep.* **2003**, *22*, 16–24. [CrossRef] [PubMed]

6. Wang, Z.; Fan, G.; Dong, Y.; Zhai, X.; Deng, M.; Zhao, Z.; Liu, W.; Cao, Y. Implications of polyploidy events on the phenotype, microstructure, and proteome of paulownia australis. *PLoS ONE* **2017**, *12*, e0172633. [CrossRef] [PubMed]

7. Ju, G.; Wang, R.; Zhou, Y.; Dajin, R. Study on the relationship between the content of vitamin C and resistance of paulownia spp. To witches' broom. *For. Res.* **1996**, *9*, 421–424.

8. Wang, R.; Wang, S.; Sun, X. Effects of hormones on the morphogenesis of witches broom of *Paulownia* spp. *Sci. Silvae Sin.* **1981**, *17*, 281–286.

9. Yue, H.; Wu, K.; Wu, Y.; Zhang, J.; Sun, R. Cloning and Characterization of Three Subunits of the Phytoplasma Sec Protein Translocation System Associated with the Paulownia Witches'-Broom. *Plant Prot.* **2009**, *35*, 25–31.

10. Wang, J.; Zhu, X.; Gao, R.; Lin, C.; Li, Y.; Xu, Q.; Piao, C.; Li, X.; Li, H.; Tian, G. Genetic and serological analyses of elongation factor EF-TU of paulownia witches'-broom phytoplasma (16sri-d). *Plant Pathol.* **2010**, *59*, 972–981. [CrossRef]

11. Lin, C.; Zhou, T.; Li, H.; Fan, Z.; Li, Y.; Piao, C.; Tian, G. Molecular characterisation of two plasmids from paulownia witches'-broom phytoplasma and detection of a plasmid-encoded protein in infected plants. *Eur. J. Plant Pathol.* **2008**, *123*, 321–330. [CrossRef]

12. Du, T.; Wang, Y.; Hu, Q.; Chen, J.; Liu, S.; Huang, W.; Lin, M. Transgenic paulownia expressing shiva-1 gene has increased resistance to paulownia witches' broom disease. *J. Integr. Plant Biol.* **2005**, *47*, 1500–1506. [CrossRef]

13. Fan, G.; Zhang, B.; Zhai, X.; Liu, F.; Ma, Y.B.; Kan, S. Effects of rifampin on the changes of morphology and plant endogenous hormones of paulownia seedlings with witches' broom. *J. Henan Agric. Univ.* **2007**, *41*, 387–391.

14. Fan, G.; Cao, X.; Niu, S.; Deng, M.; Zhao, Z.; Dong, Y. Transcriptome, microrna, and degradome analyses of the gene expression of paulownia with phytoplamsa. *BMC Genom.* **2015**, *16*, 896. [CrossRef] [PubMed]

15. Fan, G.; Cao, X.; Zhao, Z.; Deng, M. Transcriptome analysis of the genes related to the morphological changes of paulownia tomentosa plantlets infected with phytoplasma. *Acta Physiol. Plant.* **2015**, *37*, 1–12. [CrossRef]

16. Fan, G.; Xu, E.; Deng, M.; Zhao, Z.; Niu, S. Phenylpropanoid metabolism, hormone biosynthesis and signal transduction-related genes play crucial roles in the resistance of paulownia fortunei to paulownia witches' broom phytoplasma infection. *Genes Genom.* **2015**, *37*, 913–929. [CrossRef]

17. Liu, R.; Dong, Y.; Fan, G.; Zhao, Z.; Deng, M.; Cao, X.; Niu, S. Discovery of genes related to witches broom disease in paulownia tomentosa x paulownia fortunei by a de novo assembled transcriptome. *PLoS ONE* **2013**, *8*, e80238. [CrossRef] [PubMed]

18. Mou, H.Q.; Lu, J.; Zhu, S.F.; Lin, C.L.; Tian, G.Z.; Xu, X.; Zhao, W.J. Transcriptomic analysis of paulownia infected by paulownia witches'-broom phytoplasma. *PLoS ONE* **2013**, *8*, e77217. [CrossRef] [PubMed]

19. Fan, G.; Cao, Y.; Deng, M.; Zhai, X.; Zhao, Z.; Niu, S.; Ren, Y. Identification and dynamic expression profiling of micrornas and target genes of paulownia tomentosa in response to paulownia witches' broom disease. *Acta Physiol. Plant.* **2016**, *39*, 28. [CrossRef]

20. Fan, G.; Niu, S.; Xu, T.; Deng, M.; Zhao, Z.; Wang, Y.; Cao, L.; Wang, Z. Plant-pathogen interaction-related micrornas and their targets provide indicators of phytoplasma infection in paulownia tomentosa x paulownia fortunei. *PLoS ONE* **2015**, *10*, e0140590. [CrossRef] [PubMed]

21. Fan, G.; Niu, S.; Zhao, Z.; Deng, M.; Xu, E.; Wang, Y.; Yang, L. Identification of micrornas and their targets in paulownia fortunei plants free from phytoplasma pathogen after methyl methane sulfonate treatment. *Biochimie* **2016**, *127*, 271–280. [CrossRef] [PubMed]

22. Niu, S.; Fan, G.; Deng, M.; Zhao, Z.; Xu, E.; Cao, L. Discovery of micrornas and transcript targets related to witches' broom disease in paulownia fortunei by high-throughput sequencing and degradome approach. *Mol. Genet. Genom.* **2016**, *291*, 181–191. [CrossRef] [PubMed]

23. Cao, X.; Fan, G.; Dong, Y.; Zhao, Z.; Deng, M.; Wang, Z.; Liu, W. Proteome profiling of paulownia seedlings infected with phytoplasma. *Front. Plant Sci.* **2017**, *8*, 342. [CrossRef] [PubMed]

24. Rinn, J.L.; Chang, H.Y. Genome regulation by long noncoding RNAs. *Annu. Rev. Biochem.* **2012**, *81*, 145–166. [CrossRef] [PubMed]

25. Ben Amor, B.; Wirth, S.; Merchan, F.; Laporte, P.; d'Aubenton-Carafa, Y.; Hirsch, J.; Maizel, A.; Mallory, A.; Lucas, A.; Deragon, J.M.; et al. Novel long non-protein coding RNAs involved in arabidopsis differentiation and stress responses. *Genome Res.* **2009**, *19*, 57–69. [CrossRef] [PubMed]

26. Liu, J.; Jung, C.; Xu, J.; Wang, H.; Deng, S.; Bernad, L.; Arenas-Huertero, C.; Chua, N.H. Genome-wide analysis uncovers regulation of long intergenic noncoding RNAs in arabidopsis. *Plant Cell* **2012**, *24*, 4333–4345. [CrossRef] [PubMed]

27. Wang, H.; Chung, P.J.; Liu, J.; Jang, I.C.; Kean, M.J.; Xu, J.; Chua, N.H. Genome-wide identification of long noncoding natural antisense transcripts and their responses to light in arabidopsis. *Genome Res.* **2014**, *24*, 444–453. [CrossRef] [PubMed]

28. Ding, J.; Lu, Q.; Ouyang, Y.; Mao, H.; Zhang, P.; Yao, J.; Xu, C.; Li, X.; Xiao, J.; Zhang, Q. A long noncoding rna regulates photoperiod-sensitive male sterility, an essential component of hybrid rice. *Proc. Natl. Acad. Sci. USA* **2012**, *109*, 2654–2659. [CrossRef] [PubMed]

29. Zhang, Y.C.; Liao, J.Y.; Li, Z.Y.; Yu, Y.; Zhang, J.P.; Li, Q.F.; Qu, L.H.; Shu, W.S.; Chen, Y.Q. Genome-wide screening and functional analysis identify a large number of long noncoding RNAs involved in the sexual reproduction of rice. *Genome Biol.* **2014**, *15*, 512. [CrossRef] [PubMed]

30. Xin, M.; Wang, Y.; Yao, Y.; Song, N.; Hu, Z.; Qin, D.; Xie, C.; Peng, H.; Ni, Z.; Sun, Q. Identification and characterization of wheat long non-protein coding RNAs responsive to powdery mildew infection and heat stress by using microarray analysis and SBS sequencing. *BMC Plant Biol.* **2011**, *11*, 61. [CrossRef] [PubMed]

31. Zhang, H.; Chen, X.; Wang, C.; Xu, Z.; Wang, Y.; Liu, X.; Kang, Z.; Ji, W. Long non-coding genes implicated in response to stripe rust pathogen stress in wheat (*Triticum aestivum* L.). *Mol. Biol. Rep.* **2013**, *40*, 6245–6253. [CrossRef] [PubMed]

32. Fan, C.; Hao, Z.; Yan, J.; Li, G. Genome-wide identification and functional analysis of lincrnas acting as mirna targets or decoys in maize. *BMC Genom.* **2015**, *16*, 793. [CrossRef] [PubMed]

33. Lu, X.; Chen, X.; Mu, M.; Wang, J.; Wang, X.; Wang, D.; Yin, Z.; Fan, W.; Wang, S.; Guo, L.; et al. Genome-wide analysis of long noncoding RNAs and their responses to drought stress in cotton (*Gossypium hirsutum* L.). *PLoS ONE* **2016**, *11*, e0156723. [CrossRef] [PubMed]

34. Wang, M.; Yuan, D.; Tu, L.; Gao, W.; He, Y.; Hu, H.; Wang, P.; Liu, N.; Lindsey, K.; Zhang, X. Long noncoding RNAs and their proposed functions in fibre development of cotton (*Gossypium* spp.). *New Phytol.* **2015**, *207*, 1181–1197. [CrossRef] [PubMed]

35. Joshi, R.K.; Megha, S.; Basu, U.; Rahman, M.H.; Kav, N.N. Genome wide identification and functional prediction of long non-coding RNAs responsive to sclerotinia sclerotiorum infection in *Brassica napus*. *PLoS ONE* **2016**, *11*, e0158784. [CrossRef] [PubMed]

36. Tian, J.; Song, Y.; Du, Q.; Yang, X.; Ci, D.; Chen, J.; Xie, J.; Li, B.; Zhang, D. Population genomic analysis of gibberellin-responsive long non-coding RNAs in populus. *J. Exp. Bot.* **2016**, *67*, 2467–2482. [CrossRef] [PubMed]

37. Shuai, P.; Liang, D.; Tang, S.; Zhang, Z.; Ye, C.Y.; Su, Y.; Xia, X.; Yin, W. Genome-wide identification and functional prediction of novel and drought-responsive lincrnas in populus trichocarpa. *J. Exp. Bot.* **2014**, *65*, 4975–4983. [CrossRef] [PubMed]

38. Zhu, Q.H.; Stephen, S.; Taylor, J.; Helliwell, C.A.; Wang, M.B. Long noncoding RNAs responsive to fusarium oxysporum infection in *Arabidopsis thaliana*. *New Phytol.* **2014**, *201*, 574–584. [CrossRef] [PubMed]

39. Wang, J.; Yu, W.; Yang, Y.; Li, X.; Chen, T.; Liu, T.; Ma, N.; Yang, X.; Liu, R.; Zhang, B. Genome-wide analysis of tomato long non-coding RNAs and identification as endogenous target mimic for microrna in response to tylcv infection. *Sci. Rep.* **2015**, *5*, 16946. [CrossRef] [PubMed]

40. Cui, J.; Luan, Y.; Jiang, N.; Bao, H.; Meng, J. Comparative transcriptome analysis between resistant and susceptible tomato allows the identification of lncrna16397 conferring resistance to phytophthora infestans by co-expressing glutaredoxin. *Plant J.* **2017**, *89*, 577–589. [CrossRef] [PubMed]

41. Trapnell, C.; Pachter, L.; Salzberg, S.L. Tophat: Discovering splice junctions with RNA-Seq. *Bioinformatics* **2009**, *25*, 1105–1111. [CrossRef] [PubMed]

42. Trapnell, C.; Roberts, A.; Goff, L.; Pertea, G.; Kim, D.; Kelley, D.R.; Pimentel, H.; Salzberg, S.L.; Rinn, J.L.; Pachter, L. Differential gene and transcript expression analysis of RNA-seq experiments with tophat and cufflinks. *Nat. Protoc.* **2012**, *7*, 562–578. [CrossRef] [PubMed]

43. Sun, L.; Zhang, Z.; Bailey, T.L.; Perkins, A.C.; Tallack, M.R.; Xu, Z.; Liu, H. Prediction of novel long non-coding RNAs based on RNA-Seq data of mouse klf1 knockout study. *BMC Bioinform.* **2012**, *13*, 331. [CrossRef] [PubMed]

44. Li, H.; Wang, Y.; Chen, M.; Xiao, P.; Hu, C.; Zeng, Z.; Wang, C.; Wang, J.; Hu, Z. Genome-wide long non-coding RNA screening, identification and characterization in a model microorganism chlamydomonas reinhardtii. *Sci. Rep.* **2016**, *6*, 34109. [CrossRef] [PubMed]

45. Nawrocki, E.P.; Kolbe, D.L.; Eddy, S.R. Infernal 1.0: Inference of rna alignments. *Bioinformatics* **2009**, *25*, 1335–1337. [CrossRef] [PubMed]

46. Jia, H.; Osak, M.; Bogu, G.K.; Stanton, L.W.; Johnson, R.; Lipovich, L. Genome-wide computational identification and manual annotation of human long noncoding RNA genes. *RNA* **2010**, *16*, 1478–1487. [CrossRef] [PubMed]

47. Tafer, H.; Hofacker, I.L. Rnaplex: A fast tool for RNA-RNA interaction search. *Bioinformatics* **2008**, *24*, 2657–2663. [CrossRef] [PubMed]

48. Livak, K.J.; Schmittgen, T.D. Analysis of relative gene expression data using real-time quantitative PCR and the 2(-delta delta c(t)) method. *Methods (San Diego, Calif.)* **2001**, *25*, 402–408. [CrossRef] [PubMed]

49. Van Sandt, V.S.; Guisez, Y.; Verbelen, J.P.; Vissenberg, K. Analysis of a xyloglucan endotransglycosylase/hydrolase (xth) from the lycopodiophyte selaginella kraussiana suggests that xth sequence characteristics and function are highly conserved during the evolution of vascular plants. *J. Exp. Bot.* **2006**, *57*, 2909–2922. [CrossRef] [PubMed]

50. Osato, Y.; Yokoyama, R.; Nishitani, K. A principal role for atxth18 in *Arabidopsis thaliana* root growth: A functional analysis using rnai plants. *J. Plant Res.* **2006**, *119*, 153–162. [CrossRef] [PubMed]

51. Thompson, J.E.; Fry, S.C. Restructuring of wall-bound xyloglucan by transglycosylation in living plant cells. *Plant J.* **2001**, *26*, 23–34. [CrossRef] [PubMed]

52. Shimojima, M.; Hoffmann-Benning, S.; Garavito, R.M.; Benning, C. Ferredoxin-dependent glutamate synthase moonlights in plant sulfolipid biosynthesis by forming a complex with SQD1. *Arch. Biochem. Biophys.* **2005**, *436*, 206–214. [CrossRef] [PubMed]

53. Kliebenstein, D.J. Secondary metabolites and plant/environment interactions: A view through *Arabidopsis thaliana* tinged glasses. *Plant Cell Environ.* **2004**, *27*, 675–684. [CrossRef]

54. Litvak, M.E.; Monson, R.K. Patterns of induced and constitutive monoterpene production in conifer needles in relation to insect herbivory. *Oecologia* **1998**, *114*, 531–540. [CrossRef] [PubMed]

55. Ringer, K.L.; McConkey, M.E.; Davis, E.M.; Rushing, G.W.; Croteau, R. Monoterpene double-bond reductases of the (−)-menthol biosynthetic pathway: Isolation and characterization of cdnas encoding (−)-isopiperitenone reductase and (+)-pulegone reductase of peppermint. *Arch. Biochem. Biophys.* **2003**, *418*, 80–92. [CrossRef]

56. Choi, H.W.; Lee, B.G.; Kim, N.H.; Park, Y.; Lim, C.W.; Song, H.K.; Hwang, B.K. A role for a menthone reductase in resistance against microbial pathogens in plants. *Plant Physiol.* **2008**, *148*, 383–401. [CrossRef] [PubMed]

57. Finnegan, E.J.; Peacock, W.J.; Dennis, E.S. DNA methylation, a key regulator of plant development and other processes. *Curr. Opin. Genet. Dev.* **2000**, *10*, 217–223. [CrossRef]

58. Bestor, T.H.; Verdine, G.L. DNA methyltransferases. *Curr. Opin. Cell Biol.* **1994**, *6*, 380–389. [CrossRef]

59. Cao, X.; Fan, G.; Deng, M.; Zhao, Z.; Dong, Y. Identification of genes related to paulownia witches' broom by AFLP and MSAP. *Int. J. Mol. Sci.* **2014**, *15*, 14669–14683. [CrossRef] [PubMed]

60. Cao, X.; Fan, G.; Zhao, Z.; Deng, M.; Dong, Y. Morphological changes of paulownia seedlings infected phytoplasmas reveal the genes associated with witches' broom through aflp and msap. *PLoS ONE* **2014**, *9*, e112533. [CrossRef] [PubMed]

© 2017 by the authors. Licensee MDPI, Basel, Switzerland. This article is an open access article distributed under the terms and conditions of the Creative Commons Attribution (CC BY) license (http://creativecommons.org/licenses/by/4.0/).

forests

MDPI

Article

De Novo Sequencing and Assembly Analysis of Transcriptome in *Pinus bungeana* Zucc. ex Endl.

Qifei Cai [1,2,3], Bin Li [1,2,3], Furong Lin [1,2,3], Ping Huang [1,2,3], Wenying Guo [1,2,3] and Yongqi Zheng [1,2,3,*]

1   State Key Laboratory of Tree Genetics and Breeding, Chinese Academy of Forestry, Beijing 100091, China; caiqifei0416@163.com (Q.C.); libin1200@163.com (B.L.); linfr888@163.com (F.L.); pippin09@163.com (P.H.); 13801288911@139.com (W.G.)
2   Research Institute of Forestry, Chinese Academy of Forestry, Beijing 100091, China
3   Key Laboratory of Tree Breeding and Cultivation of State Forestry Administration, Chinese Academy of Forestry, Beijing 100091, China
*   Correspondence: zyq8565@126.com; Tel: +86-010-6288-8565

Received: 24 January 2018; Accepted: 19 March 2018; Published: 20 March 2018

Abstract: To enrich the molecular data of *Pinus bungeana* Zucc. ex Endl. and study the regulating factors of different morphology controled by apical dominance. In this study, *de novo* assembly of transcriptome annotation was performed for two varieties of *Pinus bungeana* Zucc. ex Endl. that are obviously different in morphology. More than 147 million reads were produced, which were assembled into 88,092 unigenes. Based on a similarity search, 11,692 unigenes showed significant similarity to proteins from *Picea sitchensis* (Bong.) Carr. From this collection of unigenes, a large number of molecular markers were identified, including 2829 simple sequence repeats (SSRs). A total of 158 unigenes expressed differently between two varieties, including 98 up-regulated and 60 down-regulated unigenes. Furthermore, among the differently expressed genes (DEGs), five genes which may impact the plant morphology were further validated by reverse transcription quantitative polymerase chain reaction (RT-qPCR). The five genes related to cytokinin oxidase/dehydrogenase (CKX), two-component response regulator ARR-A family (ARR-A), plant hormone signal transduction (AHP), and MADS-box transcription factors have a close relationship with apical dominance. This new dataset will be a useful resource for future genetic and genomic studies in *Pinus bungeana* Zucc. ex Endl.

Keywords: *De novo*; *Pinus bungeana* Zucc. ex Endl.; transcriptome; assembly analysis

---

1. Introduction

Pinus bungeana Zucc. ex Endl. is an endemic species to China, with high economic and ecological values for landscaping and afforestation. Owing to long-term over-exploitation, natural forest of *Pinus bungeana* Zucc. ex Endl. has been severely damaged and natural populations are mainly found on the top of hills with poor conditions usually isolated and fragmented. Genetic diversity in populations is losing. The plant morphology, which is the most visible difference, has been commonly used for genetic diversity studies. Additionally, its regulation and secondary metabolites should also be focused on.

Sizable numbers of molecular markers and genes data from the genome are playing an increasingly important role in population genomic studies of fine-scale genetic variation and the genetic basis of traits [1]. Nevertheless, very limited genomic data is available on *Pinus bungeana* Zucc. ex Endl. Transcriptome sequencing is an efficient means to generate functional genomic level data for the plants [2,3]. Transcriptome sequencing is a developed approach to transcriptome profiling that uses deep-sequencing technologies [4]. Studies using this method have already altered our view of the extent and complexity of the eukaryotic transcriptome [5,6].

Apical dominance occurs when the shoot apex inhibits the growth of lateral buds, leading to the vertical growth of the plant [7]. Growing upward vertically is important for the tree. In this way, it can get more sun light to maintain photosynthesis [8]. It is best demonstrated via decapitation, which leads to apical dominance. Branching control is regulated by divergent mechanisms among different species [7,9,10]. The functions of hormones in apical dominance have been included in investigations for nearly five decades. Evidence from hormonal studies suggests that apically produced auxin indirectly suppresses axillary bud outgrowth that is promoted by cytokine originating from roots/shoots [11–13]. The mechanism of branching control is still the perplexing problem which is explored by many hypotheses including the classical hypothesis, the auxin transport hypothesis, and the bud transition hypothesis [11]. Exhibiting significant involvement with other hormones, cytokinins (CK) are signaling hormonal molecules that may have an essential impact in regulating cytokinesis, growth, and development in plants [14,15]. In addition, the CKX, a flavoenzyme irresistibly degrading CK into adenine/adenosine moiety, is critical for maintaining CK homeostasis in plants. Therefore, the CKX plays a very important role by means of controlling the balance of CK. The functions of the CKX gene, including pygmyism, high yield, and anti-aging, are reported in many plants, such as maize, rice, and *Arabidopsis thaliana* (Linn.) Heynh. in Holl & Heynh. Even though the precision of hormone content analyses in tissue has greatly improved in recent years, there have been no reports on the understanding of *Pinus spp.*

Next generation transcriptome sequencing was employed to characterize the transcriptome of *Pinus bungeana* Zucc. ex Endl. and to develop genomic resources to support further studies of the species A new variety of *Pinus bungeana* Zucc. ex Endl. named "Penxian", which loses apical dominance, has also been sequenced and analyzed to characterize the architecture regulation mechanism of this species at molecular level.

2. Materials and Methods

2.1. Materials

Bud samples of two varieties of *Pinus bungeana* Zucc. ex Endl. namely 'Tree No. 9' and 'Penxian' were collected from six-year-old seedlings grown in a glasshouse in Changping, Beijing, China. The variety 'Penxian' has no main stem due to loss of apical dominance (Figure 1). Three replicate experiments were made for each sample. After sampling, the buds were immediately frozen in liquid nitrogen and homogenized with a pestle and mortar.

2.2. Methods

2.2.1. Sample Preparation for Illumina Sequencing

Total RNA was extracted using TRIzol reagent (Invitrogen, Carlsbad, CA, USA) according to the manufacturer's protocol. The RNA samples were digested using DNase I at 37 °C for 30 min to remove potential genomic DNA contamination. RNA concentration and quality were assessed using a NanoDrop ND-1000 spectrophotometer (Thermo Scientific, Wilmington, DE, USA) and Agilent 2100 Bioanalyzer (Agilent Technologies, Santa Clara, CA, USA). Poly (A) mRNA was purified from 3 µg of total RNA using Oligo (dT) magnetic beads. The mRNA was fragmented using divalent cations at an elevated temperature. Double-stranded complementary to RNA (cDNA) was synthesized using a SuperScript double-stranded cDNA synthesis kit (Invitrogen, Carlsbad, CA, USA) with random hexamer primers (Illumina). These cDNA fragments were subjected to purification, end repair, and ligation to sequencing adapters. Ligation products were purified with magnetic beads and separated by agarose gel electrophoresis. A range of cDNA fragments (200 ± 25 bp) were excised from the gel and selected for PCR amplification as templates. The cDNA library was sequenced on a paired-end flow cell using IlluminaHiSeq 2500 platform (2×100 bp read length) by Biomarker Co. (Beijing, China).

Figure 1. The plant morphology of two Pinus bungeana varieties. (**a**) The new variety named Penxian; (**b**) the ordinary variety named 'Tree No. 9'.

2.2.2. *De novo* Assembly and Annotation

The raw paired-end reads were filtered to remove those containing adapters and reads with more than 5% unknown nucleotides or a low *Q*-value ($\leq$10) base more than 20%. Then, clean data from each sample were utilized to do RNA *de novo* assembly with Trinity [16]. All the assembled unigenes were annotated against the NCBI (National Center of Biotechnology Information) protein Nr (https://www.ncbi.nlm.nih.gov/refseq/) [17], Swiss-Prot (http://www.uniprot.org/) [18], COG (Phylogenetic classification of proteins encoded in complete genomes, https://www.ncbi.nlm.nih.gov/COG/) [19], Pfam (https://pfam.xfam.org/) [20] database using BLASTX (Basic Local Alignment Search Tool, https://blast.ncbi.nlm.nih.gov/Blast.cgi) [21] with a typical *E*-value threshold of 1.0×10^{-5}. Metabolic pathway analysis was performed using the KEGG (Kyoto Encyclopedia of Genes and Genomes, http://www.kegg.jp/) [22]. The BLAST2GO (https://www.blast2go.com/) [23] program was used to get GO (Gene Ontology, http://go-database-sql.org/) [24] annotations of unigenes for describing biological processes, molecular functions, and cellular components. The dataset is available from the NCBI Short Read Archive (SRA) with the accession number: SRX2024410 (https://www.ncbi.nlm.nih.gov/sra/SRX2024410).

2.2.3. Differential Expression Analysis and Functional Enrichment

To identify DEGs between different samples, the expression level of each unigene was calculated according to the fragments per kilobase of exon per million mapped reads (FPKM) method [6]. The R statistical package software DESeq was used for differential expression analysis with default parameters. The DEGs between different samples were restricted with FDR (False Discovery Rate) $\leq$ 0.001 and the absolute value of log2 Ratio $\geq$ 1. In addition, functional-enrichment analyses including GO and KEGG were performed to identify which DEGs were significantly enriched in GO terms and metabolic pathways at a Bonferroni-corrected *p*-value $\leq$ 0.05 compared with the whole-transcriptome background. KEGG pathway functional enrichment analysis was carried out by KOBAS (http://kobas.cbi.pku.edu.cn/) [25,26].

2.2.4. Prediction of Unigene Coding Regions

The coding sequence (CDS) for unigene was predicted by BlastX [21] and getorf [27]. The unigene sequences were searched against Nr [17], COG [19], KEGG [20], and Swiss-Prot [18] protein databases using BLASTX (*e*-value < 10^{-5}) [21]. Unigenes aligned to a higher priority database will not be aligned to a lower priority database. The best alignment results were used to determine the sequence direction of unigenes. When a unigene could not be aligned to any database, the getorf program was introduced to decide its sequence orientation and protein coding region prediction.

2.2.5. Detection of SSR

The SSR detection of unigenes was performed with MicroSAtellite (MISA) software (http://pgrc.ipk-gatersleben.de/misa/misa.html). We searched for SSRs with motifs ranging from mono- to hexa-nucleotides in size. The minimum of repeat units was set as follows: ten repeat units for mono-nucleotides, six for di-nucleotides, and five for tri-, tetra-, penta-, and hexa-nucleotides.

2.2.6. Differential Expression Level Analysis by RT-qPCR

To verify the reliability and accuracy of RNA-Seq-based expression level analysis, five transcripts (Table S1.) which may impact the plant morphology were selected from the contigs database and evaluated the gene expression level in the two *Pinus bungeana* Zucc. ex Endl. varieties which were described above. RT-qPCR experiments were performed in triplicate. The methods of RNA isolation and cDNA synthesis were the same as described for RNA-Seq above. The specific primer (Table S2.) set for RT-qPCR was designed with the Oligo 7.57 software (MBI, Colorado Springs, CO, USA). The cDNA was diluted and amplified using a 7500 Fast Real-Time PCR Systems (Thermo Fisher, Waltham, MA, USA) and SYBR Premix Ex Taq kit (Thermo Fisher, Waltham, MA, USA) according to the manufacturer's instructions. Standard curves for individual unigenes were calculated using a dilution series of the pMD19-T vector containing the target gene. The relative expressions of the genes in the cold-treated seedlings compared with the control ones were determined based on comparisons with the reference gene ubiquitin-conjugating enzyme. Therefore, the amplification of the target gene was normalized with the amplification of the reference gene to correct amplification variations. The relative expression data were calculated according to the $2^{-\Delta\Delta Ct}$ method and are presented as fold change [28].

3. Results

3.1. Sequencing Data Quality and Assembly of Reads

In total, approximately 148 million paired end reads were generated by high throughput sequencing. After removing adapters, low-quality sequences, and ambiguous reads, we obtained approximately 25 million, 25 million, 25 million, 24 million, 25 million, and 23 million clean reads from the ordinary variety samples (Tree No. 9-1, Tree No. 9-2, and Tree No. 9-3), and new variety samples (Penxian-1, Penxian-2, and Penxian-3), respectively. Approximately 37.21 Gb bases were obtained with 91.56% Q30 bases with a GC content of 45.85% (Table 1). Trinity was employed for transcriptome *de novo* assembly with next-generation short-read sequences. A total of 124,692 transcripts with a mean length of 922 were assembled. Additionally, we further clustered the transcripts into 88,092 unigenes with a mean length of 758 bp and an N50 value of 1357 bp. Among the unigenes, 51,803 (58.81%) were between 200–500 bp, 16,917 (19.20%) were between 500–1000 bp, and 19,372 (21.99%) unigenes were greater than 1 kb. The shorter sequences, which were less than 200 bp in length, were excluded, as they may lack a characterized protein domain or be too short to show a sequence that matches the contigs (Table 2). The length distributions of the unigenes are shown in Figure 2.

Table 1. Statistical table of all samples sequencing data.

Samples	Read Number	Base Number	GC Content	% ≥ Q30
Tree No. 9-1	24,713,913	6,226,575,673	45.77	91.77
Tree No. 9-2	25,402,758	6,399,644,060	45.78	91.68
Tree No. 9-3	24,731,097	6,230,626,717	45.64	91.45
Penxian-1	24,088,230	6,068,870,102	46.48	91.26
Penxian-2	25,067,325	6,315,681,345	45.88	91.68
Penxian-3	23,679,887	5,966,576,473	45.53	91.51
Total/average	147,683,210	37,207,974,370	45.85	91.56

Table 2. Unigenes Quantity and Percentage of Length.

Unigenes Length	Total Number	Percentage
200–300	31,783	36.08%
300–500	20,020	22.73%
500–1000	16,917	19.20%
1000–2000	11,784	13.38%
2000+	7588	8.61%
Total Number	88,092	
Total Length	66,791,509	
N50 Length	1357	
Mean Length	758	

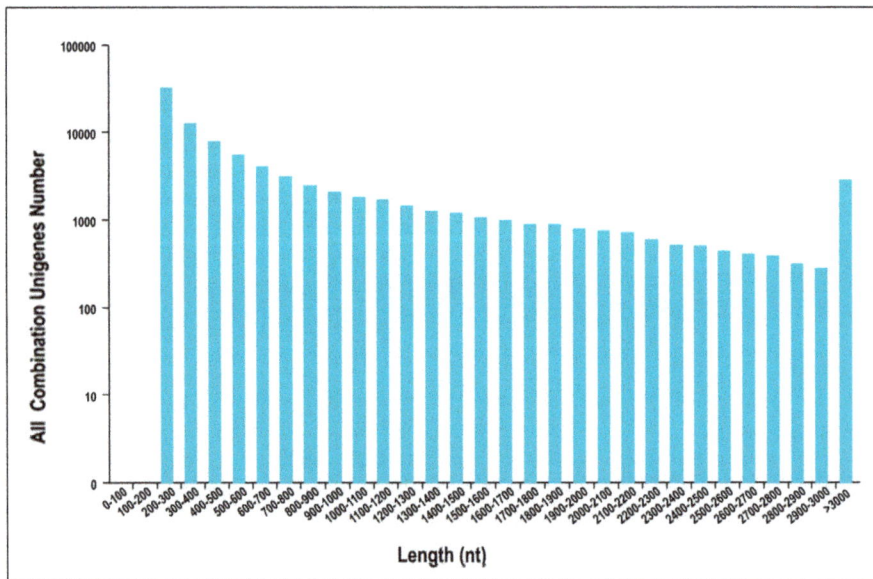

Figure 2. The length distributions of the unigenes.

3.2. Functional Annotation

Sequence annotations of all unigenes were predicted with BLAST [21] with an *E*-value threshold of 10^{-5} in the NCBI database of Nr [17], the Swiss-Prot [18] protein database, the KEGG [22] database, the COG [19] database, and the GO [24] database. Finally, after sequence annotation, 48.43% unigenes were predicted and 51.57% unigenes were still unknown. Among these annotated unigenes, 41,504 had significant matches in the Nr database, 29,287 had significant matches to the Pfam database , 25,679 had effective matches to the GO database, while 27,426 matched the Swiss-Prot database (Figure 3 and Table S3). In total, 42,667 unigenes were BLAST for homology searches, resulting in 48.43% unigenes showing similarity to known protein databases. For unigene sequences in the Nr annotations, BLAST search analysis further revealed that a total 11,692 had the most similar sequences to proteins from *Picea sitchensis*, followed by *Elaeis guineensis* Jacq., with a value of 2674, and *Amborella trichopoda* Baill., with a value of 2292 (Figure 4 and Table S4).

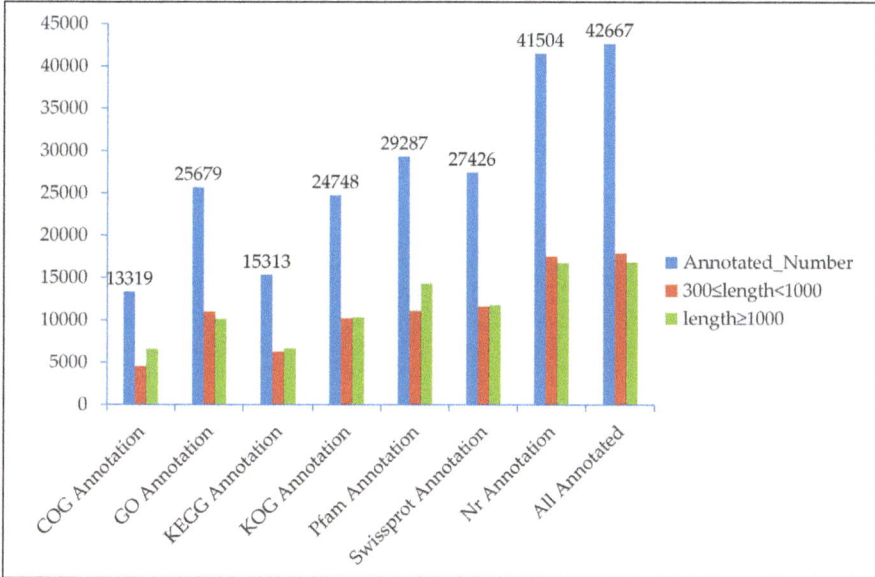

Figure 3. The Unigene Number Annotated in the Public Database Searches.

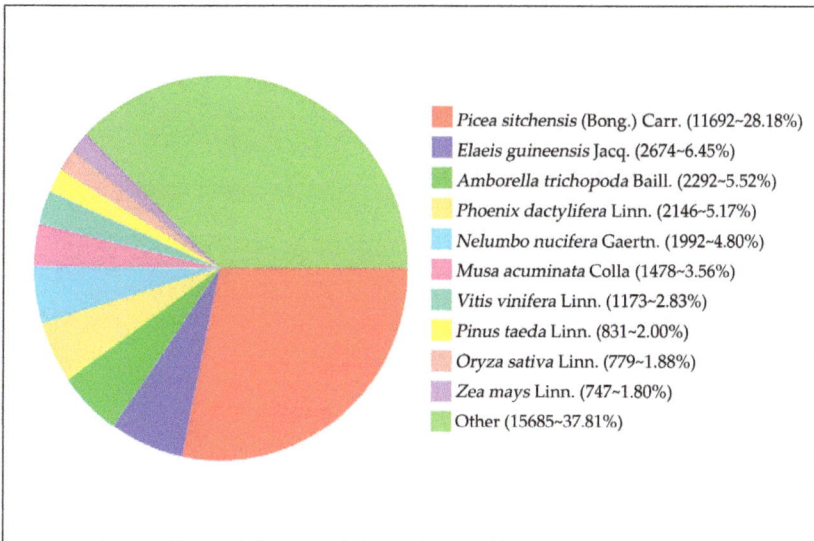

Figure 4. Nr Homologous Species Distribution.

3.3. Functional Classification

Based on the GO annotation, in total, 25,679 unigenes were assigned to 53 functional groups which were categorized into three main divisions (Figure 5 and Table S5). The Cellular components were classified into 17 groups, within which the unigenes were mostly enriched in the cell (12,836), the cell part (12,836), and the organelle (10,185). Additionally, the unique sequences were grouped into 16 groups which were included in the molecular function. The predominant groups were the catalytic activity (13,329) and binding (12,710). These were followed by transporter activity (1594) and structural molecule activity (879). In the biological process group, the matched unique sequences were categorized into 20 classes. In this group, the unigenes were highly matched to the metabolic process (17,168), followed closely by the cellular process (14,445), the single-organism process (11,988), the response to stimulus (5320), the biological regulation (4461), and the localization (3673) (Table S5).

All the assembled unigenes were searched in the COG database. A total of 18,516 matched unique sequences were categorized into 25 functional groups (Figure 6). The maximum group was replication, recombination, and repair (1773), followed by the transcription (1576), posttranslational modification (1447), translation, ribosomal structure, and biogenesis (1372), signal transduction mechanisms (1351), carbohydrate transport and metabolism (1049), and amino acid transport and metabolism (1015) (Figure 6). However, extracellular structures and nuclear structure respectively matched one and two unigenes (Figure 6). Groups with no concrete assignment, such as function unknown (536) and general function prediction only (3299), accounted for a part of unigenes (Figure 6).

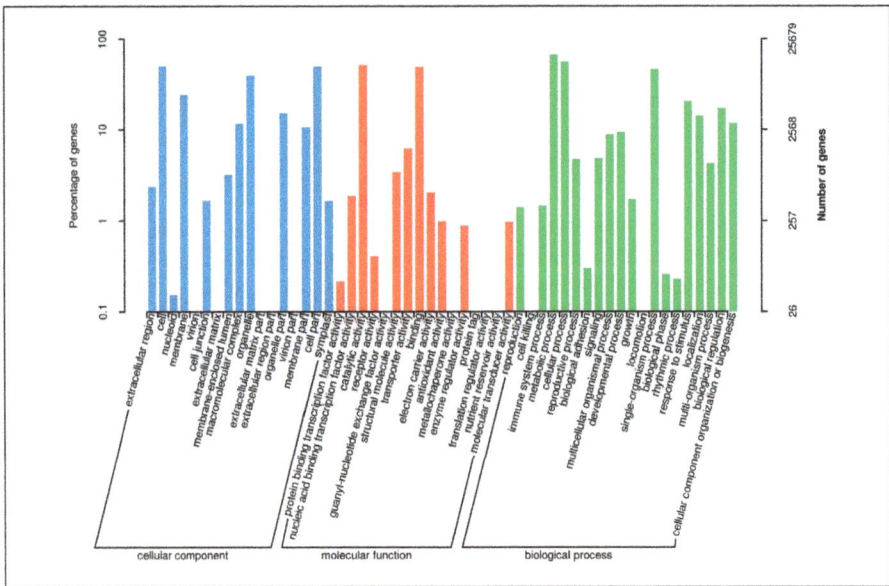

Figure 5. GO Classification of all unigenes.

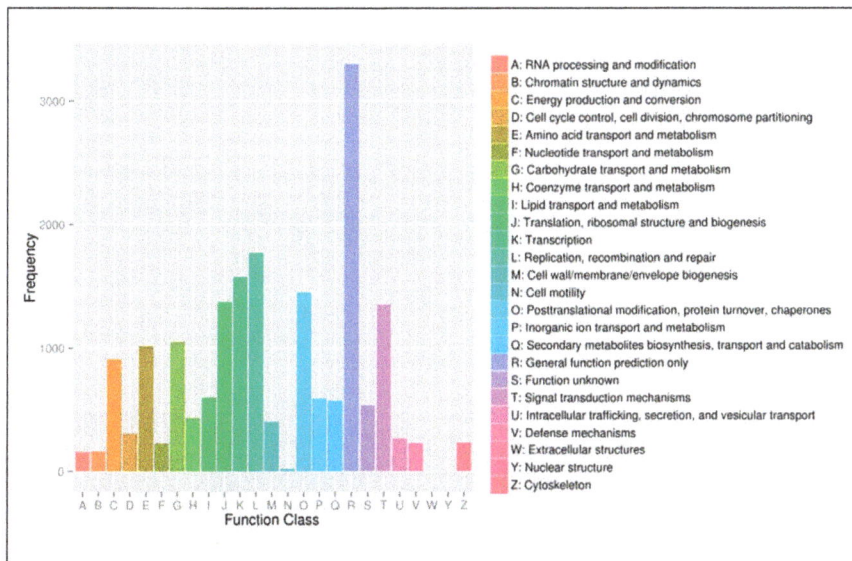

Figure 6. COG Function Classification of Consensus Sequence.

3.4. Simple Sequence Repeats (SSR) Loci Discovery

We used the generated 19,372 unigenes (≥1 kb) to exploit potential microsatellites by using MISA software. In total, this resulted in an authentication of 3323 putative microsatellites (including 1717 mononucleotides), covering 2829 nice SSR loci, and 163 loci in a compound formation. The most abundant repeat motif (excluding mononucleotides) was trinucleotide, 862, followed by dinucleotide, 646, tetra nucleotide, 76, penta nucleotide, 14, and hex nucleotide, 8 (Table 3 and Table S6). Further, we designed 2352 SSR primer pairs successfully using software Primer 3 (Table S7). These SSR loci will significantly contribute to developing polymorphic SSR markers in *Pinus* Zucc. ex Endl.

Table 3. The SSR Results.

Motif Types	Repeat Number								
	5	6	7	8	9	10	>10	Total	Percent
Dinucleotide	0	248	144	84	67	69	34	646	40.2%
Trinucleotide	562	197	86	16	1	0	0	862	53.7%
Tetra	66	8	0	0	2	0	0	76	4.7%
Penta	11	0	0	3	0	0	0	14	0.9%
Hexa	6	1	0	1	0	0	0	8	0.5%
Total	645	454	230	104	70	69	34	1606	100.0%
Percent	40.2%	28.3%	14.3%	6.5%	4.4%	4.3%	2.1%	100.0%	

3.5. DEGs between Two Samples

The analysis showed that a few of the unigenes were significantly affected between the two varieties. The Volcano plot showed the number of DEGs between the two samples (Figure 7). We identified 158 unigenes as DEGs, including 98 up-regulated and 60 down-regulated unigenes (Table 4 and Table S8).

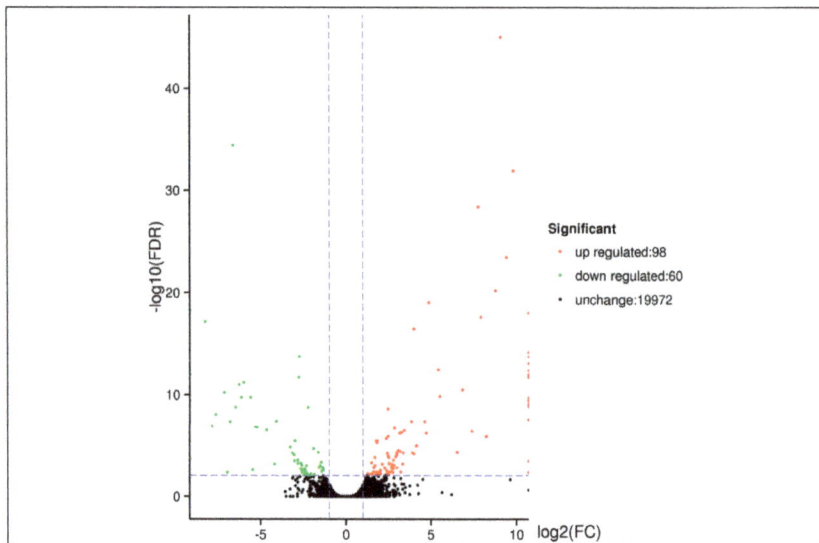

Figure 7. The Volcano Plot of DEGs.

Table 4. The DEG Unigenes Count.

DEG Set	All DEG	Up-Regulated	Down-Regulated
Tree No.9 vs. Penxian	158	98	60

3.6. Confirmation of Solexa Expression Patterns by RT-qPCR

There are different expression genes between the two varieties. As there is a difference between the two varieties in terms of morphology, we suggest that the auxin in them is different. Therefore, the genes for plant hormone, apical dominance, and genes transcriptional factor were identified in the KEGG pathways (Table 5 and Table S1) from the DEGs. The genes concerning the CKX were down-regulated. The unigenes relating to ARR-A were up-regulated. Moreover, the unigenes for AHP related up. To take the basics of DEGs of high-throughput sequencing further and confirm it, we selected five genes to validate the expression patterns of them by RT-qPCR. As it was shown in Figure 8, the expression patterns were highly consistent with the RNA-Seq analysis.

Table 5. The Unigenes in DEGs by KEGG.

Unigene Id	Pathway	Entry	Annotated Gene Name	Regulated	log2FC
c67959	ko00908	K00279	CKX	DOWN	−2.759074002
c49578	ko04075	K14492	ARR-A	UP	1.709013026
c102098	ko04075	K14490	AHP	UP	3.914710727
c121942 c91025		K09264	MADS-box transcription factor	DOWN	−1.385954738 −5.475015097

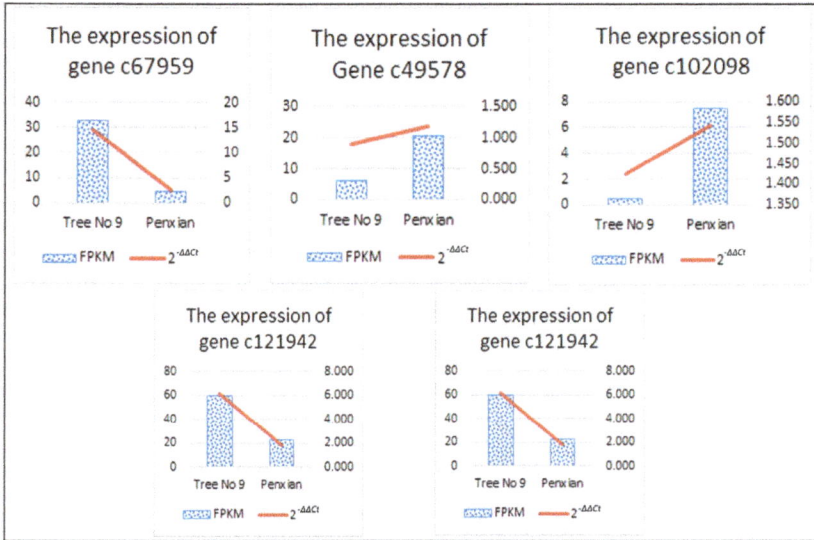

Figure 8. The expression of each gene by RT-qPCT and RNA-Seq.

4. Discussion

4.1. The Description of Pinus Bungeana Zucc. ex Endl. Transcriptome

Due to its low cost and high throughput, transcriptome sequencing is becoming a necessary method to perform gene discovery. The plants without a reference genome *de novo* transcriptome have been generated by illumine high throughput sequencing in recent years such as *Liriodendron chinense* (Hemsl.) Sarg. [29], *Corchorus capsularis* L. [30], *Camelina sativa* L. [31], and *Chlorophytum borivilianum* Santapau & R.R. Fern. [32], etc. The *Pinus spp.* are ancient species and important in modern landscape planting. In addition, the *Pinus bungeana* Zucc. ex Endl. is an endangered species [33]. Therefore, research on the new varieties of *Pinus bungeana* Zucc. ex Endl. is necessary. The new variety "Penxian" loses apical dominance completely, which is regulated by the factors related to plant hormones, including signal transduction and genes transcriptional factor. So, the transcriptome study not only enriched the lack of molecular data on *Pinus bungeana* Zucc. ex Endl., but also supplied materials for the research of apical dominance.

4.2. Transcriptome Assembly and Gene Annotation

In this paper, 42,667 (48.43%) unigenes out of 88,092 identified were annotated using BLAST searches of the public Nr [17], Swiss-Prot [18], GO [24], COG [19], and KEGG [22] databases. According to existing databases, more than 50% of unigenes were not annotated. Several reasons, such as assembly mistakes, short sequences not containing a protein domain, the deficiency of genomic information on the *Pinus app.*, the unigenes without hits probably belonging to untranslated regions, non-coding RNA, etc., may lead to this result [34]. The unigenes without hits may be considered putative novel transcribed sequences because of the lack of *Pinus app.* transcriptomic information. So, this means a large collection of unigenes of *Pinus* and expression patterns is necessary.

The annotation for each hit unigene met the GO annotation in terms of biological process, molecular function, and cellular components groups (Figure 5). The categories such as the nucleotide binding (GO: 0000166), DNA binding (GO: 0003677), zinc ion binding (GO: 0008270), protein binding (GO: 0005515), ATP binding (GO: 0005524), metal ion binding (GO: 0046872), and the structural constituent of ribosome (GO: 0003735) were well identified, indicating the need for a large number of transcripts related to various biochemical processes. These gene annotations will make contributions to the discovery of other sibling species in *Pinus app.*

4.3. Identification of Transcription Factor Involved in Pinus Bungeana Zucc. ex Endl. Architecture Regulation

Transcription factors usually relate to plant architecture regulation. Transcript profiling can be a valid tool for the characterization of apical dominance-related transcriptional factors genes. To gather information of transcription factors genes family in *Pinus bungeana* Zucc. ex Endl., we retrieved all members of transcription factors gene family from Pfam and searched against databases of *Pinus bungeana* Zucc. ex Endl. DEGs as mentioned above. In total, we found two *MADS-box* transcription factors in *Pinus bungeana* Zucc. ex Endl. (Table 5). The two transcription factors were down-regulated in "Penxian", but the opposite in "Tree No. 9". Among the transcription factors identified from our data, *MADS-box* is one of the most important transcription factors due to its roles in *Pinus bungeana* Zucc. ex Endl. architecture regulation.

4.4. DEGs between the Two Varieties

The differences between the two varieties are apparent. Also, 158 unigenes expressed differently. Actually, the previous study including the two lacebark pine varieties, using SSR markers, showed that genetic difference exists between the samples of "Penxian" and "Tree No. 9" [35]. So, this research also goes a step further on the basis of the previous study, to develop a more unambiguous understanding of the morphology of *Pinus bungeana*. Additionally, among the DEGs, CKX (K00279, EC1.5.99.12, Figure S1), ARR-A (K14492, Figure S2), and AHP (K14490, Figure S2) were related to apical dominance. The function of CKX in the regulation of plant form was reported [36]. In rice, reduced expression of CKX caused CK accumulation in inflorescence meristems and increased the number of reproductive organs [37]. In transgenic *Arabidopsis* plants, the over expression of CKX creased CK breakdown and resulted in diminished activity of the vegetative and floral apical meristems and leaf primordia, as well as the root meristematic cell activity related to plant form [38]. ARR-A and AHP are in the same pathway, plant hormone signal transduction. ARR-A increased cytokinin sensitivity and resulted in weak morphological phenotypes [39]. Furthermore, AHP is the previous process in the pathway to the result of Cell division and Shoot initiation (Ko04075, Figure S1). In this paper, the CKX was down-regulated, whilst ARR-A and AHP were up-regulated, which may be the main reason for the difference between the two varieties. However, apical dominance is very complex, which is connected with not only the auxin and CK and their interactions in controlling the buds' outgrowth, but also the buds' status of dormancy and sustained that would be easily influenced by the environmental factors, such as light, photoperiod, carbon acquisition, and nutrients [11,40]. Moreover, this paper may promote a research approach on the function of CK and ARR-A in *Pinus*.

5. Conclusions

In this study, 88,092 unigenes with an average length of 758 bp were obtained. In addition, 42,667 unigenes acquired were BLAST, resulting in 48.43% unigenes showing similarity to known protein databases. A total of 3323 putative microsatellites (including 1717 mononucleotides), covering 2829 perfect SSR loci, and 163 loci in a compound form. The DEGs related to CKX, ARR-A, AHP, and the MADS-box transcription factors between the two varieties are validated by RT-qPCR. This will provide support to for the classical hypothesis of apical dominance. The transcriptome data in this study will contribute to further gene discovery and functional exploration for *Pinus* significative characterization in deeper genetic breeding.

Supplementary Materials: The following are available online at http://www.mdpi.com/1999-4907/9/3/156/s1, Table S1: The detail sequence of the selected five transcripts, Table S2: The specific primer for RT-qPCR, Table S3: The Unigene Number Annotated in the Public Database Searches, Table S4: Nr Homologous Species Distribution, Table S5: GO Classification of all unigenes, Table S6: Frequency of identified SSR motifs, Table S7: The details of the 2352 SSR primer pairs, Table S8: The details of the 158 DEGs, Figure S1: The pathway map of ko00908,

Figure S2: The pathway map of ko04075. The dataset is available from the NCBI Short Read Archive (SRA) with an accession number SRX2024410 (https://www.ncbi.nlm.nih.gov/sra/SRX2024410).

Acknowledgments: We are grateful to the supporting funding from the National Natural Science Foundation of China (31370667), National Key Technology of China (2013BA001B06) and Inner Mongolia Hesheng Institute of Ecological Sciences & Technology (Hesheng2014).

Author Contributions: B.L. and Y.Z. conceived and designed the experiments; Q.C. performed the experiments, conducted the data analysis. and wrote the manuscript; F.L. contributed the plant material; P.H. analyzed the data; W.G. revised the manuscript.

Conflicts of Interest: The authors declare no conflict of interest.

References

1. Stinchcombe, J.R.; Hoekstra, H.E. Combining population genomics and quantitative genetics: Finding the genes underlying ecologically important traits. *Heredity* **2008**, *100*, 158–170. [CrossRef] [PubMed]
2. Andersen, J.R.; Lubberstedt, T. Functional markers in plants. *Trends Plant Sci.* **2003**, *8*, 554–560. [CrossRef] [PubMed]
3. Novaes, E.; Drost, D.R.; Farmerie, W.G.; Pappas, G.J., Jr.; Grattapaglia, D.; Sederoff, R.R.; Kirst, M. High-throughput gene and snp discovery in eucalyptus grandis, an uncharacterized genome. *BMC Genom.* **2008**, *9*, 312. [CrossRef] [PubMed]
4. Kim, J.S.; Moon, J.F.; Lim, S.H.; Kim, J.C. Applications of next generation sequencing in molecular ecology of non-model organisms. *Heredity* **2011**, *107*, 1–15. [CrossRef]
5. Wang, Z.; Gerstein, M.; Snyder, M. Rna-seq: A revolutionary tool for transcriptomics. *Nat. Rev. Genet.* **2009**, *10*, 57–63. [CrossRef] [PubMed]
6. Trapnell, C.; Williams, B.A.; Pertea, G.; Mortazavi, A.; Kwan, G.; van Baren, M.J.; Salzberg, S.L.; Wold, B.J.; Pachter, L. Transcript assembly and quantification by rna-seq reveals unannotated transcripts and isoform switching during cell differentiation. *Nat. Biotechnol.* **2010**, *28*, 511–515. [CrossRef] [PubMed]
7. Cline, M. Concepts and terminology of apical dominance. *Am. J. Bot.* **1997**, *84*, 1064. [CrossRef] [PubMed]
8. Cline, M.G. Execution of the auxin replacement apical dominance experiment in temperate woody species. *Am. J. Bot.* **2000**, *87*, 182–190. [CrossRef] [PubMed]
9. Tanaka, M.; Takei, K.; Kojima, M.; Sakakibara, H.; Mori, H. Auxin controls local cytokinin biosynthesis in the nodal stem in apical dominance. *Plant J. Cell Mol. Biol.* **2006**, *45*, 1028–1036. [CrossRef] [PubMed]
10. Reig, C.; Farina, V.; Mesejo, C.; Martínez-Fuentes, A.; Barone, F.; Agustí, M. Fruit regulates bud sprouting and vegetative growth in field-grown loquat trees (*Eriobotrya japonica* lindl.): Nutritional and hormonal changes. *J. Plant Growth Regul.* **2014**, *33*, 222–232. [CrossRef]
11. Dun, E.A.; Ferguson, B.J.; Beveridge, C.A. Apical dominance and shoot branching. Divergent opinions or divergent mechanisms? *Plant Physiol.* **2006**, *142*, 812–819. [CrossRef] [PubMed]
12. Booker, J.; Chatfield, S.; Leyser, O. Auxin acts in xylem-associated or medullary cells to mediate apical dominance. *Plant Cell* **2003**, *15*, 495–507. [CrossRef] [PubMed]
13. Tamas, I.A. *Hormonal Regulation of Apical Dominance*; Springer: Dordrecht, Netherlands, 1995; pp. 393–410. ISBN 978-0-7923-2985-5. [CrossRef]
14. Werner, T.; Motyka, V.; Strnad, M.; Schmulling, T. Regulation of plant growth by cytokinin. *Proc. Natl. Acad. Sci. USA* **2001**, *98*, 10487–10492. [CrossRef] [PubMed]
15. Werner, T.; Nehnevajova, E.; Kollmer, I.; Novak, O.; Strnad, M.; Kramer, U.; Schmulling, T. Root-specific reduction of cytokinin causes enhanced root growth, drought tolerance, and leaf mineral enrichment in arabidopsis and tobacco. *Plant Cell* **2010**, *22*, 3905–3920. [CrossRef] [PubMed]
16. Grabherr, M.G.; Haas, B.J.; Yassour, M.; Levin, J.Z.; Thompson, D.A.; Amit, I.; Adiconis, X.; Fan, L.; Raychowdhury, R.; Zeng, Q. Full-length transcriptome assembly from RNA-Seq data without a reference genome. *Nat. Biotechnol.* **2011**, *29*, 644. [CrossRef] [PubMed]
17. Deng, Y.; Jianqi, L.I.; Songfeng, W.U.; Zhu, Y.; Chen, Y.; Fuchu, H.E. Integrated nr database in protein annotation system and its localization. *Comput. Eng.* **2006**, *32*, 71–72.
18. Apweiler, R.; Bairoch, A.; Wu, C.H.; Barker, W.C.; Boeckmann, B.; Ferro, S.; Gasteiqer, E.; Huang, H.; Lopez, R.; Maqrane, M.; et al. Uniprot: The universal protein knowledgebase. *Nucleic Acids Res.* **2004**, *32*, D115–D119. [CrossRef] [PubMed]

19. Tatusov, R.L.; Galperin, M.Y.; Natale, D.A.; Koonin, E.V. The cog database: A tool for genome-scale analysis of protein functions and evolution. *Nucleic Acids Res.* **2000**, *28*, 33–36. [CrossRef] [PubMed]
20. Finn, R.D.; Coggill, P.; Eberhardt, R.Y.; Eddy, S.R.; Mistry, J.; Mitchell, A.L.; Potter, S.C.; Punta, M.; Qureshi, M.; Sangrador-Vegas, A.; et al. The pfam protein families database: Towards a more sustainable future. *Nucleic Acids Res.* **2016**, *44*, D279–D285. [CrossRef] [PubMed]
21. Altschul, S.F.; Madden, T.L.; Schaffer, A.A.; Zhang, J.; Zhang, Z.; Miller, W.; Lipman, D.J. Gapped blast and psi-blast: A new generation of protein database search programs. *Nucleic Acids Res.* **1997**, *25*, 3389–3402. [CrossRef] [PubMed]
22. Kanehisa, M.; Goto, S.; Kawashima, S.; Okuno, Y.; Hattori, M. The KEGG resource for deciphering the genome. *Nucleic Acids Res.* **2003**, *32*, D277. [CrossRef] [PubMed]
23. Conesa, A.; Gotz, S.; Garcia-Gomez, J.M.; Terol, J.; Talon, M.; Robles, M. Blast2go: A universal tool for annotation, visualization and analysis in functional genomics research. *Bioinformatics* **2005**, *21*, 3674–3676. [CrossRef] [PubMed]
24. Ashburner, M.; Ball, C.A.; Blake, J.A.; Botstein, D.; Butler, H.; Cherry, J.; Davis, A.; Dolinski, K.; Dwight, S.; Eppig, J.; et al. Gene ontology: Tool for the unification of biology. The gene ontology consortium. *Nat. Genet.* **2000**, *25*, 25–29. [CrossRef] [PubMed]
25. Xie, C.; Mao, X.; Huang, J.; Ding, Y.; Wu, J.; Dong, S.; Kong, L.; Gao, G.; Li, C.Y.; Wei, L. Kobas 2.0: A web server for annotation and identification of enriched pathways and diseases. *Nucleic Acids Res.* **2011**, *39*, W316–W322. [CrossRef] [PubMed]
26. Wu, J.; Mao, X.; Cai, T.; Luo, J.; Wei, L. Kobas server: A web-based platform for automated annotation and pathway identification. *Nucleic Acids Res.* **2006**, *34*, W720–W724. [CrossRef] [PubMed]
27. Rice, P.; Longden, I.; Bleasby, A. Emboss: The european molecular biology open software suite. *Trends Genet. TIG* **2000**, *16*, 276–277. [CrossRef]
28. Livak, K.J.; Schmittgen, T.D. Analysis of relative gene expression data using real-time quantitative PCR and the $2^{-\Delta\Delta C_T}$ method. *Methods* **2012**, *25*, 402–408. [CrossRef] [PubMed]
29. Yang, Y.; Xu, M.; Luo, Q.; Wang, J.; Li, H. De novo transcriptome analysis of *Liriodendron chinense* petals and leaves by illumina sequencing. *Gene* **2014**, *534*, 155–162. [CrossRef] [PubMed]
30. Zhang, L.; Ming, R.; Zhang, J.; Tao, A.; Fang, P.; Qi, J. De novo transcriptome sequence and identification of major bast-related genes involved in cellulose biosynthesis in jute (*Corchorus capsularis* L.). *BMC Genom.* **2015**, *16*, 1062. [CrossRef] [PubMed]
31. Mudalkar, S.; Golla, R.; Ghatty, S.; Reddy, A.R. De novo transcriptome analysis of an imminent biofuel crop, *Camelina sativa* L. Using illumina GAIIX sequencing platform and identification of SSR markers. *Plant Mol. Biol.* **2014**, *84*, 159. [CrossRef] [PubMed]
32. Kalra, S.; Puniya, B.L.; Kulshreshtha, D.; Kumar, S.; Kaur, J.; Ramachandran, S.; Singh, K. De novo transcriptome sequencing reveals important molecular networks and metabolic pathways of the plant, *Chlorophytum borivilianum*. *PLoS ONE* **2013**, *8*, e83336. [CrossRef] [PubMed]
33. Yang, Y.X.; Wang, M.L.; Liu, Z.L.; Zhu, J.; Yan, M.Y.; Li, Z.H. Nucleotide polymorphism and phylogeographic history of an endangered conifer species *Pinus bungeana*. *Biochem. Syst. Ecol.* **2016**, *64*, 89–96. [CrossRef]
34. Fu, N.; Wang, Q.; Shen, H.L. De novo assembly, gene annotation and marker development using illumina paired-end transcriptome sequences in celery (*Apium graveolens* L.). *PLoS ONE* **2013**, *8*, e57686. [CrossRef] [PubMed]
35. LI, B.; Meng, Q.; Li, Y.; Dong, W.; Zheng, Y. Identification and evaluation of germplasm resources of *Pinus bungeana*. *Hunan For. Sci. Technol.* **2016**, *43*, 1–7. [CrossRef]
36. Galuszka, P.; Frebort, I.; Sebela, M.; Sauer, P.; Jacobsen, S.; Pec, P. Cytokinin oxidase or dehydrogenase? Mechanism of cytokinin degradation in cereals. *Eur. J. Biochem.* **2001**, *268*, 450–461. [CrossRef] [PubMed]
37. Ashikari, M.; Sakakibara, H.; Lin, S.; Yamamoto, T.; Takashi, T.; Nishimura, A.; Angeles, E.R.; Qian, Q.; Kitano, H.; Matsuoka, M. Cytokinin oxidase regulates rice grain production. *Science* **2005**, *309*, 741–745. [CrossRef] [PubMed]
38. Werner, T.; Motyka, V.; Laucou, V.; Smets, R.; Van, O.H.; Schmülling, T. Cytokinin-deficient transgenic arabidopsis plants show multiple developmental alterations indicating opposite functions of cytokinins in the regulation of shoot and root meristem activity. *Plant Cell* **2003**, *15*, 2532–2550. [CrossRef] [PubMed]

39. To, J.P.; Haberer, G.; Ferreira, F.J.; Deruere, J.; Mason, M.G.; Schaller, G.E.; Alonso, J.M.; Ecker, J.R.; Kieber, J.J. Type-a arabidopsis response regulators are partially redundant negative regulators of cytokinin signaling. *Plant Cell* **2004**, *16*, 658–671. [CrossRef] [PubMed]

40. Ferguson, B.J.; Beveridge, C.A. Roles for auxin, cytokinin, and strigolactone in regulating shoot branching. *Plant Physiol.* **2009**, *149*, 1929–1944. [CrossRef] [PubMed]

© 2018 by the authors. Licensee MDPI, Basel, Switzerland. This article is an open access article distributed under the terms and conditions of the Creative Commons Attribution (CC BY) license (http://creativecommons.org/licenses/by/4.0/).

forests

Article

The Transcriptomic Responses of *Pinus massoniana* to Drought Stress

Mingfeng Du [1,2,3], **Guijie Ding** [1,2,*] **and Qiong Cai** [1,2]

1   School of Forestry Science, Guizhou University, Guiyang 550025, Guizhou, China; dmf1979@126.com (M.D.), dukecq@sina.com (Q.C.)
2   Institute for Forest Resources & Environment of Guizhou, Guizhou University, Guiyang 550025, Guizhou, China
3   School of Karst Science, Guizhou Normal University, Guiyang 550001, Guizhou, China
*   Correspondence: gjding@gzu.edu.cn; Tel.: +86-13984060976 or +86-851-88298013

Received: 29 April 2018; Accepted: 31 May 2018; Published: 4 June 2018

Abstract: Masson pine (*Pinus massoniana*) is a major fast-growing timber species planted in southern China, a region of seasonal drought. Using a drought-tolerance genotype of Masson pine, we conducted large-scale transcriptome sequencing using Illumina technology. This work aimed to evaluate the transcriptomic responses of Masson pine to different levels of drought stress. First, 3397, 1695 and 1550 unigenes with differential expression were identified by comparing plants subjected to light, moderate or severe drought with control plants. Second, several gene ontology (GO) categories (oxidation-reduction and metabolism) and Kyoto Encyclopedia of Genes and Genomes (KEGG) pathways (plant hormone signal transduction and metabolic pathways) were enriched, indicating that the expression levels of some genes in these enriched GO terms and pathways were altered under drought stress. Third, several transcription factors (TFs) associated with circadian rhythms (HY5 and LHY), signal transduction (ERF), and defense responses (WRKY) were identified, and these TFs may play key roles in adapting to drought stress. Drought also caused significant changes in the expression of certain functional genes linked to osmotic adjustment (P5CS), abscisic acid (ABA) responses (NCED, PYL, PP2C and SnRK), and reactive oxygen species (ROS) scavenging (GPX, GST and GSR). These transcriptomic results provide insight into the molecular mechanisms of drought stress adaptation in Masson pine.

Keywords: *Pinus massoniana* Lamb.; drought stress; transcriptome; transcription factor; defense response

1. Introduction

Drought is one of the world's most severe environmental stresses. It represents an increasing threat to the productivity of agriculture and forestry as it has negative impacts on plant growth and development [1]. To adapt to unfavorable conditions of water deficit, plants activate a variety of complex regulatory mechanisms by altering gene expression levels [2] and by activating complex cross-talk between biochemical and molecular processes [3]. The involved genes are typically divided into genes encoding functional proteins and those encoding regulatory proteins. Functional proteins directly protect plants and include scavengers of ROS (reactive oxygen species) [4], aquaporins [5], dehydrins [6] and others. In contrast, regulatory proteins control gene expression networks and signal transduction pathways involved in stress responses [2]. Many regulatory proteins, such as MYB and WRKY, play key roles in plant drought stress responses. In addition, ABA (abscisic acid), a crucial hormone that is often involved in signaling and stress responses, generally accumulates under drought conditions and can initiate signal transduction that results in the up-regulation of several genes involved in drought stress responses [7].

Masson pine (*Pinus massoniana*), a major coniferous tree widely distributed in southern China, is not only an economically important species that is commonly used for timber, wood pulp and rosin but also an ecologically important species in forest ecosystems [8]. Seasonal soil drought in southern China is a major natural phenomenon that constrains the production and growth of Masson pine. Therefore, it is of interest to cultivate genotypes that are resistant to drought conditions. Most studies investigating the drought tolerance of Masson pine have focused on the plant's morphology and physiology. These studies have described certain important morphological adaptations to xeric environments, such as alterations to root development [9]. Moreover, some analyses of the physiological responses of this plant have uncovered traits related to drought resistance, such as changes in MDA (malondialdehyde) and PRO (free proline) content [9] as well as changes in the activities of POD (peroxidase), SOD (superoxide dismutase) and CAT (catalase). In addition, several important drought-stress-induced genes have been identified using reverse transcription-polymerase chain reaction (RT-PCR), including F-box, Ribosomal RNA Processing 8 (RRP8), auxin response factors (ARFs), and EF1b [10]. However, despite the importance of drought resistance in Masson pine, a more comprehensive understanding of the molecular response mechanisms underlying resistance remains lacking. NGS (next-generation sequencing), a technology that provides deep sequencing sufficient to cover the entire transcriptome of an organism, has contributed greatly to studies in model and non-model plants. Expanding transcriptome information is extremely useful for the exploration of differential gene expression and key responsive factors in conifer species subjected to drought stress, such as *Pinus pinaster* [11] and *Pinus menziesii* [12]. In this study, the transcriptome of Masson pine under different drought stress conditions was evaluated using the Illumina Hi-Seq sequencing platform. The transcriptome data were used to identify genes that may be involved in the response to drought and to clarify the possible molecular mechanisms involved in Masson pine's adaptation to different drought stress conditions. The results improve our understanding of environmental acclimation mechanisms in Masson pine and will serve as an invaluable molecular-level reference to inform future work on the enhancement of drought tolerance in Masson pine.

2. Material and Methods

2.1. Plant Material and Experimental Setup

An elite pure line of Masson pine, named the 83rd family of Masson pine, obtained from the seed orchard of Guangxi Province (P. R. China) (20°36′ N, 107°28′ E) was used in this study. This line exhibited rapid growth and strong drought resistance in our previous study [9]. In April 2015, one-year-old seedlings of this elite line were cultured in pots in a ventilated nursery at the College of Forestry, Guizhou University, with a day/night room temperature of approximately 20 °C/10 °C and a light/dark photoperiod of 14 h/10 h. Each pot had a 300-mm top diameter, a 200 mm bottom diameter and a 250 mm depth and was filled with yellow soil that had developed from quaternary red clay and was collected from a Masson pine forest. The soil had a pH of approximately 5.0. Its total contents of N, P, and K were 0.16 g/kg, 0.36 g/kg and 1.50 g/kg, respectively, and its available contents of N, P, and K was 65.77 mg/kg, 10.99 mg/kg and 164.26 mg/kg, respectively. In May 2016, the two-year-old seedlings in each pot were approximately 65 cm in height. At this time, the seedlings were divided evenly into four groups, with 3 seedlings per pot and 15 pots per group. The four groups corresponding to four levels of field moisture capacity were as follows: well-watered control (CK, ≥70%), light drought (LD, 55–70%), moderate drought (MD, 45–55%), and severe drought (SD, 30–45%). The water content was controlled by potted planting [13], and the soil moisture content was measured by weighing each pot and was regulated by artificial irrigation. The seedlings were sampled after a one-month period of drought treatment, and the stem apex needles of the seedlings were selected for RNA extraction.

2.2. Total RNA Isolation, Sequencing Library Preparation and Transcriptome Assembly

RNA was extracted from four treatment seedlings with two biological replicates for each treatment and then used to construct 8 cDNA libraries. Total RNA was extracted using a Plant RNA Isolation Kit (Invitrogen, Carlsbad, CA, USA). Sequencing library construction and Illumina deep sequencing were performed using the method described by Ma et al. [14], and 150-bp paired-end reads were generated. De novo transcriptome assembly was conducted using Trinity [15]. The raw data and sequences can be found online at the NCBI Sequence Read Archive (SRA) database (accession number SRP092298) and the GenBank Transcriptome Shotgun Assembly (TSA) database (accession number GFHB00000000), respectively.

2.3. Gene Expression Quantification and Differential Expression Analysis

Gene expression was estimated using RSEM [16] for FPKM (expected number of Fragments Per Kilobase of transcript sequence per Millions base pairs sequenced) values. Differential gene expression analyses of different water conditions were conducted using the R package DESeq (http://www.bioconductor.org/packages/release/bioc/html/DESeq2.html). *p* values were adjusted to control for logFC > 1 and FDR < 0.05 using the BH (Benjamini–Hochberg) approach.

2.4. Functional Annotation and Enrichment Analysis

Gene function was annotated using the NCBI blast (http://www.ncbi.nlm.nih.gov/) [17] for Nr (NCBI non-redundant protein sequences), Nt (NCBI non-redundant nucleotide sequences), Swiss-Prot (A manually annotated and reviewed protein sequence database) and KOG/COG (Clusters of Orthologous Groups of proteins); the BLAST parameters of NR, NT and Swiss-Prot were controlled for using an e-value = 1×10^{-5}, and KOG/COG was controlled for using an e-value = 1×10^{-3}. Gene function was annotated on the Pfam (Protein family) database using hmmscan software with an e-value = 0.01; GO annotation was accomplished using blast2go software with an e-value = 1×10^{-6}; and KEGG annotation was performed using KAAS software with an e-value = 1×10^{-10} [18].

Gene ontology (GO) enrichment analysis of the differentially expressed genes (DEGs) was performed using the GOseq R package based on Wallenius' noncentral hypergeometric distribution [19]. Kyoto Encyclopedia of Genes and Genomes (KEGG) pathway enrichment analysis of the DEGs was conducted using KOBAS [20].

2.5. qRT-PCR (Quantitative Real-Time PCR) Validation

The total RNA isolated as described above was used to synthesize cDNA using the RNA LA PCR Kit (TaKaRa, Shiga, Japan) following the manufacturer's instructions. Gene-specific primers (Table S1) were designed for 9 unigenes using Primer Premier 5.0 (Premier, Canada). Three biological replicates for each reaction and three technical replicates for each biological replicate were analyzed using SYBR Premix ExTaq (TaKaRa) on a 7500 fast real-time PCR system (Applied Biosystems, Waltham, MA, USA) with the following PCR procedure parameters: 95 °C for 120 s followed by 40 cycles of 95 °C for 10 s, 61 °C for 30 s, and 72 °C for 30 s; and an additional procedure for dissociation (95 °C for 15 s, 60 °C for 60 s, and 95 °C for 15 s). qRT-PCR was performed in 20.0 μL reactions containing 10.0 μL of SYBR mix, 1.0 μL of template cDNA, 0.4 μL of forward primer (10.0 μM), 0.4 μL of reverse primer (10.0 μM), and 8.2 μL of deionized water. Amplification of three internal control genes (UBC, ubiquitin-conjugating enzyme-like protein; 18 s RNA; and GAPDH, NAD-dependent glyceraldehyde-3-phosphate dehydrogenase) [21] was used to normalize the qRT-PCR data. Quantification was achieved using comparative cycle threshold (C_t) values, and gene expression levels were calculated using the $2^{-\Delta\Delta Ct}$ method [22].

3. Results

3.1. Variations in Phenotypes during Drought Stress

Seedling phenotypes were evaluated throughout the experiment (Figure 1). Although the well-watered control seedlings displayed normal growth, the shoot tips of seedlings showed mild wilt under LD, and the wilting became increasingly severe with increasing drought stress.

Figure 1. Phenotypic variation of Masson pine seedlings under different soil moisture conditions. CK, well-watered control; LD, light drought; MD, moderate drought; and SD, severe drought.

3.2. Transcriptome Sequencing and De Novo Assembly and Annotation

A total of 390,320,648 raw reads were generated to assemble 197,612 non-redundant unigenes, which had a length range of 201–15,800 bp, an N50 of 1227 bp, and an average length of 695 bp (Figure 2).

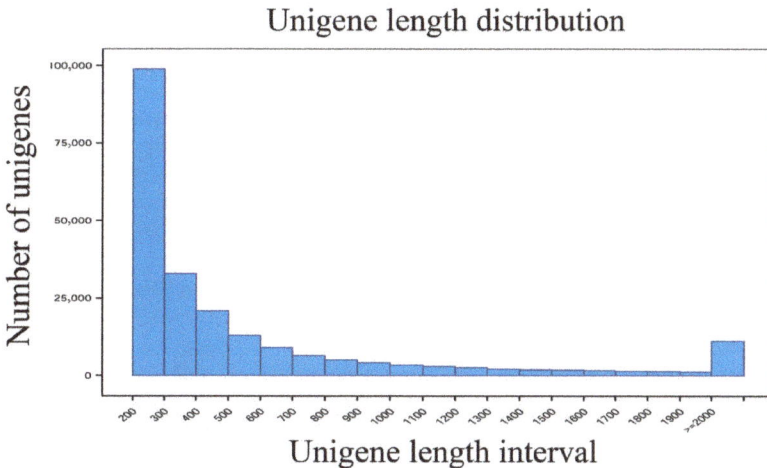

Figure 2. Distribution of assembled unigenes and the number of assembled unigenes of each length.

The sequenced unigenes were validated and annotated by alignment with public databases, including the NT (NCBI nucleotide sequences), NR (NCBI non-redundant protein sequences), Swiss-Prot (a manually annotated and reviewed protein sequence database), PFAM (protein family), GO, KEGG and KOG (EuKaryotic Orthologous Groups) databases (Table S2). Of the 197,612 unigenes, 66,825 (33.81%) and 49,085 (24.83%) had significant matches in the NR and NT databases, respectively.

In addition, 64,943 (32.86%), 35,880 (18.15%), and 30,882 (15.62%) unigenes had annotations in the GO, KOG, and KO databases, respectively. A total of 101,806 unigenes (51.51%) were successfully annotated in at least one of the above databases, and 11,874 unigenes (6%) were annotated in all seven databases.

3.3. Exploration of Gene Expression in Seedlings under Drought Stress

To elucidate the molecular activities that occurred across the different water content conditions, differential expression analyses were performed on samples collected from the three treatments (LD, MD and SD). In total, 4300 genes were differentially expressed ($q \leq 0.05$) between the samples from the three drought treatments and the well-watered control samples (Table S3). Of these, 3397, 1695 and 1550 genes were differentially expressed between the LD and CK, MD and CK, and SD and CK treatments, respectively (Table S3). Among the DEGs, 1656, 611 and 651 were found to be up-regulated ($q \leq 0.05$) in the LD, MD and SD treatments, respectively, and 1741, 1084 and 899 were found to be down-regulated ($q \leq 0.05$) in the LD, MD and SD treatments, respectively (Table S3, Figure 3).

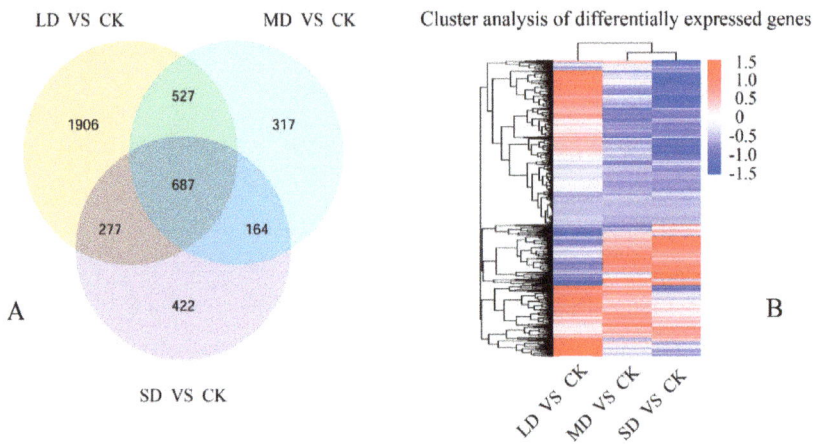

Figure 3. Venn diagram and cluster analysis of differentially expressed genes in three comparisons. (**A**) Venn diagram showing that a total of 4300 unigenes were identified as differentially expressed in the three comparisons (LD, MD and SD versus CK); the number of DEGs in each comparison are shown in each circle; the number of overlapping regions represent the 687 DEGs that were found in each comparison. (**B**) A cluster analysis of differentially expressed genes is shown in the right panel; red indicates up-regulated genes and blue indicates down-regulated genes.

The overlap among these comparisons showed that 687 genes were identified as differentially expressed under the three drought conditions (Figure 3). Of these, only 341 unigenes were identified by GO analysis, whereas 209 unigenes had no known function (Table S3). Among the former, 156 unigenes were annotated in the biological process category and were linked to signal transduction, defense response, transcriptional regulation, photosynthesis, transmembrane transport, biosynthetic processes, metabolic processes, oxidation-reduction processes, and protein phosphorylation. Our findings suggest that these biological processes may participate in the drought response.

3.4. Gene Ontology Enrichment Analysis

Enrichment analysis of GO terms derived from DEGs after drought stress was conducted to reveal the GO terms that were common among all drought samples and those that were unique to each drought sample (Table S4). Enriched ($q \leq 0.05$) GO terms were observed in the foundation categories of metabolism, oxidation-reduction, and photosynthesis, and these terms were significantly over- or underrepresented in different drought stress treatments. These results suggest that the expression of proteins associated with these GO terms was strongly affected by drought. In the category of metabolism, differential representation was found for the molecular functions of chitinases, transferases, pectinesterases, peptidases, kinases, synthases, hydrolases, peroxidases, oxidases and catalytic activity. The enriched GO terms for biological processes included hormones, single-organism, chitin, amino sugar, starch, glucose and cellular carbohydrate. It is noteworthy that lipid, cellular lipids, isoprenoids and glycosylation were affected in the lipid metabolism category, suggesting that changes in membrane lipids may have occurred. In the oxidation-reduction category, the following GO terms were significantly overrepresented: biological processes of oxidation-reduction and the molecular functions of oxidoreductase activity, including acting on peroxide as an acceptor, acting on the CH-OH group of donors, acting on paired donors, acting on single donors with the incorporation of molecular oxygen, and incorporation or reduction of molecular oxygen. These findings indicated that oxidation-reduction reactions and oxidoreductase activity were enhanced. In the categories related to photosynthesis, the down-regulated GO terms included the biological process photosynthesis and the cellular components of photosystem, photosystem I, photosystem I reaction center, photosystem II and photosystem II oxygen-evolving complex, indicating that photosynthetic functions were inhibited. Not unexpectedly, GO terms for biological processes that occur in response to stress (obsolete peroxidase reaction and response to oxidative stress) and negative regulation of catalytic activity and molecular function were highly enriched. Enriched GO terms for molecular function included chitin binding, ion binding, heme binding, ribonucleotide binding, tetrapyrrole binding and ADP binding. In the category of cellular components, the three major enriched GO terms were cell wall, apoplast, and external encapsulating structure.

Certain enriched GO terms were discovered only for a particular level of drought stress, suggesting that proteins with important specific functions are expressed in response to specific level of drought stress (Table S4). First, several phosphorylation-related GO terms were specifically enriched among the DEGs that were down-regulated under LD relative to CK, including phosphorylation, protein phosphorylation, phosphate-containing compound metabolic process, and phosphorus metabolic and modification process. Second, the GO terms for transport, including drug transport, drug transporter activity, drug transmembrane transport, and drug transmembrane transporter activity, were enriched only among the DEGs that were up-regulated under MD relative to CK. Finally, the GO terms for carbon utilization, obsolete electron transport, and electron carrier activity were enriched only among the genes that were down-regulated under SD relative to CK. Taken together, the results show that many of the GO functional categories found to be enriched were significantly inhibited in the drought-stressed seedlings.

3.5. KEGG Pathway Enrichment Analysis

The KEGG pathways that were enriched in the DEGs were analyzed to reveal the specific pathways involving the DEGs that were responsive to drought stress (Table S5). The four pathways of plant hormone signal transduction, photosynthesis, phenylalanine metabolism and phenylpropanoid biosynthesis were enriched ($q \leq 0.05$) under every drought treatment. Among these pathways, plant hormone signal transduction, phenylalanine metabolism, and phenylpropanoid biosynthesis were enriched among the up-regulated DEGs, whereas photosynthesis was enriched among the down-regulated DEGs. These results indicate that drought stress induced the signal transduction of plant hormones, which had strong effects on biosynthesis and metabolism and led to a severe decline in photosynthesis.

In addition, certain pathways enriched by DEGs occurred under the various drought treatments (Table S5). For example, enrichment of DEGs associated with the pathways of amino sugar and nucleotide sugar metabolism, circadian rhythm-plant and plant–pathogen interaction first appeared in LD samples; enrichment of DEGs associated with the pathways of NF-kappa B signaling and glutathione metabolism first appeared in MD samples; and enrichment of DEGs associated with the pathways of carbon metabolism, chemical carcinogenesis, and drug metabolism-cytochrome P450 appeared only in SD samples.

3.6. Validation by qRT-PCR

To verify the reliability of the RNA-Seq data, nine drought-responsive unigenes showing significant up- or downregulation in the drought seedlings were randomly chosen for qRT-PCR analysis (Figure 4). Among them, three unigenes (MYB (*c71819_g3*), NIP (*c85755_g1*), and MCM (*c88297_g1*)) showed constitutively down-regulated expression, and one unigene (GPX (*c92413_g2*)) showed constitutively up-regulated expression with increasing drought stress; five unigenes (DREB (*c60672_g1*), GH3 (*c94987_g2*), P450 (*c95186_g2*), P5CS (*c93699_g2*) and WRKY (*c90841_g1*)) were up-regulated under LD, MD and SD, the relative expression levels of these unigenes were higher under LD than under MD and SD. These results indicated that nine unigenes were induced by drought stress, which could assist in revealing the response to drought stress in Masson pine. For six of the unigenes (DREB, GH3, P450, GPX, MYB, and NIP), the qRT-PCR results closely matched the RNA-Seq results. The other three unigenes (P5CS, WRKY, and MCM) showed similar trends in expression, but the fold change in expression indicated by RNA-Seq was lower than that indicated by qRT-PCR. Overall, the unigene expression trends revealed by the RNA-Seq data and the qRT-PCR analysis were similar, showing that the results of the RNA-Seq analyses were valid.

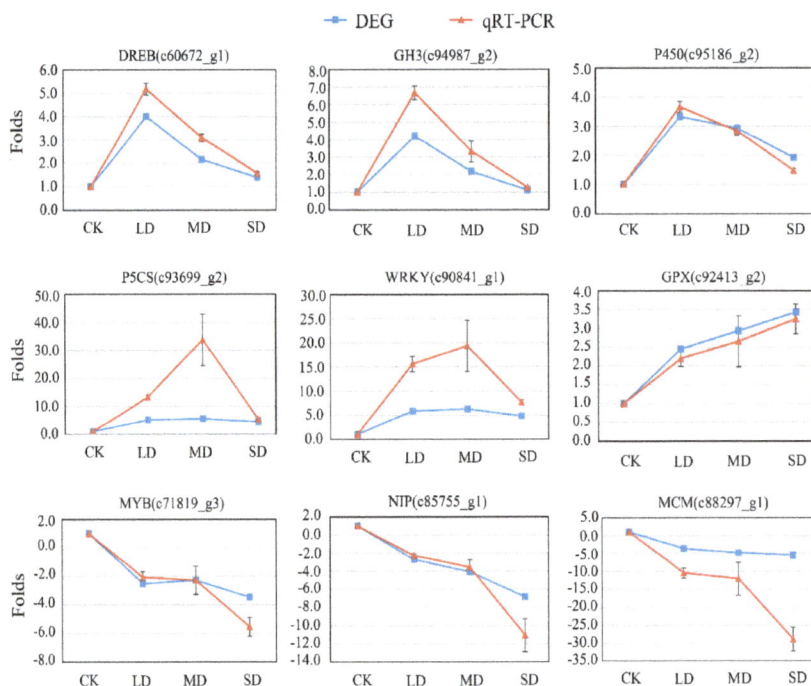

Figure 4. Expression changes of nine randomly selected unigenes as determined by qRT-PCR results and DGE sequencing data. The *x*-axis values indicate the different water content conditions. The *y*-axis values represent the change in expression under the various drought stress conditions relative to the well-watered control condition. Data represent the fold changes of expression for each unigene in the drought treatment relative to control conditions. Error bars represent standard deviations. Blue indicates the RNA-Seq results, and red indicates the qRT-PCR results.

4. Discussion

In this study, we observed that the tips of the seedlings were slightly wilted under LD. The observed phenotypic changes were considered together to assist in characterizing transcriptional responses and revealing the defense response to drought stress in conifer trees.

4.1. Resistance to Osmotic Stress at Each Level of Drought Stress

Osmotic adjustment is believed to be an adaptation to drought stress, as observed in many studies of drought-tolerance mechanisms [5]. In the present study, the P5CS (pyrroline-5-carboxylate synthase) gene (c93699_g2), which plays a role in stabilizing membranes and proteins under osmotic stress in pine [5,6], was found to be up-regulated in each drought treatment. It has been reported that the proline content is increased with increasing severity of drought conditions in this elite pure line [9]. These results suggested that Masson pine exhibits a strong capacity for osmotic adjustment via the accumulation of proline to reduce the effects of osmotic stress caused by drought.

In addition, AQPs (aquaporins) are the main membrane proteins that regulate osmotic pressure in water transport [5]. However, AQPs play complex roles due to their disparate functions and expression patterns in the response to drought stress. Under drought conditions, AQPs are up-regulated in *Phaseolus vulgaris* [23] but down-regulated in pine [5]. Similar to the pattern observed in pine, in our study, three genes encoding AQPs (NIP, c85755_g1; PIP, c91691_g1; PIP, c101640_g1) were found to be constitutively down-regulated with increasing drought stress, suggesting that drought suppresses the expression of AQPs depending on the time and degree of stress. This response reflects a mechanism of water conservation via down-regulation of AQP expression to reduce membrane permeability, resulting in the minimization of water flux and scatter in the aboveground parts of Masson pine.

4.2. Transcription Factors Responding to Stress Signals under Light Drought

TFs play a key role in regulating downstream genes involved in adversity stress responses. In this study, 142 transcription factors (TFs) were identified to be differentially expressed ($q \leq 0.05$) under drought stress, 87 were up-regulated and 55 were down-regulated (Table S6). Most of these TFs belong to the AP2/EREBP, MYB, WRKY, NAC, and HD-ZIP families. Importantly, some of the induced TFs were enriched in KEGG pathways involved in responses to environmental and physiological signals under light drought (LD).

The enriched KEGG pathway "circadian rhythm-plant" (ko04712) was linked to three TFs: one HY5 gene (c84637_g1) and two LHY genes (c85168_g1 and c91081_g2), which were down-regulated and up-regulated, respectively, under LD but showed no variation under MD and SD, indicating that HY5 and LHY were induced by light drought. HY5 is a bZIP TF that links hormone and light-signaling pathways [24], which play a part in promoting the photomorphogenesis of *A. thaliana* [25], and negatively regulates light-signaling pathways [26]. Another LHY protein, a TF that is closely related to MYB, is the central oscillator component of the light input pathway [27]. Cañas et al. [11] reported that the LHY gene of *P. pinaster*, which shows higher expression, might reflect an adaptation to light conditions rather than a transcription factor that functions to regulate diurnal rhythm. However, another analysis in *Fraxinus mandshurica* demonstrated that the LHY promoter has a pivotal role in initiating systemic responses to adverse stress [28]. According to these studies, it was suggested that HY5 and LHY might be key TFs in the light-signaling network that regulates the circadian rhythm in response to light drought stress in our study. This hypothesis remains to be validated in further studies; however, our findings provide insight into the potential mechanisms of circadian rhythmic gene expression activation associated with coniferous drought conditions.

In addition, we found that four unigenes in the "plant hormone signal transduction" pathway (ko04075), encoding the TFs ERF (c76570_g1), ARF (c83733_g1), and IAA (c77087_g1 and c92989_g1), were significantly up- or down-regulated. ERF, belonging to the AP2 family, is involved in DNA binding, and overexpression of ERF/AP2 has been confirmed to improve plant tolerance to drought in transgenic Virginia pine [29]. In our study, the expression of ERF/AP2 was constitutively up-regulated expression with increasing drought stress, with overexpression under LD versus CK, MD versus LD, and MD versus LD. These results suggest that AP2/ERF is induced by drought stress, and it might enhance drought tolerance in Masson pine. Moreover, several studies have reported that ARF regulates the expression of auxin response genes in conjunction with Aux/IAA repressors [30] and that Aux and IAA function as auxin-induced repressors and modulate the activity of DNA-binding ARFs [31]. Our results indicate that the expression of two IAA genes and an ARF gene were constitutively down-regulated expression with increasing drought stress, with marked repression in the LD versus CK and MD versus LD conditions. It appears that Aux/IAA and ARF may inhibit one another upon the onset of light drought, indicating that TFs related to growth and development in Masson pine needles begin to be inhibited upon light drought. This finding differs from a previous report showing that water stress increased IAA concentrations, thereby inducing epinastic growth in radiata pine [32]. However, the finding agrees with our previous study in which drought stress resulted in significant growth reduction in the aboveground portions of Masson pine, whereas the root growth and root–shoot ratio both significantly increased [9]. The results indicate a growth strategy to reduce aboveground growth and increase root growth, which favors water absorption from the soil and contributes to the adaptation to drought stress [33].

4.3. Defense Response of the Plant–Pathogen Interaction Pathway under Light Drought

The systematic defense response of plants under abiotic stress is an important resistance mechanism of coniferous forests [34]. In our study, the plant–pathogen interaction enriched pathway shown in Figure 5 was strongly activated in Masson pine seedlings under LD (ko04626). First, pathogenic signaling was transmitted to the cytoplasm by the recognition of FLS_2 (flagellin-sensitive 2) and EFR (EF-TU receptor). Second, the PTI response was triggered and amplified. FLS_2 and EFR were both up-regulated to activate the downstream gene encoding MEKK1 (mitogen-activated protein kinase kinase kinase 1); subsequently, MEKK1 signaling was enhanced to activate two separate pathways for the negative and positive regulation of immunity [35]. Finally, within the cell nucleus, defense-related genes, including one WRKY33 gene (*c83644_g4*), and its downstream pathogen-resistance genes NHO1 (glycerol kinase) and PR1 (pathogenesis-related protein 1) [36], were up-regulated. To date, WRKY TFs have occasionally been described as having a regulatory role in the defense of conifer species, although a regulatory role for WRKY has been more widely reported that overexpression of WRKY gene enhances the resistance to tolerance and pathogen infection to drought stress in Grapevine [37] and Horse gram [38]. Interestingly, our findings showed that the WRKY gene (*c83644_g4*) was up-regulated under LD, but showed no changes were observed under MD and SD, indicating that WRKY was induced upon light drought stress. These results provide evidence that WRKYs might play key roles in the signaling and transcriptional regulation of defense responses in Masson pine under mild drought stress.

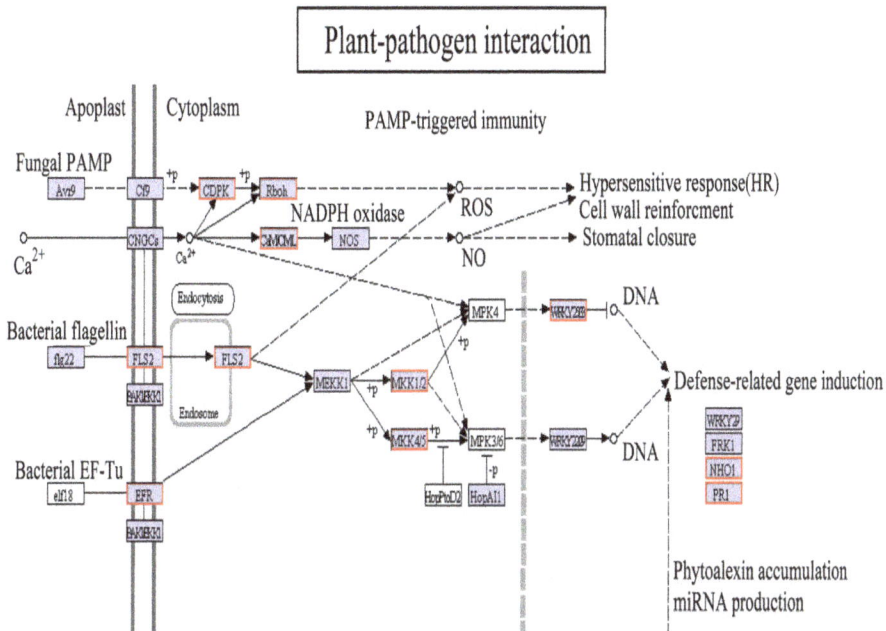

Figure 5. Unigenes inferred to be involved in the plant–pathogen interaction pathway. Blue inside the boxes indicates unigenes predicted to be involved in the pathway. White inside the boxes indicates unigenes that were not identified in the expression profile analysis. Red in the borders indicates that the genes increased expression under drought stress relative to the well-watered condition.

4.4. ABA Response under Light and Moderate Drought Stress

The plant hormone ABA is known to have a core role in the modulation of plant adaptation to drought stress [39]. Although ABA biosynthesis, signaling and responses are considered to be closely related to drought-resistance mechanisms in plants [40], information regarding the pivotal genes, specific modes and signaling pathways involved in drought resistance in conifers remains lacking. In this study, we observed that a gene encoding NCED (9-*cis*-epoxycarotenoid dioxygenase) (c71048_g1), which is a crucial enzyme for the synthesis of ABA [41] that is often overexpressed in plants under drought stress [3], was up-regulated under LD. Quan et al. [42] reported that ABAs are important hormones related to drought stress in Masson pine, with ABA content increasing with increasing drought stress. Thus, both ABA accumulation and the expression of a key gene related to ABA synthesis were found to be up-regulated under drought stress in Masson pine. It can be inferred that this up-regulation is beneficial to the development of plant drought tolerance. Moreover, three major components associated with ABA signal transduction were also found to be differentially expressed in this study. A gene encoding PYL (c87460_g1), an ABA receptor that takes part in activating ABA responses [43], was up-regulated under MD versus CK, and MD versus LD. However, a gene encoding PP2C, a type 2C protein phosphatase (c68631_g1) that is a negative regulator that inactivates SnRK2 protein kinases [44], was obviously repressed under MD versus CK, MD versus LD, and SD versus LD. In addition, two genes encoding SnRK2 (c85767_g1 and c85767_g2), which enhance drought tolerance by enhancing ABA signaling [45], were activated under LD. These results indicate that many important genes related to ABA responses, all of which are involved in the ABA signaling pathway and its double-negative regulatory system, promote the interaction between PYL and PP2C, thereby leading to PP2C inhibition and SnRK2 activation in Masson pine. These findings are consistent with those of previous studies in which drought tolerance was putatively linked to ABA signaling networks in plants [46]. In conclusion, the ABA-mediated response pathway was markedly activated under LD and MD; thus, ABA plays a central role in drought stress responses in Masson pine.

4.5. Responses to Oxidative Stress under Moderate and Severe Drought Stress

Drought stress results in the overproduction of ROS in plants [4]. The activation of many antioxidants that occurs due to drought is considered to be a protective mechanism against drought damage [47]. In our previous study, the activity of SOD was found to be markedly increased under LD [9], providing further evidence that the oxidative system might play an essential role in the response to drought stress by regulating antioxidase activity to defend against ROS damage in Masson pine.

Moreover, other ROS-scavenging enzymes were up-regulated in the enriched glutathione metabolism pathway involved in the responses to stress signals under MD and SD (ko00480) in Masson pine. Two glutathione transferases (GST) genes (*c92901_g1* and *c93725_g3*), which play a part in generic detoxification and cell adaptability under stress conditions [48], were constitutively up-regulated under different levels of drought stress in Masson pine. The expression of three GPX genes, namely, *c92413_g1*, *c92413_g2*, and *c88279_g1*, which can reduce H_2O_2 (hydrogen peroxide) and lipid hydroperoxides in the response to oxidative stress [49], was validated to be constitutively up-regulated with increasing drought stress, suggesting that GPX was activated to enhance abiotic stress tolerance. The expression of a GSR gene (*c83219_g1*), was upregulated to enhance plant tolerance to stress conditions [50] and was also upregulated under SD, indicating that GSR plays a role in the defense against ROS in Masson pine, even under severe drought conditions. Overall, these results suggest that glutathione is linked to cellular defense mechanisms against stresses caused by drought and oxidants and that GPX, GST and GSR may be positive regulators of drought tolerance in Masson pine.

Interestingly, most of the DEGs were found for the LD treatment, with fewer DEGs observed for the MD and SD treatments. These results suggest that this elite genotype of Masson pine exhibits a positive character of drought resistance in which a systemic response is rapidly activated to prevent damage under light drought stress, which then gradually returns to baseline as an adaptation to drought conditions under moderate and severe drought stress.

5. Conclusions

In this study, biological homeostasis in the Masson pine was reestablished through the collaborative action of physiological and molecular responses and growth under drought stress (Figure 6). Plant growth was slowed as an adaptation to drought stress, marked by down-regulation of the growth elements IAA and ARF. Furthermore, Masson pine exhibited an active defense and protection response that was characterized by a strong capacity for osmotic adjustment and the overexpression of genes related to ABA biosynthesis and signal transduction, and ROS scavenging, and it exhibited a rapid systemic defense against pathogenic effects. In addition, we found that drought stress is linked to the differential expression of TFs that regulate circadian rhythm, which has not previously been described in *Pinus* spp. These results will serve as a foundation for future transcriptomic research into drought tolerance in Masson pine.

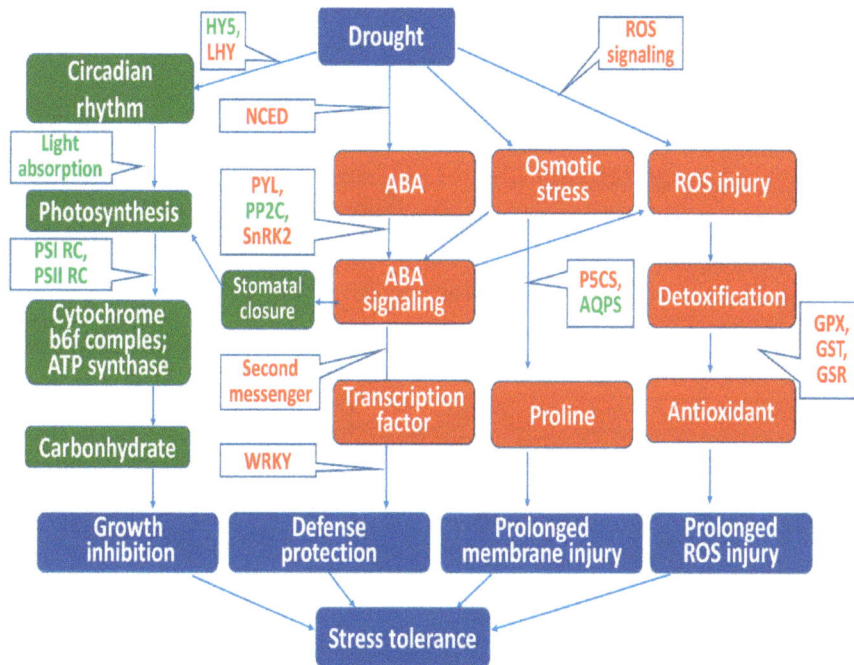

Figure 6. A model of different adaptive strategies regulated by drought-responsive genes in Masson pine. Red or green inside the white boxes indicates unigenes that showed increased or decreased expression, respectively. Blue inside the boxes indicates different adaptive strategies. Red inside the boxes indicates that the biological function was activated. Green inside the boxes indicates that the biological function was repressed.

Supplementary Materials: The following are available online at www.mdpi.com/1999-4907/9/6/326/s1, Table S1: Primers used in qRT-PCR. (XLS), Table S2: Statistical analysis of unigenes annotated in public databases. (XLS), Table S3: Unigenes differentially expressed between different water stress conditions. (XLS), Table S4: GO enrichment results for DEGs in drought-stressed versus control seedlings. (XLS), Table S5: KEGG pathway enrichment results for DEGs in drought-stressed versus control seedlings. (XLS), Table S6: Transcription factors differentially expressed in drought-stressed versus control seedlings. (XLS).

Author Contributions: M.D. and G.D. conceived and designed the research. M.D. conducted the experiments. G.D. contributed new reagents and analytical tools. M.D. and Q.C. analyzed the data. M.D. wrote the manuscript. All authors read and approved the manuscript.

Funding: The work was supported by grants from the National Natural Science Foundation of China (31660200), National science and technology support project (2015BAD09B0102), Special Core Program of Guizhou Province, P.R. China (20126001), and the Science and Technology Support Project of Guizhou Province (20172525).

Acknowledgments: We would like to thank Novogene Corporation (Beijing, China) for assistance with sequencing services.

Conflicts of Interest: The authors have no conflicts of interest to declare.

References

1. Bray, E.A. Plant responses to water deficit. *Trends Plant Sci.* **1997**, *2*, 48–54. [CrossRef]
2. Golldack, D.; Li, C.; Mohan, H.; Probst, N. Tolerance to drought and salt stress in plants: unraveling the signaling networks. *Front. Plant Sci.* **2014**, *5*, 151. [CrossRef] [PubMed]
3. Yamaguchi-Shinozaki, K.; Shinozaki, K. Transcriptional regulatory networks in cellular responses and tolerance to dehydration and cold stresses. *Annu. Rev. Plant Biol.* **2006**, *57*, 781–803. [CrossRef] [PubMed]

4. Salo, H.M.; Sarjala, T.; Jokela, A.; Häggman, H.; Vuosku, J. Moderate stress responses and specific changes in polyamine metabolism characterize *Scots pine* somatic embryogenesis. *Tree Physiol.* **2016**, *36*, 392–402. [CrossRef] [PubMed]

5. Lorenz, W.W.; Alba, R.; Yu, Y.-S.; Bordeaux, J.M.; Simões, M.; Dean, J.F.D. Microarray analysis and scale-free gene networks identify candidate regulators in drought-stressed roots of loblolly pine (*P. taeda* L.). *BMC Genom.* **2011**, *12*, 264. [CrossRef] [PubMed]

6. Perdiguero, P.; Soto, Á.; Collada, C. Comparative analysis of *Pinus pinea* and *Pinus pinaster* dehydrins under drought stress. *Tree Genet. Genomes* **2015**, *11*, 70. [CrossRef]

7. Li, H.; Yao, W.; Fu, Y.; Li, S.; Guo, Q. De novo assembly and discovery of genes that are involved in drought tolerance in Tibetan *Sophora moorcroftiana*. *PLoS ONE* **2015**, *10*, e111054. [CrossRef] [PubMed]

8. Ding, G.; Zhou, Z.; Wang, Z. *Cultivation and Utilization of Pulpwood Stand for Pinus massoniana*; China Forestry Publishing House: Beijing, China, 2006.

9. Du, M.; Ding, G.; Zhao, X. Responses to continuous drought stress and drought resistance comprehensive evaluation of different *Masson pine* families. *Sci. Silvae Sin.* **2017**, *53*, 21–29.

10. Chen, H.; Yang, Z.; Hu, Y.; Tan, J.; Jia, J.; Xu, H.; Chen, X. Reference genes selection for quantitative gene expression studies in *Pinus massoniana* L. *Trees* **2016**, *30*, 685–696. [CrossRef]

11. Cañas, R.A.; Feito, I.; Fuente-Maqueda, J.F.; Ávila, C.; Majada, J.; Cánovas, F.M. Transcriptome-wide analysis supports environmental adaptations of two *Pinus pinaster* populations from contrasting habitats. *BMC Genom.* **2015**, *16*, 909. [CrossRef] [PubMed]

12. Müller, T.; Ensminger, I.; Schmid, K.J. A catalogue of putative unique transcripts from Douglas-fir (*Pseudotsuga menziesii*) based on 454 transcriptome sequencing of genetically diverse, drought stressed seedlings. *BMC Genom.* **2012**, *13*, 673. [CrossRef] [PubMed]

13. Wang, Y.; Ding, G.J. Physiological responses of mycorrhizal *Pinus massoniana* seedlings to drought stress and drought resistance evaluation. *Ying Yong Sheng Tai Xue Bao* **2013**, *24*, 639–645. [PubMed]

14. Ma, X.; Wang, P.; Zhou, S.; Sun, Y.; Liu, N.; Li, X.; Hou, Y. De novo transcriptome sequencing and comprehensive analysis of the drought-responsive genes in the desert plant *Cynanchum komarovii*. *BMC Genom.* **2015**, *16*, 753. [CrossRef] [PubMed]

15. Grabherr, M.G.; Haas, B.J.; Yassour, M.; Levin, J.Z.; Thompson, D.A.; Amit, I.; Adiconis, X.; Fan, L.; Raychowdhury, R.; Zeng, Q.; et al. Full-length transcriptome assembly from RNA-Seq data without a reference genome. *Nat. Biotechnol.* **2011**, *29*, 644–652. [CrossRef] [PubMed]

16. Li, B.; Dewey, C. RSEM: Accurate transcript quantification from RNA-Seq data with or without a reference genome. *BMC Bioinform.* **2011**. [CrossRef] [PubMed]

17. Götz, S.; García-Gómez, J.M.; Terol, J.; Williams, T.D.; Nagaraj, S.H.; Nueda, M.J.; Robles, M.; Talón, M.; Dopazo, J.; Conesa, A. High-throughput functional annotation and data mining with the Blast2GO suite. *Nucleic Acids Res.* **2008**, *36*, 3420–3435. [CrossRef] [PubMed]

18. Moriya, Y.; Itoh, M.; Okuda, S.; Yoshizawa, A.C.; Kanehisa, M. KAAS: An automatic genome annotation and pathway reconstruction server. *Nucleic Acids Res.* **2007**, *35* (Suppl. 2), W182–W185. [CrossRef] [PubMed]

19. Young, M.D.; Wakefield, M.J.; Smyth, G.K.; Alicia, O. Gene ontology analysis for RNA-Seq: Accounting for selection bias. *Gen. Biol.* **2010**. [CrossRef] [PubMed]

20. Mao, X.; Cai, T.; Olyarchuk, J.G.; Wei, L. Automated genome annotation and pathway identification using the KEGG Orthology (KO) as a controlled vocabulary. *Bioinformatics* **2005**, *21*, 3787–3793. [CrossRef] [PubMed]

21. Fan, F.; Cui, B.; Zhang, T.; Qiao, G.; Ding, G.; Wen, X. The Temporal Transcriptomic Response of *Pinus massoniana* Seedlings to Phosphorus Deficiency. *PLoS ONE* **2014**, *8*, e105068. [CrossRef] [PubMed]

22. Livak, K.J.; Schmittgen, T.D. Analysis of Relative Gene Expression Data Using Real-Time Quantitative PCR and the $2-\Delta\Delta CT$ Method. *Methods* **2001**, *25*, 402–408. [CrossRef] [PubMed]

23. Aroca, R.; Ferrante, A.; Vernieri, P.; Chrispeels, M.J. Drought, abscisic acid and transpiration rate effects on the regulation of PIP aquaporin gene expression and abundance in *Phaseolus vulgaris* plants. *Ann. Bot.* **2006**, *98*, 1301–1310. [CrossRef] [PubMed]

24. Li, J.; Li, G.; Gao, S.; Martinez, C.; He, G.; Zhou, Z.; Huang, X.; Lee, J.H.; Zhang, H.; Shen, Y.; et al. Arabidopsis transcription factor ELONGATED HYPOCOTYL5 plays a role in the feedback regulation of phytochrome a signaling. *Plant Cell Online* **2010**, *22*, 3634–3649. [CrossRef] [PubMed]

25. Yamawaki, S.; Yamashino, T.; Nakanishi, H.; Mizuno, T. Functional characterization of HY5 homolog genes involved in early light-signaling in *Physcomitrella patens*. *Biosci. Biotechnol. Biochem.* **2011**, *75*, 1533–1539. [CrossRef] [PubMed]

26. Prasad, B.R.V.; Kumar, S.V.; Nandi, A.; Chattopadhyay, S. Functional interconnections of HY1 with MYC2 and HY5 in *Arabidopsis* seedling development. *BMC Plant Biol.* **2012**, *12*, 37. [CrossRef] [PubMed]

27. Song, H.-R. Interaction between the Late Elongated hypocotyl (LHY) and Early flowering 3 (ELF3) genes in the *Arabidopsis* circadian clock. *Genes Genom.* **2012**, *34*, 329–337. [CrossRef]

28. Wang, X.; Zeng, F.S.; Zhan, Y.G.; He, Z.L. Cloning and functional analysis of the circadian gene LHY promoter in fraxinus mandshurica. *Plant Physiol. J.* **2014**, *50*, 1675–1682. [CrossRef]

29. Tang, W.; Charles, T.M.; Newton, R.J. Overexpression of the pepper transcription factor CaPF1 in transgenic virginia pine (Pinus Virginiana Mill.) confers multiple stress tolerance and enhances organ growth. *Plant Mol. Biol.* **2005**, *59*, 603–617. [CrossRef] [PubMed]

30. Guilfoyle, T.J.; Hagen, G. Auxin response factors. *Curr. Opin. Plant Biol.* **2007**, *10*, 453–460. [CrossRef] [PubMed]

31. Liscum, E.; Reed, J.W. Genetics of Aux/IAA and ARF action in plant growth and development. *Plant Mol. Biol.* **2002**, *49*, 387–400. [CrossRef] [PubMed]

32. De Diego, N.; Perez-Alfocea, F.; Cantero, E.; Lacuesta, M.; Moncalean, P. Physiological response to drought in radiata pine: Phytohormone implication at leaf level. *Tree Physiol.* **2012**, *32*, 435–449. [CrossRef] [PubMed]

33. Wu, J.; Zhang, Y.; Yin, L.; Qu, J.; Lu, J. Linkage of cold acclimation and disease resistance through plant-pathogen interaction pathway in *Vitis amurensis* grapevine. *Funct. Integr. Genom.* **2014**, *14*, 741–755. [CrossRef] [PubMed]

34. Bonello, P.; Gordon, T.R.; Herms, D.A.; Wood, D.L.; Erbilgin, N. Nature and ecological implications of pathogen-induced systemic resistance in conifers: A novel hypothesis. *Physiol. Mol. Plant Pathol.* **2006**, *68*, 95–104. [CrossRef]

35. Kong, Q.; Qu, N.; Gao, M.; Zhang, Z.; Ding, X.; Yang, F.; Li, Y.; Dong, O.X.; Chen, S.; Li, X.; et al. The MEKK1-MKK1/MKK2-MPK4 kinase cascade negatively regulates immunity mediated by a mitogen-activated protein kinase kinase kinase in *Arabidopsis*. *Plant Cell* **2012**, *24*, 2225–2236. [CrossRef] [PubMed]

36. Li, Z.T.; Dhekney, S.A.; Gray, D.J. PR-1 gene family of grapevine: a uniquely duplicated PR-1 gene from a Vitis interspecific hybrid confers high level resistance to bacterial disease in transgenic tobacco. *Plant Cell Rep.* **2011**, *30*, 1–11. [CrossRef] [PubMed]

37. Haider, M.S.; Kurjogi, M.M.; Khalil-Ur-Rehman, M.; Fiaz, M.; Pervaiz, T.; Jiu, S.T.; Jia, H.F.; Chen, W.; Fang, J.G. Grapevine immune signaling network in response to drought stress as revealed by transcriptomic analysis. *Plant Physiol. Biochem.* **2017**, *121*, 187–195. [CrossRef] [PubMed]

38. Bhardwaj, J.; Chauhan, R.; Swarnkar, M.K.; Chahota, R.K.; Singh, A.K.; Shankar, R.; Yadav, S.K. Comprehensive transcriptomic study on horse gram (*Macrotyloma uniflorum*): De novo assembly, functional characterization and comparative analysis in relation to drought stress. *BMC Genom.* **2013**, *14*, 647. [CrossRef] [PubMed]

39. Min, H.; Chen, C.; Wei, S.; Shang, X.; Sun, M.; Xia, R.; Liu, X.; Hao, D.; Chen, H.; Xie, Q. Identification of drought tolerant mechanisms in maize seedlings based on transcriptome analysis of recombination inbred lines. *Front. Plant Sci.* **2016**, *7*, 1080. [CrossRef] [PubMed]

40. Agarwal, P.K.; Jha, B. Transcription factors in plants and ABA dependent and independent abiotic stress signalling. *Biol. Plant.* **2010**, *54*, 201–212. [CrossRef]

41. Wang, Z.-Y.; Xiong, L.; Li, W.; Zhu, J.-K.; Zhu, J. The plant cuticle is required for osmotic stress regulation of abscisic acid biosynthesis and osmotic stress tolerance in *Arabidopsis*. *Plant Cell* **2011**, *23*, 1971–1984. [CrossRef] [PubMed]

42. Quan, W.; Ding, G. Dynamic of Volatiles and Endogenous Hormones in *Pinus massoniana* Needles under Drought Stress. *Sci. Silvae Sin.* **2017**, *53*, 49–54.

43. Ma, Y.; Szostkiewicz, I.; Korte, A.; Moes, D.; Yang, Y.; Christmann, A.; Grill, E. Regulators of PP2C phosphatase activity function as abscisic acid sensors. *Science* **2009**, *324*, 1064–1068. [CrossRef] [PubMed]

44. Liu, L.; Hu, X.; Song, J.; Zong, X.; Li, D.; Li, D. Over-expression of a *Zea mays* L. protein phosphatase 2C gene (ZmPP2C) in *Arabidopsis thaliana* decreases tolerance to salt and drought. *J. Plant Physiol.* **2009**, *166*, 531–542. [CrossRef] [PubMed]

45. Zhang, H.; Mao, X.; Wang, C.; Jing, R. Overexpression of a common wheat gene TaSnRK2.8 enhances tolerance to drought, salt and low temperature in *Arabidopsis*. *PLoS ONE* **2010**, *5*, e16041. [CrossRef] [PubMed]

46. Gao, S.-Q.; Chen, M.; Xu, Z.-S.; Zhao, C.-P.; Li, L.; Xu, H.-J.; Tang, Y.-M.; Zhao, X.; Ma, Y.-Z. The soybean GmbZIP1 transcription factor enhances multiple abiotic stress tolerances in transgenic plants. *Plant Mol. Biol.* **2011**, *75*, 537–553. [CrossRef] [PubMed]

47. Dziri, S.; Hosni, K. Effects of cement dust on volatile oil constituents and antioxidative metabolism of *Aleppo pine* (*Pinus halepensis*) needles. *Acta Physiol. Plant.* **2012**, *34*, 1669–1678. [CrossRef]

48. Zeng, Q.-Y.; Lu, H.; Wang, X.-R. Molecular characterization of a glutathione transferase from *Pinus tabulaeformis* (Pinaceae). *Biochimie* **2005**, *87*, 445–455. [CrossRef] [PubMed]

49. Noctor, G.; Gomez, L.; Vanacker, H.; Foyer, C.H. Interactions between biosynthesis, compartmentation and transport in the control of glutathione homeostasis and signalling. *J. Exp. Bot.* **2002**, *53*, 1283–1304. [CrossRef] [PubMed]

50. Rao, A.S.V.C.; Reddy, A.R. Glutathione reductase: A putative redox regulatory system in plant cells. In *Sulfur Assimilation and Abiotic Stress in Plants*; Khan, N.A., Singh, S., Umar, S., Eds.; Springer: Berlin, Germany, 2008; pp. 111–147.

© 2018 by the authors. Licensee MDPI, Basel, Switzerland. This article is an open access article distributed under the terms and conditions of the Creative Commons Attribution (CC BY) license (http://creativecommons.org/licenses/by/4.0/).

forests

MDPI

Article

Comprehensive Analysis of the Cork Oak (*Quercus suber*) Transcriptome Involved in the Regulation of Bud Sprouting

Ana Usié [1,2], Fernanda Simões [3], Pedro Barbosa [1], Brígida Meireles [1], Inês Chaves [1], Sónia Gonçalves [1,4], André Folgado [1], Maria H. Almeida [5], José Matos [3,6] and António M. Ramos [1,2,*]

[1] Centro de Biotecnologia Agrícola e Agro-alimentar do Alentejo (CEBAL), Instituto Politécnico de Beja (IPBeja), Rua Pedro Soares, s.n.-Campus IPBeja/ESAB, Apartado 6158, 7801-908 Beja, Portugal; ana.usie@cebal.pt (A.U.); pedro.barbosa@cebal.pt (P.B.); brigida.meireles@cebal.pt (B.M.); ines.chaves@cebal.pt (I.C.); sonia.goncalves@cebal.pt (S.G.); andrejoaofolgado@gmail.com (A.F.)
[2] Instituto de Ciências Agrárias e Ambientais Mediterrânicas (ICAAM), Universidade de Évora, Núcleo da Mitra, Apartado 94, 7006-554 Évora, Portugal
[3] Instituto Nacional de Investigação Agrária e Veterinária, I.P. (INIAV), Av. da República, Quinta do Marquês (edifício sede), 2780-157 Oeiras, Portugal; fernanda.simoes@iniav.pt (F.S.); jose.matos@iniav.pt (J.M.)
[4] Wellcome Trust Sanger Institute, Wellcome Genome Campus Hinxton, Cambridge CB10 1SA, UK
[5] Centro de Estudos Florestais, Instituto Superior de Agronomia, Universidade de Lisboa, Tapada da Ajuda 1349-017 Lisboa, Portugal; nica@isa.ulisboa.pt
[6] Centre for Ecology, Evolution and Environmental Changes (cE3c), Faculdade de Ciências da Universidade de Lisboa, Edifício C2, 5° Piso, Sala 2.5.46 Campo Grande, 1749-016 Lisboa, Portugal
* Correspondence: marcos.ramos@cebal.pt; Tel.: +351-284-314-399

Received: 7 October 2017; Accepted: 29 November 2017; Published: 6 December 2017

Abstract: Cork oaks show a high capacity of bud sprouting as a response to injury, which is important for species survival when dealing with external factors, such as drought or fires. The characterization of the cork oak transcriptome involved in the different stages of bud sprouting is essential to understanding the mechanisms involved in these processes. In this study, the transcriptional profile of different stages of bud sprouting, namely (1) dormant bud and (2) bud swollen, vs. (3) red bud and (4) open bud, was analyzed in trees growing under natural conditions. The transcriptome analysis indicated the involvement of genes related with energy production (linking the TCA (tricarboxylic acid) cycle and the electron transport system), hormonal regulation, water status, and synthesis of polysaccharides. These results pinpoint the different mechanisms involved in the early and later stages of bud sprouting. Furthermore, some genes, which are involved in bud development and conserved between species, were also identified at the transcriptional level. This study provides the first set of results that will be useful for the discovery of genes related with the mechanisms regulating bud sprouting in cork oak.

Keywords: cork oak; transcriptome; bud development; gene expression

1. Introduction

Cork oak (*Quercus suber* L.) plays an important environmental, social, and economic role in the Mediterranean ecosystems known as "Montado" and "Dehesa", in Portugal and Spain, respectively. The ability of cork oak to produce cork in a sustainable manner is the basis for an industry that is unique in the world. However, over the last 20 years, the Iberian Peninsula has witnessed a reduction in the number of trees—due to drought, extreme temperatures, pests, and fires, among other factors, which threatens the rural economy in this part of Europe and increases the vulnerability to wildfires [1].

The vegetative bud phenology of long-lived species is crucial to their productivity, adaptability, and distribution [2]. The frequency of droughts in the Mediterranean is increasing significantly due to climate change, which suggests an aggravation of environmental conditions that are likely to increase the severity of water stress in plants [3]. Global warming is expected to modify the length of the growing season and distribution of forest tree species, changing the timing of phenological events and possibly causing frost or drought injuries, or even a failure to produce mature fruits and seeds. It was already reported that long-term exposure of young cork oak trees to contrasting temperatures impacts the leaf metabolites and gene expression profiles of key enzymes of phenolic metabolism [4]. Most Mediterranean plant species, including cork oak, have a good sprouting capacity after disturbance, displaying the ability to re-sprout from basal buds when stems or crowns are severely damaged, which is of great importance for species survival. However, cork oak is the only oak able to quickly and effectively re-sprout after fire from epicormic buds, which are positioned underneath the bark, showing a competitive advantage over coexisting woody plants. Thus, cork oak is one of the best-adapted trees persisting in ecosystems with recurrent fire-exposure, making it one of the best candidates for reforestation programs.

In pedunculate oak (*Quercus robur*), several QTL (quantitative trait loci) for bud burst and height growth have been identified [5], using the approach described by Saitagne and colleagues [6], based on a double-pseudo-testcross mapping strategy. Moreover, in *Populus*, several genes (*PHYB1*, *PHYB2*, *ABI3*, and *ABI1B*) mapped to positions where QTL for bud set and/or bud burst were identified [7], providing further support for their involvement in the regulation of bud sprouting. The *PHYB1* and *PHYB2* genes are phytochromes photoreceptors, which absorb both red light and far-red light and act as a biological switch to activate/deactivate plant growth. The *ABI3* and *ABI1B* genes are required for the establishment of dormancy, both being involved in signal transduction.

Analysis of differential expression for genes involved in bud burst was performed in sessile oak (*Quercus petraea*), which resulted in the identification of a set of relevant candidate genes for signaling the pathway of bud burst as well as hundreds of expressed-sequence-tags (ESTs) [8]. Recently, an oak gene expression atlas was also generated for two sympatric oak species, *Quercus robur* (pedunculated oak) and *Quercus petraea* [9], which identified genes associated with vegetative bud phenology and contributed relevant information for the annotation of the pedunculated oak's genome. The gene expression studies performed in cork oak have mainly targeted the identification of genes involved in cork formation, and several candidate genes have been revealed [10].

Furthermore, a multiple tissue transcriptome database was compiled, covering multiple developmental stages and physiological conditions [11]. Several studies have reported differentially expressed transcripts in different stages of development, such as acorn development [12].

Despite the knowledge that has been produced for oak species—the molecular mechanisms in cork oak underlying bud set, bud dormancy, and bud burst—still remain unclear. Additional information is needed to understand the genetic mechanisms underlying signaling and regulation in the transition from dormant to active bud development, as well as to characterize the genetic response to phenological events. Thus, in order to reveal the mechanisms involved in bud sprouting of cork oak, a whole transcriptome approach was carried out using data generated using the 454 sequencing platform. A total of four different stages of bud sprouting development, from bud dormant to bud burst, were analyzed, in order to assess the differences in gene expression over the stages of bud sprouting development.

2. Materials and Methods

2.1. Sample Collection

Bud samples were collected from eight *Quercus suber*—individuals from different origins. The eight different genotypes were selected in order to capture the wider species-level transcriptomic mechanisms associated with bud sprouting development. These trees are part of a provenance assay growing under natural conditions, established in Portugal in 1998 at Monte de Fava (Ermidas do Sado, Portugal). Samples were collected in different phases of bud development. Buds were cut from selected branches presenting one of the following development stages: (1) dormant bud, (2) bud swollen, (3) red bud, and (4) open bud (Figure 1). The samples were immediately immersed in RNA Later (AMBION, Ambion, Inc., Austin, TX, USA) and, upon arrival at the lab, stored at −80 °C for RNA extraction.

Figure 1. Development stages of bud sprouting in *Quercus suber*. Earlier stages: (**1**) dormant bud and (**2**) bud swollen. Later stages: (**3**) red bud and (**4**) open bud.

2.2. Total RNA Extraction and Sequencing

Bud samples were homogenized with a speed mill (AnalyticJena, Jena, Germany). The total RNA was extracted using the RNAQUEOUS extraction kit (AMBION, Ambion, Inc., Austin, TX, USA) and analyzed for quality using a Bioanalyzer. Reverse transcription was performed with the Mint cDNA Synthesis kit from Evrogen (Evrogen JSC, Moscow, Russia), using oligo d(T) Primer. Four cDNA libraries were constructed (two from pooled stages 1 and 2, and two from pooled stages 3 and 4). One library from each pooled stage was normalized by the Duplex-Specific Nuclease-technology, which resulted in a total of four cDNA libraries (two non-normalized and two normalized). The constructed libraries were sequenced using the 454 GS FLX Titanium platform (Roche, Basel, Switzerland). The sequence data analyzed in this manuscript are available from the NCBI Sequence Read Archive under the accession numbers SRX2642267, SRX2642266, ERX143072, and ERX133073.

2.3. Sequence Data and Transcriptome Assembly

The 454 single-end read sequences from both library types, normalized and non-normalized, were pre-processed, according to different criteria, using a pipeline combining a custom Perl script (https://github.com/anausie/cebal/blob/master/preProcessing454.pl) and open-source tools, namely Mothur [13], for quality trimming, and Sequence Cleaner (https://sourceforge.net/projects/seqclean/). These procedures were performed in order to remove adaptors, barcodes, and poly-A or poly-T tails from the read sequences as well as remove/trim from the dataset reads with low average quality and a certain number of undetermined nucleotides (N's). All the pre-processed reads were then used to perform a de novo transcriptome assembly using MIRA (Mimicking Intelligent Read Assembly)

4.9.5_2 [14], with the following parameters: -job = denovo, est, accurate, 454; -GE:not10:amm = no:mps = 100:kpmf = 90; -NW: cmrnk = warn; -SK:not = 10:mmhr = 10; -AS:mrpc = 2.

2.4. Read Mapping and Differential Expression Analysis

The pre-processed reads from each non-normalized library were aligned to the assembled contigs using BWA-mem (BWA: Burrows-Wheeler Aligner) with default parameters [15]. The mapping results were processed with Samtools [16], and only the reads that mapped to a unique location (UMR—uniquely mapped reads) were kept for further analyses. In addition, coding regions within assembled contigs were predicted with Transdecoder [17].

The featureCounts tool [18] was used to create a table with the counts for the UMRs for each transcript, required for edgeR, a Bioconductor software package [19], used to perform the differential expression analysis. Genes with low counts across the dataset were discarded, in line with edgeR guidelines (minimum of 6 reads as cutoff), and a TMM (Trimmed mean of M-values) normalization was applied to normalize library sizes before integrating them in the statistical model. EdgeR works on replicated data considering either the biological and technical variability between conditions. This variability is referred to as the Biological Coefficient Variation (BCV). Given that no replicates were available for this dataset, we considered the BCV to be 0.1, in line with edgeR recommendations. From the total list of differentially expressed genes, we filtered out those with an FDR (False Discovery Rate) value $\leq$0.05.

2.5. Validation of Differentially Expressed Genes by Quantitative Real-Time PCR Analysis

To confirm the differential expression results obtained with edgeR, quantitative real-time PCR (qPCR) assays were performed using 9 randomly chosen genes. The primers were designed using Primer3Plus software [20] (Table S1). Reverse transcription was performed using the QuantiTect Reverse Transcription Kit (Qiagen, Hilden, Germany) with 1 µg of Total RNA following the manufacturer's instructions. Relative expression quantification was performed with an iQ5 system (BioRad, USA, Hercules, CA,) using the SsoAdvanced Universal SYBR Green Supermix (BioRad, Hercules, CA, USA) and 250 nM of each primer in a final volume of 20 µL. All samples were run in triplicate and a no template control (NTC) was used for every primer pair.

The following program was used for all reactions: 95 °C for 10 min, 45 cycles at 95 °C for 10 s, 60 °C for 15 s, 72 °C for 15 s. A melting curve was generated for each reaction to assure specificity of the primers and the presence of primer-dimer. Primers' efficiencies were assessed using a serial dilution of cDNA stock. The change fold was calculated using the mathematical model described by Pfaffl [21] using four reference genes (*Act, CACs, EF-1α*, and *β-Tub*) previously reported for *Quercus suber* L. [22].

2.6. Functional Annotation

The predicted coding sequences (CDS) from Transdecoder 2.01 were functionally annotated, using a blast against the non-redundant protein plant sequence database from NCBI, with an e-value of 1×10^{-5} [23]. InterproScan 5.18.57 was used to find the protein domains and identify the Gene Ontology (GO) terms. Additionally, it also assigned KEGG (Kyoto Encyclopedia of Genes and Genomes) information to the sequences by identifying enzymes' EC numbers and the corresponding KEGG pathways [24]. All these results were analyzed with Cytoscape 3.3 and CateGOrizer [25,26].

3. Results

3.1. Preprocessing of the Sequence Data and De Novo Transcriptome Assembly

Four libraries (two normalized and two non-normalized) were constructed using RNA extracted from *Q. suber* buds were sampled at different development stages. A total of 2,245,526 reads were produced using the 454 GS FLX Titanium platform, ranging from 49 to 1201 bp (base-pairs) in length. By the end of the pre-processing step, a total of 1,772,150 reads remained in the dataset, which represented 78.9% of the initial number of reads. These results are displayed in Table 1.

Table 1. Summary statistics of the 454 sequence data preprocessing step.

Library	Raw Reads		Processed Reads	
	Number	\<AL\>	Number	\<AL\>
Non-normalized 1	566,726	574.1	452,455	519.9
Non-normalized 2	513,382	577.8	404,242	522.9
Normalized 1	595,388	568.4	464,611	523.8
Normalized 2	570,030	560.7	450,842	518.1
Total	2,245,526	-	1,772,150	-

\<AL\>: average length of reads in base pairs.

A de novo assembly of the cork oak transcriptome involved in the regulation of bud sprouting was generated with MIRA 4.9.5_2. A total of 117,094 contigs were generated, of which 115,935 were larger than 200 bp.

3.2. Read Mapping, Differential Expression

The pre-processed reads from each non-normalized library were aligned to the assembled contigs using the BWA-mem algorithm. A total of 815,287 reads were mapped, which represented 95.2% of the total, while the number of UMR was 358,785. These results are indicated in Table 2.

Table 2. Results obtained for read mapping, with BWA, against the transcriptome assembly.

Library	Reads Used	Mapped Reads (%)	UMR (%)
Non-normalized 1	452,455	430,935 (95.2%)	197,521 (43.7%)
Non-normalized 2	404,242	384,352 (95.1%)	161,264 (39.9%)
Total	856,697	815,287 (95.2%)	358,785 (41.9%)

As described above, the coding regions within transcripts were predicted, and a total of 57,034 coding regions (predicted genes) were obtained. Following the procedures for the differential expression analysis, we obtained a total of 58 differentially expressed genes between the two pools of bud development stages: 38 down-regulated—genes with a higher expression level or only expressed in early stages (pool 1)—and 20 up-regulated—genes with a higher expression level or only expressed in later stages (pool 2).

3.3. Functional Annotation

The set of protein sequences where coding regions were predicted by Transdecoder (57,034 predicted genes) was annotated with blastp (protein blast) against the non-redundant NCBI plants database, which resulted in 90.7% (51,728) of the predicted genes having at least one hit. Taking into account the best hit for each predicted gene, a distribution of occurrences over the annotated species was performed, yielding a total number of 692 plant species (Figure 2). The majority of the hits identified were against *Vitis vinifera* (8.9% of the best hits). Cork oak, as well as most of the species found in the top 10 of the best hits, belong to rosids, one of the major clades of order.

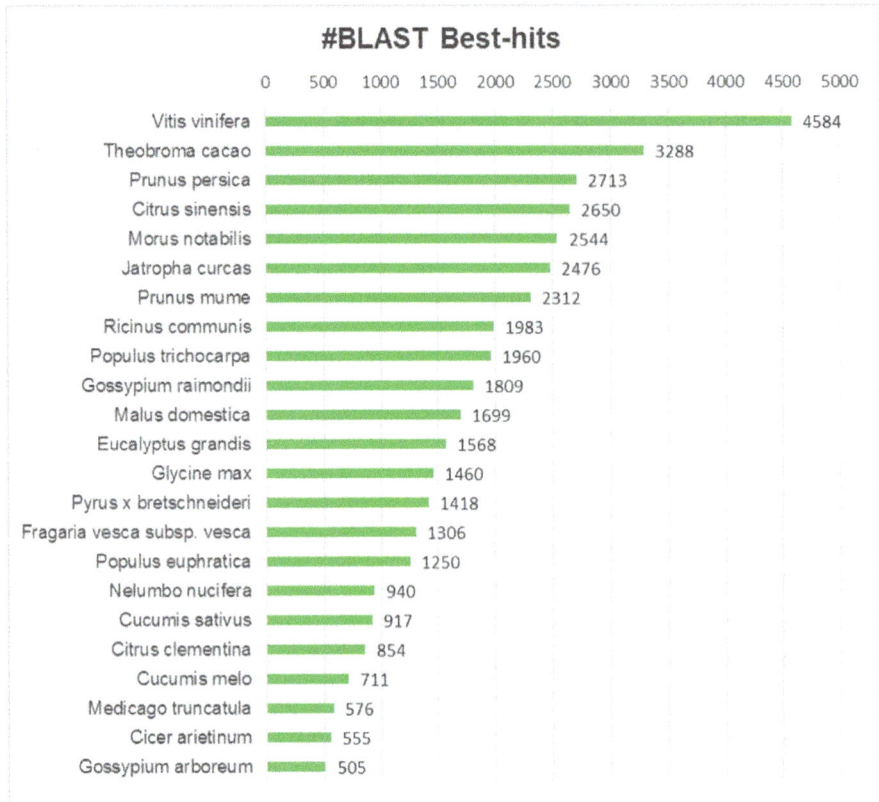

Figure 2. Blast top hits by species. The most representative species with a minimum of 500 hits are represented.

Additionally, InterPro was used to identify protein domains, functional classes of GOs (gene ontology) and KEGG pathways associated with the sequences of the predicted genes. A total of 48,022 sequences had at least one protein domain, and 29,383 sequences mapped to at least one GO term, covering a total of 1602 different GO terms, of which 40.4% belonged to Biological Process (BP), 47.1% to Molecular Function (MF) and 12.5% to Cellular Components (CC). Moreover, 3680 sequences were associated with at least one KEGG pathway, for a total of 114 KEGG pathways and 387 different enzymes (Figure 3).

The GO terms were further analyzed with CateGOrizer which mapped the GOs against the Plant GOSlim database. We obtained a total of 41 subcategories of GO terms belonging to BP, 24 to MF, and 24 to CC categories. BP, MF, and CC categories contained a total of 19,881, 11,169, and 25,441 gene sequences, respectively.

Regarding the 58 differentially expressed genes, 36 mapped to at least one GO over the total number of 52 different GO terms identified. CateGOrizer identified 17 subcategories of GO terms belonging to BP, 12 to MF, and 7 to CC, over 22, 2622, and 8 different GOs, respectively (Figures 4–6). With respect to KEGG pathways, only four genes, codifying a different enzyme each, were associated with eight KEGG pathways.

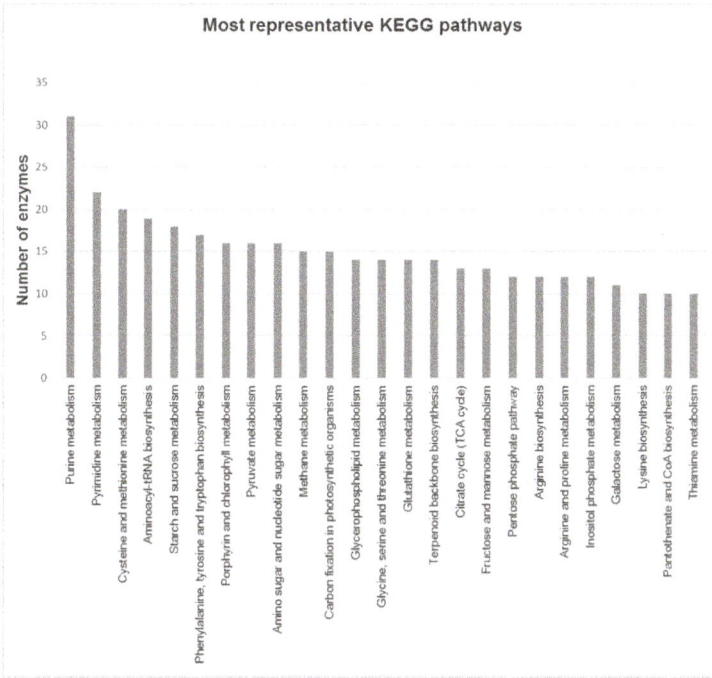

Figure 3. The most representative KEGG pathways associated with all predicted genes. Only pathways with at least eight associated enzymes are represented.

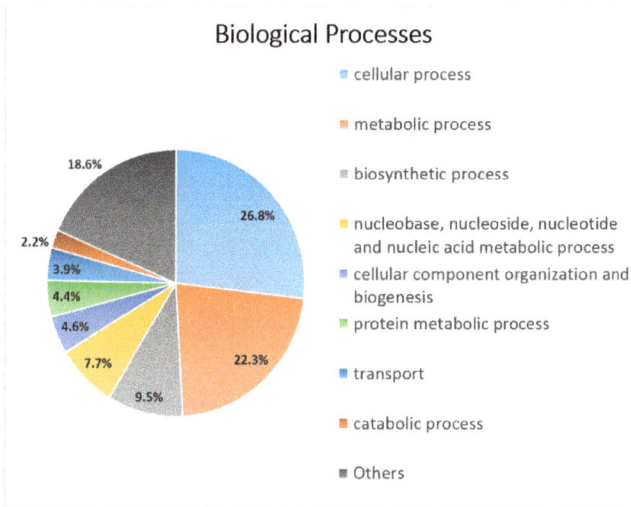

Figure 4. Subcategories of GOs identified by CateGOrizer within the Biological Processes over the whole set of predicted genes.

Molecular Function

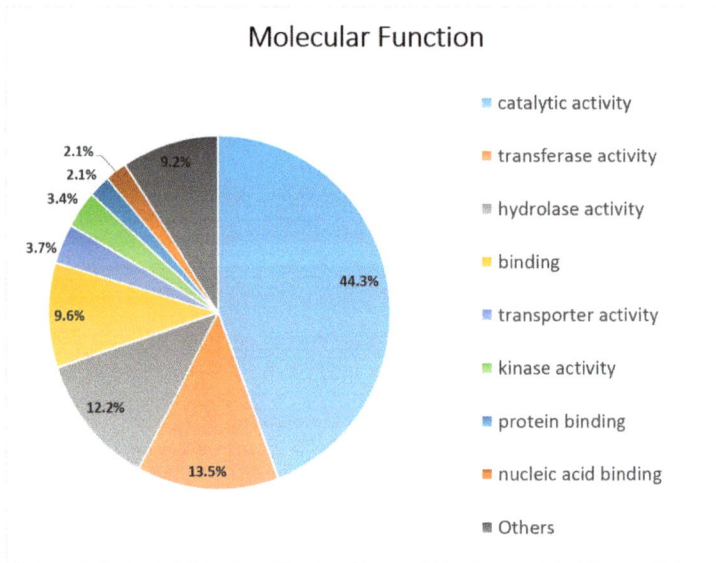

- catalytic activity
- transferase activity
- hydrolase activity
- binding
- transporter activity
- kinase activity
- protein binding
- nucleic acid binding
- Others

Figure 5. Subcategories of GOs identified by CateGOrizer within the Molecular Function over the whole set of predicted genes.

Cellular Component

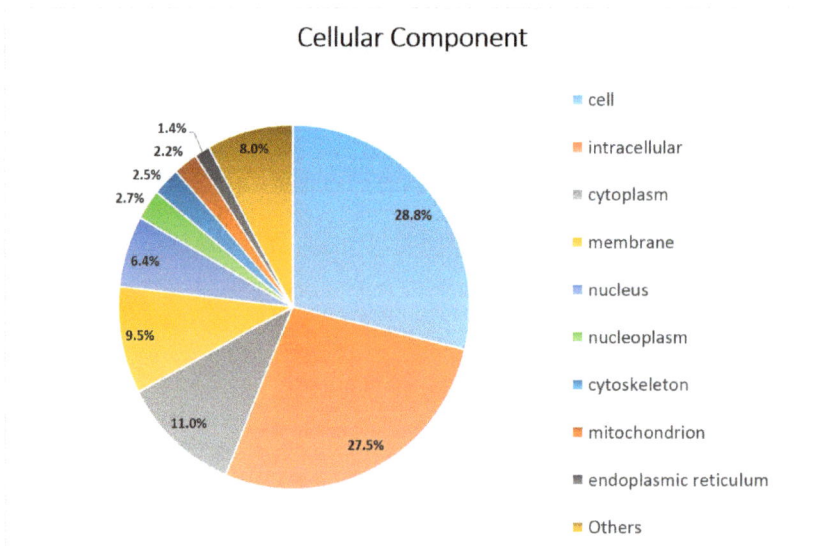

- cell
- intracellular
- cytoplasm
- membrane
- nucleus
- nucleoplasm
- cytoskeleton
- mitochondrion
- endoplasmic reticulum
- Others

Figure 6. Subcategories of GOs identified by CateGOrizer within the Cellular Components over the whole set of predicted genes.

The vast majority of differentially expressed genes associated with the MF GO terms were down-regulated and/or annotated as core histones (Figure 7). The DE (differentially expressed) genes associated with the CC GO terms with mostly down-regulated and/or annotated as r-proteins and histones (Figure 8).

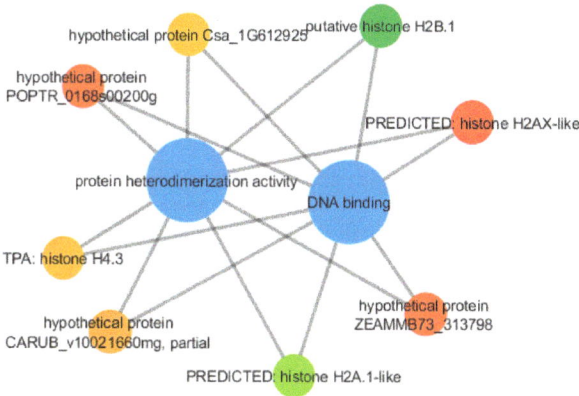

Figure 7. Differentially expressed genes associated with some Molecular Function GO terms identified. Blue nodes represent the MF GO terms while the other nodes represent the associated DE genes. The color of each DE gene node follows an RGB color scale, going from the most down-regulated in red to the most up-regulated in green.

Briefly, the results obtained with the differential expression analysis revealed several candidate genes related with bud sprouting, from which a subset is displayed in Table 3.

Table 3. Subset of differentially expressed genes identified between the earlier and later stages. A gene with a positive logFC (logarithm fold change) value is (more) expressed in the later stages, while a gene with a negative logFC value is (more) expressed in the earlier stages.

Annotation	LogFC	*p*-Value	Gene Identifier ID
unnamed protein product	8.48	1.22×10^{-12}	296082254
Ribosomal protein S3Ae	7.88	4.37×10^{-5}	976900419
putative histone H2B.1	7.83	7.13×10^{-5}	703085592
PREDICTED: 60S ribosomal protein L8-3	7.45	1.67×10^{-6}	449434174
basic blue copper family protein	7.38	3.17×10^{-6}	224054286
serin/threonine protein kinase	7.38	3.17×10^{-6}	38343920
hypothetical protein PRUPE_ppa012332mg	7.3	6.03×10^{-6}	595797137
PREDICTED: pentatricopeptide repeat-containing protein At1g74750-like	7.3	6.03×10^{-6}	470127288
40S ribosomal protein S17C	7.13	2.20×10^{-5}	313586437
60S ribosomal L38	−7.15	1.15×10^{-5}	728829564
PREDICTED: succinate dehydrogenase	−3.34	3.39×10^{-5}	449455896
hypothetical protein EUGRSUZ_G02560 (myo-inositol oxynase)	−4.57	1.55×10^{-7}	629099270
Translation elongation factor 1 alpha	−5.2	1.53×10^{-10}	110224776
PREDICTED: uncharacterized protein	−7.22	6.03×10^{-6}	645275588
PREDICTED: polyubiquitin	−7.22	6.03×10^{-6}	743820616
PREDICTED: protein NUCLEAR FUSION DEFECTIVE 2	−7.22	6.03×10^{-6}	470125885
aquaporin TIP2;2	−7.22	6.03×10^{-6}	383479044
Ribosomal protein L31e family protein	−7.48	4.66×10^{-7}	590724775
RecName: Full = Agglutinin; AltName: Full = CCA	−7.7	7.00×10^{-8}	48428322
PREDICTED: histone H2AX-like	−7.94	3.11×10^{-9}	658055874
agglutinin isoform	−8.1	2.68×10^{-10}	85376265
chlorophyll a/b-binding protein	−8.84	9.97×10^{-15}	2804572

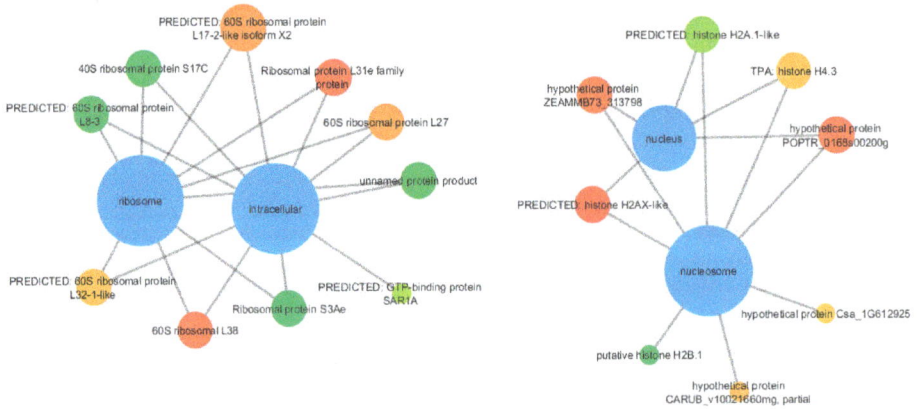

Figure 8. Differentially expressed genes associated with some Cellular Components (CC) GO terms identified. Blue nodes represent the CC GO terms while the other nodes represent the associated DE genes. The color of each DE gene node follows an RGB color scale, going from the most down-regulated in red to the most up-regulated in green.

3.4. qPCR Validation

In order to validate the results obtained by the bioinformatics analyses, a total of nine DE genes were randomly selected to assess their expression by qPCR. The qPCR assay for the selected transcripts shows an expression pattern similar to the one obtained by the bioinformatics analyses. For several of the genes tested, the change fold value is higher in the RNA-seq data when compared to the qPCR results (Figure 9), which may be explained by the amplification step required for sequencing of the transcriptome as well as some degradation of the stored RNA.

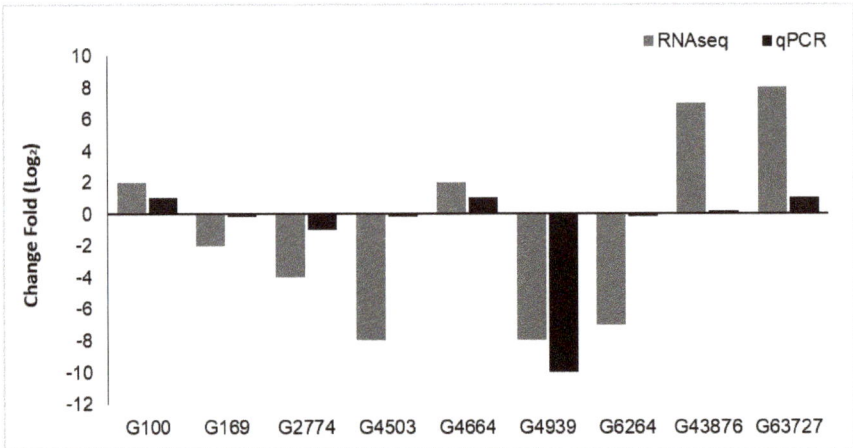

Figure 9. Change fold for each of the transcripts tested for validation. The values are represented as the Log2 of the normalized expression value of both RNA-seq and qPCR data.

4. Discussion

Cork oak is an important natural resource with economic and ecological impact in the Mediterranean area. This work adds relevant information to be used in cork oak biology, with the potential to enable genetics and genomics approaches targeting a deeper molecular knowledge of genes intervening in processes dealing with the ecosystem responses to external factors, which can affect bud sprouting.

In previous studies, several stages of bud sprouting development—such as bud burst, bud endodormancy, bud ecodormancy, and bud dormancy—were examined in other *Quercus* species [8,9,27] using approaches based on suppression subtract hybridization (SSH), microarrays, and ESTs. These studies identified some candidate genes involved in the regulation of bud sprouting, but the molecular mechanisms associated with bud sprouting development in cork oak remain uncharacterized.

In order to tackle this limitation, we performed a transcriptome characterization analysis in four stages, grouped as an early stage (dormant bud and bud swollen stages) and a later stage (red bud and open bud stages), after which a group of candidate genes was identified. Evidence for a differential expression pattern of genes involved in key mechanisms of cork oak bud sprouting was detected, even though the risk of false-positive results was increased by the fact that biological replicates were not available.

In order to characterize the transcriptome involved in bud sprouting development, a classification by KEGG Orthology (KO) of the most representative KEGG pathways (Figure 3) associated with the assembled transcripts was performed. The KOs identified included nucleotide metabolism, amino acid metabolism, carbohydrate metabolism, energy metabolism, and lipid metabolism. The results indicated that the transcriptome of bud sprouting development is mostly focused on the production of essential precursors and metabolites for the synthesis of macromolecules, such as polysaccharides, which represents most of the cell-wall biomass and energy production. Thus, at the initial stages, a significant effort is made to supply the plant with the means required for growth, including the considerable amount of energy needed.

4.1. A High Input of Energy Is Required during Bud Sprouting in Cork Oak

In plants, the presence of the cyclic flux of the tricarboxylic acid (TCA) cycle is not necessary if the ATP demand is low or other sources of ATP are available, such as through photosynthesis [28]. In the cork oak bud sprouting transcriptome, the cyclic flux of the TCA cycle is active since all of the enzymes involved in this cyclic flux are expressed, indicating that a high energy production is required in the development of bud sprouting.

The succinate dehydrogenase enzyme (EC 1.3.5.1), also known as complex II, plays an important role linking the TCA cycle and the electron transport system (ETS) by catalyzing the oxidation of succinate to fumarate and the reduction of ubiquinone to ubiquinol. Both the TCA cycle and ETS are involved in the production of energy for the cell via the aerobic respiratory chain [29]. In the earlier stages of development, a high consumption of energy is expected and required for growth [30]. Hence, the over-expression pattern found for this gene in the early stages is in accordance with the metabolic state and needs of the plant during the initial phases of bud burst.

Moreover, a basic blue copper family protein, belonging to the cupredoxins family, was more significantly expressed in the later stages. It is well known that cupredoxins serve as mobile electron carriers in a variety of charge transport systems [31]. This suggests that, besides linking the TCA cycle to ETS via succinate dehydrogenase in the early stages, the ETS is fully functional due to the expression of the blue copper family protein in the later stages of development.

The main role of Acetyl-CoA is to deliver the acetyl group to the TCA cycle for energy production. However, it is also used by the Acetyl-CoA carboxylase enzyme to produce malonyl-CoA. It has been generally accepted that this enzyme, expressed in the transcriptome analyzed in this study, is almost exclusively responsible for the production of malonyl-CoA. This molecule is the precursor for the formation of flavonoids, interacting with Coumoraloyl-CoA, a product of the phenylpropanoid pathway [32]. Essential enzymes required for the transformation of phenylalanine into coumoroyl-CoA are expressed in the cork oak bud sprouting transcriptome, including phenylanine ammonia lyase (EC 4.3.1.24), cinnamate 4-hydroxylase (EC 1.14.13.11), and 4-Coumarate-CoA ligase (EC 6.2.1.12), as well as several other important enzymes for flavonoid biosynthesis, such as chalcone synthase (EC 2.3.1.74), chalcone isomerase (EC 5.5.1.6), and flavanone 3-hydrolase (EC 1.14.11.9), all of which are involved in the early steps of this pathway.

Flavonoids are secondary metabolites associated with several biological functions in plants, including the signaling of plant growth and development [33] as well as the capacity to regulate the activity of proteins responsible for cell growth. The interaction of flavonoids with auxins has been established for some time, where flavonoids control the distribution of auxins influencing developmental processes [32,34–37].

Large amounts of nitrogen are required for the synthesis of nucleic acids and proteins, thus making nitrogen a limiting resource for plant growth. Arginine amino acid chemical properties make it especially suitable to store nitrogen. Several enzymes involved in the arginine biosynthesis were expressed in the present study transcriptome. El Zein and colleagues (2011), in a sessile oak study, observed that, during leaf growth, most of the nitrogen used was provided from the stored nitrogen. Therefore, during bud sprouting development, the plant seems to be storing nitrogen via arginine synthesis, in order to use it when the leaf starts to grow [38].

The role played by the translation elongation factor 1α gene in protein and actin cytoskeleton synthesis, and possibly also in plant development, has been demonstrated in a number of eukaryotes [39,40]. A previous study in tobacco plants showed that its expression was higher in younger, developing tissues than in older or more mature tissues [41]. Moreover, the same gene was also highlighted as a candidate gene for proteomic analysis of shoot apical meristem transition from dormancy to activation in *Cunninghamia lanceolata (Lamb.) Hook* [42]. In that study, the comparison of the reactivating and active stages with the dormant stage of shoot apical meristem demonstrated that higher levels of proteins involved in translation were present, results that provide further support for the involvement of the translation elongation factor 1α gene in the processes associated with bud burst and sprouting. Additionally, in this work, this gene was over-expressed in the early stages, results that are consistent with the ones determined in those studies.

4.2. Cork Oak Bud Burst and Development Is under Tight Hormonal Regulation

Plant growth regulators can be divided into five main groups: auxins, cytokinins, gibberellins (GAs), ethylene, and abscisic acid (ABA). Auxins, cytokinins, and GAs are the ones that largely affect plant growth [43]. Several genes codifying these hormones were expressed in the transcriptome analysis. Auxins usually act together with, or in opposition to, other plant hormones, such as GAs and cytokinins. The interaction between these three hormones plays a crucial role in the regulation of bud dormancy and bud burst in many plant species [44,45].

In our work, two superfamilies of auxins were present in the transcriptome, including auxin-binding protein (ABP) 19 and ABP20. An analysis of the ABP19 and ABP20 expression pattern in the two stages of bud sprouting suggests that ABP19 is more important in the early stage, whereas ABP20's influence is larger in the later stage. In a previous work [46], ABP20 displayed high expression levels in buds, contrasting with the extremely low expression of ABP19, which provided evidence towards the involvement of ABP20 in the differentiation and development of both floral and vegetative buds. The identification of high expression genes codifying ABP20 in the later stage, despite the absence of statistical significance for differential expression, suggests that ABP20 plays a role in cork oak bud development.

Strigolactone, which is a carotenoid derivate hormone, displayed a pattern of higher expression in the early stages. This hormone regulates bud growth as it has more influence on the inhibition of axillary bud growth [47]. One relevant protein in strigolactone signaling is MAX2/RMS4 [48], which was codified by genes in the transcriptome. This protein is an F-box protein and regulates multiple targets at different stages of development in order to optimize plant growth and development [49]. The presence in the cork oak bud transcriptome of genes codifying enzymes involved in the ubiquitin-proteasome system, such as ubiquitin E3-ligase, expressed in the early stages of bud sprouting, could be linked with the involvement of MAX2 in strigolactone or ABA signaling, since the number of E3 ligases involved is more than any other hormone [50]. In the early stages, genes codifying indole-3-acetic acid (IAA), which, together with ubiquitin E3-ligase, plays a crucial role in auxin signaling [51], were also identified. Considering that auxins are capable of regulating the activity of strigolactones, which can inhibit bud growth, and that cytokinins can stimulate bud growth [52], it is essential to identify the intermediate steps to understand the mechanism of this antagonistic control.

The expression of the gibberellin 3-beta-dioxygenase 4 enzyme in the cork oak bud sprouting transcriptome, which produces GA1 or GA4 bioactive GAs, indicates that gibberellin synthesis is active during bud sprouting. In fact, it is well known that GAs influence varied development processes, such as stem elongation, germination, and dormancy. After dormancy release, the number of GAs increases when the sprouting stage begins. The chitin-inducible gibberellin-responsive protein (CIGR), codified by several genes expressed in this transcriptome, belongs to the GRAS family, which plays a role in plant growth and development. Specifically, this protein is associated with the regulation of plant height as well as leaf size, the latter of which contains young elongation tissues. The relation of CIGR to the regulation of leaf size may indicate the relevance of this protein in the last stages of the bud sprouting development where the leaf is starting to grow [53].

Cytokinins are fundamental in the regulation of the cell division and elongation processes that occur in buds and are considered to be promoters of sprouting [54]. In fact, their direct interaction with auxins influences bud burst, since sprouting begins when the cytokinin concentrations at the bud are higher than auxins. Additionally, there is evidence that these hormones can also be involved in tuber dormancy release [45].

ABA hormones are also related to plant growth. The expression of ABA-insensitive (ABIB) genes in the studied transcriptome is consistent with the association of ABA with bud dormancy, dormancy release, bud set, and bud break. For instance, in *Populus*, the ABIB1 gene was associated with bud set and bud break [7]. Moreover, there is evidence that ABA hormones regulate the synthesis of flavonoids, which, as mentioned, are also involved in plant growth and development [55].

The light-harvesting complex a/b binding protein (LHCB) is one of the most abundant membrane proteins in nature. This protein plays a fundamental role in plant adaptation to environmental stresses, protection against light stress, and control of chloroplast functions. In this study, the chlorophyll a/b binding protein was over-expressed in the early stages. The expression of the LHCB genes is regulated by multiple factors, such as the light, oxidative stress, and ABA. The expression of chlorophyll a/b binding protein can be associated with the maturation of chloroplasts, since the levels of LHCB were reported to generally increase during cold stress [56], which is a concomitant abiotic stress factor to which Mediterranean *Quercus suber* buds are exposed.

4.3. Genes Related to Water Status Play a Key Role in Cork Oak Bud Sprouting

The growth rate of a cell depends on the amount of water uptake and the capacity of the cell wall to extend. Extensive water requirements are needed for flushing and shoot elongation because the rehydration of meristems, cell expansion, and metabolic pathway recovery are indispensable for plant growth. Consequently, genes related to water stress, such as aquaporins (AQPs) and dehydrins (DHNs), have a fundamental role in the regulation of bud burst timing processes.

The transport of water across the membrane is a key factor for expansion growth. AQPs function as membrane channels that selectively transport water at the cellular level. They can be divided into two main groups or subfamilies, which include the plasma membrane intrinsic proteins (PIPs) and the tonoplast intrinsic proteins (TIPs) [57,58]. A relation between the PIPs and TIPs expression and cell expansion has been observed in several plant tissues [59,60]. However, the functional relation between water transport at the whole plant level and PIPs/TIPs expression still remains unclear, despite some studies demonstrating the role of various AQPs in root water transport [61,62]. Additionally, abiotic stresses, such as cold temperatures, induce changes in AQPs expression, mainly in PIPs and TIPs [63,64].

The accumulation of DHNs occurs in response to drought stress and low temperatures in plant tissues, therefore playing an important role in the winter dormancy, providing protection, as well as freezing tolerance and cold acclimation.

Thus, considering that the early stages of bud burst coincide with the period of the year when the coldest temperatures are observed, it is likely that both AQPs (such as TIP2,2, which was found only expressed in the early stages of bud sprouting) and DHNs would be crucial for survival of the buds in the initial stages of their development.

4.4. An Alternative Pathway for the Synthesis of Polysaccharides

In plants, UDP-Glucoronic Acid (GlcA) acts as an important precursor in the synthesis of many different polysaccharides, such as xylose, arabinose, and GalA, for cell-wall biomass [65]. UDP-GlcA is synthesized via two independent routes, which include the myo-inositol oxidation pathway (MIOP) and the nucleotide sugar oxidation pathway. Karkonen and colleagues [66] suggested that the predominant route can differ among species, tissues, and different development stages. It has been demonstrated that the MIOP is present in a variety of plant tissues since this pathway was first proposed about 40 years ago [67]. The MIOP is initialized by the catalyzation of the thermodynamically irreversible oxygenative cleavage of myo-inositol to GlcA by the myo-inositol oxynase (EC 1.13.99.1). The gene coding for this enzyme was found expressed in the transcriptome analyzed, providing support for the activation of this alternative pathway for the synthesis of polysaccharides. Moreover, in this work, this enzyme was over-expressed in the early stage, suggesting that MIOP activates in the earlier stages of development, which is in accordance with the findings of previous studies [68]. The production of UDP-GlcA is also required to generate a number of other sugars, which represent a large amount of cell-wall biomass.

4.5. Stress Response and Development Related Proteins Are Expressed in the Cork Oak Bud Transcriptome

Some genes involved in the response to biotic and abiotic stresses were differentially expressed in this work, such as the polyubiquitin and agglutinin genes. The former plays a relevant role in wound response, while the latter is responsible for the immobilization of non-pathogenic organisms. The polyubiquitin genes belong to the ubiquitin gene family and are also present in a wide variety of plants. The ubiquitin pathway is involved in the senescence and stress responses in plants, playing a relevant role in the modification of protein behavior during wound response. This gene is involved in protein degradation, in the chromatin structure, ribosome biogenesis, and cellular receptor proteins [69,70]. The agglutinin is considered a glycoprotein and is involved in the immobilization of non-pathogenic bacteria in the different plant tissues [71]. It was previously identified in a white oak study [53], where two stages of bud dormancy were compared. Since, in this work, the dormant bud is represented in the early stages, the over-expression of the gene codifying this protein is consistent with the results determined previously.

The protein kinases and phosphatases are involved in the signaling pathways, playing fundamental roles in stress signal transduction [72]. The serine/threonine protein kinase was over-expressed in the later stages of bud sprouting. This protein receives the information from the receptors that sense the changes of the environmental conditions and convert it into a response, altering the metabolism or the cell growth and division [70].

During plant growth and development, increased levels of protein synthesis are indispensable and the synthesis of new ribosomes is crucial to growth. In this work, different ribosomal proteins were over-expressed in the two stages of bud sprouting [73]. Similar to ribosomal proteins, histones were also over-expressed in the two stages. Both ribosomal proteins and histones were expressed in vegetative and reproductive meristems, those being the two types of proteins that are more associated with plant growth than they are with plant dormancy [74].

4.6. Common Sets of Genes Used by Different Species in the Regulation of Bud Burst and Development

An additional analysis of the genes expressed in the cork oak transcriptome studied in this work revealed the presence of genes that were previously associated with bud burst, bud dormancy, bud sprouting, and flower development in other species, such as the sessile oak (*Quercus petraea*) [8,9,75–77]. In the two studies involving the *Quercus* species, the set of genes for which differential expression was found encompassed genes that were also expressed in our study. These included genes such as galactinol-synthase (important for tolerance to drought, high salinity, and cold stress), SKP1 (which is described as a regulator of seed germination, dormancy or bud burst) and LEA (whose protein products act as protection from desiccation and temperature stress) in *Quercus petraea* [8], as well as several genes associated with ribosome biogenesis (60S ribosomal protein L14, ribosomal protein S24e, and 60S ribosomal protein L12), response to water deprivation (Bax inhibitor 1 and CBL interacting protein kinase 6) and response to auxin (auxin influx transporter) in *Quercus robur* [9].

Moreover, genes involved in the regulation of bud sprouting and flower development in *Arabidopsis thaliana*, such as FRIGIDA, SOC1, and SVP [75,76], as well as genes implicated in the transition from endodormancy to ecodormancy of leaf buds in *Pyrus pyrifolia* such as expansin, cellulose synthase, polygalacturonase, arabinogalactan (all of them related with the plant cell wall) and germin-like protein [77], were also expressed in the cork oak transcriptome analyzed in this study. These results highlight a set of genes whose expression during bud development and sprouting is common in different species of the *Quercus* genus, and likewise in *Arabidopsis thaliana* and *Pyrus pyrifolia*, despite the pattern of differential expression not being detected in this work, which could be due to biological and/or technical limitations.

5. Conclusions

In this study, a set of candidate genes for bud development in *Quercus suber* was identified, revealing for the first time some of the mechanisms involved in the genetic regulation of cork oak's phenological events. The vast amount of GO terms and pathways related to bud sprouting confirms that this is a complex genetic and molecular development process. The KOs identified indicate that the transcriptome of bud sprouting development is mostly focused on the production of essential precursors and metabolites for the synthesis of macromolecules as well as energy production and where water transport seems to play an important role. The data presented here substantially increase our understanding of the complex global cellular mechanisms of bud sprouting in the particular cork producing oak *Quercus suber*. Indeed, these results might be useful in molecular-assisted selection, when the selection aims for pest, frost, or drought tolerance, or to avoid scenarios of an advancing spring phenology that may render young plants vulnerable to late spring frost damages in a climate change scenario.

Supplementary Materials: The following are available online at www.mdpi.com/1999-4907/8/12/486/s1, Table S1: Sequences of the primers used in the qPCR validation assays.

Acknowledgments: This project was funded by "Fundação para a Ciência e a Tecnologia" (FCT) through the projects SOBREIRO/0039/2009: "Consórcio de ESTs de sobreiro-abrolhamento e desenvolvimento foliar" and UID/AGR/00115/2013. Financial support for A. Usié, P. Barbosa, B. Meireles and A.M. Ramos was provided by Investigador FCT project IF/00574/2012/CP1209/CT0001: "Genetic characterization of national animal and plant resources using next-generation sequencing".

Author Contributions: José Matos and Fernanda Simões conceived and designed the study. Fernanda Simões, André Folgado and Sónia Gonçalves performed the laboratorial experiments. Ana Usié performed bioinformatics analyses of the data. Ana Usié, Brígida Meireles, Pedro Barbosa, and Inês Chaves contributed in the bioinformatics analyses. A. M. Ramos coordinated the bioinformatics analyses. José Matos, Fernanda Simões, Ana Usié, Brígida Meireles, Maria H. Almeida, and A. M. Ramos interpreted the results. Ana Usié and A. M. Ramos wrote the manuscript. Brígida Meireles, Fernanda Simões, Sónia Gonçalves, and Maria H. Almeida revised the manuscript. All authors have read and approved this version of the manuscript.

Conflicts of Interest: The authors declare no conflict of interest.

References

1. Berrahmouni, N.; Regato, P.; Stein, C. *Beyond Cork—A Wealth of Resources for People and Nature, Lessons from the Mediterranean*; WWF Mediterranean: Rome, Italy, 2007; Volume 118.
2. Chuine, I.; Beaubien, E.G. Phenology is a major determinant of tree species range. *Ecol. Lett.* **2001**, *4*, 500–510. [CrossRef]
3. Pereira, H. *Cork: Biology, Production and Uses*; Elsevier: Amsterdam, The Netherlands, 2007; ISBN 9780444529671.
4. Chaves, I.; Passarinho, J.A.P.; Capitão, C.; Chaves, M.M.; Fevereiro, P.; Ricardo, C.P.P. Temperature stress effects in *Quercus suber* leaf metabolism. *J. Plant Physiol.* **2011**, *168*, 1729–1734. [CrossRef] [PubMed]
5. Scotti-Saintagne, C.; Bodénès, C.; Barreneche, T.; Bertocchi, E.; Plomion, C.; Kremer, A. Detection of quantitative trait loci controlling bud burst and height growth in *Quercus robur* L. *Theor. Appl. Genet.* **2004**, *109*, 1648–1659. [CrossRef] [PubMed]
6. Saintagne, C.; Bodénès, C.; Barreneche, T.; Pot, D.; Plomion, C.; Kremer, A. Distribution of genomic regions differentiating oak species assessed by QTL detection. *Heredity* **2004**, *92*, 20–30. [CrossRef] [PubMed]
7. Chen, T.H.H.; Howe, G.T.; Bradshaw, H.D., Jr. Molecular genetic analysis of dormancy-related traits in poplars. *Weed Sci.* **2002**, *50*, 232–240. [CrossRef]
8. Derory, J.; Léger, P.; Garcia, V.; Schaeffer, J.; Hauser, M.-T.; Salin, F.; Luschnig, C.; Plomion, C.; Glössl, J.; Kremer, A. Transcriptome analysis of bud burst in sessile oak (*Quercus petraea*). *New Phytol.* **2006**, *170*, 723–738. [CrossRef] [PubMed]
9. Lesur, I.; Le Provost, G.; Bento, P.; Da Silva, C.; Leplé, J.-C.; Murat, F.; Ueno, S.; Bartholomé, J.; Lalanne, C.; Ehrenmann, F.; et al. The oak gene expression atlas: Insights into Fagaceae genome evolution and the discovery of genes regulated during bud dormancy release. *BMC Genom.* **2015**, *16*, 112. [CrossRef] [PubMed]
10. Teixeira, R.T.; Fortes, A.M.; Pinheiro, C.; Pereira, H. Comparison of good- and bad-quality cork: Application of high-throughput sequencing of phellogenic tissue. *J. Exp. Bot.* **2014**, *65*, 4887–4905. [CrossRef] [PubMed]
11. Pereira-Leal, J.B.; Abreu, I.A.; Alabaça, C.S.; Almeida, M.H.; Almeida, P.; Almeida, T.; Amorim, M.I.; Araújo, S.; Azevedo, H.; Badia, A.; et al. A comprehensive assessment of the transcriptome of cork oak (*Quercus suber*) through EST sequencing. *BMC Genom.* **2014**, *15*, 371. [CrossRef] [PubMed]
12. Miguel, A.; de Vega-Bartol, J.; Marum, L.; Chaves, I.; Santo, T.; Leitão, J.; Varela, M.C.; Miguel, C.M. Characterization of the cork oak transcriptome dynamics during acorn development. *BMC Plant Biol.* **2015**, *15*, 158. [CrossRef] [PubMed]
13. Schloss, P.D.; Westcott, S.L.; Ryabin, T.; Hall, J.R.; Hartmann, M.; Hollister, E.B.; Lesniewski, R.A.; Oakley, B.B.; Parks, D.H.; Robinson, C.J.; et al. Introducing mothur: Open-source, platform-independent, community-supported software for describing and comparing microbial communities. *Appl. Environ. Microbiol.* **2009**, *75*, 7537–7541. [CrossRef] [PubMed]
14. Chevreux, B.; Pfisterer, T.; Drescher, B.; Driesel, A.J.; Müller, W.E.G.; Wetter, T.; Suhai, S. Using the miraEST assembler for reliable and automated mRNA transcript assembly and SNP detection in sequenced ESTs. *Genome Res.* **2004**, *14*, 1147–1159. [CrossRef] [PubMed]
15. Li, H.; Durbin, R. Fast and accurate short read alignment with Burrows-Wheeler transform. *Bioinformatics* **2009**, *25*, 1754–1760. [CrossRef] [PubMed]
16. Li, H.; Handsaker, B.; Wysoker, A.; Fennell, T.; Ruan, J.; Homer, N.; Marth, G.; Abecasis, G.; Durbin, R. The Sequence Alignment/Map format and SAMtools. *Bioinformatics* **2009**, *25*, 2078–2079. [CrossRef] [PubMed]
17. Haas, B.J.; Papanicolaou, A.; Yassour, M.; Grabherr, M.; Blood, P.D.; Bowden, J.; Couger, M.B.; Eccles, D.; Li, B.; Lieber, M.; et al. De novo transcript sequence reconstruction from RNA-seq using the Trinity platform for reference generation and analysis. *Nat. Protoc.* **2013**, *8*, 1494–1512. [CrossRef] [PubMed]
18. Liao, Y.; Smyth, G.K.; Shi, W. featureCounts: An efficient general purpose program for assigning sequence reads to genomic features. *Bioinformatics* **2014**, *30*, 923–930. [CrossRef] [PubMed]
19. Robinson, M.D.; McCarthy, D.J.; Smyth, G.K. edgeR: A Bioconductor package for differential expression analysis of digital gene expression data. *Bioinformatics* **2009**, *26*, 139–140. [CrossRef] [PubMed]
20. Untergasser, A.; Nijveen, H.; Rao, X.; Bisseling, T.; Geurts, R.; Leunissen, J.A.M. Primer3Plus, an enhanced web interface to Primer3. *Nucleic Acids Res.* **2007**, *35*, W71–W74. [CrossRef] [PubMed]

21. Pfaffl, M.W. A new mathematical model for relative quantification in real-time RT-PCR. *Nucleic Acids Res.* **2001**, *29*, e45. [CrossRef] [PubMed]

22. Marum, L.; Miguel, A.; Ricardo, C.P.; Miguel, C. Reference gene selection for quantitative real-time PCR normalization in *Quercus suber*. *PLoS ONE* **2012**, *7*, e35113. [CrossRef]

23. Altschul, S.F.; Gish, W.; Miller, W.; Myers, E.W.; Lipman, D.J. Basic local alignment search tool. *J. Mol. Biol.* **1990**, *215*, 403–410. [CrossRef]

24. Zdobnov, E.M.; Apweiler, R. InterProScan—An integration platform for the signature-recognition methods in InterPro. *Bioinformatics* **2001**, *17*, 847–848. [CrossRef] [PubMed]

25. Hu, Z.-L.; Bao, J.; Reecy, J. CateGOrizer: A web-based program to batch analyze gene ontology classification categories. *Online J. Bioinform.* **2008**, *9*, 108–112.

26. Shannon, P.; Markiel, A.; Ozier, O.; Baliga, N.S.; Wang, J.T.; Ramage, D.; Amin, N.; Schwikowski, B.; Ideker, T. Cytoscape: A software Environment for integrated models of biomolecular interaction networks. *Genome Res.* **2003**, *13*, 2498–2504. [CrossRef] [PubMed]

27. Derory, J.; Scotti-Saintagne, C.; Bertocchi, E.; Le Dantec, L.; Graignic, N.; Jauffres, A.; Casasoli, M.; Chancerel, E.; Bodénès, C.; Alberto, F.; et al. Contrasting relationships between the diversity of candidate genes and variation of bud burst in natural and segregating populations of European oaks. *Heredity* **2010**, *104*, 438–448. [CrossRef] [PubMed]

28. Sweetlove, L.J.; Beard, K.F.M.; Nunes-Nesi, A.; Fernie, A.R.; Ratcliffe, R.G. Not just a circle: Flux modes in the plant TCA cycle. *Trends Plant Sci.* **2010**, *15*, 462–470. [CrossRef] [PubMed]

29. Pieczenik, S.R.; Neustadt, J. Mitochondrial dysfunction and molecular pathways of disease. *Exp. Mol. Pathol.* **2007**, *83*, 84–92. [CrossRef] [PubMed]

30. Jardim-Messeder, D.; Caverzan, A.; Rauber, R.; de Souza Ferreira, E.; Margis-Pinheiro, M.; Galina, A. Succinate dehydrogenase (mitochondrial complex II) is a source of reactive oxygen species in plants and regulates development and stress responses. *New Phytol.* **2015**, *208*, 776–789. [CrossRef] [PubMed]

31. Nersissian, A.M.; Immoos, C.; Hill, M.G.; Hart, P.J.; Williams, G.; Herrmann, R.G.; Selverstone Valentine, J. Uclacyanins, stellacyanins, and plantacyanins are distinct subfamilies of phytocyanins: Plant-specific mononuclear blue copper proteins. *Protein Sci.* **1998**, *7*, 71915–71929. [CrossRef] [PubMed]

32. Peer, W.A.; Brown, D.E.; Tague, B.W.; Muday, G.K.; Taiz, L.; Murphy, A.S. Flavonoid accumulation patterns of transparent testa mutants of arabidopsis. *Plant Physiol.* **2001**, *126*, 536–548. [CrossRef] [PubMed]

33. Broun, P. Transcriptional control of flavonoid biosynthesis: A complex network of conserved regulators involved in multiple aspects of differentiation in *Arabidopsis*. *Curr. Opin. Plant Biol.* **2005**, *8*, 272–279. [CrossRef] [PubMed]

34. Jacobs, M.; Rubery, P.H. Naturally occurring auxin transport regulators. *Science* **1988**, *241*, 346–349. [CrossRef] [PubMed]

35. Bernasconi, P. Effect of synthetic and natural protein tyrosine kinase inhibitors on auxin efflux in zucchini (Cucurbita pepo) hypocotyls. *Physiol. Plant.* **1996**, *96*, 205–210. [CrossRef]

36. Murphy, A.; Peer, W.A.; Taiz, L. Regulation of auxin transport by aminopeptidases and endogenous flavonoids. *Planta* **2000**, *211*, 315–324. [CrossRef] [PubMed]

37. Brown, D.E.; Rashotte, A.M.; Murphy, A.S.; Normanly, J.; Tague, B.W.; Peer, W.A.; Taiz, L.; Muday, G.K. Flavonoids act as negative regulators of auxin transport in vivo in *Arabidopsis*. *Plant Physiol.* **2001**, *126*, 524–535. [CrossRef] [PubMed]

38. El Zein, R.; Breda, N.; Gerant, D.; Zeller, B.; Maillard, P. Nitrogen sources for current-year shoot growth in 50-year-old sessile oak trees: An in situ 15N labeling approach. *Tree Physiol.* **2011**, *31*, 1390–1400. [CrossRef] [PubMed]

39. Linz, J.E.; Sypherd, P.S. Expression of three genes for elongation factor 1 alpha during morphogenesis of *Mucor racemosus*. *Mol. Cell. Biol.* **1987**, *7*, 1925–1932. [CrossRef] [PubMed]

40. Hovemann, B.; Richter, S.; Walldorf, U.; Cziepluch, C. Two genes encode related cytoplasmic elongation factors 1 alpha (EF-1 alpha) in *Drosophila melanogaster* with continuous and stage specific expression. *Nucleic Acids Res.* **1988**, *16*, 3175–3194. [CrossRef] [PubMed]

41. Ursin, V.M.; Irvine, J.M.; Hiatt, W.R.; Shewmaker, C.K. Developmental analysis of elongation factor-1 alpha expression in transgenic tobacco. *Plant Cell* **1991**, *3*, 583–591. [CrossRef] [PubMed]

42. Xu, H.; Cao, D.; Chen, Y.; Wei, D.; Wang, Y.; Stevenson, R.A.; Zhu, Y.; Lin, J. Gene expression and proteomic analysis of shoot apical meristem transition from dormancy to activation in *Cunninghamia lanceolata (Lamb.) Hook. Sci. Rep.* **2016**, *6*, 19938. [CrossRef] [PubMed]

43. Davies, P.J. The plant hormones: Their nature, occurrence, and functions. In *Plant Hormones*; Springer: Dordrecht, The Netherlands, 2010; pp. 1–15.

44. O'Neill, D.P.; Davidson, S.E.; Clarke, V.C.; Yamauchi, Y.; Yamaguchi, S.; Kamiya, Y.; Reid, J.B.; Ross, J.J. Regulation of the gibberellin pathway by auxin and DELLA proteins. *Planta* **2010**, *232*, 1141–1149. [CrossRef] [PubMed]

45. Bajji, M.; M'Hamdi, M.; Gastiny, F.; Rojas-Beltran, J.A.; du Jardin, P. Catalase inhibition accelerates dormancy release and sprouting in potato (*Solanum tuberosum* L.) tubers. *Biotechnol. Agron. Soc. Environ.* **2007**, *11*, 121–131.

46. Ohmiya, A. Characterization of ABP19/20, sequence homologues of germin-like protein in Prunus persica L. *Plant Sci.* **2002**, *163*, 683–689. [CrossRef]

47. Ongaro, V.; Leyser, O. Hormonal control of shoot branching. *J. Exp. Bot.* **2008**, *59*, 67–74. [CrossRef] [PubMed]

48. Santner, A.; Estelle, M. The ubiquitin-proteasome system regulates plant hormone signaling. *Plant J.* **2010**, *61*, 1029–1040. [CrossRef] [PubMed]

49. Shen, H.; Luong, P.; Huq, E. The F-Box Protein MAX2 Functions as a positive regulator of photomorphogenesis in Arabidopsis. *Plant Physiol.* **2007**, *145*, 1471–1483. [CrossRef] [PubMed]

50. Kelley, D.; Estelle, M. Ubiquitin-mediated control of plant hormone signaling. *Plant Physiol.* **2012**, *160*, 47–55. [CrossRef] [PubMed]

51. Mayzlish-Gati, E.; LekKala, S.; Resnick, N.; Wininger, S.; Bhattacharya, C.; Lemcoff, J.H.; Kapulnik, Y.; Koltai, H. Strigolactones are positive regulators of light-harvesting genes in tomato. *J. Exp. Bot.* **2010**, *61*, 3129–3136. [CrossRef] [PubMed]

52. Brewera, P.B.; Koltaib, H.; Beveridge, C.A. Diverse roles of strigolactones in plant development. *Mol. Plant* **2013**, *6*, 18–28. [CrossRef] [PubMed]

53. Ueno, S.; Klopp, C.; Leplé, J.C.; Derory, J.; Noirot, C.; Léger, V.; Prince, E.; Kremer, A.; Plomion, C.; Le Provost, G. Transcriptional profiling of bud dormancy induction and release in oak by next-generation sequencing. *BMC Genom.* **2013**, *14*, 236. [CrossRef] [PubMed]

54. Meier, A.R.; Saunders, M.R.; Michler, C.H. Epicormic buds in trees: A review of bud establishment, development and dormancy release. *Tree Physiol.* **2012**, *32*, 565–584. [CrossRef] [PubMed]

55. Zheng, C.; Halaly, T.; Acheampong, A.K.; Takebayashi, Y.; Jikumaru, Y.; Kamiya, Y.; Or, E. Abscisic acid (ABA) regulates grape bud dormancy, and dormancy release stimuli may act through modification of ABA metabolism. *J. Exp. Bot.* **2015**, *66*, 1527–1542. [CrossRef] [PubMed]

56. Dhanaraj, A.L.; Slovin, J.P.; Rowland, L.J. Analysis of gene expression associated with cold acclimation in blueberry floral buds using expressed sequence tags. *Plant Sci.* **2004**, *166*, 863–872. [CrossRef]

57. Kaldenhoff, R.; Fischer, M. Functional aquaporin diversity in plants. *Biochim. Biophys. Acta Biomembr.* **2006**, *1758*, 1134–1141. [CrossRef] [PubMed]

58. Maurel, C.; Verdoucq, L.; Luu, D.-T.; Santoni, V. Plant Aquaporins: Membrane channels with multiple integrated functions. *Annu. Rev. Plant Biol.* **2008**, *59*, 595–624. [CrossRef] [PubMed]

59. Ludevid, D.; Hofte, H.; Himelblau, E.; Chrispeels, M.J. The expression pattern of the tonoplast intrinsic protein gamma-tip in *Arabidopsis-Thaliana* is correlated with cell enlargement. *Plant Physiol.* **1992**, *100*, 1633–1639. [CrossRef] [PubMed]

60. Phillips, A.L.; Huttly, A.K. Cloning of two gibberellin-regulated cDNAs from *Arabidopsis thaliana* by subtractive hybridization: Expression of the tonoplast water channel, γ-TIP, is increased by GA3. *Plant Mol. Biol.* **1994**, *24*, 603–615. [CrossRef] [PubMed]

61. Chaumont, F.; Moshelion, M.; Daniels, M.J. Regulation of plant aquaporin activity. *Biol. Cell* **2005**, *97*, 749–764. [CrossRef] [PubMed]

62. Chen, G.P.; Wilson, I.D.; Kim, S.H.; Grierson, D. Inhibiting expression of a tomato ripening-associated membrane protein increases organic acids and reduces sugar levels of fruit. *Planta* **2001**, *212*, 799–807. [CrossRef] [PubMed]

63. Afzal, Z.; Howton, T.; Sun, Y.; Mukhtar, M. The roles of aquaporins in plant stress responses. *J. Dev. Biol.* **2016**, *4*, 9. [CrossRef]

64. Ahamed, A.; Murai-Hatano, M.; Ishikawa-Sakurai, J.; Hayashi, H.; Kawamura, Y.; Uemura, M. Cold stress-induced acclimation in rice is mediated by root-specific aquaporins. *Plant Cell Physiol.* **2012**, *53*, 1445–1456. [CrossRef] [PubMed]

65. Reiter, W.-D. Biochemical genetics of nucleotide sugar interconversion reactions. *Curr. Opin. Plant Biol.* **2008**, *11*, 236–243. [CrossRef] [PubMed]

66. Kärkönen, A. Biosynthesis of UDP-GlcA: Via UDPGDH or the myo-inositol oxidation pathway? *Plant Biosyst. Int. J. Deal. Asp. Plant Biol.* **2005**, *139*, 46–49. [CrossRef]

67. Loewus, F.A.; Kelly, S.; Neufeld, E.F. Metabolism of myo-inositol in plants: Conversion to pectin, hemicellulose, d-xylose and sugar acids. *Proc. Natl. Acad. Sci. USA* **1962**, *48*, 421–425. [CrossRef] [PubMed]

68. Roberts, R.M.; Loewus, F. The conversion of d-Glucose-6-C to cell wall polysaccharide material in *Zea mays* in presence of high endogenous levels of myoinositol. *Plant Physiol.* **1973**, *52*, 646–650. [CrossRef] [PubMed]

69. Belknap, W.R.; Garbarino, J.E. The role of ubiquitin in plant senescence and stress responses. *Trends Plant Sci.* **1996**, *10*, 331–335. [CrossRef]

70. Garbarino, J.E.; Rockhold, D.R.; Belknap, W.R. Expression of stress-responsive ubiquitin genes in potato tubers. *Plant Mol. Biol.* **1992**, *20*, 235–244. [CrossRef] [PubMed]

71. Janse, J.D. Chapter: 3 Disease and symptoms caused by plant pathogenic bacteria. In *Phytobacteriology: Principles and Practice*; Janse, J.D., Ed.; CABI: Wallingford, Oxon, UK; New York, NY, USA, 2005; pp. 91–93.

72. Kulik, A.; Wawer, I.; Krzywińska, E.; Bucholc, M.; Dobrowolska, G. SnRK2 Protein Kinases—Key regulators of plant response to abiotic stresses. *OMICS* **2011**, *15*, 859–872. [CrossRef] [PubMed]

73. Stafstrom, J.P. Regulation of growth and dormancy in pea axillary buds. In *Dormancy in Plants: From Whole Plant Behaviour to Cellular Control*; Viémont, J.D., Crabbé, J., Eds.; CABI: Wallingford, Oxon, UK; New York, NY, USA, 2000; pp. 337–339.

74. Devitt, M.; Stafstrom, J. Cell cycle regulation during growth-dormancy cycles in pea axillary buds. *Plant Mol. Biol.* **1995**, *29*, 255–265. [CrossRef] [PubMed]

75. Horvath, D. Common mechanisms regulate flowering and dormancy. *Plant Sci.* **2009**, *177*, 523–531. [CrossRef]

76. Kryvych, S.; Nikiforova, V.; Herzog, M.; Perazza, D.; Fisahn, J. Gene expression profiling of the different stages of *Arabidopsis thaliana* trichome development on the single cell level. *Plant Physiol. Biochem.* **2008**, *46*, 160–173. [CrossRef] [PubMed]

77. Nishitani, C.; Saito, T.; Ubi, B.E.; Shimizu, T.; Itai, A.; Saito, T.; Yamamoto, T.; Moriguchi, T. Transcriptome analysis of *Pyrus pyrifolia* leaf buds during transition from endodormancy to ecodormancy. *Sci. Hortic.* **2012**, *147*, 49–55. [CrossRef]

© 2017 by the authors. Licensee MDPI, Basel, Switzerland. This article is an open access article distributed under the terms and conditions of the Creative Commons Attribution (CC BY) license (http://creativecommons.org/licenses/by/4.0/).

![forests logo] *forests*

MDPI

Article

Genome-Wide Analysis of Gene and microRNA Expression in Diploid and Autotetraploid *Paulownia fortunei* (Seem) Hemsl. under Drought Stress by Transcriptome, microRNA, and Degradome Sequencing

Zhenli Zhao †, Suyan Niu †, Guoqiang Fan *, Minjie Deng and Yuanlong Wang

Institute of Paulownia, Henan Agricultural University, 95 Wenhua Road, Jinshui District, 450002 Zhengzhou, China; zhaozhl2006@126.com (Z.Z.); suyanniu_happy@126.com (S.N.); dengmj1980@126.com (M.D.); xiaoyunxia2012@126.com (Y.W.)
* Correspondence: fanguoqiangdr@163.com; Tel.: +86-0371-6355-8605
† These authors are the co-first authors (Suyan Niu and Zhenli Zhao contributed equally to this work).

Received: 14 January 2018; Accepted: 9 February 2018; Published: 13 February 2018

Abstract: Drought is a common and recurring climatic condition in many parts of the world, and it can have disastrous impacts on plant growth and development. Many genes involved in the drought response of plants have been identified. Transcriptome, microRNA (miRNA), and degradome analyses are rapid ways of identifying drought-responsive genes. The reference genome sequence of *Paulownia fortunei* (Seem) Hemsl. is now available, which makes it easier to explore gene expression, transcriptional regulation, and post-transcriptional in this species. In this study, four transcriptome, small RNA, and degradome libraries were sequenced by Illumina sequencing, respectively. A total of 258 genes and 11 miRNAs were identified for drought-responsive genes and miRNAs in *P. fortunei*. Degradome sequencing detected 28 miRNA target genes that were cleaved by members of nine conserved miRNA families and 12 novel miRNAs. The results here will contribute toward enriching our understanding of the response of *Paulownia fortunei* trees to drought stress and may provide new direction for further experimental studies related the development of molecular markers, the genetic map construction, and other genomic research projects in Paulownia.

Keywords: *Paulownia fortunei*; gene expression; transcriptome; microRNA; degradome; drought

1. Introduction

Drought is a common form of environmental stress that constrains plant growth and development and crop production. During their evolution, plants have evolved processes to respond to adverse stresses, including signal transduction and hormonal regulation [1]. Under drought stress conditions, cell turgor can be maintained at a balanced level through osmotic adjustment, and cell elasticity and size can be adjusted via protoplasmic tolerance, which allows plants to tolerate drought stress. The drought stress response has been studied in various plants, including crops, vegetables, and trees.

Paulownia is a fast-growing woody plant native to China, where it has been used widely for landscaping and silviculture; however, the ecological benefits of Paulownia are being severely affected by the increasingly dry climate. The *Paulownia fortunei* (Seem) Hemsl. genome has been published recently, which will help in better understanding the drought tolerance of Paulownia species. To enlarge the germplasm and breed a drought-resistant variety, autotetraploids ($4n = 80$) have been obtained from diploid parent plants ($2n = 40$) using colchicines [2]. Polyploidy and whole genome duplication lay a foundation for the innovation of genetic material [3]. The altered characters of polyploidy—for instance, drought tolerance or pathogen resistance—make these plants more adapted to rough environmental conditions [4,5]. Different genotypes may cause plants to display different response patterns to drought stress. Therefore, breeding plants with enhanced genetic characteristics requires an in-depth understanding of the regulation mechanisms of gene expression under drought stress.

Next-generation sequencing techniques have made the analysis of genome-wide transcriptome available and affordable. To date, a large number of genes related to drought stress have been reported [6–9]. MicroRNAs (miRNAs) are the kind of non-coding RNAs that play vital roles in regulation of gene expression at the post-transcriptional level. They regulate gene expression either by cleaving mRNA or by translational repression [10,11]. Plant miRNAs were implicated to play important roles in plants' responses to adversity stresses [12–16]. Functional analyses have demonstrated that miRNAs are essential to the ability of plants to resist environment stress [17]. However, in the Paulownia species, the networks that underlie these stress responses have not been fully characterized.

Plant hormones play pivotal roles in plant growth and stress tolerance. ABA (phytohormone abscisic acid) is known as a stress hormone because it can be sensitively and rapidly accumulated under drought stress condition in order to reduce or avoid the inhibition of plant growth and development and it can be rapidly decreased and degraded when the stress is removed [18–21]. ABA regulates the expression of many genes, which leads to changes in important physiological and biochemical indexes that help plants to survive under environment stress [22]. During this process, drought stress increases endogenous ABA levels and induces ABA-dependent and ABA-independent transcriptional regulatory networks [22,23]. It has been shown that genes involved in the regulatory networks are active under drought condition, especially transcription factors (TFs), which are employed to enhance drought tolerance in plants [24,25]. For instance, an increasing number of reports have suggested that the overexpression of TFs such as the MYB (transcription factor MYB), NAC (NAM, ATAF, and CUC transcription factor), WRKY (transcription factor WRKY), ZFP (Zinc finger protein), bHLH (basic helix-loop-helix), ZF-HD (zinc finger homeodomain protein), and AP2 (APETALA 2) families plays a critical role in regulating the expression of specific stress-related genes and plant stress signal transduction, therefore improving the resistance of *Arabidopsis* and rice plants to drought and salinity stresses [26–29]. Moreover, it has been shown that drought stress affects miRNAs through regulating their target genes, which encode the MYB, AP2, and ZF-HD TFs that are involved in ABA-dependent and ABA-independent pathways, and that these changes can allow plants to better adapt to water deficit. For instance, ABA treatment and drought stress have been shown to induce the accumulation of miR159, which targets the encoding of the MYB TFs that positively modulate ABA responses during seed germination in *Arabidopsis* [30]. Although the complex interplay between transcriptional and posttranscriptional regulation of drought response in *Arabidopsis* has been reported, this type of interplay is not well characterized in Paulownia. To understand the molecular mechanisms associated with the responses of diploid and autotetraploid *P. fortunei* to drought stress, we sequenced four transcriptome, small RNA, and degradome libraries by high-throughput sequencing. As a result, we obtained information on a large number of genes and biological processes that might be involved in molecular mechanisms of drought responses in Paulownia. The results will help researchers explore the gene expression differences between diploid and autotetraploid *P. fortunei* under drought stress at the transcriptome and miRNA level, which is of fundamental importance to agricultural production.

2. Materials and Methods

2.1. Plant Materials

The tissue culture samples were harvested from the Institute of Paulownia, Henan Agricultural University (Zhengzhou, Henan Province, China). The autotetraploid clone of *Paulownia fortunei* was obtained by in vitro treatment of a diploid *P. fortunei* clone with colchicine from a previous study [2]. This diploid and its derived autotetraploid clone were the same genome with differences at the ploidy level. The samples were treated according to the method of Zhang et al. [2]. Briefly, the *P. fortunei* tissue culture seedlings (diploid and autotetraploid) were cultured for 30 days. Samples were transferred into nutrition blocks containing normal garden soil for 30 days. Then, samples with the same growth consistency were transferred individually into nutrition pots (30 cm in diameter) containing the same weight of ordinary garden soil. After 50 days, the samples with the same growth consistency were used to carry out the drought treatment. Diploid and autotetraploid *P. fortunei* plantlets were treated with 75% (control) and 25% (drought treatment) relative soil water contents and named PF2 and PF2T, and PF4 and PF4T, respectively. After 6, 9, and 12 days of treatment, the leaves (second pairs) were picked from the drought-treated plants. At least three biological duplicates with the same growing consistency under each condition were chosen and the leaves were harvested. The obtained leaves were immediately frozen in liquid nitrogen for total RNA extraction.

2.2. Physiological Responses of Diploid and Autotetraploid Paulownia fortunei to Drought Stress

Physiological characteristics and biochemical indexes including chlorophyll, proline, soluble sugar, soluble protein content, malondialdehyde concentration, superoxide dismutase activity, relative water content, and relative electrical conductivity of the well-watered control samples (PF2 and PF4) and drought treatment samples (PF2T_12D and PF4T_12D) were calculated in this study. The chlorophyll content was calculated following the methods of Bojović et al. [31] and Arnon [32]. The proline content was calculated by using the method of Bates et al. [33]. The soluble protein content, soluble sugar content, malondialdehyde concentration, and superoxide dismutase activity were measured by the method described previously [34]. The relative water content of the leaf was calculated by using the method described previously [35]. The relative electrical conductivity was calculated according to the methods by Li [34]. The ANOVA Duncan analysis was performed using SPSS 19.0 software (SPASS, Inc., Chicago, IL, USA).

2.3. Construction, Processing, and Annotation of the P. fortunei cDNA Libraries

Total RNA was extracted from each sample using Trizol reagent following the manufacturer's instructions (Invitrogen, Carlsbad, CA, USA). Then, the samples (PF2, PF2T_12D, PF4, and PF4T_12D) were treated with DNase I, and magnetic beads with Oligo (dT) were used to isolate the mRNA. The mRNA was then mixed with fragmentation buffer to obtain short fragments. The cDNA was synthesized using the mRNA fragments as templates. Briefly, the short fragments had been purified and suspended with elution buffer in a TruSeq™ RNA sample preparation kit (Catalog No. RS-930-2001) for end reparation and single nucleotide A (adenine) addition (Illumina, San Diego, CA, USA). The short fragments were then connected with adapters in the TruSeq™ RNA sample preparation kit. Poly (A) fragments were selected for connected adapters in TruSeq™ RNA sample preparation kit. Suitable fragments were used as templates for PCR amplification. After the quantification and qualification of the sample libraries were performed, these libraries were sequenced on an Illumina HiSeq™ 2000 (Illumina, San Diego, CA, USA).

After the paired-end sequencing was conducted, the raw data for these libraries were obtained and then deposited in the NIH SRA (Short Read Archive) database (http://www.ncbi.nlm.nih.gov/sra) under accession number SRP030466. The reads (about 110 bp length) from either end of cDNA fragment were yielded for each sequencing sample. We used in-house Perl scripts to process the raw data by removing low-quality reads, reads containing adapters, and reads containing poly (N) sequences in order to obtain high-quality reads. Quality control and other downstream analyses were performed on the clean data. The clean reads were mapped to the *P. fortunei* genome reference and gene reference sequences with Bowtie2 program (http://bowtie-bio.sourceforge.net/bowtie2/index.shtml.) [36]. The transcript expression levels were quantified using the RSEM (RSEM v1.3.0) software package (https://deweylab.github.io/RSEM/). The fragments per kilo base of transcript (FPKM) was used to identify DEGs (differentially expressed genes) from comparisons between two libraries [37]. Statistically significant differences in gene expression levels were determined using the Audic-Claverie algorithm [38]. The *p*-values were calculated according to previously established methods. The threshold *p*-value in multiple tests was calculated using the false discovery rate (FDR) method of Benjamini and Hochberg [39]. Threshold FDR < 0.001 and | log2Ratio | > 1 were used to judge the significance of differences in gene expression. Gene functional annotation was based on BLAST searches (E-value cutoff of 10^{-5}) against NCBI's non-redundant protein sequence (nr) database, and Kyoto Encyclopaedia of Genes and Genomes (KEGG) pathway (http://www.genome.jp/kegg/) and Gene Ontology (GO) databases annotations (http://www.geneontology.org/).

2.4. Construction, Processing, and Annotation of the P. fortunei Small RNA Libraries

The same total RNA samples as those used for transcriptome sequencing were also subjected to build the small RNA (sRNA) libraries and then sequenced on an Illumina GAIIx platform. Briefly, sRNA fragments (18 to 30 nt) were separated and purified by polyacrylamide gel electrophoresis. Then, using the T4 RNA ligase, sRNA fragments were ligated to 5′ and 3′ adaptors. The adaptor-ligated sRNAs were transcribed to single-stranded cDNA and amplified by 12 PCR cycles. The final products were used to construct the libraries for sequencing.

After low quality reads, adapters, and contaminated reads were filtered, we obtained the clean reads. The length distribution of the clean reads was analyzed, and the reads were then mapped to the *P. fortunei* genome and gene databases. After removing repeat sRNAs, sequences that matched rRNAs, tRNAs, snRNAs, and snoRNAs were removed. The remaining unique sequences were compared against the plant known miRNA sequences in miRBase (Release 21.0; http://www.mirbase.org/) with a maximum of two mismatches allowed [40] in order to identify the conserved miRNAs in *P. fortunei*. MIREAP (http://sourceforge.net/projects/mireap/) and RNAfold (http://rna.tbi.univie.ac.at/cgi-bin/RNAfold.cgi) software were used to predict novel miRNAs from the remaining unknown sRNAs under the basic criteria described by Meyers et al. [41].

2.5. Differential Expression Analysis of Conserved and Novel miRNAs

The abundance of miRNAs from four libraries was normalized to one million by total clean reads of miRNAs in each sample. Fold changes between any two libraries were calculated as: fold change = log2 (sample 1/sample 2). The miRNAs with fold changes >1 or <−1 and with $p \leq 0.05$ were considered likely to be differentially expressed in response to drought stress. The *p*-values were evaluated according to previously established methods [38].

p-value:

$$P(x|y) = \left(\frac{N_2}{N_1}\right)\frac{(x+y)!}{x!y!(1+\frac{N_2}{N_1})^{(x+y+1)}} \tag{1}$$

$$C(y \leq y_{\max}|x) = \sum_{y=0}^{y \leq y_{\min}} p(x|y) \tag{2}$$

$$D(y \geq y_{\max}|x) = \sum_{y \geq y_{\max}}^{\infty} p(x|y) \tag{3}$$

where *N*1and *N*2 represent the total number of reads in PTF2 and PTF2H (or PTF4 and PTF4H; PTF2H and PA4H; or PTF2 and PTF4), respectively; x and y represent the number of reads surveyed in PTF2 and PTF2H (or PTF4 and PTF4H; or PTF2H and PTF4H; or PTF2 and PTF4), respectively; C and D can be regarded as the probability discrete distribution of the *p*-value inspection.

2.6. Identification of miRNA Target Genes Associated with Drought Stress

To understand the function of target genes in regulating drought stress, four corresponding degradome libraries were built as described elsewhere [42,43]. Poly (A) RNA was isolated from total RNA using an Oligotex kit (QIAGEN, Santa Clarita, CA, USA) and ligated to an RNA adapter that had a 3′ *Mme*I recognition site. First strand cDNA was obtained using oligo (dT) and the PCR product was digested with *Mme*I. The digested products were ligated with degenerate nucleotides at the 3′ ends using T4 DNA ligase. The ligation product was amplified with 20 PCR cycles and then sequenced on an Illumina HiSeq™ 2000 system.

After initial processing, quality sequences 20–22 nt in length were collected for subsequent analysis. The unique sequence signatures were match to the *P. fortunei* genome and gene databases using SOAP software (http://soap.genomics.org.cn). Sequences with less than six mismatches and with zero mismatches between the 10th and 11th nucleotides were considered as potential targets. The CleaveLand pipeline was used to identify the miRNA gene targets. According to the abundance of tags and cleavage sites on the transcript, the cleaved targets were grouped into five categories. Perfectly matched sequences were kept selected and extended to approximately 35 nt in length by adding 15 nt of the upstream sequences. All resulting reads were reverse complemented and mapped to the miRNAs identified in this study. T-plots were built according to the distribution of signatures along the transcripts. The identified targets were subjected to BLASTX analysis and assigned GO annotations (http://www.geneontology.org/) based on their alignment with known sequences. Furthermore, pathway analyses of the targets were performed based on Blastall hits (*E*-value threshold < 10^{-5}) against the KEGG database (http://www.genome.jp/kegg/).

2.7. Quantitative Real-Time Polymerase Chain Reaction (qRT-PCR) for Verification of the Sequencing Data

Potential genes and miRNAs and their corresponding targets related to drought response were selected randomly for qRT-PCR validation. The RNA samples from the leaves of the PF2, PF2T_6D, PF2T_9D, PF2T_12D, PF4, PF4T_6D, PF4T_9D, PF4T_12D samples were extracted with Trizol (Sangon, Shanghai, China). These samples were harvested at the same time as those used for transcriptome, miRNAs, and degradome sequencing. The purified RNA was denatured and first-strand cDNAs were synthesized using the PrimeScript RT reagent Kit (Takara, Dalian, China). The primers for genes (contained six miRNA targets) were designed by Beacon Designer (version 7.7, Premier Biosoft International, Ltd., Palo Alto, CA, USA) (Table S1). The specific stem-loop primers for the miRNAs were designed according to the method reported previously [44] (Table S1). The cDNAs were then amplified in the Bio-Rad CFX96TM Real-Time System (Bio-Rad, Hercules, CA, USA) under the PCR parameters defined by Fan et al. [45]. Three biological replicates were carried out for each gene and miRNA. The average threshold cycle (Ct) was normalized and the relative expression changes were calculated using the $2^{-\Delta\Delta Ct}$ method [46]. The Paulownia 18S rRNA and U6 snRNA were chosen as internal reference genes for normalization [45].

3. Results

3.1. Physiological Responses of Diploid and Autotetraploid Paulownia fortunei to Drought Stress

The physiological responses of two *P. fortunei* genotypes to drought stress tolerance were determined in this present study. Figure 1 shows that the chlorophyll and relative water contents were decreased while the proline, malondialdehyde, soluble sugar, and relative electrical conductivity were increased in the drought stress treatment plants (autotetraploids and diploids). The protein contents and superoxide dismutase activity showed increased trends in the drought stress treatment plants (autotetraploids and diploids) but differences did not reach a significant level. Moreover, these physiological and biochemical indexes also exhibited differences between diploid and autotetraploid plants. The autotetraploid plants always stored higher levels of soluble sugar contents and lower relative water contents than the corresponding diploid plants both in well-watered and drought stress conditions (Figure 1). These results are consistent with the previous results presented by other scholars [2,47].

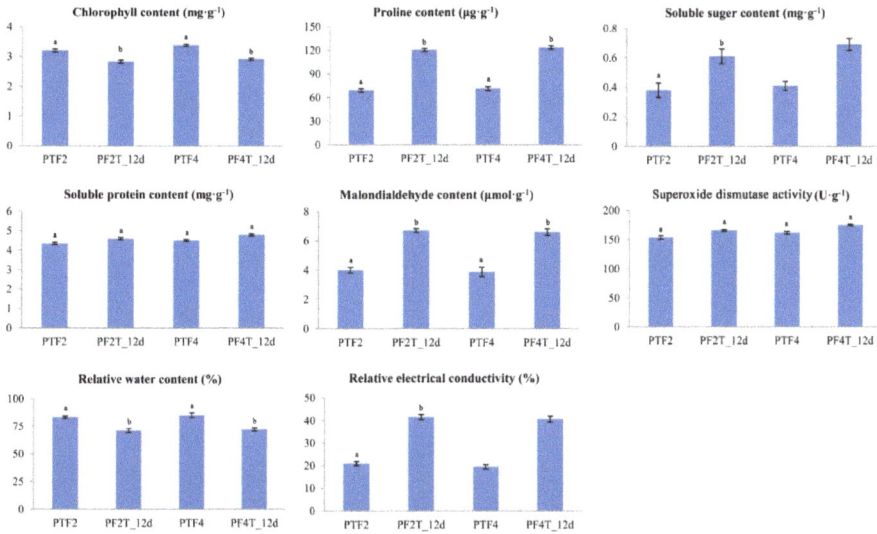

Figure 1. Effects of drought stress on *P. fortunei* (Seem) Hemsl. physiology. PF2, well-watered diploid; PF2T_12d, 12 days drought-treated diploid; PF4, well-watered autotetraploid; PF4T_12d, 12 days drought-treated autotetraploid. The different letters within a physiological and biochemical index indicate significant difference, while the same letters within a physiological and biochemical index indicate no significant differences ($p < 0.05$).

3.2. Transcriptome Data Mapping to the P. fortunei Genome

To get an overview of the diploid and autotetraploid *P. fortunei* transcriptomes under drought stress, we generated four cDNA libraries from leaves from PF2, PF2T_12D, PF4, and PF4T_12D plants. A total of 54,049,716 clean reads were mapped to the *P. fortunei* genome sequence, accounting for 84.10% of all the transcripts in the PF2 library. Moreover, 31,664,626 transcripts, accounting for 49.27% of the transcripts in the four libraries, were mapped to the *P. fortunei* gene database. The alignment statistics for the transcripts in the PPF2, PF2T_12D, PF4, and PF4T_12D libraries are shown in Table 1. Among these reads, 39,381,687 (PF2), 40,236,126 (PF2T_12D), 33,117,090 (PF4), and 34,095,108 (PF4T_12D) were matched perfectly to the Paulownia reference genome, while about 23,034,911 (PF2), 22,138,954 (PF2T_12D), 19,896,750 (PF4), and 18,735,185 (PF4T_12D) reads were perfectly matched to the reference genes (Table 1). The numbers of unique reads that were matched to the reference genome were 46,383,583 (PF2), 47,298,908 (PF2T_12D), 47,201,196 (PF4), and 47,540,030 (PF4T_12D), and the numbers of unique reads that matched to the Paulownia reference genes were 21,725,268 (PF2), 20,998,174 (PF2T_12D), 21,955,604 (PF4), and 21,592,724 (PF4T_12D). Altogether, 54,049,716 (PF2), 55,463,664 (PF2T_12D), 55,033,798 (PF4), and 56,335,628 (PF4T_12D) reads were matched to the reference genome, and 31,664,626 (PF2), 30,623,746 (PF2T_12D), 33,230,102 (PF4), and 31,595,018 (PF4T_12D) reads were matched to the reference genes. However, because of the significant sequencing depth that can be obtained by Illumina sequencing and the incomplete annotation of the *P. fortunei* genome, 14,668,029 (PF2), 15,227,538 (PF2T_12D), 21,916,708 (PF4), and 22,240,520 (PF4T_12D) reads were not matched to the reference genome, and 8,629,715 (PF2), 8,484,792 (PF2T_12D), 13,333,352 (PF4), and 12,859,833 (PF4T_12D) reads were not matched to the reference genes.

Table 1. The statistics of clean reads mapped to reference gene and genome.

		Map to Genome		Map to Gene	
	Category	Reads Number	Percent	Reads Number	Percent
PF2	Total Reads	64,267,312	100.00%	64,267,312	100.00%
	Total Base Pairs	5,784,058,080	100.00%	5,784,058,080	100.00%
	Total Mapped Reads	54,049,716	84.10%	31,664,626	49.27%
	Perfect Match	39,381,687	61.28%	23,034,911	35.84%
	Mismatch	14,668,029	22.82%	8,629,715	13.43%
	Unique Match	46,383,583	72.17%	21,725,268	33.80%
	Multi-position Match	7,666,133	11.93%	9,939,358	15.47%
	Total Unmapped Reads	10,217,596	15.90%	32,602,684	50.73%
PF2T_12D	Total Reads	66,551,876	100.00%	66,551,876	100.00%
	Total Base Pairs	5,989,668,840	100.00%	5,989,668,840	100.00%
	Total Mapped Reads	55,463,664	83.34%	30,623,746	46.01%
	Perfect Match	40,236,126	60.46%	22,138,954	33.27%
	Mismatch	15,227,538	22.88%	8,484,792	12.75%
	Unique Match	47,298,908	71.07%	20,998,174	31.55%
	Multi-position Match	8,164,756	12.27%	9,625,572	14.46%
	Total Unmapped Reads	11,088,212	16.66%	35,928,128	53.99%
PF4	Total Reads	67,073,816	100.00%	67,073,816	100.00%
	Total Base Pairs	6,036,643,440	100.00%	6,036,643,440	100.00%
	Total Mapped Reads	55,033,798	82.05%	33,230,102	49.54%
	Perfect Match	33,117,090	49.37%	19,896,750	29.66%
	Mismatch	21,916,708	32.68%	13,333,352	19.88%
	Unique Match	47,201,196	70.37%	21,955,604	32.73%
	Multi-position Match	7,832,602	11.68%	11,274,498	16.81%
	Total Unmapped Reads	12,040,018	17.95%	33,843,712	50.46%
PF4T_12D	Total Reads	68,805,144	100.00%	68,805,144	100.00%
	Total Base Pairs	6,192,462,960	100.00%	6,192,462,960	100.00%
	Total Mapped Reads	56,335,628	81.88%	31,595,018	45.92%
	Perfect Match	34,095,108	49.55%	18,735,185	27.23%
	Mismatch	22,240,520	32.32%	12,859,833	18.69%
	Unique Match	47,540,030	69.09%	21,592,724	31.38%
	Multi-position Match	8,795,598	12.78%	10,002,294	14.54%
	Total Unmapped Reads	12,469,516	18.12%	37,210,124	54.08%

3.3. Comparison of Gene Expression Profiles between the Two P. fortunei Genotypes

To further explore the differences between diploid and autotetraploid *P. fortunei* under drought stress, we compared gene expression levels for four comparisons (Figure 2a) and detected 7526 (PF2 vs. PF2T_12D), 5205 (PF4 vs. PF4T_12D), 6103 (PF2 vs. PF4), and 3590 (PF2T_12D vs. PF4T_12D) genes that were differentially expressed (Tables S2–S5). In the PF2 vs. PF2T_12D comparison, 3270 genes were upregulated and 4256 were downregulated, while in comparing PF4 vs. PF4T_12D, 1885 and 3320 genes were upregulated and downregulated, respectively (Figure 2b). Moreover, a total of 9591 DEGs were identified in these two *P. fortunei* genotypes, and 3140 DEGs were common in the PF2 vs. PF2T_12D and PF4 vs. PF4T_12D comparisons. These common DEGs indicated a general drought response. Among the differentially expressed genes (DEGs), 622, 402, 509, and 356 were annotated as transcription factors (TFs) in PF2 vs. PF2T_12D, PF4 vs. PF4T_12D, PF2 vs. PF4, and PF2T vs. PF4T_12D, respectively (Tables S2–S5 for details).

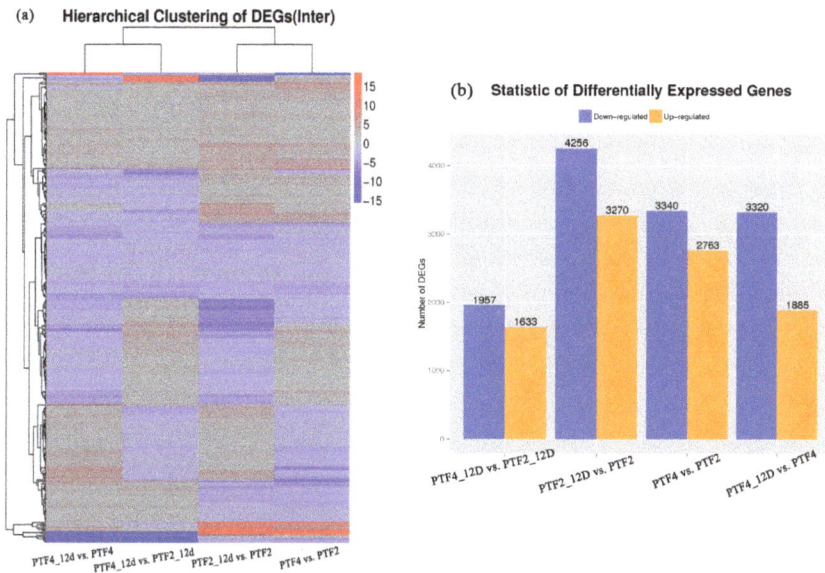

Figure 2. Statistics of gene expression in four libraries; (**a**) The cluster chart of genes expression; (**b**) The differently expressed genes in the four comparisons of PTF4_12d vs. PTF2_12d, PTF2_12d vs. PTF2, PTF4 vs. PTF2, and PTF4_12d vs. PTF4. DEGs: differentially expressed genes.

To assign the DEGs to pathways, we performed a KEGG pathway analysis. Among these obtained DEGs, 4606 (61.20%) and 3121 (59.96%) were mapped to 128 and 126 KEGG pathways for PF2 vs. PF2T and PF4 vs. PF4T_12D comparisons, respectively. The metabolic pathway was the most represented, with 1149 (24.95%) DEGs (PF2 vs. PF2T_12D) and 837 (26.82%) DEGs (PF4 vs. PF4T_12D), followed by secondary metabolites biosynthesis, plant-pathogen interaction, and plant hormone signal transduction in two comparisons (Tables S6 and S7). The DEGs were also annotated with GO terms. In the PF2 vs. PF2T and PF4 vs. PF4T comparisons, under the category of cellular component, cell and cell part were the most represented terms, with 2860 DEGs (83.0%) and 1956 DEGs (82.3%), followed by intracellular (71.8% and 70.8%), and intracellular part (71.6% and 70.5%). The terms assigned under the biological process and molecular function categories are list in Supplementary Figures S1 and S2.

3.4. Drought Responsive Genes in the Two P. fortunei Genotypes

According to the expression patterns in the four comparisons, the DEGs were classified into eight types: A (upregulated in both PF2 vs. PF2T_12D and PF4 vs. PF4T_12D), B (downregulated in PF2 vs. PF2T_12D and upregulated in PF4 vs. PF4T_12D), C (downregulated in both PF2 vs. PF2T_12D and PF4 vs. PF4T_12D), D (upregulated in PF2 vs. PF2T and downregulated in PF4 vs. PF4T_12D) (Table S8); E (upregulated in both PF2 vs. PF4 and PF2T_12D vs. PF4T_12D), F (downregulated in PF2 vs. PF4 and upregulated in PF2T_12D vs. PF4T_12D), G (downregulated in both PF2 vs. PF4 and PF2T_12D vs. PF4T_12D), and H (upregulated in PF2 vs. PF4 and downregulated in PF2T_12D vs. PF4T_12D) (Table S9). Finally, we detected co-expressed DEGs in types A and E, a and G, C and E, and C and G, and 258 co-expressed DEGs (as the type I) were selected as the drought responsive candidate genes (Table S10). Among these co-expressed DEGS, 24 of them coding 17 transcription factor (TF) families were identified (Table 2). Moreover, the correlation coefficient of these DEGs between these two genotypes was analyzed, with the correlation coefficient being 0.736 for PF2 vs. PF2T_12D and PF4 vs. PF4T_12D comparisons (Figure S3). These results provided further evidence that these 258 co-expressed DEGs were the drought responsive genes. Furthermore, based on the GO and KEGG pathway annotations, 172 and 147 DEGs were used to perform GO and KEGG analysis, respectively (Figures 3 and 4 and Table S11).

Table 2. The co-expressed DEGs coding the transcription factors.

Gene ID	TF Family	Included Domain	Excluded Domain
PAU000179.1	C3H	zf-CCCH	PHD, SNF2_N, zf-C2H2
PAU000931.1	MYB	Myb_DNA-binding	ARID, Response_reg, SNF2_N
PAU000931.1	Trihelix	trihelix	-
PAU001653.1	GRAS	GRAS	-
PAU001788.1	MYB	Myb_DNA-binding	ARID, Response_reg, SNF2_N
PAU001788.1	G2-like	G2-like	Response_reg
PAU002069.1	GRAS	GRAS	-
PAU003364.1	NAC	NAM	-
PAU003475.1	GRAS	GRAS	-
PAU006678.1	HSF	HSF_DNA-bind	-
PAU007281.2	ABI3VP1	B3	AP2, Auxin_resp
PAU008439.1	mTERF	mTERF	-
PAU010113.2	AP2-EREBP	AP2	-
PAU011010.1	LOB	DUF260	-
PAU012184.1	bHLH	HLH	-
PAU012899.1	MYB-related	Myb_DNA-binding	ARID, G2-like, Response_reg, SNF2_N,trihelix
PAU012899.1	MYB	Myb_DNA-binding	ARID, Response_reg, SNF2_N
PAU017462.1	NAC	NAM	-
PAU018191.1	MYB-related	Myb_DNA-binding	ARID, G2-like, Response_reg, SNF2_N,trihelix
PAU018191.1	MYB	Myb_DNA-binding	ARID, Response_reg, SNF2_N
PAU019436.2	NAC	NAM	-
PAU021504.1	NAC	NAM	-
PAU021792.1	C2C2-GATA	GATA	-
PAU022916.1	ARF	Auxin_resp	-
PAU023755.1	ARF	Auxin_resp	-
PAU025363.1	WRKY	WRKY	-
PAU029472.1	LIM	LIM	-
PAU030808.2	GRAS	GRAS	-

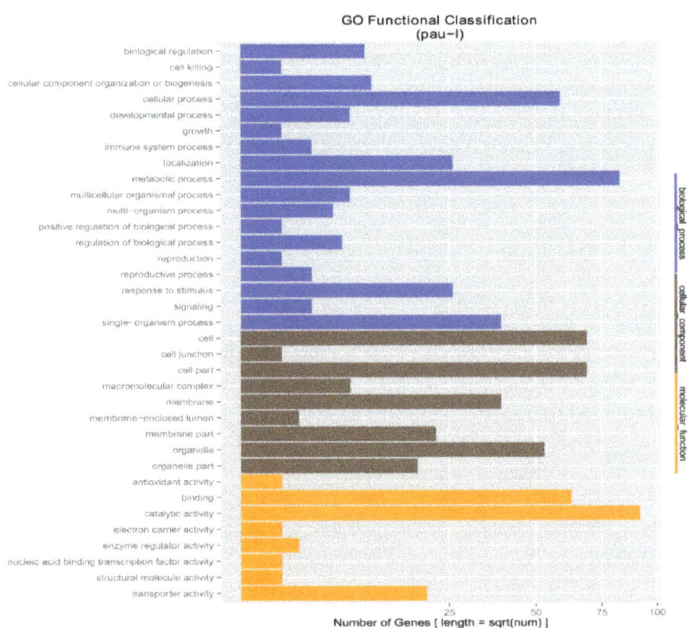

Figure 3. Gene Ontology (GO) for the co-expressed DEGs in two *P. fortunei* genotypes.

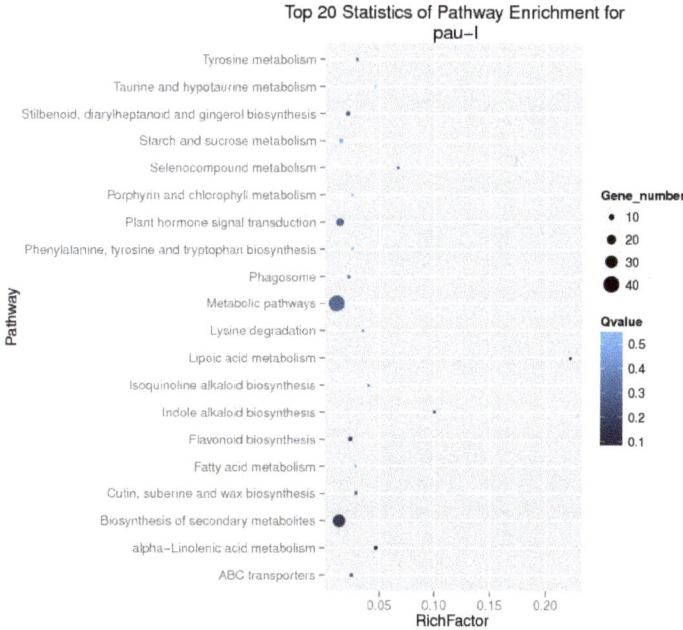

Figure 4. Scatter chart of top 20 pathway enrichment for the co-expressed DEGs.

3.5. Analysis of Small RNA Library Data

We obtained a total of 15,707,321 (PF2), 14,828,766 (PF2T_12D), 14,153,413 (PF4), and 16,151,854 (PF4T_12D) raw reads from the four libraries. After initial processing, 15,534,913 (PF2), 14,759,868 (PF2T_12D), 14,024,334 (PF4), and 16,067,026 (PF4T_12D) clean reads were obtained. The sRNAs were mapped to the *P. fortunei* genome sequence, and classified and annotated as miRNAs, rRNAs, snRNAs, and snoRNAs by searches against the GenBank, Rfam, and miRbase 21.0 databases. A total of 22,521 unique miRNAs were obtained from the PF2 library, accounting for 0.74% of the unique sRNAs in PF2 (Table 3). Among the four libraries, the largest number of miRNA sequences that mapped to the genome was from the PF2 library, possibly because the reference genome in this present research was from the diploid *P. fortunei* species.

Table 3. Annotation of small RNA (sRNA) sequences in four libraries.

Category	Species	Unique sRNAs	Percent%	Reads Number	Percent%
miRNA	PF2	22,521	0.47%	2,882,549	18.56%
	PF2T-12D	19,768	0.42%	2,788,081	18.89%
	PF4	19,454	0.39%	2,608,853	18.60%
	PF4T-12D	21,358	0.46%	2,317,495	14.42%
rRNA	PF2	51,685	1.09%	718,451	4.62%
	PF2T-12D	27,242	0.59%	271,635	1.84%
	PF4	38,987	0.78%	280,929	2%
	PF4T-12D	38,819	0.83%	604,494	3.76%
repeat	PF2	257,952	5.44%	658,140	4.24%
	PF2T-12D	243,489	5.23%	605,304	4.10%
	PF4	196,217	3.94%	469,620	3.35%
	PF4T-12D	175,370	3.74%	511,910	3.19%

Table 3. Cont.

Category	Species	Unique sRNAs	Percent%	Reads Number	Percent%
snRNA	PF2	1846	0.04%	3375	0.02%
	PF2T-12D	1316	0.03%	2821	0.02%
	PF4	1729	0.03%	3247	0.02%
	PF4T-12D	1630	0.03%	3582	0.02%
snoRNA	PF2	647	0.01%	1257	0.01%
	PF2T-12D	487	0.01%	1083	0.01%
	PF4	635	0.01%	1170	0.01%
	PF4T-12D	697	0.01%	1124	0.01%
tRNA	PF2	15,176	0.32%	569,369	3.67%
	PF2T-12D	7542	0.16%	293,078	1.99%
	PF4	10,605	0.21%	292,476	2.09%
	PF4T-12D	10,210	0.22%	669,113	4.16%
unann	PF2	4,394,575	92.63%	10,701,772	68.89%
	PF2T-12D	4,353,098	93.56%	10,797,866	73.16%
	PF4	4,711,849	94.62%	10,368,085	73.93%
	PF4T-12D	4,436,314	94.71%	11,959,262	74.43%

3.6. Identification of Conserved and Novel miRNAs

Unique reads that mapped to the mature plant miRNAs in miRBase 21.0 with less than three mismatches were considered to be the conserved miRNAs. We identified 146 precursors that fulfilled the criteria, giving 63 unique mature miRNAs, which have also been identified as mature miRNAs in other plant species. These 63 mature miRNAs are produced by 137 precursors and are distributed in 30 families (Table S12). The miR156 family was the largest, with 16 members, followed by the miR395, miR164, and miR166 families. Pfo-miR166a-3p was the most abundant miRNA, with 1,131,274 (PF2), 869,592 (PF2T_12D), 866,819 (PF4), and 1,080,900 (PF4T_12D) clean reads in these four libraries, respectively.

To identify novel miRNAs, the unannotated unique sRNA sequences were mapped to the *P. fortunei* genome sequence. Then, the secondary structures of the mapped sequences were predicted to identify sRNAs that could form the characteristic hairpin structure. As a result, we identified 100 novel miRNAs belonging to 79 miRNA families. Among these novel miRNA candidates, except four novel miRNA candidates (Pfo-mir31b, Pfo-mir31c, Pfo-mir32, and Pfo-mir542), the miRNA* for 96 of them were identified, supporting their identification as novel miRNAs. The minimal folding free energies for the precursor pre-miRNA sequences varied from -153 to -18 kcal mol^{-1} (average: -52.3 kcal mol^{-1}) in these identified novel miRNAs. The minimal free energies of plant miRNA precursors are commonly below -18 kcal mol^{-1} [48]. Furthermore, RNA sequences with average minimal free energies close to -46 kcal mol^{-1} are considered to be true miRNAs, compared with the higher values typical for other categories of RNAs [49,50]. Therefore, we concluded that these novel *P. fortunei* miRNAs are likely to be real miRNAs. Detailed information about the miRNA sequences is shown in Supplementary Table S13.

3.7. Analysis of miRNA Expression Profiles in the Two P. fortunei Genotypes

The abundances of the miRNAs in the four sRNA libraries were normalized in order to calculate the fold changes and *p*-values. If the fold-changes were >1 or <−1 and *p*-values were <0.05, the miRNAs were considered to be differentially expressed miRNAs. A total of 212 miRNAs in the four libraries were identified as being differentially expressed. Among them, we found 109 differentially expressed miRNAs in the PF2 vs. PF2T_12D comparison, including 50 (10 upregulated and 40 downregulated) that were conserved miRNAs and 59 (31 upregulated and 28 downregulated) that were novel miRNAs. In the PF4 vs. PF4T_12D comparison, 77 miRNAs were differentially expressed; among them,

34 (12 upregulated and 22 downregulated) were conserved miRNAs and 43 (23 upregulated and 20 downregulated) were novel miRNAs. In addition, 32 miRNAs in common were differentially expressed in both of the two *P. fortunei* genotypes under drought stress. Of these miRNAs, 21 (eight upregulated and 13 downregulated) were expressed in the same expression form in the two *P. fortunei* genotypes under drought stress. Moreover, we also found that 110 and 138 miRNAs were differentially expressed in the PF2 vs. PF4 and PF2T_12D vs. PF4T_12D comparisons, respectively. According to the same manner that was used to select the drought responsive genes, we found 11 miRNAs likely to be the drought responsive miRNAs (Tables S12 and S13).

3.8. Identification of miRNA Target Genes Associated with Drought Stress

To understand the potential regulatory roles of theses identified differentially expressed miRNAs, we constructed and sequenced four degradome libraries. After removing low-quality adapter and redundant sequences, 20,229,141 (PF2), 18,274,196 (PF2T_12D), 22,134,487 (PF4), and 17,221,529 (PF4T_12D) clean reads were obtained. A total of 17,140,999 (84.73%), 15,376,217 (84.14%), 17,032,997 (76.95%), and 13,537,235 (78.61%) tags were mapped to the *P. fortunei* genome (Table S14). The CleaveLand pipeline [51] was used to detect candidate target genes. The target genes were classed into four categories (Figure 5) according to the relative abundances of the degradome tags at the transcript sites. Twenty-eight potential targets were cleaved by miRNAs from nine conserved miRNA families and 12 novel miRNAs (Table S14). Among these target genes, the largest category was 'Category 2'. The target genes were annotated by the GO database to better understand the roles of these identified miRNAs and also to illustrate the correlation between miRNA and mRNA.

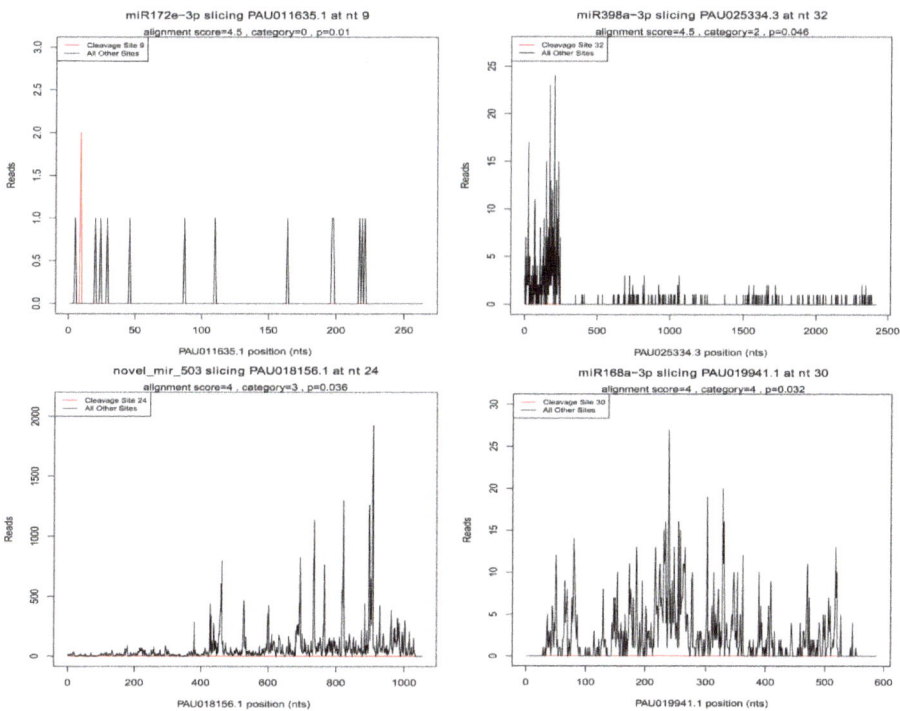

Figure 5. Target plots (t-plots) of miRNA targets in different categories confirmed by degradome sequencing.

3.9. Confirmation of Genes, miRNAs, and miRNA Targets by qRT-PCR

To confirm the Illumina sequencing data, 12 genes, 14 miRNAs, and six target genes from diploid and autotetraploid samples were monitored by qRT-PCR assays at 0, 6, 9, and 12 days after drought treatment. Figure 6 shows that the expression profiles of the genes tested by qRT-PCR exhibited similar trends to the deep sequencing except the genes PAU000881.1 (uncharacterized protein) and PAU017533.1 (4-beta-*N*-acetylglucosaminyltransferase-like isoform 1), which were down and up slightly in the PF4T_12D and PF2T_12D, respectively, compared to their control samples PF4 and PF2. For the miRNAs validation, the expressions of Pfo-miR166a-3p and Pfo-miR408 in the PF4T_12D samples showed inverse trends with the results of the deep sequencing but showed the same trends with that in the PF2T_12D samples compared to their corresponding control samples. The rest of the miRNAs exhibited the same trends as the results determined by the deep sequencing (Figure 7). We also found that, whether for genes or miRNAs, the expressions displayed the different trends under different time of drought treatments (Figures 6 and 7).

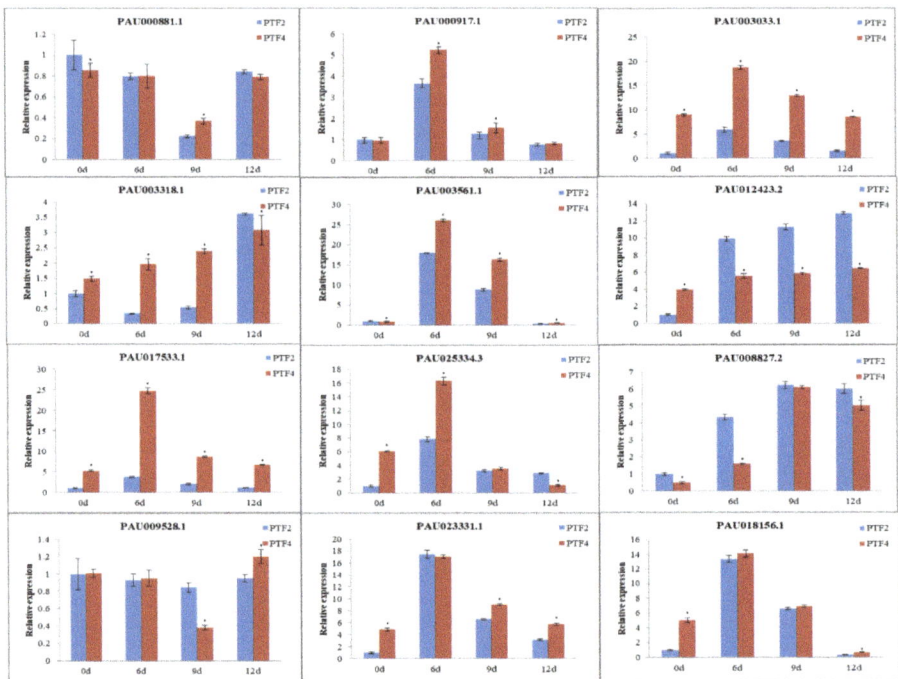

Figure 6. Relative expression levels of candidate drought response genes in two *P. fortunei* genotypes. 18S rRNA was used as the endogenous reference gene. Bar = Standard deviation. The difference between the diploid and autotetraploid *P. fortunei* within each drought treatment time point was significant (*, $p < 0.05$) according to the *t*-test. 0d, well-watered diploid and autotetraploid plants; 6d, 6 days drought-treated diploid and autotetraploid plants; 9d, 9 days drought-treated diploid and autotetraploid plants; 12d, 12 days drought-treated diploid and autotetraploid plants.

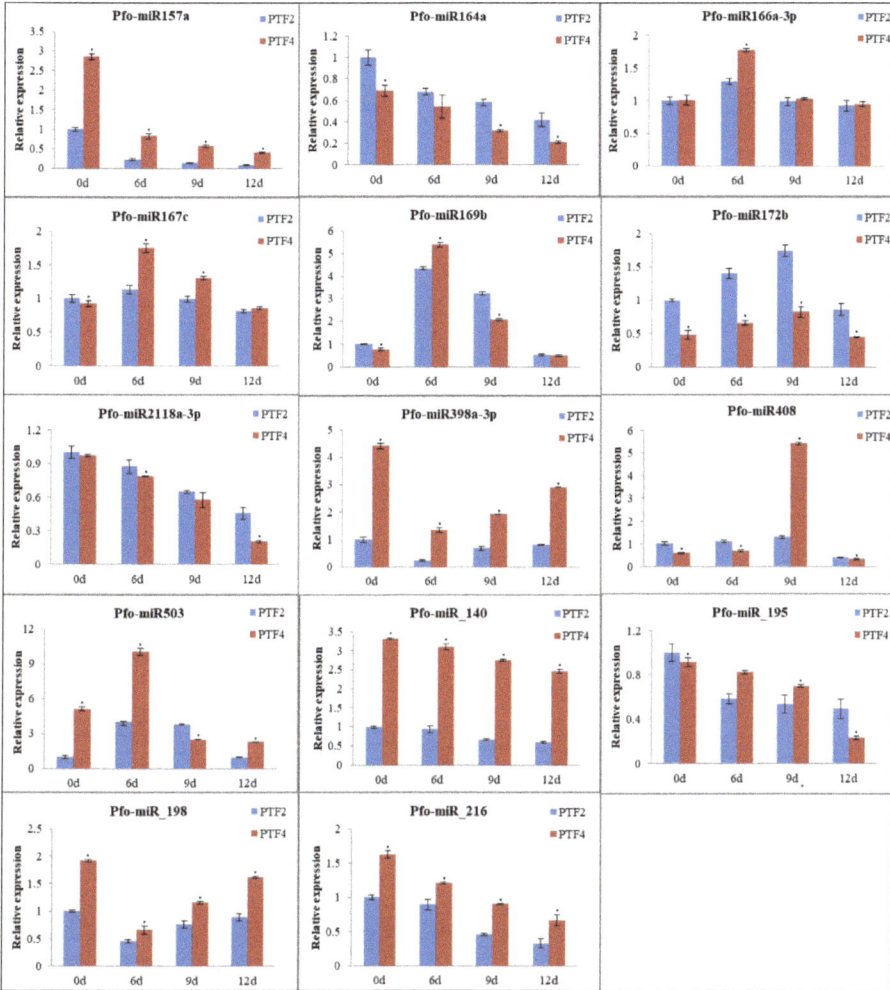

Figure 7. Relative expression levels of drought response miRNAs in two *P. fortunei* genotypes. U6 snRNA was used as the endogenous reference gene. Bar = Standard deviation. The difference between the diploid and autotetraploid *P. fortunei* within each drought treatment time point was significant (*, $p < 0.05$) according to the *t*-test. 0d, well-watered diploid and autotetraploid plants; 6d, 6 days drought-treated diploid and autotetraploid plants; 9d, 9 days drought-treated diploid and autotetraploid plants; 12d, 12 days drought-treated diploid and autotetraploid plants.

Furthermore, the potential correlation between miRNAs and targets was also investigated. As shown in Figure 8, the expressions of three target genes (PAU012423.2, PAU003033.1, and PAU017533.1) had negative correlations with the corresponding miRNAs (Pfo-miR157a, Pfo-miR166a-3p, and Pfo-miR172b), and other two target genes (PAU003561.1 and PAU003561.1) showed positive correlations with miRNAs (Pfo-miR408 and Pfo-mir198). Moreover, the rest of the targets showed mixed correlations with their miRNAs (Figure 4). The same results have been reported in other plant species such as *Phalaenopsis aphrodite* Rchb.f. [52] and cotton [53]. Meanwhile, the results implied that the expression patterns between miRNAs and their target genes were very complex.

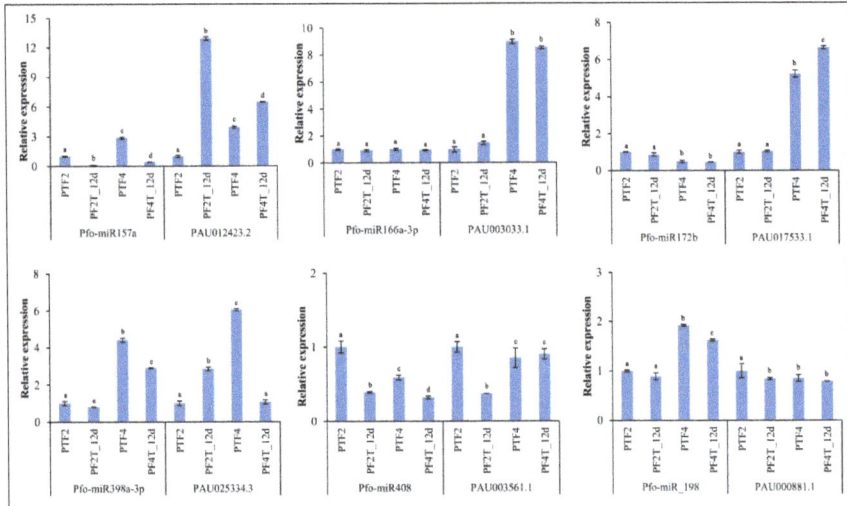

Figure 8. Relative expression of the target genes of six *P. fortunei* miRNAs by Quantitative Real-Time Polymerase Chain Reaction (qRT-PCR) assay. PF2, well-watered diploid; PF2T_12d, 12 days drought-treated diploid; PF4, well-watered autotetraploid; PF4T_12d, 12 days drought-treated autotetraploid. The different letters within a miRNA expression or a gene expression indicate significant difference, while the same letters within a miRNA expression or a gene expression indicate no significant differences ($p < 0.05$).

4. Discussion

Transcriptome sequencing is a convenient way of rapidly obtaining genomic information when a reference genome is not available, while a transcript that might be overlapping refers to a reference gene in genome. Genome-wide transcriptome analysis can capture an unbiased view of the complete RNA transcript profile, which allows the transcriptional level of each gene to be monitored. In this study, based on the Paulownia reference genome, we used transcriptome, miRNA, and degradome sequencing to identify the genes and miRNAs related to the response of Paulownia plants to drought stress. As a result, 258 DEGs and 11 miRNAs related to drought stress response were identified from the two *P. fortunei* genotypes. Comparing the results from Dong et al. [47] and Fan et al. [54], we found that less DEGs and miRNAs were identified in this study due to different experimental designs (including the screening strategies for drought responsive genes and reference genome for gene identification) between the study by Dong et al. [47] and this present research.

Molecular markers are closely related to several different agronomic characters and abiotic and disease resistance traits. These molecular markers include hybridization-based and PCR-based markers that have been reported in many crop species and tree species [55–60]. Recently, with the advent of next-generation sequencing, the transcriptome sequencing technology not only can capture the genes expressed at the transcriptional level but can also provide the opportunity to detect nucleotide variations in the model or non-model species. Thus, the sequence-based markers, single nucleotide polymorphism (SNP) markers, were generated rapidly. The SNP markers had been previously developed for many model or non-model species such as rice, *Arabidopsis*, *Hemarthria*, eggplant, and legumes [61–67]. Furthermore, Han et al. [63] previously reported the discovery of the novel molecular markers based on the transcription factors in legumes species. In this study, we found 622 and 402 DEGs were annotated as transcription factors in PF2 vs. PF2T_12D and PF4 vs. PF4T_12D comparisons, respectively. These TFs may act as excellent targets for developing drought stress-related molecular markers in Paulownia species in the future.

Transcriptional regulation is mediated by the interplay between TFs and specific-regulatory regions of the genome, leading to gene activation and/or other changes [68]. Therefore, some TFs could be considered as the master regulators, because they are involved in the regulation of other TFs and downstream genes, protein phosphatases, protein kinases, and enzymes involved in phospholipid metabolism, and the components of calcium-coupled phosphoprotein cascades [69]. In this study, we focused on TF genes that were differentially expressed in response to drought stress in *P. fortunei*. Several putative TF families were identified in the leaves of *P. fortunei*, including MYB, WRKY, bHLH, zinc finger, NAC, HSF, and AP2/EREBP.

The MYB family of TFs contains a conserved MYB domain, which is present in all eukaryotes. In *Arabidopsis*, the flavonoids were induced by stress, which was implied to be a positive response to drought stress [70]. Overexpression of MYB12, a regulator of flavonol synthesis, resulted in less water loss [71]. In *Arabidopsis*, MYB41 was induced by drought and may be involved in regulating cuticle deposition and cell expansion under drought stress [72]. Many R2R3-MYB proteins in *Arabidopsis* have been reported to be involved in drought responses. For example, for MYB60, a regulator of stomatal movement, its overexpression resulted in hypersensitivity to water deficiency, and it was found to be downregulated under drought stress [73]. In this study, 39 and 26 MYB TFs were upregulated in PF2 vs. PF2T and PF4 vs. PF4T, respectively. We found that six of these MYB TFs were upregulated in both comparisons, while three were downregulated in PF2 vs. PF2T and upregulated in PF4 vs. PF4T.

The phytohormone ABA was involved in plant adaptation under drought stress [18]. It was known that genes (including TFs) involved in regulatory networks are active under drought conditions (Figure S4) [24]. In this study, we identified WRKY and NAC TFs that were differentially expressed in the drought-stressed plants. Previous studies have shown that WRKY18 and WRKY60 were the positive regulators of ABA signaling during stress response and seed germination, while WRKY40 was a negative regulator of ABA signaling. WRKYs18 and 60 are weak transcriptional activators, whereas WRKY40 binds to the promoters of stress-induced TF genes, such as DREB2A, DREB1A/CBF3, and MYB2, and represses the expression of these genes [74,75]. ABA signal perception leads to the induction of WRKYs18 and 40 and their products can bind to the W-box in the WRKY60 promoter sequence, thereby inducing it [74]. In this study, 14 and 13 WRKY TFs were found to be upregulated in PF2 vs. PF2T and PF4 vs. PF4T, respectively. Overexpression of several stress-responsive NAC TFs in *Arabidopsis* and rice has been reported to impart drought tolerance in transgenic plants. For example, transgenic plants overexpressing NAC TFs were shown to have enhanced drought tolerance and increased sensitivity to ABA [76,77]. Similarly, overexpression of ATAF1, another NAC TF, improved drought tolerance [78]. In the present study, we found that lots of NAC TFs were involved in ABA signaling and that their biosynthesis was significantly upregulated or downregulated in response to drought stress. This finding is consistent with our expectation that ABA plays key roles in drought stress responses and plant growth and development.

Several candidate genes encoding chlorophyll a/b binding proteins were identified in this study. Typically, this protein family contains a chlorophyll-binding CAB domain. There are some indications that these genes are involved in high light protection and encode light-harvesting-like proteins. All organisms that perform oxygenic photosynthesis also contain light-induced-like proteins (LILs), which contain a CAB domain and encode the light-harvesting antenna found in higher plants [79]. The LILs are negatively regulated by the light-harvesting complex (LHC) proteins; for example, under high light conditions, when the expression of LILs is increased, the expression of LHCs is repressed. In *Arabidopsis* [80] and rice [81] exposed to drought stress, the genes encoding LILs were found to be upregulated, indicating the general importance of these genes in plants under drought stress.

It was known that ploidy variation induces a certain type of gene expression at various levels because of epigenetic interaction between redundant genes [82,83]. In this study, all the clean sRNA reads were mapped to the *P. fortunei* genome. We found that the percentage of reads that mapped to the *P. fortunei* genome was lower in the autotetraploid library compared with the diploid library even though the diploid *P. fortunei* and its derived autotetraploid *P. fortunei* clone used in the present study were the same genome with the difference at the ploidy level. These findings indicated that whole-genome duplication did not simply increase gene expression by two-fold; indeed, the evolution of genes in genome duplication perhaps transformed the gene structure, leading to different expression patterns in the two genotypes. Furthermore, most miRNA families had more numbers than reported in a previous study, probably because of the availability of a reference genome sequence, which contains more information than the transcriptome data. In addition, many sequences were filtered out when the conserved miRNAs were identified. In this study, 63 conserved miRNAs were identified and 26 of them were identified for the first time in Paulownia, including highly conserved miRNAs such as, miR164, miR171, miR172, miR396, miR398, and miR408. Moreover, we found that 60 conserved miRNAs and 64 novel miRNAs were upregulated or downregulated predominantly in only one of the genotypes after drought treatment; for example, pfo-miR160a/b/c-3p, pfo-miR164i, and pfo-miR166b-3p were downregulated only in diploid clones, and pfo-miR530b and pfo-miR5720c-3p were upregulated only in autotetraploid clones under drought stress. The different expression patterns of the above miRNAs in the two comparisons suggested that these miRNAs may be related to variations in the capacity of the two Paulownia genotypes to adapt to drought stress.

It has been shown that plant response to drought stress involves a large number of genes in regulatory networks, which are active at the transcriptional and post-transcriptional levels [84]. Degradome analysis combined with transcriptome analysis has proven to be helpful in revealing the potential relationships between DEGs, miRNAs, and target genes. In this study, the transcriptome and degradome analyses results identified eight and six target genes that were upregulated and downregulated in PF2 vs. PF2T and PF4 vs. PF4T, respectively. Among those target genes, the gene targeted by pfo-miR157 was differentially expressed in the PF2 vs. PF2T_12D and PF4 vs. PF4T_12D comparisons. MiR157 negatively targets the Squamosa promoter binding protein-like (SPL) TFs in plants, and miR157 overexpression has been found to result in the delayed onset of flowering and adult traits in *Arabidopsis* [85]. Current research on miR157 has focused mainly on its role in morphology changes and regulation of blooming. Here, the transcriptome and degradome sequencing data have provided evidence that drought stress can disturb the expression of a miR157 target gene, indicating a novel role for miR157 and its target in response to drought stress. We also found that pfo-miR408-3p positively regulated its corresponding targets genes which encoded the blue basic protein. The RNA-Seq results showed that the target gene blue basic protein was downregulated in the PF2 vs. PF2T_12D and PF4 vs. PF4T_12D comparisons, while the expression of pfo-miR408-3p was also downregulated in two comparisons by the qRT-PCR analysis. However, pfo-miR6262 and its target gene were both upregulated in PF2 vs. PF2T_12D. In addition, the miR160 family was reported to target auxin response factors, which are essential for plant growth and development. Under normal conditions, the auxin-responsive genes, which are regulated by miR160, were sufficient for plant growth and development. Under drought stress conditions, miR160 was found to be upregulated, which caused a reduction in the transcript levels of the genes encoding auxin response factors, resulting in the attenuation of plant growth and development. In *Medicago truncatula* Gaertn. and pea, miR398 was downregulated by drought treatment [24,86]. The target gene encoded the potassium channel KAT1, which may be regulated by ABA to keep the homeostasis of potassium. Additionally, it is important for osmotic adjustment, biofilm formation, and regulation of seedlings growth [87,88]. In the current study, we found that pfo-miR160e-3p was upregulated in both of the two genotypes under the drought stress, whereas pfo-miR398a/b-3p was downregulated in PF2 vs. PF2T, and slightly downregulated in PF4 vs.PF4T. These results suggested these miRNAs may play important roles in *Paulownia fortunei* plants under dehydration stress and implied that the correlation between miRNAs and their target genes are more complex than we previously thought.

5. Conclusions

In this research, we integrated transcriptome, miRNAs, and degradome data to investigate the drought-responsive genes and miRNAs in *P. fortunei* at the genome-wide. Fortunately, we have discovered 258 genes and 11 miRNAs related to drought stress response from the two *P. fortunei* genotypes. Based on GO and KEGG pathway annotation of these drought responsive genes and miRNAs, we found most of these genes were predicted to participate in ABA signaling and biosynthesis, photosynthesis, flavonol synthesis, and auxin signal transduction pathways. The results provide valuable information for drought tolerance molecular mechanisms that are necessary to understand in order to enhance yields under environment stresses. Furthermore, we also found that TFs, especially for the MYB, WRKY, and NCA families, play an irreplaceable role in tolerance to drought stress in Paulownia trees. These TFs may be potential targets for developing molecular markers in Paulownia species in the future.

Supplementary Materials: The following are available online at http://www.mdpi.com/1999-4907/9/2/88/s1, Table S1: Quantitative RT-PCR primers for genes and miRNAs. Table S2: Comparison of the differentially expressed genes in PF2 vs. PF2T_12D. Table S3: Comparison of the differentially expressed genes in PF4 vs. PF4T_12D. Table S4: Comparison of the differentially expressed genes in PF2 vs. PF4. Table S5: Comparison of the differentially expressed genes in PF2T_12D vs. PF4T_12D. Table S6: Comparison of the KEGG annotation of DGEs in PF2 vs. PF2T_12D. Table S7: Comparison of the KEGG annotation of DGEs in PF4 vs. PF4T_12D. Table S8: Comparisons of the consistently differentially expressed genes in PF2 vs. PF2T_12D and PF4 vs. PF4T_12D. Table S9: Comparisons of the consistently differentially expressed genes in PF2 vs. PF4 and PF2T_12D vs. PF4T_12D. Table S10: The DEGs of type I in two *P. fortunei* genotypes. Table S11: KEGG annotation for the type I DEGs in two *P. fortunei* genotypes. Table S12: The conserved miRNAs identified in two *P. fortunei* genotypes. Table S13: The novel miRNAs identified in two *P. fortunei* genotypes. Table S14: Identified miRNA target genes in *P. fortunei* by degradome analysis. Figure S1: Gene Ontology of the DEGs in PF2 vs. PF2T_12D. Figure S2: Gene Ontology of the DEGs in PF4 vs. PF4T_12D. Figure S3: The correlation coefficient of DEGs between two *P. fortunei* genotypes. Figure S4: The schematic representation of transcription factors involved in drought-stress-responses.

Acknowledgments: We thank Beijing Genomics Institute-Shenzhen (BGI-Shenzhen) for helping us with the throughput RNA-seq and sRNA-seq analyses. We also thank Edanz China for assistance with the language proofing.

Author Contributions: G.F. conceived and designed the experiments; Z.Z. performed the experiments; S.N. and Z.Z. analyzed the data; M.D. contributed reagents and analysis tools; S.N. and Y.W. wrote the paper.

Conflicts of Interest: The authors declare no conflict of interest.

References

1. Huang, G.T.; Ma, S.L.; Bai, L.P.; Zhang, L.; Ma, H.; Jia, P.; Liu, J.; Zhong, M.; Guo, Z.F. Signal transduction during cold, salt, and drought stresses in plants. *Mol. Biol. Rep.* **2012**, *39*, 969–987. [CrossRef] [PubMed]

2. Zhang, X.; Liu, R.; Fan, G.; Zhao, Z.; Deng, M. Study on the physiological response of tetraploid paulownia to drought. *J. Henan Agric. Univ.* **2013**, *47*, 543–551.

3. Raes, J. Duplication and divergence: The evolution of new genes and old ideas. *Ann. Rev. Genet.* **2004**, *38*, 615–643.

4. Jackson, J.A.; Tinsley, R.C. Parasite infectivity to hybridising host species: A link between hybrid resistance and allopolyploid speciation? *Int. J. Parasitol.* **2003**, *33*, 137–144. [CrossRef]

5. Ramsey, J.; Schemske, D.W. Neopolyploidy in flowering plants. *Ann. Rev. Ecol. Syst.* **2002**, *33*, 589–639. [CrossRef]

6. Ni, Z.; Hu, Z.; Jiang, Q.; Zhang, H. Gmnfya3, a target gene of mir169, is a positive regulator of plant tolerance to drought stress. *Plant Mol. Biol.* **2013**, *82*, 113–129. [CrossRef] [PubMed]

7. Yang, J.; Zhang, N.; Mi, X.; Wu, L.; Ma, R.; Zhu, X.; Yao, L.; Jin, X.; Si, H.; Wang, D. Identification of mir159s and their target genes and expression analysis under drought stress in potato. *Comput. Biol. Chem.* **2014**, *53 Pt B*, 204–213. [CrossRef] [PubMed]

8. Jovanović, Ž.; Stanisavljević, N.; Mikić, A.; Radović, S.; Maksimović, V. Water deficit down-regulates mir398 and mir408 in pea (*Pisum sativum* L.). *Plant Physiol. Biochem.* **2014**, *83*, 26–31. [CrossRef] [PubMed]

9. Sorin, C.; Declerck, M.; Christ, A.; Blein, T.; Ma, L.; Lelandais-Brière, C.; Njo, M.F.; Beeckman, T.; Crespi, M.; Hartmann, C. A mir169 isoform regulates specific nf-ya targets and root architecture in arabidopsis. *New Phytol.* **2014**, *202*, 1197–1211. [CrossRef] [PubMed]

10. Bartel, D.P. Micrornas: Target recognition and regulatory functions. *Cell* **2009**, *136*, 215–233. [CrossRef] [PubMed]

11. Brodersen, P.; Sakvarelidze-Achard, L.; Bruun-Rasmussen, M.; Dunoyer, P.; Yamamoto, Y.Y.; Sieburth, L.; Voinnet, O. Widespread translational inhibition by plant mirnas and sirnas. *Science* **2008**, *320*, 1185–1190. [CrossRef] [PubMed]

12. Covarrubias, A.A.; Reyes, J.L. Post-transcriptional gene regulation of salinity and drought responses by plant micrornas. *Plant Cell Environ.* **2010**, *33*, 481–489. [CrossRef] [PubMed]

13. Wang, Y.-G.; An, M.; Zhou, S.-F.; She, Y.-H.; Li, W.-C.; Fu, F.-L. Expression profile of maize micrornas corresponding to their target genes under drought stress. *Biochem. Genet.* **2014**, *52*, 474–493. [CrossRef] [PubMed]

14. Lv, D.K.; Xi, B.; Yong, L.; Ding, X.D.; Ying, G.; Hua, C.; Wei, J.; Wu, N.; Zhu, Y.M. Profiling of cold-stress-responsive mirnas in rice by microarrays. *Gene* **2010**, *459*, 39–47. [CrossRef] [PubMed]

15. Nigam, D.; Kumar, S.; Mishra, D.C.; Rai, A.; Smita, S.; Saha, A. Synergistic regulatory networks mediated by micrornas and transcription factors under drought, heat and salt stresses in *Oryza sativa* spp. *Gene* **2015**, *555*, 127–139. [CrossRef] [PubMed]

16. Wang, Y.; Zhao, Z.; Deng, M.; Liu, R.; Niu, S.; Fan, G. Identification and functional analysis of micrornas and their targets in platanus acerifolia under lead (pb) stress. *Int. J. Mol. Sci.* **2015**, *16*, 7098–7111. [CrossRef] [PubMed]

17. Jover-Gil, S.; Candela, H.; Ponce, M.R. Plant micrornas and development. *Int. J. Dev. Biol.* **2005**, *49*, 733–744. [CrossRef] [PubMed]

18. Singh, D.; Laxmi, A. Transcriptional regulation of drought response: A tortuous network of transcriptional factors. *Front. Plant Sci.* **2015**, *6*, 895. [CrossRef] [PubMed]

19. Iuchi, S.; Kobayashi, M.; Taji, T.; Naramoto, M.; Seki, M.; Kato, T.; Tabata, S.; Kakubari, Y.; Yamaguchi-Shinozaki, K.; Shinozaki, K. Regulation of drought tolerance by gene manipulation of 9-cis-epoxycarotenoid, a key in abscisic acid biosynthesis in arabidopsis. *Plant J.* **2001**, *27*, 325–333. [CrossRef] [PubMed]

20. Zhang, J.; Jia, W.; Yang, J.; Ismail, A.M. Role of aba in integrating plant responses to drought and salt stresses. *Field Crop Res.* **2006**, *97*, 111–119. [CrossRef]

21. Yoshida, T.; Mogami, J.; Yamaguchi-Shinozaki, K. Aba-dependent and aba-independent signaling in response to osmotic stress in plants. *Curr. Opin. Plant Biol.* **2014**, *21*, 133–139. [CrossRef] [PubMed]

22. Taishi, U.; Kazuo, N.; Takuya, M.; Takashi, K.; Masaru, T.; Kazuo, S.; Kazuko, Y.S. Molecular basis of the core regulatory network in aba responses: Sensing, signaling and transport. *Plant Cell Physiol.* **2010**, *51*, 1821–1839.

23. Shinozaki, K.; Yamaguchi-Shinozaki, K. Gene networks involved in drought stress response and tolerance. *J. Exp. Bot.* **2007**, *58*, 221–227. [CrossRef] [PubMed]

24. Shinozaki, K.; Yamaguchi-Shinozaki, K.; Seki, M. Regulatory network of gene expression in the drought and cold stress responses. *Curr. Opin. Plant Biol.* **2003**, *6*, 410–417. [CrossRef]

25. Hussain, S.S.; Kayani, M.A.; Amjad, M. Transcription factors as tools to engineer enhanced drought stress tolerance in plants. *Biotechnol. Progr.* **2011**, *27*, 297–306. [CrossRef] [PubMed]

26. Abe, H.; Urao, T.; Ito, T.; Seki, M.; Shinozaki, K.; Yamaguchi-Shinozaki, K. Arabidopsis atmyc2 (bhlh) and atmyb2 (myb) function as transcriptional activators in abscisic acid signaling. *Plant Cell* **2003**, *15*, 63–78. [CrossRef] [PubMed]

27. Zheng, X.; Chen, B.; Lu, G.; Han, B. Overexpression of a NAC transcription factor enhances rice drought and salt tolerance. *Biochem. Biophys. Res. Commun.* **2009**, *379*, 985–989. [CrossRef] [PubMed]

28. Hu, H.; Dai, M.; Yao, J.; Xiao, B.; Li, X.; Zhang, Q.; Xiong, L. Overexpressing a NAM, ATAF, and CUC (NAC) transcription factor enhances drought resistance and salt tolerance in rice. *Proc. Natl. Acad. Sci. USA* **2006**, *103*, 12987–12992. [CrossRef] [PubMed]

29. Aharoni, A.; Dixit, S.; Jetter, R.; Thoenes, E.; van Arkel, G.; Pereira, A. The shine clade of ap2 domain transcription factors activates wax biosynthesis, alters cuticle properties, and confers drought tolerance when overexpressed in arabidopsis. *Plant Cell* **2004**, *16*, 2463–2480. [CrossRef] [PubMed]

30. Reyes, J.; Chua, N. Aba induction of mir159 controls transcript levels of two myb factors during. *Plant J. Cell Mol. Biol.* **2007**, *49*, 592–606. [CrossRef] [PubMed]

31. Bojović, B.; Stojanović, J. Chlorophyll and carotenoid content in wheat cultivars as a function of mineral nutrition. *Arch. Biol. Sci.* **2005**, *57*, 283–290. [CrossRef]

32. Arnon, D.I. Copper enzymes in isolated chloroplasts. Polyphenoloxidase in beta vulgaris. *Plant Phys.* **1949**, *24*, 1–15. [CrossRef]

33. Bates, L.; Waldren, R.; Teare, I. Rapid determination of free proline for water-stress studies. *Plant Soil* **1973**, *39*, 205–207. [CrossRef]

34. Li, H. *Principle and Technology of Plant Physiology and Biochemistry*; Higher Education Press: Beijing, China, 2000.

35. Liu, J.; Liu, X. *Tutorial of Plant Physiology Experiments*; Higher Education Press: Beijing, China, 2011.

36. Langmead, B.; Salzberg, S.L. Fast gapped-read alignment with bowtie 2. *Nat. Method* **2012**, *9*, 357–359. [CrossRef] [PubMed]

37. Mortazavi, A.; Williams, B.A.; McCue, K.; Schaeffer, L.; Wold, B. Mapping and quantifying mammalian transcriptomes by RNA-seq. *Nat. Meth.* **2008**, *5*, 621–628. [CrossRef] [PubMed]

38. Audic, S.; Claverie, J.M. The significance of digital gene expression profiles. *Genom. Res.* **1997**, *7*, 986–995. [CrossRef]

39. Benjamini, Y.; Yekutieli, D. The control of the false discovery rate in multiple testing under dependency. *Ann. Stat.* **2001**, *29*, 1165–1188.

40. Kozomara, A.; Griffiths-Jones, S. Mirbase: Integrating microrna annotation and deep-sequencing data. *Nucleic Acid Res.* **2010**, *39*, D152–D157. [CrossRef] [PubMed]

41. Meyers, B.C.; Axtell, M.J.; Bonnie, B.; Bartel, D.P.; David, B.; Bowman, J.L.; Xiaofeng, C.; Carrington, J.C.; Xuemei, C.; Green, P.J. Criteria for annotation of plant micrornas. *Plant Cell* **2008**, *20*, 3186–3190. [CrossRef] [PubMed]

42. Addo-Quaye, C.; Eshoo, T.W.; Bartel, D.P.; Axtell, M.J. Endogenous sirna and mirna targets identified by sequencing of the arabidopsis degradome. *Curr. Biol.* **2008**, *18*, 758–762. [CrossRef] [PubMed]

43. German, M.A.; Pillay, M.; Jeong, D.H.; Hetawal, A.; Luo, S.; Janardhanan, P.; Kannan, V.; Rymarquis, L.A.; Nobuta, K.; German, R.; et al. Global identification of microrna-target RNA pairs by parallel analysis of RNA ends. *Nat. Biotechnol.* **2008**, *26*, 941–946. [CrossRef] [PubMed]

44. Chen, C.; Ridzon, D.A.; Broomer, A.J.; Zhou, Z.; Lee, D.H.; Nguyen, J.T.; Barbisin, M.; Xu, N.L.; Mahuvakar, V.R.; Andersen, M.R.; et al. Real-time quantification of micrornas by stem-loop rt-pcr. *Nucleic Acid Res.* **2005**, *33*, e179. [CrossRef] [PubMed]

45. Fan, G.; Niu, S.; Zhao, Z.; Deng, M.; Xu, E.; Wang, Y.; Yang, L. Identification of micrornas and their targets in paulownia fortunei plants free from phytoplasma pathogen after methyl methane sulfonate treatment. *Biochimie* **2016**, *127*, 271–280. [CrossRef] [PubMed]

46. Livaka, K.J.; Schmittgenb, T.D. Analysis of relative gene expression data using real-time quantitative PCR and the 2(-delta delta c(t)) method. *Methods* **2011**, *25*, 402–408. [CrossRef] [PubMed]

47. Dong, Y.; Fan, G.; Zhao, Z.; Deng, M. Compatible solute, transporter protein, transcription factor, and hormone-related gene expression provides an indicator of drought stress in paulownia fortunei. *Funct. Integr. Genomics* **2014**, *14*, 479–491. [CrossRef] [PubMed]

48. Han, J.; Xie, H.; Kong, M.L.; Sun, Q.P.; Li, R.Z.; Pan, J.B. Computational identification of mirnas and their targets in phaseolus vulgaris. *Genet. Mol. Res.* **2014**, *13*, 310–322. [CrossRef] [PubMed]

49. Baohong, Z.; Xiaoping, P.; Cannon, C.H.; Cobb, G.P.; Anderson, T.A. Conservation and divergence of plant microrna genes. *Plant J. Cell Mol. Biol.* **2006**, *46*, 243–259.

50. Zhang, B.H.; Pan, X.P.; Cox, S.B.; Cobb, G.P.; Anderson, T.A. Evidence that mirnas are different from other rnas. *Cell. Mol. Life Sci.* **2006**, *63*, 246–254. [CrossRef] [PubMed]

51. Addo-Quaye, C.; Miller, W.; Axtell, M.J. Cleaveland: A pipeline for using degradome data to find cleaved small RNA targets. *Bioinformatics* **2009**, *25*, 130–131. [CrossRef] [PubMed]

52. An, F.M.; Chan, M.T. Transcriptome-wide characterization of mirna-directed and non-mirna-directed endonucleolytic cleavage using degradome analysis under low ambient temperature in phalaenopsis aphrodite subsp. Formosana. *Plant Cell Physiol.* **2012**, *53*, 1737–1750. [CrossRef] [PubMed]

53. Wei, M.; Wei, H.; Wu, M.; Song, M.; Zhang, J.; Yu, J.; Fan, S.; Yu, S. Comparative expression profiling of mirna during anther development in genetic male sterile and wild type cotton. *BMC Plant Biol.* **2013**, *13*, 66. [CrossRef] [PubMed]

54. Fan, G.; Niu, S.; Li, X.; Wang, Y.; Zhao, Z.; Deng, M.; Dong, Y. Functional analysis of differentially expressed micrornas associated with drought stress in diploid and tetraploid paulownia fortunei. *Plant Mol. Biol. Rep.* **2017**, *35*, 389–398. [CrossRef]

55. Xiaoshen, Z.; Guoqiang, F.; Zhenli, Z.; Xibing, C.; Gaili, Z.; Minjie, D.; Yanpeng, D. Analysis of diploid and its autotetraploid paulownia tomentosa× p. Fortunei with aflp and msap. *Sci. Silv. Sin.* **2013**, *10*, 026.

56. Zhang, Y.; Cao, X.; ZhaiI, X.; Fan, G. Study on DNA extraction of aflp reaction system for paulownia plants. *J. Henan Agric. Univ.* **2009**, *6*, 007.

57. De la Rosa, R.; Angiolillo, A.; Guerrero, C.; Pellegrini, M.; Rallo, L.; Besnard, G.; Bervillé, A.; Martin, A.; Baldoni, L. A first linkage map of olive (*Olea europaea* L.) cultivars using rapd, aflp, rflp and ssr markers. *Theor. Appl. Genet.* **2003**, *106*, 1273–1282. [CrossRef] [PubMed]

58. Marques, C.; Araujo, J.; Ferreira, J.; Whetten, R.; O'malley, D.; Liu, B.-H.; Sederoff, R. AFLP genetic maps of eucalyptus globulus and e. Tereticornis. *Theor. Appl. Genet.* **1998**, *96*, 727–737. [CrossRef]

59. Phillips, R.; Vasil, I. *DNA-Based Markers in Plants, Advances in Cellular and Molecular Biology of Plants*; Kluwer: Dordrecht, the Netherlands, 2001.

60. Gupta, P.K.; Varshney, R. The development and use of microsatellite markers for genetic analysis and plant breeding with emphasis on bread wheat. *Euphytica* **2000**, *113*, 163–185. [CrossRef]

61. Huang, X.; Yan, H.-D.; Zhang, X.; Zhang, J.; Frazier, T.P.; Huang, D.; Lu, L.; Huang, L.; Liu, W.; Peng, Y.; et al. De novo transcriptome analysis and molecular marker development of two hemarthria species. *Front. Plant Sci.* **2016**, *7*, 496. [CrossRef] [PubMed]

62. Gramazio, P.; Blanca, J.; Ziarsolo, P.; Herraiz, F.J.; Plazas, M.; Prohens, J.; Vilanova, S. Transcriptome analysis and molecular marker discovery in solanum incanum and s. Aethiopicum, two close relatives of the common eggplant (*Solanum melongena*) with interest for breeding. *BMC Genom.* **2016**, *17*, 300. [CrossRef] [PubMed]

63. Han, Y.; Khu, D.-M.; Torres-Jerez, I.; Udvardi, M.; Monteros, M.J. Plant transcription factors as novel molecular markers for legumes. In *Sustainable Use of Genetic Diversity in Forage and Turf Breeding*; Springer: Berlin, Germany, 2010; pp. 421–425.

64. Lu, T.; Lu, G.; Fan, D.; Zhu, C.; Li, W.; Zhao, Q.; Feng, Q.; Zhao, Y.; Guo, Y.; Li, W. Function annotation of the rice transcriptome at single-nucleotide resolution by rna-seq. *Genom. Res.* **2010**, *20*, 1238–1249. [CrossRef] [PubMed]

65. Zhai, R.; Feng, Y.; Zhan, X.; Shen, X.; Wu, W.; Yu, P.; Zhang, Y.; Chen, D.; Wang, H.; Lin, Z. Identification of transcriptome snps for assessing allele-specific gene expression in a super-hybrid rice xieyou9308. *PLoS ONE* **2013**, *8*, e60668. [CrossRef] [PubMed]

66. Gan, X.; Stegle, O.; Behr, J.; Steffen, J.G.; Drewe, P.; Hildebrand, K.L.; Lyngsoe, R.; Schultheiss, S.J.; Osborne, E.J.; Sreedharan, V.T. Multiple reference genomes and transcriptomes for arabidopsis thaliana. *Nature* **2011**, *477*, 419. [CrossRef] [PubMed]

67. Chekanova, J.A.; Gregory, B.D.; Reverdatto, S.V.; Chen, H.; Kumar, R.; Hooker, T.; Yazaki, J.; Li, P.; Skiba, N.; Peng, Q. Genome-wide high-resolution mapping of exosome substrates reveals hidden features in the arabidopsis transcriptome. *Cell* **2007**, *131*, 1340–1353. [CrossRef] [PubMed]

68. Zhang, D.F.; Li, B.; Jia, G.Q.; Zhang, T.F.; Dai, J.R.; Li, J.S.; Wang, S.C. Isolation and characterization of genes encoding grf transcription factors and gif transcriptional coactivators in maize (*Zea mays* L.). *Plant Sci.* **2008**, *175*, 809–817. [CrossRef]

69. Nadgauda, R.S.; Parasharami, V.A.; Mascarenhas, A.F. Precocious flowering and seeding behaviour in tissue-cultured bamboos. *Nature* **1990**, *344*, 335–336. [CrossRef]

70. Thorsen, S.; Rugulies, R.; Løngaard, K.; Borg, V.; Thielen, K.; Bjorner, J.B. Edl3 is an f-box protein involved in the regulation of abscisic acid signalling in arabidopsis thaliana. *Int. Arch. Occup. Environ. Health* **2011**, *62*, 5547–5560.

71. Nakabayashi, R.; Yonekura-Sakakibara, K.; Urano, K.; Suzuki, M.; Yamada, Y.; Nishizawa, T.; Matsuda, F.; Kojima, M.; Sakakibara, H.; Shinozaki, K.; et al. Enhancement of oxidative and drought tolerance in arabidopsis by overaccumulation of antioxidant flavonoids. *Plant J.* **2014**, *77*, 367–379. [CrossRef] [PubMed]

72. Cominelli, E.; Sala, T.; Calvi, D.; Gusmaroli, G.; Tonelli, C. Over-expression of the arabidopsis atmyb41 gene alters cell expansion and leaf surface permeability. *Plant J.* **2008**, *53*, 53–64. [CrossRef] [PubMed]

73. Oh, J.; Kwon, Y.; Kim, J.; Noh, H.; Hong, S.-W.; Lee, H. A dual role for myb60 in stomatal regulation and root growth of arabidopsis thaliana under drought stress. *Plant Mol. Biol.* **2011**, *77*, 91–103. [CrossRef] [PubMed]

74. Chen, H.; Lai, Z.; Shi, J.; Xiao, Y.; Chen, Z.; Xu, X. Roles of arabidopsis wrky18, wrky40 and wrky60 transcription factors in plant responses to abscisic acid and abiotic stress. *BMC Plant Biol.* **2010**, *10*, 443–462. [CrossRef] [PubMed]

75. Shang, Y.; Yan, L.; Liu, Z.Q.; Cao, Z.; Mei, C.; Xin, Q.; Wu, F.Q.; Wang, X.F.; Du, S.Y.; Jiang, T. The mg-chelatase h subunit of arabidopsis antagonizes a group of wrky transcription repressors to relieve aba-responsive genes of inhibition. *Plant Cell* **2010**, *22*, 1909–1935. [CrossRef] [PubMed]

76. Miki, F.; Yasunari, F.; Kyonoshin, M.; Motoaki, S.; Keiichiro, H.; Masaru, O.T.; Lam-Son Phan, T.; Kazuko, Y.S.; Kazuo, S. A dehydration-induced NAC protein, rd26, is involved in a novel aba-dependent stress-signaling pathway. *Plant J.* **2004**, *39*, 863–876.

77. Tran, L.S.; Nakashima, K.; Sakuma, Y.; Simpson, S.D.; Fujita, Y.; Maruyama, K.; Fujita, M.; Seki, M.; Shinozaki, K.; Yamaguchi-Shinozaki, K. Isolation and functional analysis of arabidopsis stress-inducible NAC transcription factors that bind to a drought-responsive cis-element in the early responsive to dehydration stress 1 promoter. *Plant Cell* **2004**, *16*, 2481–2498. [CrossRef] [PubMed]

78. Wu, Y.; Deng, Z.; Lai, J.; Zhang, Y.; Yang, C.; Yin, B.; Zhao, Q.; Zhang, L.; Li, Y.; Yang, C.; et al. Dual function of arabidopsis ATAF1 in abiotic and biotic stress responses. *Cell Res.* **2009**, *19*, 1279–1290. [CrossRef] [PubMed]

79. Neilson, J.A.D.; Durnford, D.G. Evolutionary distribution of light-harvesting complex-like proteins in photosynthetic eukaryotes. *Genome* **2010**, *53*, 68–78. [CrossRef] [PubMed]

80. Xu, Y.H.; Liu, R.; Yan, L.; Liu, Z.Q.; Jiang, S.C.; Shen, Y.Y.; Wang, X.F.; Zhang, D.P. Light-harvesting chlorophyll a/b-binding proteins are required for stomatal response to abscisic acid in arabidopsis. *J. Exp. Bot.* **2011**, *63*, 1095–1106. [CrossRef] [PubMed]

81. Minh-Thu, P.T.; Hwang, D.J.; Jeon, J.S.; Nahm, B.H.; Kim, Y.K. Transcriptome analysis of leaf and root of rice seedling to acute dehydration. *Rice* **2013**, *6*, 1–18. [CrossRef] [PubMed]

82. Li, X.-Q. Natural attributes and agricultural implications of somatic genome variation. *Curr. Issue Mol. Biol.* **2016**, *20*, 29–46.

83. Comai, L. The advantages and disadvantages of being polyploid. *Nat. Rev. Genet.* **2005**, *6*, 836–846. [CrossRef] [PubMed]

84. Wang, T.; Lei, C.; Zhao, M.; Tian, Q.; Zhang, W.H. Identification of drought-responsive micrornas in medicago truncatula by genome-wide high-throughput sequencing. *BMC Genom.* **2011**, *12*, 367. [CrossRef] [PubMed]

85. Park, W.; Scheffler, B.E.; Bauer, P.J.; Campbell, B.T. Identification of the family of aquaporin genes and their expression in upland cotton (*Gossypium hirsutum* L.). *BMC Plant Biol.* **2010**, *10*, 142. [CrossRef] [PubMed]

86. Formey, D.; Iñiguez, L.P.; Peláez, P.; Li, Y.F.; Sunkar, R.; Sánchez, F.; Reyes, J.L.; Hernández, G. Genome-wide identification of the phaseolus vulgaris srnaome using small RNA and degradome sequencing. *BMC Genom.* **2015**, *16*, 1–17. [CrossRef] [PubMed]

87. Berkowitz, G.; Zhang, X.; Mercier, R.; Leng, Q.; Lawton, M. Co-expression of calcium-dependent protein kinase with the inward rectified guard cell k+ channel kat1 alters current parameters in xenopus laevis oocytes. *Plant Cell Physiol.* **2000**, *41*, 785–790. [CrossRef] [PubMed]

88. Maathuis, F.J.; Sanders, D. Mechanisms of potassium absorption by higher plant roots. *Physiol. Plant.* **1996**, *96*, 158–168. [CrossRef]

© 2018 by the authors. Licensee MDPI, Basel, Switzerland. This article is an open access article distributed under the terms and conditions of the Creative Commons Attribution (CC BY) license (http://creativecommons.org/licenses/by/4.0/).

forests

MDPI

Article

Tamarix microRNA Profiling Reveals New Insight into Salt Tolerance

Jianwen Wang, Meng Xu, Zhiting Li, Youju Ye, Hao Rong and Li-an Xu *

Co-Innovation Center for Sustainable Forestry in Southern China, Nanjing Forestry University,
Nanjing 210037, China; 13390780572@189.cn (J.W.); mengxu412@126.com (M.X.); lizhiting93@163.com (Z.L.);
yeyj9403@163.com (Y.Y.); mrronghao@outlook.com (H.R.)
* Correspondence: laxu@njfu.edu.cn; Tel.: +86-025-8542-7882

Received: 27 February 2018; Accepted: 27 March 2018; Published: 3 April 2018

Abstract: The halophyte tamarisk (*Tamarix*) is extremely salt tolerant, making it an ideal material for salt tolerance-related studies. Although many salt-responsive genes of *Tamarix* were identified in previous studies, there are no reports on the role of post-transcriptional regulation in its salt tolerance. We constructed six small RNA libraries of *Tamarix chinensis* roots with NaCl treatments. High-throughput sequencing of the six libraries was performed and microRNA expression profiles were constructed. We investigated salt-responsive microRNAs to uncover the microRNA-mediated genes regulation. From these analyses, 251 conserved and 18 novel microRNA were identified from all small RNAs. From 191 differentially expressed microRNAs, 74 co-expressed microRNAs were identified as salt-responsive candidate microRNAs. The most enriched GO (gene ontology) terms for the 157 genes targeted by differentially expressed microRNAs suggested that transcriptions factors were highly active. Two hub microRNAs (miR414, miR5658), which connected by several target genes into an organic microRNA regulatory network, appeared to be the key regulators of post-transcriptional salt-stress responses. As the first survey on the tamarisk small RNAome, this study improves the understanding of tamarisk salt-tolerance mechanisms and will contribute to the molecular-assisted resistance breeding.

Keywords: *Tamarix*; salt tolerance; salt-responsive microRNA; regulatory network

1. Introduction

MicroRNAs (miRNAs or miRs) are endogenous small (~22 nt) noncoding RNAs that play important roles in plant development and morphogenesis through post-transcriptional regulation. For plant miRNAs, mRNA degradation is the major regulatory approach compared with translational inhibition because the complementation of miRNAs and target genes in plants is much greater than in animals [1–3]. Over the past decade, many plant miRNAs have been identified as stress-responsive or developmental regulatory molecules, and some have been functionally confirmed in genetic complementation experiments [4–6]. Dozens of miRNAs like miR167, miR169, miR319 and miR393, which are regulators of plant growth or development, have important effects in abiotic stress responses [7,8]. Many previous studies have suggested that miRNAs play important roles in abiotic stress and their potential functions in the regulation of stress responses are indicated by the increasing number of miRNAs expression profile studies [9–11].

Salinity is a major abiotic stress that impedes crop breeding. Hundreds of salt-tolerant genes and transcription factors (TFs), such as v-myb avian myeloblastosis viral oncogene homolog (MYB), WRKY and nuclear factor-Y (NF-Y) have been identified [12,13]. Recently, several new genes had been verified for increasing tolerance of transformed plants by T-DNA tagging overexpression, e.g., *Oryza sativa* salt tolerance activation 2-dominant [14], *Oryza sativa* intermediate filament like protein [15–17]. Some verified microRNA-target interactions have provided insights into miRNAs regulatory mechanisms of salt-stress responses. For example, the subtle regulation of the genes targeted by the miR398 family, Cu/Zn superoxide dismutases and downstream genes, have been investigated in many plant abiotic stress studies [18,19]. Recent studies revealed nearly forty conserved miRNAs responding to salt stress in different plants [20,21]. The salt-responsive miRNAs could dominate the plant growth attenuation and organ morphogenesis under salt stress. All these findings imply the miRNAs are potential key regulatory factors of the complex salt-tolerance traits.

The halophyte tamarisk (*Tamarix*) distributes naturally in the salinity-rich areas, such as seashore beaches, saline-alkali deserts, and delta-estuaries. As one of the most salt-tolerant trees, it is an important plant material for high salt-tolerance studies. Many *Tamarix* salt-responsive genes, such as aquaporin, peroxidase (POD), basic leucine zipper (bZIP), and NAC-domain transcription factor (NAC), have been identified by transgenic plant studies or microarray studies [22–24]. Although the studies of individual genes or several members of a gene family have gradually clarified the salt-tolerance mechanisms, they may be inefficient and time-consuming for uncovering the complex gene regulatory network responsible for salt stress responses, especially for deciphering the signaling crosstalk of secondary stress induced by high salinity. Several gene expression profile studies of *Tamarix* using transcriptome sequencing (RNA-seq) or microarrays have identified thousands of differentially expressed genes (DEGs) as salt-tolerant candidate genes [22,25,26]. Based on the previous studies, a genome-wide survey of miRNAs could outline the upstream regulators of salt-tolerant genes and illustrate an overview of post-transcriptional regulation under salt stress. Some miRNA high-throughput sequencing studies of halophytes predicted a number of microRNAs and target genes (miRNA-target pairs), which improved our understanding of gene regulatory networks on salt stress responses [27,28]. These studies could be basic references for experimental design of other halophytes' miRNA profile studies.

As the first research on the sRNAome survey of *Tamarix chinensis* (*T. chinensis*), this study aimed to build miRNA profiles (salt treatment for 0.5 h, 1 h and control group) to identify salt-responsive candidate microRNAs. We investigated differentially expressed miRNAs and their important target genes by comprehensive analysis. The results indicated that two microRNAs and four target genes could be important factors of the post-transcriptional regulatory network responsible for high salt tolerance.

2. Materials and Methods

2.1. Plant Materials and Experimental Design

The 5-year-old *T. chinensis* trees selected as the ortet were planted in a nursery of Danyang (Zhenjiang, Jiangsu, China). The annual semi-lignified shoots of the *T. chinensis* ortet were collected as cutting material. Identical cuttings (7 cm in length and 1 cm in diameter) were cultured in sand substrate under the conditions of light 16 h, dark 8 h and temperature of 25 °C. The roots of one-month-old ramets were immersed in NaCl solution of three gradient concentrations (0.8%, 1.4%, 2% M/V) to estimate critical concentration of salt-resistance limit. The control group was immersed in water to estimate water stress effect. According to plant growing situation observations, we decided on 2% NaCl solution for the salt-stress treatment. Five time intervals (0.5 h, 1 h, 2 h, 4 h and 8 h) and a control group were set up for one-month-old (Group I) and two-month-old (Group II) ramets to decide on salt-responsive time points. After that, the roots were fully harvested and stored at −80 °C for POD enzymatic activity assay and RNA extraction for high-throughput sequencing.

2.2. Peroxidase (POD) Enzyme Activity

In preparation for POD activity assay, 100 mg roots were weighed for each sample of the different treatments (salt treatment for 0.5 h, 1 h, 2 h, 4 h and 8 h × Group I, II and two control groups). Plant POD Assay kits (A084-3, Jiancheng, Nanjing, China) were applied for POD activity assay according to the manufacturer's instructions based on absorbency of the maximum absorption wave length (420 nm).

2.3. Small RNA Deep Sequencing

Six samples (two time points and one control group of Group I, II) were selected for RNA libraries construction, as shown in Table 1. Total RNA was isolated according to slightly improved CTAB-LiCl protocol [29]. RNA co ncentration was measured in a Qubit 2.0 Fluorometer (Life Technologies, Carlsbad, CA, USA) using the Qubit® RNA Assay Kit (Life Technologies, Carlsbad, CA, USA). RNA integrity was assessed according to RNA Integrity Number (RIN) of the Agilent Bioanalyzer 2100 system (Agilent Technologies, Santa Clara, CA, USA). Small RNA libraries were constructed with 3 µg total RNA of six samples. Before reverse transcriptase PCR, NEB 3′ and 5′ SR Adaptors from NEBNext® Multiplex Small RNA Library Prep Set (NEB, Ipswich, MA, USA) were specifically ligated to tags of six libraries After PCR amplification, DNA fragments corresponding to 140~160 bp (the length of cDNA of small noncoding RNA plus the 3′ and 5′ adaptors) were screened out as target fragments. Library quality was assessed on the Agilent Bioanalyzer 2100 System to ensure correct length of insert fragments. After cluster generation using TruSeq SR Cluster Kit v3-cBot-HS on the cBot Cluster Generation System (Illumina, San Diego, CA, USA), library preparations were sequenced on the Illumina Hiseq 2500 platform (Illumina, San Diego, CA, USA) and 50 bp single-end reads were generated. The libraries construction and high-throughput sequencing were performed at the Beijing Novo Gene Genomics Institute, China. The raw data have been uploaded to the Sequence Read Archive of National Center for Biotechnology Information (NCBI) and await Sequence Read Archive (SRA) data accession numbering.

Table 1. The samples and treatments in the study.

Samples ID	Tissue	Age	Treatment
IC I-0.5h I-1h	primary roots from cuttings of different age	Group I: One-month-old cuttings	water for 1 h, control group 2% salt for 0.5 h 2% salt for 1 h
IIC II-0.5h II-1h		Group II: Two-month-old cuttings	Water for 1 h, control group 2% salt for 0.5 h 2% salt for 1 h

2.4. Conserved and Novel MicroRNAs Identification

Clean reads were obtained by removing reads (1) containing poly-N; (2) with 5′ adapter contaminants; (3) without 3′ adapter or the insert tag; (4) containing poly-A or T or G or C; and (5) of low quality from raw data. Then, clean reads in the range of 18–30 nt were screened out as the sRNA tags. The sRNA tags were mapped to reference transcriptome [30] by Bowtie [31] without mismatch. Mapped sRNA tags were used to identify known miRNA, and miRBase 20.0 [32] was used as reference. Modified software miRdeep2 [33] and srna-tools-cli [34] were used to obtain the conserved miRNA and draw the precursors' secondary structures. The precursors less than three reads were rejected for the correctness of miRNA identification.

sRNA tags were mapped to RepeatMasker [35] and Rfam [36] database to eliminate irrelevant tags. The protein-coding genes (tags), repeat sequences, rRNA, tRNA, snRNA, and snoRNA were identified and erased from source reads. Then transcripts which were mapped by these novel microRNA candidate tags were collected from the reference transcriptome. According to the hairpin structure of precursor, novel miRNAs were identified by the combined prediction analysis of the softwares miREvo [37] and miRdeep2.

2.5. Differentially Expressed MicroRNA Identification

The miRNA expression levels were estimated by transcripts per million (TPM). Differential expression analysis was performed using the DEGseq R package [38]. *p*-Value was adjusted using *q*-value [39]. "*q*-value < 0.01" and " | log2(fold change) | > 1" were set as the threshold of significant differentially expressed miRNA identification.

2.6. Target Genes Prediction

Coding Sequence (CDS) of all Unigenes of transcriptome [30] were predicted according to the BLAST hits in NR (NCBI non-redundant database) and Unigenes with no hits were performed on ESTscan 3.0 [40]. Unigenes containing "minus" open reading fragment (ORF) were converted to reverse complementary sequence. A total of 59,331 Unigenes was used as target genes source. The miRNA target genes prediction was performed by psRobot [41] and psRNATarget [42]. Finally, target genes predicted by the two tools were intersected to reject ambiguous results.

2.7. MicroRNA Expression Validation by qRT-PCR

For the study, 100 ng total RNA was reverse-transcribed using miRNA specific stem-loop primers (Table S1) with PrimeScript™ RT Master Mix Kit (Takara, Dalian, China). The qRT-PCR using gene-specific primers and universal primer (Table S1) were performed on Viia 7 Real-Time PCR System (ABI, Carlsbad, CA, USA). The reaction component and program were set according to PowerUp™ SYBR™ Green Master Mix (ABI, Carlsbad, CA, USA) manufacturer recommendations with three technical replications. Reference gene TIFY (unpublished) was used as control for normalization of $2^{-\Delta\Delta Ct}$ methods to calculate relative expression of miRNAs. Student's *t*-test of correlation analysis of high-throughput sequencing and qRT-PCR was performed using the basic R package.

3. Results

3.1. Selection of NaCl Treatment Method

The extent and timing of NaCl stress in this study were determined by culturing rooted cuttings. The cuttings were considered more sensitive to salt than tamarisk tree used in our previous transcriptome study [30]. Unexpectedly, based on naked-eye observation of the shoots, cuttings under 0% (M/V), 0.8% (136 mM), 1.4% (239 mM), and 2% (342 mM) continuous NaCl stress did not show any abnormal morphological characteristics over a 7-d period. Until the eighth day, some twigs of cuttings treated by 2% NaCl became mildly dehydrated (turned yellow). Taking into account our observation and previous study of tamarisk salt-tolerance limit [25], we selected 2% NaCl as the salt concentration for our experiment. POD enzyme activity demonstrates the stress response intensity [43]. Here, the POD enzyme activity increased after 0.5 h of treatment, and reached the peak at 1 h. Then, it decreased to a stable level during 2–8 h (Figure 1). The response pattern of one-month-old cuttings (Group I) was similar to two-month-old cuttings (Group II). Thus, the appropriate salt treatment times were 0.5 h and 1 h. Therefore, roots of Group I and II with 2% NaCl treatment for 0.5 h and 1 h were used for the following libraries construction (Table 1).

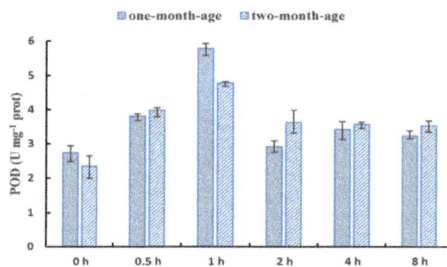

Figure 1. POD enzyme activity dynamic changes under salt stress. The *x*-axis represents the NaCl treatment time and *y*-axis represents the peroxidase (POD) activity.

3.2. Small RNAs Generated from Deep Sequencing

Six small RNA (sRNA) libraries (IC, I-0.5h, I-1h, IIC, II-0.5h and II-1h; Table 1) yielded 129,101,552 redundant reads (6.629 Gb data) with 12–27 million reads per library. Then, 7–15 million unique reads of 18–30 nt per library were selected for sRNA reads mapping and 45.69–68.88% reads mapped to the reference transcriptome. Only these mapped reads were selected for the microRNAs identification. All of the libraries had similar distribution of noncoding RNA families including rRNA and tRNA (~16–18%), snRNA and snoRNA (~0.28–0.46%). Detailed information, including reads numbers and percentages of the six libraries, are listed in Table 2. The 24-nt sRNAs were the most abundant sRNAs (~10% of all reads) based on the size distribution of sRNA (Figure 2). Approximately 80% of sRNAs (80.47%, 81.02%, 81.04%, 81.68%, 79.83% and 82.69% per library) returned no hits in BLAST algorithm-based search and the percentage was similar with sRNAomes of other plants [10,11,44]. The high percentage of unannotated sRNAs suggested they had not been fully appreciated in former studies.

Table 2. Small RNA annotation and reads number.

Small RNA Library	Roots of 2% NaCl Treated Cuttings						Type of Analysis
	Group I			Group II			
	IC	I-0.5h	I-1h	IIC	II-0.5h	II-1h	
Clean reads	20,304,895 (1.038 G)	12,017,139 (0.626 G)	27,490,082 (1.426 G)	21,379,294 (1.093 G)	24,151,563 (1.231 G)	23,758,579 (1.215 G)	Deep sequencing
Clean reads of small RNA	12,268,690	7,656,408	15,286,485	12,991,605	14,999,466	14,327,078	Length selection
Mapped small RNA reads	8,450,784 (68.88%)	3,535,105 (46.17%)	6,984,424 (45.69%)	7,002,413 (53.9%)	8,045,446 (53.64%)	7,872,758 (54.95%)	Map to reference
rRNA and tRNA snRNA and snoRNA	1,531,042 (18.12%) 30,146 (0.36%)	631,272 (17.86%) 9131 (0.26%)	1,352,577 (19.37%) 23,888 (0.34%)	1,242,869 (17.75%) 23,057 (0.33%)	1,378,558 (17.13%) 37,014 (0.46%)	1,294,286 (16.44%) 22,183 (0.28%)	BLAST
Conserved miRNA reads	41,979 (0.50%)	29,239 (0.83%)	31,756 (0.45%)	100,486 (1.44%)	57,517 (0.71%)	45,494 (0.58%)	BLAST
Conserved unique miRNA	180	158	167	207	187	170	
Novel miRNA reads	609 (0.01%)	491 (0.01%)	522 (0.01%)	1282 (0.02%)	848 (0.01%)	642 (0.01%)	Prediction
Novel miRNA	12	12	14	15	13	14	

Group I: one-month-old plants; Group II: two-month-old plants; IC, I-0.5h, I-1h: control group, salt treatment 0.5 h, salt treatment 1 h of Group I; IIC, II-0.5h, II-1h: control group, salt treatment 0.5 h, salt treatment 1 h of Group II. The percentage of Mapped small RNA reads is the ratio of (Mapped small RNA reads/Clean reads of small RNA). The other percentages are ratio of (corresponding reads/Mapped small RNA reads).

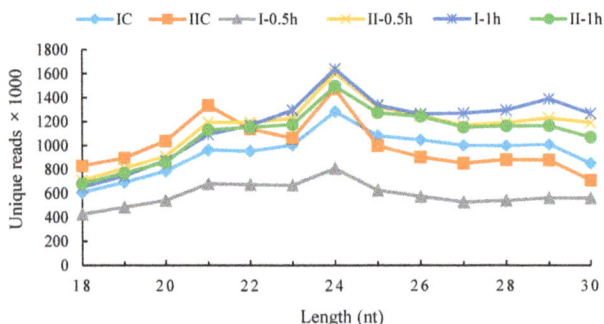

Figure 2. Size distribution of small RNAs of the six libraries. The *x*-axis represents sequence lengths. The *y*-axis represents the number of unique reads (unique small RNAs). The six libraries are distinguished by different colors. IC, I-0.5h, I-1h: control group, salt treatment 0.5 h, salt treatment 1 h of one-month-old plants; IIC, II-0.5h, II-1h: control group, salt treatment 0.5 h, salt treatment 1 h of two-month-old plants.

3.3. Conserved and Novel MicroRNA Identification

In total, 558 known miRNA precursors (pre-miRNAs) and 28 novel pre-miRNAs were identified from the RNA reads. Of the 586 pre-miRNAs, 521 were classified into 46 miRNA families. The top-3 families, MIR159, 156, 166 had 56, 52, 42 identified pre-miRNAs respectively (Table S2). Their distributions were consistent with those of all plant pre-miRNAs (Table S2).

Of the 558 known pre-miRNAs, 307 were rejected due to lack of reads mapped in microRNA mature sequences. Thus, 251 known and 28 novel pre-miRNAs were selected for microRNA identification. Here, 251 conserved miRNAs were identified from the former and 18 novel miRNAs together with 19 putative miRNAs* (miRNA star strand) of the precursor reverse arm (−3p) were identified from the latter. All the 279 pre-miRNAs had canonical stem-loop structures by RNA secondary structural analyses. The numbers of conserved and novel miRNAs per library were 180–207 and 12–15 respectively, and the corresponding reads are listed in Table 2.

The nucleotide bias indicated that the first 5′ nucleotide with the highest frequency was U. The 5′ end nucleotides of miRNAs [45] are important for the binding stability of Argonaute (AGO). AGO has a strong preference in combining with miRNAs starting in U, as seen with rice AGO 1, 2 and 3 [46]. In addition to the U bias of first nucleotide, the C bias of 19th nucleotide seemed to be specific to *Tamarix* (Figure 3a).

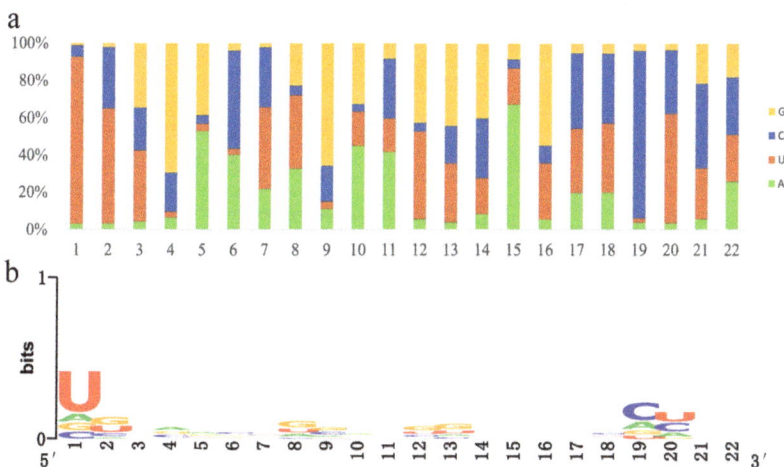

Figure 3. Sequence characteristics of miRNAs. (a) The nucleotide bias of miRNAs. The distributions of four bases are illustrated with different colored columns. The *x*-axis represents miRNA nucleotides from 5′ to 3′. The *y*-axis represents the percentage. (b) The nucleotide conservation of miRNAs. The nucleotide conservation is indicated by the overall height of the stack of symbols, while the height of symbols within the stack indicates the relative frequencies of each nucleotide (U, A, C, or G). The *x*-axis represents miRNA nucleotides from 5′ to 3′. The *y*-axis represents the conservation bits. The graphics were made using WebLogo [47].

The sequence logo [47] showed the highest conservation of the first nucleotide and a high conservation of the 19th and 20th nucleotides, which is consistent with nucleotide bias (Figure 3b). However, other non-conservative nucleotides did not correspond with the bias, e.g., the G bias of 4th, 9th nucleotide and the A bias of 15th nucleotide (Figure 3a). The nonconformity between nucleotide bias and conservation resulted from the abundance difference of microRNAs. The high-abundance microRNAs had greater weight for nucleotides bias but no effect on nucleotide conservation.

3.4. MicroRNA Profile Building and Time-Course Analysis

The miRNA profiles of six libraries (Table S3) were constructed based on expression level. The expression level was quantified using transcripts per million (TPM). Here, 122 (45.41%), 35 (12.95%) and 112 (41.64%) miRNAs were distributed in 0–15, 15–60 and 60– TPM intervals respectively. The intervals represented the low-, middle- and high-expression levels of miRNA abundance (Table 3).

The TPM probability density distributions reflected the overview of the miRNA abundance (Figure 4). The probability density distribution of Group I changed from peak value (IC) to valley value (I-0.5h, I-1h) suggesting a significant shift occurred in interval 0–1.5 (TPM = 1–30). For Group II, on the contrary, only a minor change occurred in interval 1–2.5 (TPM = 10–315). The profiles' changes were in conformity with the POD enzyme activity to some degree, and the difference between Groups I and II meets the expectation that younger tissues are more sensitive to salt stress.

Table 3. MicroRNA numbers and percentages in different TPM intervals.

TPM Interval	IC	IIC	I-0.5h	II-0.5h	I-1h	II-1h	Mean
0–15	128 (47.58%)	109 (40.52%)	125 (46.47%)	125 (46.47%)	128 (47.58%)	118 (42.87%)	122 (45.41%)
15–60	28 (10.41%)	43 (15.99%)	30 (11.15%)	30 (11.15%)	35 (13.01%)	43 (15.99%)	35 (12.95%)
>60	113 (42.01%)	117 (43.49%)	114 (42.38%)	114 (42.38%)	106 (39.41%)	108 (40.15%)	112 (41.64)

IC, I-0.5h, I-1h: control group, salt treatment 0.5 h, salt treatment 1 h of one-month-old plants; IIC, II-0.5h, II-1h: control group, salt treatment 0.5 h, salt treatment 1 h of two-month-old plants. TPM, transcripts per million.

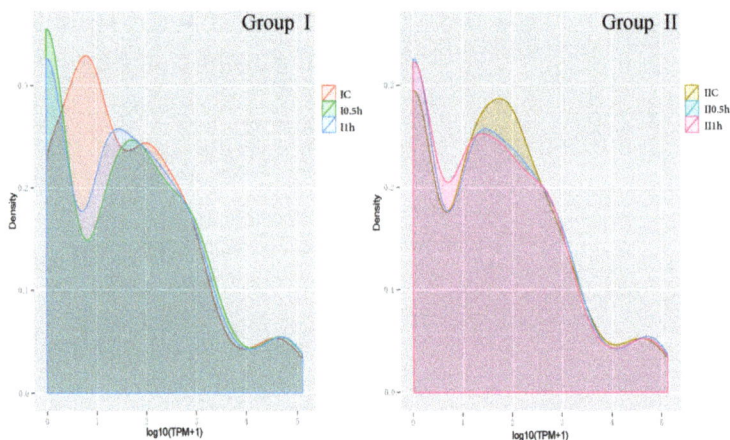

Figure 4. The density distribution profiles of miRNAs. Libraries of each group (**I,II**) are presented with different colors. The *x*-axis represents the $\log_{10}$ (TPM + 1). The *y*-axis represents the probability value of the TPM distribution. TPM: transcript per million; Group I: one-month-old plants; Group II: two-month-old plants. IC, I-0.5h, I-1h: control group, salt treatment 0.5 h, salt treatment 1 h of Group I; IIC, II-0.5h, II-1h: control group, salt treatment 0.5 h, salt treatment 1 h of Group II.

A short time-series expression miner (STEM) analysis [48] clustered miRNAs according to their temporal expression patterns. Profiles for Groups I (Figure 5a) and II (Figure 5b) showed dynamic changes of miRNA abundance. In total, 239 miRNAs and 237 miRNAs generated STEM profiles for Group I and II respectively. Three significantly enriched temporal expression profiles, clusters 0, 2 and 6, represented the main miRNA expression patterns. Cluster 0 was enriched with the most miRNAs (45 miRNAs of Group I and 40 miRNAs of Group II) and showed a "decrease to steady level" pattern. These miRNAs were down-regulated to a low abundance under salt stress. Cluster 2 (Group I) was enriched with 23 "down to up" regulated miRNAs and showed an opposite pattern to Cluster 6

(Group II) with 39 miRNAs. The other clusters were not significantly enriched, and Clusters 3 and 8 represented continuously decreasing and increasing patterns, indicating that those miRNAs had ultra (highest or lowest) abundance levels after 1 h of salt stress.

Stem-loop quantitative Real-Time PCR (qRT-PCR) is a reliable method for measuring miRNA expression levels. We used this technology to validate the expression patterns of ten miRNAs randomly selected from corresponding ten STEM clusters (Figure 5c). The Pearson correlation coefficient (R^2 = 0.50854) and two-tailed *t*-test (*p*-value = 0.000808) indicated the significant positive correlations between sRNA-seq and qRT-PCR (Figure 5d). The comparisons of profiles showed that the expression patterns generated by qRT-PCR in each cluster were in high agreement with the sRNA-seq at 0.5 h. Additionally, microRNA profiles of Group II had higher consistency levels with sRNA-seq on average compared with Group I (Figure 5c).

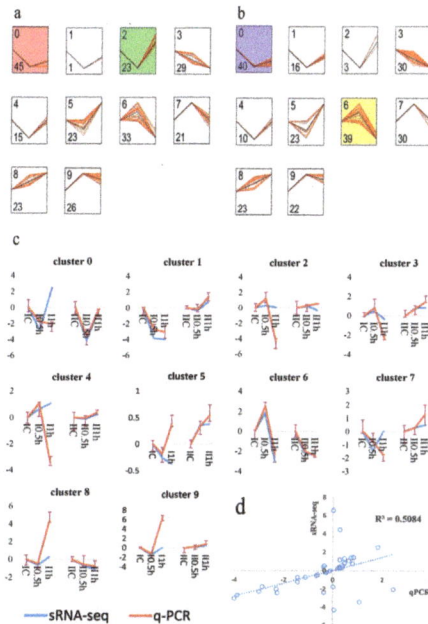

Figure 5. Time-course expression and validation of miRNAs. Short Time-series Expression Miner (STEM) [48] analysis of (**a**) Group I and (**b**) Group II. Group I: one-month-old plants; Group II: two-month-old plants. The cluster ID and the number of miRNAs assigned to the corresponding cluster are indicated by the two figures at the upper- and lower-left corner, respectively. Each STEM profile is made up of ten miRNAs clusters sorted in decreasing order of miRNAs number. In every cluster, the *x*-axis represents the time series of 0, 0.5 and 1 h and the *y*-axis represents relative expression level with log normalization. The black line represents the model profile and the red lines represent individual miRNA expression profiles. Besides, the background color indicates corresponding clusters were statistically significantly enriched based on permutation test. The colored profiles had a statistically significant number of genes assigned. Non-white profiles of the same color represent profiles grouped into a single cluster. (**c**) Comparisons of the profiles between sRNA-seq and qRT-PCR. Ten miRNAs were randomly selected from corresponding clusters of STEM profiles. The *x*-axis represents six small RNA libraries and the *y*-axis represents relative expression level of log normalization (log2). Blue and red lines represent expression profiles of high throughput sequencing and qPCR, respectively. The error bars of qRT-PCR technical replicates are represented in I—shaped lines. (**d**) Pearson correlation analysis between qPCR and high throughput sequencing data. The correlation coefficient is 0.5084 (*p* = 0.000808).

3.5. Differentially Expressed MicroRNAs Identification

Overall, 191 miRNAs including 62 Group I-specific miRNAs, 55 Group II-specific miRNAs and 74 co-expressed miRNAs, were screened as the differentially expressed miRNAs (Figure 6a). According to the up/down-regulation of 74 co-expressed miRNAs, 41 miRNAs maintained a monotone regulation in time series. In the 41 miRNAs, 26 (11 up-regulated and 15 down-regulated) miRNAs had consistent regulatory patterns between two groups but 15 other miRNAs were conversely regulated. Considering the microRNAs differentially expressed in both groups, the 74 co-expressed miRNAs were selected as salt-responsive candidate microRNAs.

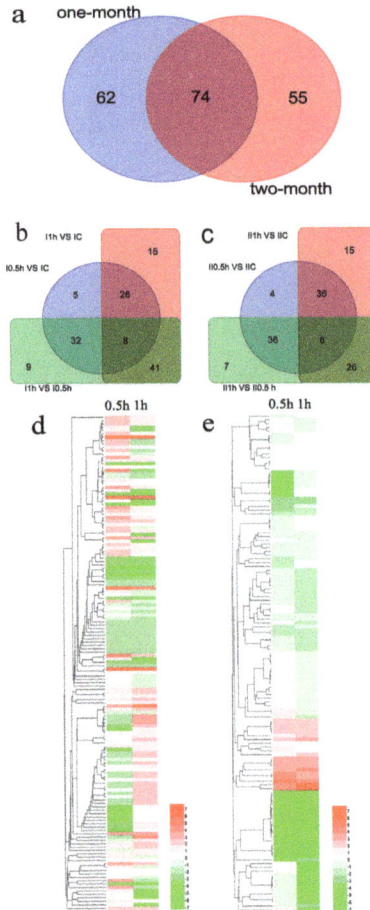

Figure 6. The differentially expressed miRNAs. (**a**) Venn diagram of Group I vs. II comparison. (**b,c**) Venn diagram of three comparisons in Groups I and II. (**d,e**) Heatmaps of Group I and II. The log$_2$ [fold change] is demonstrated in the corresponding color gradation. Group I: one-month-old plants; Group II: two-month-old plants. IC, I-0.5h, I-1h: control group, salt treatment 0.5 h, salt treatment 1 h of Group I; IIC, I-0.5h, II-0.5h, II-1h: control group, salt treatment 0.5 h, salt treatment 1 h of Group II.

In Group I (Figure 6b), 136 differentially expressed miRNAs consisted of 71, 90 and 90 miRNAs from the three libraries comparison of I-0.5h vs. IC, I-1h vs. IC, I-1h vs. I-0.5h. Additionally, 34, 40 and 49 miRNAs were differentially expressed in each pairwise comparison. In Group II (Figure 6c),

129 differentially expressed miRNAs consisted of 82, 83 and 75 miRNAs from II-0.5h vs. IIC, II-1h vs. IIC, II-1h vs. II-0.5h. Additionally, 42, 42 and 36 miRNAs were differentially expressed in each pairwise comparison. In view of the fact that these differentially expressed miRNAs come from only a single set of biological replicates, any individual microRNA differential expression must be validated prior to follow-up study.

The heatmaps of Group I (136 miRNAs, Figure 6d) and Group II (129 miRNAs, Figure 6e) illustrated the fold changes of the differentially expressed miRNAs. Although the numbers of differentially expressed miRNAs in two groups were similar, the higher percentage of extremely differentially expressed (fold change >2 or <−2) miRNAs in Group I indicated that an intense salt stress response occurred in younger roots. In addition, in Group II, the higher percentage of down-regulated miRNAs indicated the older roots may respond to salt stress by down-regulating miRNAs to promote the expression of targeted salt-tolerant genes.

3.6. Target Genes Prediction and MicroRNA Regulatory Network

A total of 706 Unigenes were predicted as targets of 219 miRNAs. Fifty miRNAs had no targets due to conflicting predictions between psRoboot and psRNATarget. The 1475 putative miR-target pairs indicated that every miRNA regulated 7 targets and every predicted target was regulated by 2 miRNAs on average. In the 706 putative targets, 523 targets were function-annotated in the NR database, including 86 TFs such as squamosa promoter binding protein-like, MYB, NAC, APETALA2-like (AP2), homeobox-leucine zipper protein (HD-ZIP) and TCP domain transcription factor. The 510 predicted targets of all annotated target genes were divided into 46 gene ontology (GO) terms (Table S4) including the top three terms: binding (GO: 0005488), cellular process (GO: 0009987) and metabolic process (GO: 0008152). The 266 targets were annotated in 23 clusters of orthologous groups for eukaryotic complete genomes (KOG) function categories (Table S4) including the top three categories: General function prediction only (R), Transcription (K) and Post-translational modification, protein turnover, chaperones (O). The 107 targets were annotated in 89 Kyoto Encyclopedia of Genes and Genomes (KEGG) pathways (Table S4) including the top three pathways: Ubiquitin mediated proteolysis (ko04120), Spliceosome (ko03040) and RNA degradation (ko03018).

The 157 targets and 151 differentially expressed miRNAs constructed 454 miR-target pairs. To focus on the biological progress of salt responses, we performed the GO enrichment analysis using agriGO v2.0 [49], which suggested 88 target genes were enriched in 11 GO terms, including four terms of Cellular Component and seven terms of Molecular Function (Table S5). The GO directed acyclic hierarchical graph (Figure 7) made up a logical framework of enriched terms. The top three most enriched GO terms, chromatin binding (GO: 0003682), DNA binding (GO: 0003677) and nucleic acid binding (GO: 0003676), revealed the process of regulatory elements binding to DNA was highly active under salt stress. Considering TFs are the main targets of plant miRNAs, the results illustrated that miRNAs affect downstream responsive gene expression by regulating their TFs.

To demonstrate the interaction network under salt stress, we selected the 74 co-expressed miRNAs from all the differentially expressed microRNAs as salt responsive candidate miRNAs. In addition, the DEGs of RNA-seq were selected from all the predicted target genes. In total, 70 miR-target pairs (38 salt responsive miRNAs and 41 DEGs) and 8 proteins interactions (7 DEGs) were involved in the microRNA regulatory network on the Cytoscape platform [50]. The interactions between proteins were predicted according to the STRING database using poplar as the reference. The regulatory manner of cleavages or inhibition was predicted by psRNATarget. The false discovery rate and gene function annotation of DEGs were acquired from transcriptome data. Four genes use the Unigene IDs (e.g., compXXXX_c0) [30] as names because of their lack of hits in the function annotation. Eight sub-networks constituted the whole regulatory network. The sub-network mediated by ath-miR5658 (ath-miR414, ath-824-5p, miR172 and osa-miR395u) was the most complex. The other seven sub-networks were mediated by only one miRNA or several miRNAs of the same family.

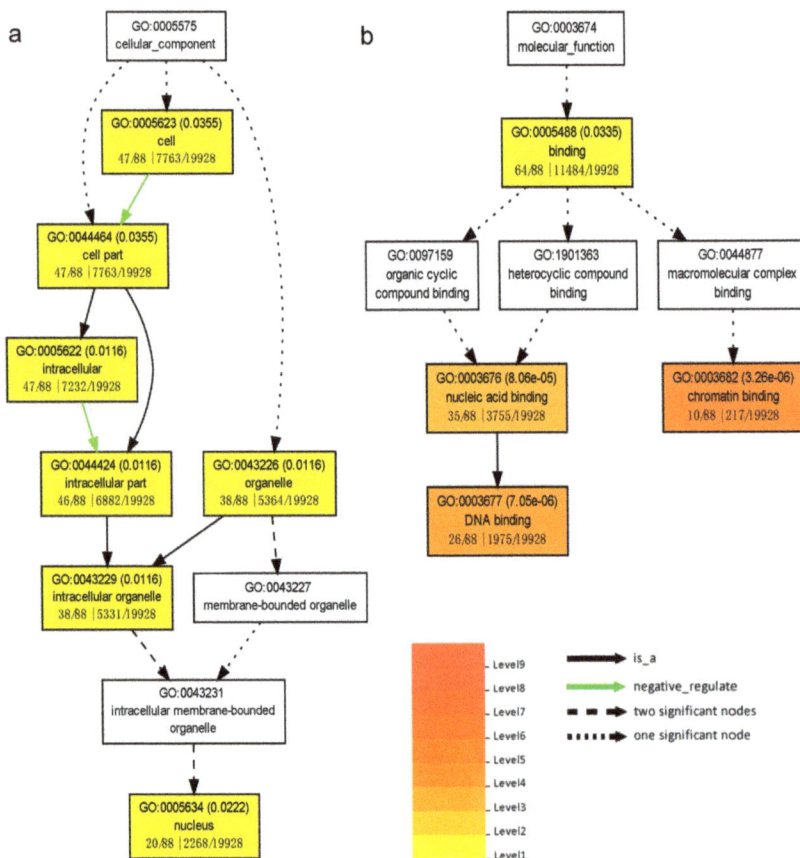

Figure 7. Directed acyclic hierarchical graph (DAG) of enriched GO terms. (**a**) DAG of enriched GO terms in Cellular Component; (**b**) DAG of enriched GO terms in Molecular Function. The DAG shows the GO structural networks, which are represented by boxes with GO numbers and names in the first and second lines, respectively. The third line of each box represents the numbers of targets associated with the given GO term, the enriched targets, the Unigenes associated with the given GO term and all the enriched Unigenes. The color gradation represents the enrichment significance level which is corresponding to Q value (the number in brackets) based on multiple testing correction of *p* value [39]. Relationship types of terms are represented with colored arrows.

4. Discussion

4.1. Strategy for Time-Course Experimental Design

High-quality expression profiles depend on the optimal strategy of experimental design, especially in high-throughput sequencing experiments. Based on previous reports [51,52], NaCl concentrations higher than 2% would induce salt shock symptoms of *Tamarix*, including curled or yellow leaves, branch exsiccation and individual death. Our 2% saline solution culture indicated the limit of *T. chinensis* salt-tolerance should be higher than the experiential concentration 2%. So 2% NaCl concentration was decided as the salt treatment on the prerequisite of salt shock on plants. Time points and intervals are the core of time-course experimental design [27,53,54]. In our study, plant materials, the young adventitious roots of cuttings, were so sensitive to abiotic stress that the time intervals and time points could dramatically affect the identification of differentially expressed miRNAs. The POD enzymes activity is a representative physiological index of salt stress responses. We set 0 h, 0.5 h, 1 h, 2 h, 4 h and 8 h as the time course because the transcriptome of roots quickly responded to salt stress. We decided on the choice of the time point of peak (1 h) and the preceding one (0.5 h) because the strongest and second strongest responses happened in the time points. The TPM distribution significantly change in I-0.5h and I-1h (Figure 4). It was consistent with the trend change of POD enzyme activity in 0.5 h and 1 h. This proved the rationality of time-point choice. The differentially expressed microRNAs that were co-expressed both in Group I and II were considered a strategy of salt-responsive microRNA identification for compensation of experimental error control [55–59]. Based on this, all of the 74 co-expressed miRNAs were selected as salt-responsive candidate microRNAs.

4.2. The miRNA Regulation Network of Tamarix under Stress

Among the 41 putative target genes in the microRNA regulatory network, 13 TFs were responsive to abiotic stress, like *Ethylene-responsive transcription factor (ERF)*, *HD-ZIP and MYB*. Most of the other 28 putative targets were important signaling genes such as BR-signaling kinase (BSK), Receptor-like protein 12 and Protein tyrosine kinase. Interestingly, three important TFs were putatively targeted by multiple miRNA family members. The three TFs were Scarecrow-like protein 15 (SCL15), no apical meristem protein (NAM or CUC1 NAC) and AP2. Functions of the three TFs were well studied in model plant species. SCL15 is an important target of the miR171 family [60], and 14 miR171 members were identified in this study. The miR171-*SCL* interaction was identified in target predictions [61] and southern hybridization experiments in Arabidopsis [62]. Interactions of *AP2* with miR5658 and miR172 family members were only computationally predicted in *Arabidopsis* [63]. In addition, miR172-AP2 and miR164-*NAM* pairs have been identified in other high-throughput sequencing studies [26,64]. In our study, *NAM* was putatively targeted by six up-regulated miR164 family members. SCL15 (*AP2*) were putatively regulated by the members of miR171 (miR172) with reverse expression patterns (Figure 8, label colors). The three TFs' functions and regulatory patterns indicated that microRNAs' (miR171, miR172, miR5658) mediated targets regulation was intricate and influenced by multiple factors.

In the seven protein-encoding genes, five genes consisted a putative small interaction network (Figure 8, dotted lines), including HD-ZIP, SLD (delta8-fatty-acid desaturase), PUB31 (U-box domain-containing protein 31), PERK10 (proline-rich receptor-like protein kinase) and CRK25 (cysteine-rich receptor-like protein kinase 25). The HD-ZIP associated with ABA-mediated salt tolerance was significantly differentially expressed in many abiotic stress profiles studies [26,65,66]. SLD encodes an important desaturase of lipids synthesis that enhances the salt tolerance of overexpressed Arabidopsis and other plants [67–69]. PUB plays a key role in ubiquity-26s proteasome pathway-mediated stress tolerance by regulating hormonal signaling [67]. The two kinase PERK10 and CRK25 are receptor-like protein kinases that could be the candidate sensors of salt stress, and their importance had been discussed in our previous study [30]. miR5658, miR414, miR395, miR824 and other targets were connected in a complex sub-network of proteins interactions. In this sub-network, the highest connectivity (four edges) of four targets (HD-ZIP, PUB31, SLD, PERK10) suggested that these hub genes could influence multiple pathways compared with other targets. Because some hub genes were directly (like AP2) or indirectly (like HD-ZIP) regulated by two or more microRNAs, their regulation of downstream salt responsive genes would be more sensitive and controllable.

In the differentially expressed miRNAs' regulatory network, the two hubs ath-miR5658 and ath-miR414 (Figure 8) appeared to be more important regulators of salt-responsive genes than other microRNAs. Although mRNA cleavage is the major mechanism of miRNA-introduced silencing, several target genes were predicted to be regulated by translational inhibition (Figure 8, lines with T arrow) like "ctr-miR171—comp25618_c0", "ptc-miR482d-3p—RGA4 (Leucine-rich repeat)". However, the reliability of the translational inhibition predicted computationally based on mismatches in the paring regions, needs to be experimentally validated. Especially, both hub miRNAs regulated targets through translational inhibition and cleavage, indicating their flexible regulatory modes for different salt responsive genes. ath-miR5658 mediated the regulation of 22 targets, including the direct regulation of salt tolerance genes like BSK, inactive receptor kinase (RKL), as well as the indirect regulation based on "miR-TF-downstream genes" pathway. The two regulatory strategies were also used in the ath-miR414 mediated pathway. For the network connection, between miR56568 and miR414, the proteins Ligand-gated ion channel, Transcription factor HEX, SLD and HD-ZIP connected the two hub miRNAs in a complex organic network. This organic network is regulated in a complicated manner that involves miRNA-directed target and target-directed miRNA interactions. Thus, the two miRNAs could affect the targets of other miRNAs. As indicated earlier, ath-miR5658 and ath-miR414 may regulate *Tamarix* responses to environmentally complicated and changeable salt stress factors by the subtle and comprehensive post-transcriptional regulation of different target genes.

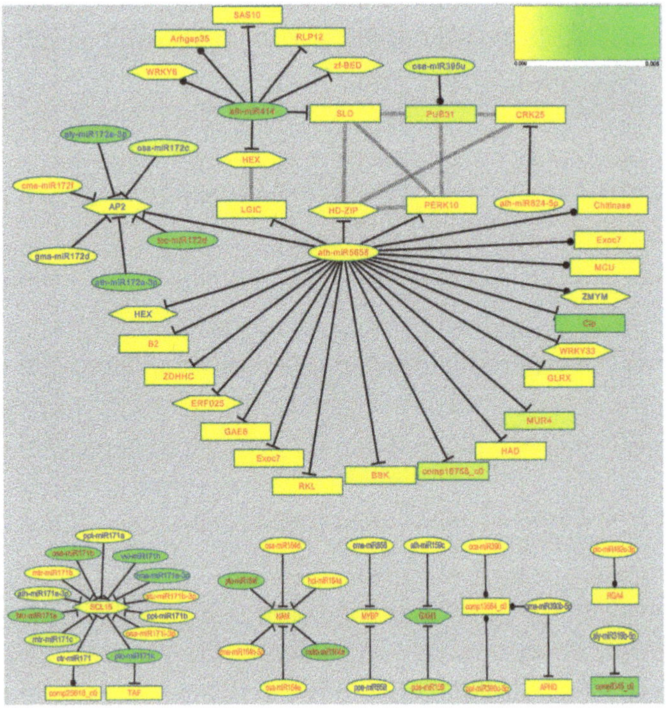

Figure 8. MicroRNA regulatory network. The colors of nodes, based on the color gradation, represent differential expression's false discovery rate (FDR) between 0–0.005 (scale in the top right, the lower the value the more accurate the data). The miRNAs are represented with circle nodes. Transcription factors are represented with hexagon nodes, other genes with rectangle nodes. The edges represent the putative regulation styles. Cleavage (prediction) is represented by solid lines with "T" arrow, inhibition (prediction) by solid lines with circle arrow. Interactions between proteins (prediction) are represented with dotted lines. Up-regulation is represented with red labels, down-regulation with blue labels.

5. Conclusions

In summary, this research investigated the sRNAome of *T. chinensis* and identified 192 differentially expressed microRNAs from the time series expression profiles. In addition, 74 co-expressed microRNAs were identified as salt-responsive candidate microRNAs. Four putative target genes (HD-ZIP, PUB31, SLD, PERK10) acted as the bridges in the microRNA regulatory network. Two microRNAs miR5658 (ath-miR5658) and miR414 (ath-miR414), appeared to be the potential key regulators of the post-transcriptional regulatory network under salt stress. These findings will aid in understanding of *T. chinensis* salt tolerance. Further studies should focus on the experimental validation of these microRNAs' functions and evaluate their potential as candidate molecules in salt-resistance breeding.

Supplementary Materials: The following are available online at www.mdpi.com/1999-4907/9/4/180/s1, Table S1: qRT-PCR primers, Table S2: MicroRNA families of pre-microRNAs, Table S3: MicroRNAs expression profile, Table S4: Target genes function classification, Table S5: GO enriched analysis of target genes of differentially expressed microRNAs.

Acknowledgments: This research was financially supported by the Science and Technology Support Program of Jiangsu Province (CN), BE2011321, the Priority Academic Program Development of Jiangsu Higher Education Institutions (PAPD), and the Doctorate Fellowship Foundation of Nanjing Forestry University.

Author Contributions: L.X. and M.X. conceived and designed the project. J.W. undertook the enzyme activity and molecular biology experiment. J.W., Z.L., Y.Y. and H.R. participated in the data analysis. J.W. drafted the manuscript. L.X. and M.X. modified the manuscript. All authors have read and approved the manuscript for publication.

Conflicts of Interest: The authors declare no conflict of interest.

References

1. Bartel, D.P. MicroRNAs: Target recognition and regulatory functions. *Cell* **2009**, *136*, 215–233. [CrossRef] [PubMed]
2. Brousse, C.; Liu, Q.; Beauclair, L.; Deremetz, A.; Axtell, M.J.; Bouche, N. A non-canonical plant microRNA target site. *Nucleic Acids Res.* **2014**, *42*, 5270–5279. [CrossRef] [PubMed]
3. Li, J.; Reichel, M.; Li, Y.; Millar, A.A. The functional scope of plant microRNA-mediated silencing. *Trends Plant Sci.* **2014**, *19*, 750–756. [CrossRef] [PubMed]
4. Li, C.; Zhang, B. MicroRNAs in control of plant development. *J. Cell Physiol.* **2016**, *231*, 303–313. [CrossRef] [PubMed]
5. Jones-Rhoades, M.W.; Bartel, D.P.; Bartel, B. MicroRNAs and their regulatory roles in plants. *Annu. Rev. Plant Biol.* **2006**, *57*, 19–53. [CrossRef] [PubMed]
6. Sunkar, R.; Li, Y.; Jagadeeswaran, G. Functions of microRNAs in plant stress responses. *Trends Plant Sci.* **2012**, *17*, 196–203. [CrossRef] [PubMed]
7. Khraiwesh, B.; Zhu, J.; Zhu, J. Role of miRNAs and siRNAs in biotic and abiotic stress responses of plants. *Biochim. Biophys. Acta (BBA) Gene Regul. Mech.* **2012**, *1819*, 137–148. [CrossRef] [PubMed]
8. Shukla, L.I.; Chinnusamy, V.; Sunkar, R. The role of microRNAs and other endogenous small RNAs in plant stress responses. *BBA-Gene Regul. Mech.* **2008**, *1779*, 743–748. [CrossRef] [PubMed]
9. Ding, D.; Zhang, L.; Wang, H.; Liu, Z.; Zhang, Z.; Zheng, Y. Differential expression of miRNAs in response to salt stress in maize roots. *Ann. Bot.-Lond.* **2009**, *103*, 29–38. [CrossRef] [PubMed]
10. Xie, F.; Wang, Q.; Sun, R.; Zhang, B. Deep sequencing reveals important roles of microRNAs in response to drought and salinity stress in cotton. *J. Exp. Bot.* **2015**, *66*, 789–804. [CrossRef] [PubMed]
11. Xie, F., Jr.; Stewart, C.N.; Taki, F.A.; He, Q.; Liu, H.; Zhang, B. High-throughput deep sequencing shows that microRNAs play important roles in switchgrass responses to drought and salinity stress. *Plant Biotechnol. J.* **2014**, *12*, 354–366. [CrossRef] [PubMed]
12. Muchate, N.S.; Nikalje, G.C.; Rajurkar, N.S.; Suprasanna, P.; Nikam, T.D. Plant salt stress: Adaptive responses, tolerance mechanism and bioengineering for salt tolerance. *Bot. Rev.* **2016**, *82*, 371–406. [CrossRef]
13. Zhu, J.K. Salt and drought stress signal transduction in plants. *Annu. Rev. Plant Biol.* **2002**, *53*, 247–273. [CrossRef] [PubMed]
14. Kumar, M.; Choi, J.; An, G.; Kim, S. Ectopic expression of OsSta2 enhances salt stress tolerance in rice. *Front. Plant Sci.* **2017**, *8*, 316. [CrossRef] [PubMed]
15. Soda, N.; Sharan, A.; Gupta, B.K.; Singlapareek, S.L.; Pareek, A. Evidence for nuclear interaction of a cytoskeleton protein (OsIFL) with metallothionein and its role in salinity stress tolerance. *Sci. Rep.-UK* **2016**, *6*, 34762. [CrossRef] [PubMed]
16. Xie, R.; Jin, Z.; Ma, Y.; Pan, X.; Dong, C.; Pang, S.; He, S.; Deng, L.; Yi, S.; Zheng, Y. Combined analysis of mRNA and miRNA identifies dehydration and salinity responsive key molecular players in citrus roots. *Sci. Rep.-UK* **2017**, *7*, 42094. [CrossRef] [PubMed]

17. Fu, R.; Zhang, M.; Zhao, Y.; He, X.; Ding, C.; Wang, S.; Feng, Y.; Song, X.; Li, P.; Wang, B. Identification of salt tolerance-related microRNAs and their targets in maize (*Zea mays* L.) using high-throughput sequencing and degradome analysis. *Front. Plant Sci.* **2017**, *8*, 864. [CrossRef] [PubMed]

18. Dugas, D.V.; Bartel, B. Sucrose induction of Arabidopsis miR398 represses two Cu/Zn superoxide dismutases. *Plant. Mol. Biol.* **2008**, *67*, 403–417. [CrossRef] [PubMed]

19. Sunkar, R.; Kapoor, A.; Zhu, J.K. Posttranscriptional induction of two Cu/Zn superoxide dismutase genes in Arabidopsis is mediated by downregulation of miR398 and important for oxidative stress tolerance. *Plant Cell* **2006**, *18*, 2051–2065. [CrossRef] [PubMed]

20. Kawa, D.; Testerink, C. Regulation of mRNA decay in plant responses to salt and osmotic stress. *Cell. Mol. Life Sci.* **2017**, *74*, 1165–1176. [CrossRef] [PubMed]

21. Kumar, V.; Khare, T.; Shriram, V.; Wani, S.H. Plant small RNAs: The essential epigenetic regulators of gene expression for salt-stress responses and tolerance. *Plant Cell Rep.* **2018**, *37*, 61–75. [CrossRef] [PubMed]

22. Gao, C.; Wang, Y.; Liu, G.; Yang, C.; Jiang, J.; Li, H. Expression profiling of salinity-alkali stress responses by large-scale expressed sequence tag analysis in *Tamarix hispid*. *Plant Mol. Biol.* **2008**, *66*, 245–258. [CrossRef] [PubMed]

23. Ji, X.; Liu, G.; Liu, Y.; Zheng, L.; Nie, X.; Wang, Y. The bZIP protein from *Tamarix hispida*, ThbZIP1, is ACGT elements binding factor that enhances abiotic stress signaling in transgenic Arabidopsis. *BMC Plant Biol.* **2013**, *13*, 151. [CrossRef] [PubMed]

24. Yang, G.; Wang, C.; Wang, Y.; Guo, Y.; Zhao, Y.; Yang, C.; Gao, C. Overexpression of ThVHAc1 and its potential upstream regulator, ThWRKY7, improved plant tolerance of Cadmium stress. *Sci. Rep.-UK* **2016**, *6*, 18752. [CrossRef] [PubMed]

25. Yang, C.P.; Wang, Y.C.; Liu, G.F.; Jiang, J.; Zhang, G.D. Study on expression of genes in *Tamarix androssowii* under NaHCO₃ stress using gene chip technology. *Sheng Wu Gong Cheng Xue Bao* 2005, *21*, 220–226. [PubMed]

26. Wang, C.; Gao, C.; Wang, L.; Zheng, L.; Yang, C.; Wang, Y. Comprehensive transcriptional profiling of NaHCO₃-stressed *Tamarix hispida* roots reveals networks of responsive genes. *Plant Mol. Biol.* **2014**, *84*, 145–157. [CrossRef] [PubMed]

27. Yang, C.; Wei, H. Designing microarray and RNA-Seq experiments for greater systems biology discovery in modern plant genomics. *Mol. Plant* **2015**, *8*, 196–206. [CrossRef] [PubMed]

28. Yaish, M.W.; Sunkar, R.; Zheng, Y.; Ji, B.; Al-Yahyai, R.; Farooq, S.A. A genome-wide identification of the miRNAome in response to salinity stress in date palm (*Phoenix dactylifera* L.). *Front. Plant Sci.* **2015**, *6*, 946. [CrossRef] [PubMed]

29. Xu, M.; Zang, B.; Yao, H.S.; Huang, M.R. Isolation of high quality RNA and molecular manipulations with various tissues of *Populus*. *Russ. J. Plant Physl.* **2009**, *56*, 716–719. [CrossRef]

30. Wang, J.; Xu, M.; Gu, Y.; Xu, L. Differentially expressed gene analysis of *Tamarix chinensis* provides insights into NaCl-stress response. *Trees-Struct. Funct.* **2017**, *31*, 645–658. [CrossRef]

31. Langmead, B. Aligning short sequencing reads with Bowtie. *Curr. Protoc. Bioinform.* **2010**, *32*, 11–17.

32. Griffiths-Jones, S. *miRBase: The microRNA Sequence Database*; Humana Press: New York, NY, USA, 2006; Volume 342, pp. 129–138.

33. Friedlander, M.R.; Mackowiak, S.D.; Li, N.; Chen, W.; Rajewsky, N. miRDeep2 accurately identifies known and hundreds of novel microRNA genes in seven animal clades. *Nucleic Acids Res.* **2012**, *40*, 37–52. [CrossRef]

34. Moxon, S.; Schwach, F.; Dalmay, T.; MacLean, D.; Studholme, D.J.; Moulton, V. A toolkit for analysing large-scale plant small RNA datasets. *Bioinformatics* **2008**, *24*, 2252–2253. [CrossRef] [PubMed]

35. Saha, S.; Bridges, S.; Magbanua, Z.V.; Peterson, D.G. Empirical comparison of ab initio repeat finding programs. *Nucleic Acids Res.* **2008**, *36*, 2284–2294. [CrossRef] [PubMed]

36. Kalvari, I.; Argasinska, J.; Quinones-Olvera, N.; Nawrocki, E.P.; Rivas, E.; Eddy, S.R.; Bateman, A.; Finn, R.D.; Petrov, A.I. Rfam 13.0: Shifting to a genome-centric resource for non-coding RNA families. *Nucleic Acids Res.* **2017**. [CrossRef] [PubMed]

37. Wen, M.; Shen, Y.; Shi, S.; Tang, T. miREvo: An integrative microRNA evolutionary analysis platform for next-generation sequencing experiments. *BMC Bioinform.* **2012**, *13*, 140. [CrossRef] [PubMed]

38. Wang, L.; Feng, Z.; Wang, X.; Wang, X.; Zhang, X. DEGseq: An R package for identifying differentially expressed genes from RNA-seq data. *Bioinformatics* **2010**, *26*, 136–138. [CrossRef] [PubMed]

39. Storey, J.D. The positive false discovery rate: A Bayesian interpretation and the q-value. *Ann. Stat.* **2003**, *31*, 2013–2035. [CrossRef]

40. Iseli, C.; Jongeneel, C.V.; Bucher, P. ESTScan: A program for detecting, evaluating, and reconstructing potential coding regions in EST sequences. *Proc. Int. Conf. Intell. Syst. Mol. Biol.* **1999**, *99*, 138–148.

41. Wu, H.J.; Ma, Y.K.; Chen, T.; Wang, M.; Wang, X.J. PsRobot: A web-based plant small RNA meta-analysis toolbox. *Nucleic Acids Res.* **2012**, *40*, W22–W28. [CrossRef] [PubMed]

42. Dai, X.; Zhao, P.X. psRNATarget: A plant small RNA target analysis server. *Nucleic Acids Res.* **2011**, *39*, W155–W159. [CrossRef] [PubMed]

43. He, K.; Guo, C. Effects of salt stress on SOD and POD activities in three bamboos. *J. Jiangsu Forestryence Technol.* **1995**, *22*, 11–14.

44. Wan, L.; Wang, F.; Guo, X.; Lu, S.; Qiu, Z.; Zhao, Y.; Zhang, H.; Lin, J. Identification and characterization of small non-coding RNAs from Chinese fir by high throughput sequencing. *BMC Plant Biol.* **2012**, *12*, 146. [CrossRef] [PubMed]

45. Hutvagner, G.; Simard, M.J. Argonaute proteins: Key players in RNA silencing. *Nat. Rev. Mol. Cell Biol.* **2008**, *9*, 22–32. [CrossRef] [PubMed]

46. Wu, L.; Zhang, Q.; Zhou, H.; Ni, F.; Wu, X.; Qi, Y. Rice microRNA effector complexes and targets. *Plant Cell* **2009**, *21*, 3421–3435. [CrossRef] [PubMed]

47. Crooks, G.E.; Hon, G.; Chandonia, J.M.; Brenner, S.E. WebLogo: A sequence logo generator. *Genome Res.* **2004**, *14*, 1188–1190. [CrossRef] [PubMed]

48. Ernst, J.; Bar-Joseph, Z. STEM: A tool for the analysis of short time series gene expression data. *BMC Bioinform.* **2006**, *7*, 191. [CrossRef] [PubMed]

49. Tian, T.; Liu, Y.; Yan, H.; You, Q.; Yi, X.; Du, Z.; Xu, W.; Su, Z. agriGO v2.0: A GO analysis toolkit for the agricultural community, 2017 update. *Nucleic Acids Res.* **2017**, *45*, W122–W129. [CrossRef] [PubMed]

50. Shannon, P.; Markiel, A.; Ozier, O.; Baliga, N.S.; Wang, J.T.; Ramage, D.; Amin, N.; Schwikowski, B.; Ideker, T. Cytoscape: A software environment for integrated models of biomolecular interaction networks. *Genome Res.* **2003**, *13*, 2498–2504. [CrossRef] [PubMed]

51. Yin, J.; Li, W.; Zhang, P.; Zhang, J.; Liu, L. Study on the salt resistance of conifer Tamarisk in hydroponic experiment. *J. Tianjin Univ. Technol.* **2016**, *32*, 42–45.

52. Liu, K.; Deng, C.; Hao, J. Physiological response of *Tamarix austromongolica* to NaCl stress. *Guangdong Agric. Sci.* **2012**, *10*, 38–42.

53. Hansen, K.D.; Wu, Z.; Irizarry, R.A.; Leek, J.T. Sequencing technology does not eliminate biological variability. *Nat. Biotechnol.* **2011**, *29*, 572–573. [CrossRef] [PubMed]

54. Liu, Y.; Zhou, J.; White, K.P. RNA-seq differential expression studies: More sequence or more replication? *Bioinformatics* **2014**, *30*, 301–304. [CrossRef] [PubMed]

55. Schurch, N.J.; Schofield, P.; Gierlinski, M.; Cole, C.; Sherstnev, A.; Singh, V.; Wrobel, N.; Gharbi, K.; Simpson, G.G.; Owen-Hughes, T.; et al. How many biological replicates are needed in an RNA-seq experiment and which differential expression tool should you use? *RNA* **2016**, *22*, 839–851. [CrossRef] [PubMed]

56. Gierliński, M.; Cole, C.; Schofield, P.; Schurch, N.J.; Sherstnev, A. Statistical models for RNA-seq data derived from a two-condition 48-replicate experiment. *Bioinformatics* **2015**, *31*, 3625–3630. [CrossRef] [PubMed]

57. Nagalakshmi, U.; Wang, Z.; Waern, K.; Shou, C.; Raha, D. The transcriptional landscape of the yeast genome defined by RNA sequencing. *Science* **2008**, *320*, 1344–1349. [CrossRef] [PubMed]

58. McIntyre, L.M.; Lopiano, K.K.; Morse, A.M.; Amin, V.; Oberg, A.L.; Young, L.J.; Nuzhdin, S.V. RNA-seq: Technical variability and sampling. *BMC Genom.* **2011**, *12*, 293. [CrossRef] [PubMed]

59. Wang, X.; Li, X.; Zhang, S.; Korpelainen, H.; Li, C. Physiological and transcriptional responses of two contrasting *Populus* clones to nitrogen stress. *Tree Physiol.* **2016**, *36*, 628–642. [CrossRef] [PubMed]

60. Rhoades, M.W.; Reinhart, B.J.; Lim, L.P.; Burge, C.B.; Bartel, B.; Bartel, D.P. Prediction of plant microRNA targets. *Cell* **2002**, *110*, 513–520. [CrossRef]

61. Hwang, E.; Shin, S.; Yu, B.; Byun, M.; Kwon, H. miR171 family members are involved in drought response in *Solanum tuberosum*. *J. Plant Biol.* **2011**, *54*, 43–48. [CrossRef]

62. Llave, C.; Xie, Z.; Kasschau, K.D.; Carrington, J.C. Cleavage of Scarecrow-like mRNA targets directed by a class of Arabidopsis miRNA. *Science* **2002**, *297*, 2053–2056. [CrossRef] [PubMed]

63. Breakfield, N.W.; Corcoran, D.L.; Petricka, J.J.; Shen, J.; Sae-Seaw, J.; Rubio-Somoza, I.; Weigel, D.; Ohler, U.; Benfey, P.N. High-resolution experimental and computational profiling of tissue-specific known and novel miRNAs in Arabidopsis. *Genome Res.* **2012**, *22*, 163–176. [CrossRef] [PubMed]

64. Feng, H.; Duan, X.; Zhang, Q.; Li, X.; Wang, B.; Huang, L.; Wang, X.; Kang, Z. The target gene of tae-miR164, a novel NAC transcription factor from the NAM subfamily, negatively regulates resistance of wheat to stripe rust. *Mol. Plant Pathol.* **2014**, *15*, 284–296. [CrossRef]

65. Chen, X.; Chen, Z.; Zhao, H.; Zhao, Y.; Cheng, B.; Xiang, Y. Genome-wide analysis of soybean HD-Zip gene family and expression profiling under salinity and drought treatments. *PLoS ONE* **2014**, *9*, e87156. [CrossRef] [PubMed]

66. Fujita, Y.; Fujita, M.; Shinozaki, K.; Yamaguchi-Shinozaki, K. ABA-mediated transcriptional regulation in response to osmotic stress in plants. *J. Plant Res.* **2011**, *124*, 509–525. [CrossRef] [PubMed]

67. Lyzenga, W.J.; Stone, S.L. Abiotic stress tolerance mediated by protein ubiquitination. *J. Exp. Bot.* **2012**, *63*, 599–616. [CrossRef] [PubMed]

68. Rodriguez-Vargas, S.; Sanchez-Garcia, A.; Martinez-Rivas, J.M.; Prieto, J.A.; Randez-Gil, F. Fluidization of membrane lipids enhances the tolerance of *Saccharomyces cerevisiae* to freezing and salt stress. *Appl. Environ. Microb.* **2006**, *73*, 110–116. [CrossRef] [PubMed]

69. Zhang, J.; Liu, H.; Sun, J.; Li, B.; Zhu, Q.; Chen, S.; Zhang, H. Arabidopsis fatty acid desaturase FAD2 is required for salt tolerance during seed germination and early seedling growth. *PLoS ONE* **2012**, *7*, e30355. [CrossRef] [PubMed]

© 2018 by the authors. Licensee MDPI, Basel, Switzerland. This article is an open access article distributed under the terms and conditions of the Creative Commons Attribution (CC BY) license (http://creativecommons.org/licenses/by/4.0/).

![forests logo] *forests*

MDPI

Article

De Novo Transcriptome Sequencing in *Passiflora edulis* Sims to Identify Genes and Signaling Pathways Involved in Cold Tolerance

Sian Liu [1,2], Anding Li [3,*], Caihui Chen [1,2], Guojun Cai [3], Limin Zhang [3], Chunyan Guo [4] and Meng Xu [1,2,*]

1   Co-Innovation Center for Sustainable Forestry in Southern China, Nanjing Forestry University, Nanjing 210037, China; sianliu@foxmail.com (S.L.); chencaihui0110@163.com (C.C.)
2   College of Forestry, Nanjing Forestry University, Nanjing 210037, China
3   Institute of Mountain Resources, Guizhou Academy of Science, Guiyang 550001, China; gzu_gjcai@163.com (G.C.); zhanglimin563406@163.com (L.Z.)
4   Guizhou Botanical Garden, Guizhou Academy of Science, Guiyang 550001, China; gchy_922@sina.com
*   Correspondence: anndynlee@126.com (A.L.); xum@njfu.edu.cn (M.X.);
    Tel.: +86-139-8405-9737 (A.L.); +86-150-9430-7586 (M.X.)

Received: 22 September 2017; Accepted: 6 November 2017; Published: 12 November 2017

Abstract: The passion fruit (*Passiflora edulis* Sims), also known as the purple granadilla, is widely cultivated as the new darling of the fruit market throughout southern China. This exotic and perennial climber is adapted to warm and humid climates, and thus is generally intolerant of cold. There is limited information about gene regulation and signaling pathways related to the cold stress response in this species. In this study, two transcriptome libraries (KEDU_AP vs. GX_AP) were constructed from the aerial parts of cold-tolerant and cold-susceptible varieties of *P. edulis*, respectively. Overall, 126,284,018 clean reads were obtained, and 86,880 unigenes with a mean size of 1449 bp were assembled. Of these, there were 64,067 (73.74%) unigenes with significant similarity to publicly available plant protein sequences. Expression profiles were generated, and 3045 genes were found to be significantly differentially expressed between the KEDU_AP and GX_AP libraries, including 1075 (35.3%) up-regulated and 1970 (64.7%) down-regulated. These included 36 genes in enriched pathways of plant hormone signal transduction, and 56 genes encoding putative transcription factors. Six genes involved in the ICE1–CBF–COR pathway were induced in the cold-tolerant variety, and their expression levels were further verified using quantitative real-time PCR. This report is the first to identify genes and signaling pathways involved in cold tolerance using high-throughput transcriptome sequencing in *P. edulis*. These findings may provide useful insights into the molecular mechanisms regulating cold tolerance and genetic breeding in *Passiflora* spp.

Keywords: passion fruit; cold tolerance; RNA sequencing; DEG; ICE1–CBF–COR

1. Introduction

Passiflora is the largest genus of the Passifloraceae family, with more than 500 species [1]. *Passiflora* species are distributed throughout Latin America, and Brazil and Colombia are the centers of diversity for this genus [2], and many of these species are widely cultivated for their edible fruit, medicinal efficacy, and ornamental properties. In the early 20th century, *Passiflora edulis* Sims as an edible fruit was introduced to China, mainly in Taiwan, Guangdong, Guangxi and Fujian. *P. edulis* is known for its taste, is used in Brazilian traditional folk medicine and is included in pharmacopoeias of several countries [3,4]. Leaf extracts of *P. edulis* are considered to treat alcoholism, anxiety and insomnia [5]. The flower has been used for the treatment of cough and bronchitis, and the seed oil as a lubricant and massage oil [6].

Low temperatures represent a major abiotic constraint to plant growth, productivity and distribution [7]. To ensure optimal growth and survival, plants must respond and adapt to cold stress, by implementing a wide range of biochemical and physiological processes. Currently, the most thoroughly understood cold-signaling pathway is the ICE1–CBF–COR pathway. C-repeat binding factors (CBFs) can activate the expression of numerous downstream cold-responsive (*COR*) genes by binding to the cis-acting elements [8,9]. Overexpression of *AtCBF1* or *AtCBF3* enhanced chilling, drought and salt stress tolerance in many species, including *Brassica* spp. [10], wheat (*Triticum aestivum* L.) [11], tomato (*Solanum lycopersicum* L.) [12] and rice (*Oryza sativa* L.) [13]. Inducer of CBF expression 1(ICE1) belongs to the bHLH transcription factor (TF) family, which control *CBF* genes [14]. ICE1 is affected by ambient temperature and is in an inactive state in a warm environment; however, it is activated upon exposure to cold and induces *CBF* expression. Mitogen-activated protein kinase (MAPK) kinase 2 (MKK2) is activated by cold and controls COR expression to improve plant tolerance to freezing [15]. Recent studies have shown that chilling stress activates *SNF1-related protein kinases 2.6/open stomata 1* (*SnRK2.6/OST1*), and then *SnRK2.6* interacts with and phosphorylates ICE1 to activate the *CBF–COR* gene-expression cascade and increase cold tolerance [14,16]. In *Arabidopsis thaliana* (L.) Heynh, only 12% of *COR* genes were regulated by *CBF* [17], which indicated that other regulatory pathways may also be activated in response to cold stress. The abscisic acid (ABA) signal transduction pathway played a key role in plant cold tolerance, with approximately 10% of the ABA response genes involved in cold stress [18], and ABA signaling pathways may be involved in regulating *COR* expression [19,20].

The passion fruit (*P. edulis*), also known as the purple granadilla, is widely cultivated as the new darling of the fruit market throughout southern China. This exotic and perennial climber is adapted to warm and humid climates, and thus is generally intolerant of cold. Fortunately, we have developed a cold-tolerant *P. edulis* variety ('Pingtang 1', KEDU), which is widely cultivated in the south of Guizhou. From breeding and popularization, this variety has undergone testing of several rounds of extreme low temperature, especially in the 2008 Snow Disaster in the south of China; it showed significantly higher cold tolerance in the field than the other varieties mainly cultivated in Guangdong and Guangxi [21]. However, there is limited information about gene regulation and signaling pathways related to the cold stress response in this species. Here, two transcriptome libraries (KEDU_AP vs. GX_AP) were constructed from the aerial parts of cold-tolerant and cold-susceptible varieties of *P. edulis*, respectively. De novo transcriptome sequencing was carried out to identify genes and signaling pathways involved in cold tolerance. This report is the first to identify genes and signaling pathways involved in cold tolerance using high-throughput transcriptome sequencing in *P. edulis*.

2. Materials and Methods

2.1. Plant Materials

In a previous field investigation in Pingtang County, Guizhou Province, we found that after the 2008 Snow Disaster, several *P. edulis* ('Pingtang 1') survived and that they could survive winter temperatures below −2°C [21]. We carried out conservation and cutting propagation, and in a series of experiments found that 'Pingtang 1' had significantly higher cold tolerance than 'Purple Fragrance 1' [21]. The two varieties of *P. edulis*, cold-tolerant 'Pingtang 1' and cold-susceptible 'Purple Fragrance 1' were planted in Pingtang County, Guizhou Province (25°44'24.77″ N, 106°48'45.44″ E; altitude 884 m). The aerial parts (stem, leaf, flower and fruit) of 'Pingtang 1' and 'Purple Fragrance 1' were sampled for RNA extraction in the winter of 2016 (November 9, temperature of 4–6 °C). All *P. edulis* samples were stored at −80 °C in an ultra-low temperature freezer.

2.2. Total RNA Extraction, cDNA Library Preparation and Transcriptome Sequencing

Total RNA was extracted from samples according to the protocol of RNeasy Plant Mini Kit (Qiagen, Hilden, Germany). RNA integrity was monitored by agarose gel electrophoresis (1%) and using an Agilent 2100 Bioanalyzer (Agilent Technologies, Palo Alto, CA, USA). The cDNA library of 'Pingtang 1' was named KEDU_AP and that of 'Purple Fragrance 1' named GX_AP. Following the manufacturer's instructions, two cDNA libraries were generated using a NEBNext® Ultra™ RNA Library Prep Kit for Illumina® (NEB, Boston, Massachusetts, USA). According to the manufacturer's recommendations, clustering of the index-coded samples was performed on a cBot Cluster Generation System using TruSeq PE Cluster Kit v3-cBot-HS (Illumina, San Diego, CA, USA). The prepared libraries were sequenced, and 150-bp paired-end reads were generated (Illumina Hiseq 2500, San Diego, CA, USA). The raw images were transformed using CASAVA base-calling into the FASTQ format of raw reads (raw data). To get clean reads, low-quality reads and adapter sequences were removed using SeqPrep (https://github.com/jstjohn/SeqPrep) and Cutadapt [22].

2.3. Transcriptome Assembly and Function Annotation

Transcriptome assembly was achieved using Trinity version r20140413p1 [23] based on the left.fq and right.fq, with the min_kmer_cov set as 2 by default and all other parameters set as their defaults. For function annotation, the longest transcript of each gene was defined as the 'unigene'. All assembled unigenes were BLASTed in Nr, Nt, Pfam, KOG/COG, Swiss-Prot, KEGG ortholog database (KO) and Gene ontology (GO) databases using BLAST2GO of version 2.5 with a cut-off E-value of 10^{-6} [24].

2.4. Differential Expression Analysis

Gene expression level of all samples was estimated by mapping clean reads to the Trinity transcripts assembly using RSEM version 1.2.15 [25], with the bowtie2 parameter set at mismatch 0. Then, read counts for each gene were obtained from the mapping results. Prior to Differentially expressed genes (DEG) analysis, the read counts were normalized using the edgeR program package with the Trimmed Mean of *M*-values method [26,27]. The DEGseq R package was used to analyze differential expression of two *P. edulis* samples. The *p*-value was adjusted using *q*-value [28]. The significant differential expression threshold was set as *q*-value < 0.005 and $|\log_2(\text{foldchange})| > 1$ [29]. The identified DEGs were used for GO enrichment analyses, which were performed using the GOseq R package (version 1.10.0), based on the Wallenius non-central hypergeometric distribution [30]. Kyoto encyclopedia of genes and genomes (KEGG) enrichment analysis was performed using KOBAS version 2.0.12 [31].

2.5. Quantitative Real-Time PCR (qRT-PCR)

Total RNA was extracted from various samples with the RNeasy Plant Mini Kit (Qiagen, Hilden, Germany) and treated with RNAse-free DNase I (Ambion, Austin, Texas, USA). Of the DNase-treated RNA, 1 µg was reverse transcribed using random hexamer primers. The resulting cDNA was diluted, and 2 µL of the diluted cDNA was used. Specific primers (Tm, 58–61 °C) were designed to generate PCR products of 70–150 bp (Table S1). The qRT-PCR was performed on an ABI ViiA 7 Real-time PCR platform. FastStart Universal SYBR Green Master (Rox) was used for qRT-PCR assays according to the manufacturer's protocol. For each sample, three replicates were performed in a final volume of 20 µL containing 10 µL of SYBR Premix Ex Taq (2×), 0.4 µL of 50 × ROX Reference Dye II, 0.4 µL (10 µM) of each primer, 2 µL of cDNA, and 6.8 µL of dH2O. Thermo-cycling conditions were as follows: initial denaturation at 95 °C for 30 s, followed by 40 cycles of 95 °C for 5 s, and 60 °C annealing and extension for 34 s. All reactions were performed in triplicate. The specificity of the PCR amplification procedures was checked with a heat dissociation protocol after the final cycle of the PCR to ensure that each amplicon was a single product. Relative expression was calculated as the difference in delta-Ct between the target gene and the internal control, *histone H3.3* (*HIS*) gene.

3. Results

3.1. Transcriptome Sequencing and De Novo Assembly

In total, there were 60,881,198 raw reads generated from KEDU_AP and 71,719,162 from GX_AP. The sequencing raw data were submitted to the Short Reads Archive (SRA) database under the accession number SRP106510. Of the raw reads from KEDU_AP, more than 95.87% bases had a *q* value ≥ 20, and for GX_AP 95.15%. The GC-contents were 44.84% and 45.20% for KEDU_AP and GX_AP, respectively. After removing low-quality reads, 57,840,324 clean reads from KEDU_AP and 68,443,694 from GX_AP were obtained. These were used for de novo assembly.

The Trinity software generated 127,474 transcripts with an average length of 1077 bp and an N50 of 2057 bp (Table 1). In total, 86,880 unigenes were obtained in the range of 201–13,397 bp with an N50 length of 2222 bp. Of these, 20,947 (24.11%) were 200–500 bp, 21,684 (24.96%) were 500–1000 bp, 22,051 (25.38%) were 1–2 kb and the remaining 22,198 (25.55%) were >2 kb (Table 1).

Table 1. Length distribution of unigenes and transcripts.

Nucleotide Length	Transcripts	Unigenes
200–500 bp	60,598	20,947
0.5–1k bp	22,605	21,684
1–2k bp	22,073	22,051
>2k bp	22,198	22,198
Total	127,474	86,880
Min length (bp)	201	201
Mean length (bp)	1077	1449
Max length (bp)	13,397	13,397
N50 (bp)	2057	2222

3.2. Function Annotation and Classification

All the 86,880 assembled unigenes were annotated to the seven databases using the BLAST algorithm (Table 2). In total, 64,067 unigenes were annotated, accounting for 73.74% (Table 2). There were 10,457 (12.03%) unigenes successfully annotated in all seven databases. Analyses showed that 60,028 (69.09%) unigenes had high homology with sequences in the Nr database and 46,093 (53.05%) in the Nt database. The details of other database proportions are shown in Table 2.

Table 2. Unigenes were annotated to the seven databases.

Component	Number of Unigenes	Percentage (%)
Annotated in NCBI non-redundant protein sequences (NR)	60,028	69.09
Annotated in NCBI non-redundant nucleotide sequences (NT)	46,093	53.05
Annotated in KEGG ortholog database (KO)	24,312	27.98
Annotated in Swiss-Prot	47,717	54.92
Annotated in Pfam	45,544	52.42
Annotated in Gene ontology (GO)	46,754	53.81
Annotated in EuKaryotic Orthologous Groups (KOG)	17,742	20.42
Annotated in all databases	10,457	12.03
Annotated in at least one database	64,067	73.74
Total unigenes	86,880	

There were 46,754 unigenes divided into three functional GO categories: biological process (BP), cellular component (CC) and molecular function (MF) (Figure 1). In the BP category, these matched unigenes were annotated to 25 GO terms, with the three top terms being 'cellular process' (27,925), 'metabolic process' (26,692) and 'single-organism process' (21,336). For CC and MF, these unigenes were clustered into 20 and 10 GO terms, respectively, with 'cell' (16,101) and 'binding' (27,248) as the largest subcategories. There were 17,742 unigenes divided into 26 groups for KOG analysis (Figure 2): the largest group was R (general functional prediction only), followed by O (post-translational modification, protein turnover and chaperon) and then J (translation, ribosomal structure and biogenesis).

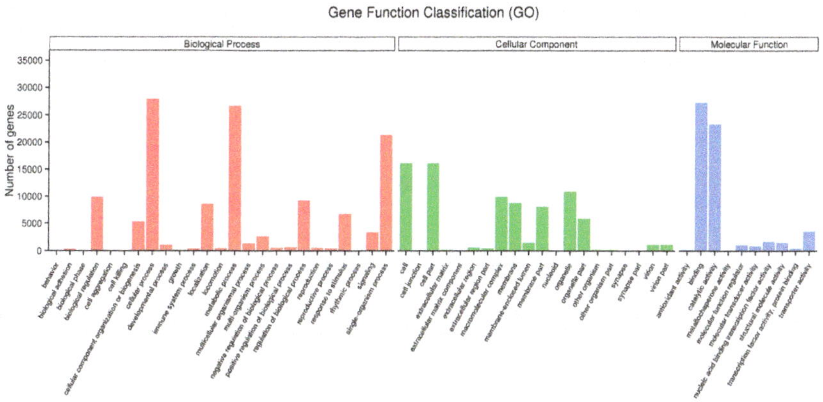

Figure 1. Gene ontology (GO) classification of unigenes. All the annotated unigenes were divided into three functional GO categories: biological process (BP), cellular component (CC) and molecular function (MF).

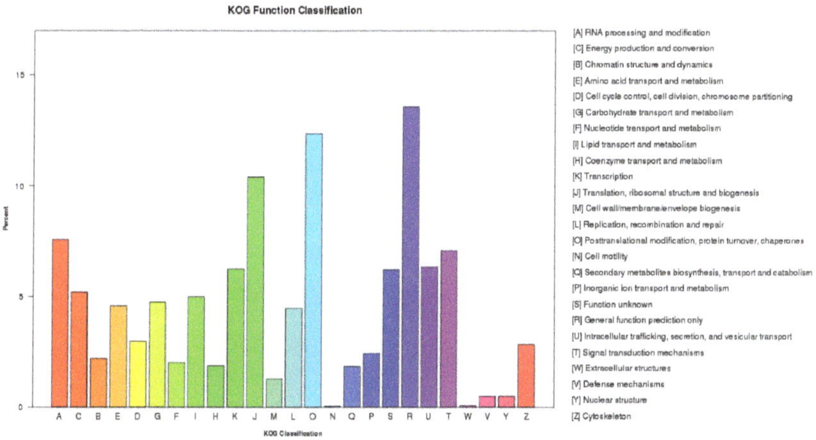

Figure 2. EuKaryotic Orthologous Groups (KOG) annotation of putative proteins. The *x*-axis indicates the names of the 25 groups of KOG. The *y*-axis indicates the percentage of the number of genes annotated to the group out of the total number of genes annotated.

A total of 18,366 unigenes were assigned to 19 metabolic pathways in the KEGG database (Figure 3). These 19 KEGG pathways were clustered into five branches: cellular processes (A) of 1164 unigenes, environmental information processing (B) of 999 unigenes, genetic information processing (C) of 4935 unigenes, metabolism (D) of 10,430 unigenes and organismal systems (E) of 838 unigenes. The metabolic pathways with the largest number of unigenes were 'carbohydrate metabolism' (2101), followed by 'translation' (2087) and 'folding, sorting and degradation' (1528).

KEGG Classification

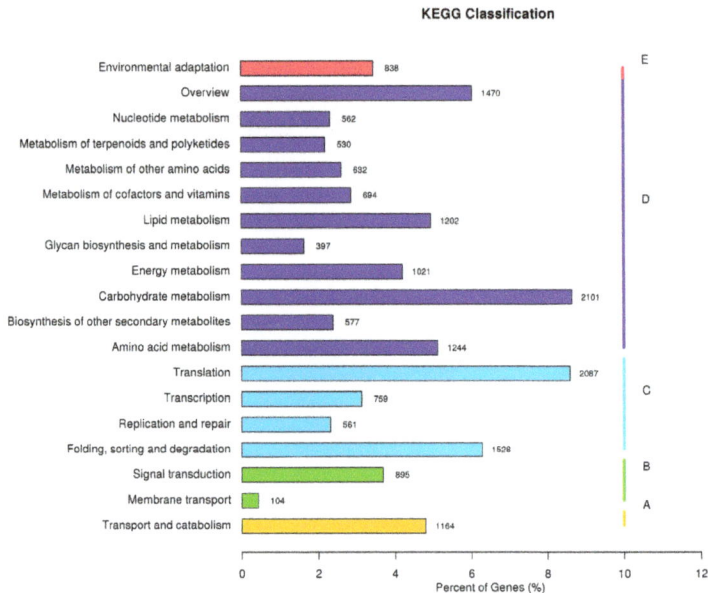

Figure 3. Kyoto encyclopedia of genes and genomes (KEGG) annotation of putative proteins. The *x*-axis indicates the percentage of the number of genes annotated to the pathway out of the total number of genes annotated. The *y*-axis indicates the name of the KEGG metabolic pathway. The genes were divided into five branches according to the KEGG metabolic pathway: Cellular Processes (A), Environmental Information Processing (B), Genetic Information Processing (C), Metabolism (D) and Organismal Systems (E).

3.3. DEG Identification, GO and KEGG Enrichment Analysis

A total of 80,810 unigenes (Fragments per kilobase of exon per million fragments mapped (FPKM) > 0.3) in two groups were identified, with 57,213 unigenes in common. There were 3045 significant DEGs identified between the KEDU_AP and GX_AP libraries. For these DEGs, if $\log_2$Foldchange >1, the DEG was considered as up-regulated but if $\log_2$Fold change <−1, it was considered as down-regulated. The 3045 DEGs included 1075 up-regulated and 1970 down-regulated DEGs (Figure 4).

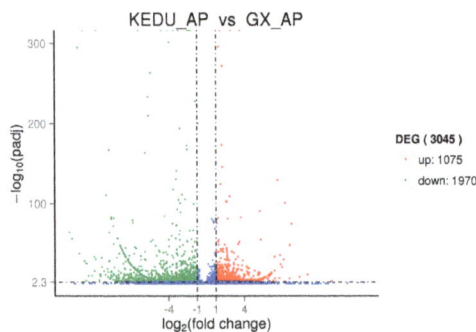

Figure 4. Differentially expressed genes (DEGs) identified between KEDU_AP and GX_AP. The red dots represent significant up-regulated genes and the green dots represent down-regulated genes.

Between the two libraries of enriched GO terms, the analysis showed that 'catalytic activity' and 'single-organism metabolic process' had the highest degree of enrichment (Figure 5). In the KEGG enrichment analysis (Figure 6), the most significantly enriched pathway was 'flavonoid biosynthesis', which mainly consisted of up-regulated DEGs. The second most significantly enriched pathway was 'phenylpropanoid biosynthesis', which also mainly consisted of up-regulated DEGs. Down-regulated DEGs were dominant in the 'photosynthesis-antenna proteins' and 'amino sugar and nucleotide sugar metabolism' (Figure S1). The greatest numbers of DEGs were involved in 'starch and sucrose metabolism', followed by 'phenylpropanoid biosynthesis'.

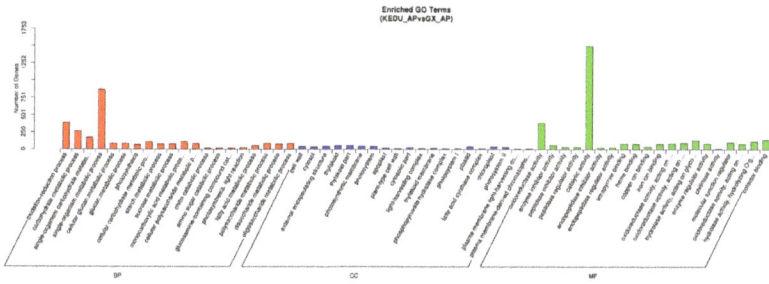

Figure 5. Functional gene ontology classification of DEGs. The *y*-axis indicates the number of DEGs in a category.

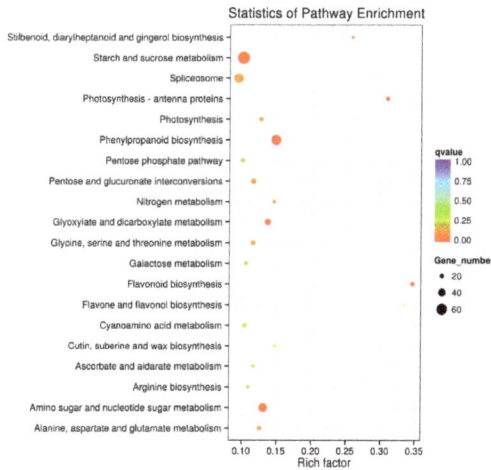

Figure 6. KEGG pathway enrichment of DEGs. The *y*-axis indicates the pathway name, and the *x*-axis indicates the enrichment factor corresponding to the pathway. The *q*-value is represented by the color of the dot. The number of DEGs is represented by the size of the dots.

3.4. DEGs Involved in the Plant Hormone Signal Transduction Pathways

Plant hormones play an important role in plant cold tolerance. A total of 36 DEGs and 677 background unigenes were in the enriched pathway of plant hormone signal transduction (Figure 7). Plant hormone signal transduction contained eight sub-paths. In the auxin signal transduction sub-pathway, there were two up-regulated and nine down-regulated genes. In the cytokinin signal transduction sub-pathway, there were three up-regulated and four down-regulated genes. There was just one up-regulated gene in the gibberellin signal transduction sub-pathway. Only one down-regulated gene was related to jasmonic acid signal transduction. There were up-regulated and down-regulated genes in the brassinolide and salicylic acid signal transduction pathways, which of them all contained three up-regulated genes and one down-regulated gene. Ethylene and ABA have vital functions in cold-stress signaling. In the ethylene signal transduction sub-pathway, there were two up-regulated genes and one down-regulated gene. In the ABA signal transduction sub-pathway, there were two up-regulated genes and one down-regulated gene for PYR/PYL, two down-regulated genes for PP2C, and one up-regulated and two down-regulated genes for SnRK2.

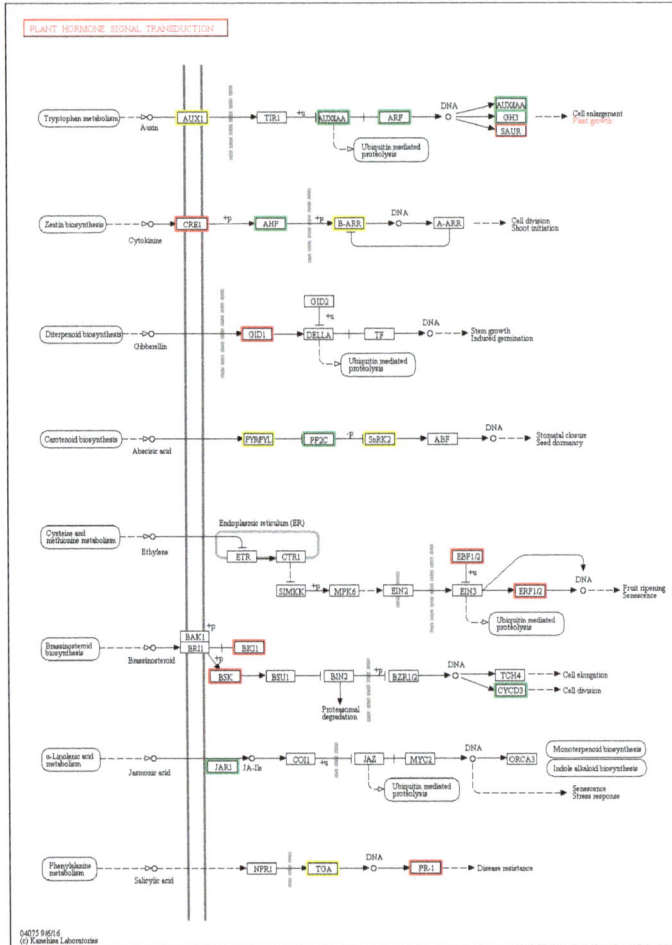

Figure 7. DEGs involved in the plant hormone signal transduction pathway. The enriched KEGG ortholog database (KO) terms are colored according to DEG regulation: red indicates up-regulation, green indicates down-regulation, and yellow indicates both up- and down-regulation.

3.5. Verification by qRT-PCR

In the ICE1–CBF–COR pathway, we identified one ICE1, one COR, two ICE1-like and two CBF genes. These genes were up-regulated in cold-tolerant variety 'Pingtang 1' (KEDU_AP) by qRT-PCR verification (Figure 8). To verify the reliability of the transcriptome data, 11 DEGs were randomly selected and examined using qRT-PCR at the transcriptional level (Figure S2). The expression patterns of the 11 DEGs were consistent with the transcriptome data (R^2 = 0.81321, *p*-value = 0.002334). These results indicate that our transcriptomic analysis was highly reproducible and reliable.

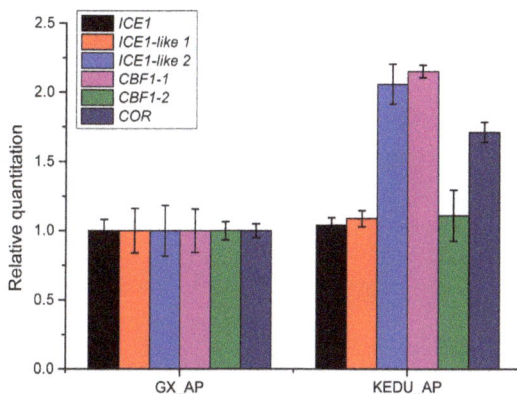

Figure 8. The expression of six genes involved in the ICE1–CBF–COR pathway. *ICE*, Cluster-5862.10114; *ICE-like 1*, Cluster-5862.30769; *ICE-like 2*, Cluster-5862.1181; *CBF1-1*, Cluster-5862.30492; *CBF1-2*, Cluster-5862.29181; and *COR*, Cluster-5862.23126.

4. Discussion

As a key messenger, the cytosolic Ca^{2+} occupies an important role in response to cold stress [32,33]. Research shows that cold-induced Ca^{2+} influx is positively correlated with accumulation of cold-induced transcripts in *Arabidopsis* [34] and alfalfa (*Medicago sativa* L.) [35]. Cold stress increases the cytosolic Ca^{2+} level, and this is then sensed by Ca^{2+} sensor proteins such as calmodulin (CaM) and calcineurin B-like proteins (CBLs) [36,37]. The increased levels of Ca^{2+} in plant cells affect the expression level of CBF and COR genes in the cold signaling pathway [7,38]. In this study, 332 unigenes were annotated as CaMs and CBLs, and included 14 DEGs. Among these DEGs, seven (four CaM and three CBL) genes were up-regulated in KEDU_AP, and seven (six CaM and one CBL) were down-regulated (Figure S3 and Table 3).

Table 3. Identified DEGs, unigene numbers and ratios of DEG/unigene.

Component	Unigenes	DEGs	Ratio (%)
AP2/EREBP	216	8	3.7
WRKY	161	12	7.5
bZIP	181	8	4.4
MYB	387	14	3.6
NAC	221	11	5.0
B-ARR	20	3	15
CaM/CBLs	332	14	4.2
MAPK	93	11	11.8
LEA	38	4	10.5

Plant protein kinases, such as Mitogen-activated protein kinases (MAPKs), play a central role in cellular signaling. MKK2 is a member of the MAPK family, and is upstream of MPK4 and MPK6, which are activated by low temperatures [14]. Previous studies showed that the influx of calcium is involved in cold-stress regulation of MAPKs [39]. The receptor-like kinase CRLK1 may associate calcium signaling with the MAPK cascade by binding to calcium, and CaM interacts with MEK kinase 1 (MEKK1) [40] (Figure 9). We identified 93 MAPK unigenes, which included 11 DEGs: nine up-regulated and two down-regulated (Figure S3 and Table 3).

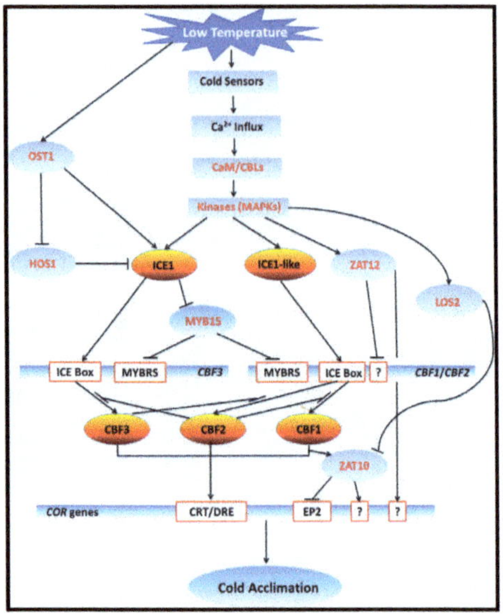

Figure 9. Diagram of cold-responsive regulatory networks [8,14]. The arrows indicate activation, whereas lines ending with a bar indicate negative regulation; the "?" indicates unknown cis-elements. Abbreviations: CaM, calmodulin; CBLs, calcineurin B-like proteins; MAPKs, mitogen-activated protein kinases; ICE1, inducer of CBF expression 1; MYB, V-MYB avian myeloblastosis viral oncogene homolog; MYBRS, MYB transcription factor recognition sequence; OST1, *open stomata 1 (SnRK2.6)*; HOS1, *high expression of osmotically responsive genes1*; CBF, C-repeat binding factor; COR, *cold-responsive genes*; CRT, C-repeat elements; DRE, dehydration-responsive elements; LOS2, *low expression of osmotically responsive genes 2*; ZAT10 and ZAT12 are two C2H2 zinc finger transcription factors.

In this work, 56 differentially expressed TFs were identified, and were divided into six TF families: AP2/EREBP, WRKY, bZIP, MYB, NAC and B-ARR (Figure S4 and Table 3). Most of them have been reported to be associated with cold tolerance in plants [41–44]. The TFs of CBFs belong to the AP2/EREBP family [45]. Three CBF genes are known in Arabidopsis, and are induced with plant exposure to cold, and regulate COR gene accumulation [46,47] (Figure 9). Overexpression of OsMYB4 could improve freezing tolerance in Arabidopsis by increasing COR gene expression, and OsMYBS3 and OsMYB2 enhanced cold resistance in rice [48–50]. We identified six genes in the ICE1-CBF-COR pathway of *P. edulis*, and their expression levels in the cold-tolerant variety 'Pingtang 1' were higher than that in cold-susceptible 'Purple Fragrance 1'.

In addition to the TFs, other important genes involved in plant cold tolerance include late-embryogenesis-abundant proteins (LEAs), antifreeze proteins (AFPs), superoxide dismutase (SOD) and proline [51–53]. LEAs are highly hydrophilic and provide protection for plants during cold stress. We also found four differentially expressed LEAs (Figure S3 and Table 3), with a high expression level in KEDU_AP.

In this study, we identified candidate genes and signaling pathways associated with cold tolerance by transcriptome sequencing of cold-tolerant variety 'Pingtang 1' and cold-susceptible 'Purple Fragrance 1'. The heatmap of DEGs showed that many genes related to cold-tolerance had higher expression levels in the cold-tolerant variety, indicating that they played a role in the chilling stress response. This report is the first to identify genes and signaling pathways involved in cold tolerance using high-throughput transcriptome sequencing in *P. edulis*. These findings may provide useful insights into the molecular mechanisms regulating cold tolerance and genetic breeding in *Passiflora* spp.

5. Conclusions

P. edulis is mainly distributed in tropical and subtropical regions and has difficulty surviving at low temperatures. To elucidate the molecular mechanisms of cold tolerance, we collected the aboveground parts of cold-tolerant variety 'Pingtang 1' and cold-susceptible 'Purple Fragrance 1' and subjected them to high-throughput sequencing. In total, 26,284,018 clean reads were obtained, and 86,880 unigenes with a mean size of 1449 bp were assembled. There were 3045 significant differentially expressed genes (DEGs) identified between 'Pingtang 1' (KEDU_AP) and 'Purple Fragrance 1' (GX_AP). To provide reference for the cold-tolerance breeding of *P. edulis*, we screened many DEGs, constructed their expression profiles and analyzed their functions; some potentially vital cold-tolerance genes and transcription factors were identified, and the expression levels of the two varieties were compared and analyzed. As the first report on the high-throughput sequencing of cold-tolerant *P. edulis*, this study should provide novel insights into cold-tolerance genes for *P. edulis* and be a valuable molecular basis for study in *Passiflora spp.*

Supplementary Materials: The following are available online at www.mdpi.com/1999-4907/8/11/435/s1, Table S1: Primer sequences; Figure S1: KEGG pathway enrichment of up-regulated and down-regulated DEGs; Figure S2: Comparison of expression results between RNA sequencing (RNA-seq) and qRT-PCR; Figure S3: Heatmap of important DEGs associated with cold-tolerance; Figure S4: Heatmap of differentially expressed transcription factors (TFs) associated with cold-tolerance.

Acknowledgments: This project was supported by the Key Science and Technology Program of Guizhou Province ((2016)2525), the National Key Research and Development Program of China (2016YFC050260404), the Key Science and Technology Program of Guangzhou Province (201604020006), the Applied Basic Research Program of Guizhou Province ((2014)200208), the Priority Academic Program Development of Jiangsu Higher Education Institutions (PAPD) and the Collaborative Innovation Plan of Jiangsu Higher Education (CIP).

Author Contributions: S.L. and M.X. analyzed the data and wrote the manuscript. S.L. and C.C. performed the experiments. G.C., L.Z. and C.G. participated in data analysis. A.L. provided helpful suggestions in design of the project. M.X. conceived and designed the project. All authors read and approved the final manuscript.

Conflicts of Interest: The authors declare no conflict of interest.

Abbreviations

Nr	NCBI non-redundant protein sequences
Nt	NCBI non-redundant nucleotide sequences
Pfam	Protein family
COG	Clusters of orthologous groups of proteins
KOG	EuKaryotic Orthologous Groups
Swiss-Prot	A manually annotated and reviewed protein sequence database
KEGG	Kyoto encyclopedia of genes and genomes
KO	KEGG ortholog database
GO	Gene ontology
FPKM	Fragments per kilobase of exon per million fragments mapped
DEGs	Differentially expressed genes

References

1. Rodriguez, R.R.; Dallas, M.P. Aislamientoy cultivo de protoplastos enmaracuyá isolation and cultive of protoplast in passion fruit. *Acta Biol. Colomb.* **2004**, *9*, 35–46.
2. Benincá, J.P.; Montanher, A.B.; Zucolotto, S.M.; Schenkel, E.P.; Fröde, T.S. Evaluation of the anti-inflammatory efficacy of *Passiflora edulis*. *Food Chem.* **2007**, *104*, 1097–1105. [CrossRef]
3. Bernacci, L.C.; Soares-Scott, M.D.; Junqueira, N.T.V.; Passos, I.R.D.S.; Meletti, L.M.M. *Passiflora edulis* sims the correct taxonomic way to cite the yellow passionfruit (and of others colors). *Rev. Bras. Frutic.* **2008**, *30*, 566–576. [CrossRef]
4. Rudnicki, M.; de Oliveira, M.R.; Veiga Pereira, T.d.; Reginatto, F.H.; Dal-Pizzol, F.; Fonseca Moreira, J.C. Antioxidant and antiglycation properties of *Passiflora alata* and *Passiflora edulis* extracts. *Food Chem.* **2007**, *100*, 719–724. [CrossRef]

5. Benigni, R.; Capra, C.; Cattorini, P. Plante medicinale; chimica, farmacologia e terapia. *Milan Inverni Della Beffa* **1962**, *2*, 1012–1022.

6. Zibadi, S.; Watson, R.R. Passion fruit (*Passiflora edulis*) composition, efficacy and safety. *Evid. Based Integr. Med.* **2004**, *1*, 183–187. [CrossRef]

7. Chinnusamy, V.; Zhu, J.K.; Sunkar, R. Gene regulation during cold stress acclimation in plants. *Methods Mol. Biol.* **2010**, *639*, 39–55. [PubMed]

8. Chinnusamy, V.; Zhu, J.; Zhu, J.K. Cold stress regulation of gene expression in plants. *Trends Plant Sci.* **2007**, *12*, 444–451. [CrossRef] [PubMed]

9. Thomashow, M.F. Plant cold acclimation: Freezing tolerance genes and regulatory mechanisms. *Annu. Rev. Plant Biol.* **1999**, *50*, 571–599. [CrossRef] [PubMed]

10. Jaglo, K.R.; Kleff, S.; Amundsen, K.L.; Zhang, X.; Haake, V.; Zhang, J.Z.; Deits, T.; Thomashow, M.F. Components of the *Arabidopsis* C-repeat/dehydration-responsive element binding factor cold-response pathway are conserved in *Brassica napus* and other plant species. *Plant Physiol.* **2001**, *127*, 910–917. [CrossRef] [PubMed]

11. Pellegrineschi, A.; Reynolds, M.; Pacheco, M.; Brito, R.M.; Almeraya, R.; Yamaguchi-Shinozaki, K.; Hoisington, D. Stress-induced expression in wheat of the *Arabidopsis thaliana* DREB1a gene delays water stress symptoms under greenhouse conditions. *Genome* **2004**, *47*, 493–500. [CrossRef] [PubMed]

12. Hsieh, T.H.; Lee, J.T.; Yang, P.T.; Chiu, L.H.; Charng, Y.Y.; Wang, Y.C.; Chan, M.T. Heterology expression of the arabidopsis C-repeat/dehydration response element binding factor 1 gene confers elevated tolerance to chilling and oxidative stresses in transgenic tomato. *Plant Physiol.* **2002**, *129*, 1086–1094. [CrossRef] [PubMed]

13. Oh, S.J.; Song, S.I.; Kim, Y.S.; Jang, H.J.; Kim, S.Y.; Kim, M.; Kim, Y.K.; Nahm, B.H.; Kim, J.K. Arabidopsis CBF3/DREB1a and ABF3 in transgenic rice increased tolerance to abiotic stress without stunting growth. *Plant Physiol.* **2005**, *138*, 341–351. [CrossRef] [PubMed]

14. Zhu, J.K. Abiotic stress signaling and responses in plants. *Cell* **2016**, *167*, 313–324. [CrossRef] [PubMed]

15. Teige, M.; Scheikl, E.; Eulgem, T.; Doczi, R.; Ichimura, K.; Shinozaki, K.; Dangl, J.L.; Hirt, H. The MKK2 pathway mediates cold and salt stress signaling in *Arabidopsis*. *Mol. Cell* **2004**, *15*, 141–152. [CrossRef] [PubMed]

16. Ding, Y.; Li, H.; Zhang, X.; Xie, Q.; Gong, Z.; Yang, S. Ost1 kinase modulates freezing tolerance by enhancing ice1 stability in arabidopsis. *Dev. Cell* **2015**, *32*, 278–289. [CrossRef] [PubMed]

17. Fowler, S. *Arabidopsis* transcriptome profiling indicates that multiple regulatory pathways are activated during cold acclimation in addition to the CBF cold response pathway. *Plant Cell* **2002**, *14*, 1675–1690. [CrossRef] [PubMed]

18. Kreps, J.A.; Wu, Y.; Chang, H.S.; Zhu, T.; Wang, X.; Harper, J.F. Transcriptome changes for *Arabidopsis* in response to salt, osmotic, and cold stress. *Plant Physiol.* **2002**, *130*, 2129–2141. [CrossRef] [PubMed]

19. Mantyla, E.; Lang, V.; Palva, E.T. Role of abscisic acid in drought-lnduced freezing tolerance, cold acclimation, and accumulation of lti78 and rabi8 proteins in *Arabidopsis thaliana*. *Plant Physiol.* **1995**, *107*, 141–148. [CrossRef] [PubMed]

20. Gilmour, S.J.; Thomashow, M.F. Cold acclimation and cold-regulated gene expression in ABA mutants of *Arabidopsis thaliana*. *Plant Mol. Biol.* **1991**, *17*, 1233–1240. [CrossRef] [PubMed]

21. Dong, W.P.; Luo, C.; Long, X.Q.; Hu, J.; Li, Y. Effects of low temperature stress on physiological indexes of cold resistance of *Passiflora edulis*. *Plant Physiol. J.* **2015**, *51*, 771–777.

22. Martin, M. Cutadapt removes adapter sequences from high-throughput sequencing reads. *EMBnet J.* **2011**, *17*, 10–12. [CrossRef]

23. Grabherr, M.G.; Haas, B.J.; Yassour, M.; Levin, J.Z.; Thompson, D.A.; Amit, I.; Adiconis, X.; Fan, L.; Raychowdhury, R.; Zeng, Q.; et al. Full-length transcriptome assembly from RNA-seq data without a reference genome. *Nat. Biotechnol.* **2011**, *29*, 644–652. [CrossRef] [PubMed]

24. Götz, S.; García-Gómez, M.J.; Terol, J.; Williams, T.D.; Nagaraj, S.H.; Nueda, M.J.; Robles, M.; Talón, M.; Dopazo, J.; Conesa, A. High-throughput functional annotation and data mining with the blast2go suite. *Nucleic Acids Res.* **2008**, *36*, 3420–3435. [CrossRef] [PubMed]

25. Li, B.; Dewey, C.N. Rsem: Accurate transcript quantification from RNA-seq data with or without a reference genome. *BMC Bioinform.* **2011**, *12*, 323. [CrossRef] [PubMed]

26. Robinson, M.; Oshlack, A. A scaling normalization method for differential expression analysis of RNA-seq data. *Genome Biol.* **2010**, *11*, R25. [CrossRef] [PubMed]

27. Robinson, M.; McCarthy, D.; Smyth, G. Edger: A bioconductor package for differential expression analysis of digital gene expression data. *Bioinformatics* **2010**, *26*, 139–140. [CrossRef] [PubMed]

28. Storey, J.D.; Tibshirani, R. Statistical significance for genomewide studies. *Proc. Natl. Acad. Sci. USA* **2003**, *100*, 9440–9445. [CrossRef] [PubMed]

29. Anders, S.; Huber, W. Differential expression analysis for sequence count data. *Genome Biol.* **2010**, *11*, R106. [CrossRef] [PubMed]

30. Young, M.D.; Wakefield, M.J.; Smyth, G.K.; Oshlack, A. Gene ontology analysis for RNA-seq: Accounting for selection bias. *Genome Biol.* **2010**, *11*, R14. [CrossRef] [PubMed]

31. Mao, X.; Cai, T.; Olyarchuk, J.G.; Wei, L. Automated genome annotation and pathway identification using the kegg orthology (ko) as a controlled vocabulary. *Bioinformatics* **2005**, *21*, 3787–3793. [CrossRef] [PubMed]

32. Reddy, A.S.N.; Ali, G.S.; Celesnik, H.; Day, I. Coping with stresses roles of calcium and calciumcalmodulin-regulated gene expression. *Plant Cell* **2011**, *23*, 2010–2032. [CrossRef] [PubMed]

33. Janska, A.; Marsık, P.; Zelenkova, S.; Ovesna, J. Cold stress and acclimation—What is important for metabolic adjustment. *Plant Biol.* **2010**, *12*, 395–405. [CrossRef] [PubMed]

34. Henriksson, K.N.; Trewavas, A.J. The effect of short-term low-temperature treatments on gene expression in arabidopsis correlates with changes in intracellular Ca^{2+} levels. *Plant Cell Environ.* **2003**, *26*, 485–496. [CrossRef]

35. Reddy, V.S.; Reddy, A.S.N. Proteomics of calcium-signaling components in plants. *Phytochemistry* **2004**, *65*, 1745–1776. [CrossRef] [PubMed]

36. Tuteja, N.; Mahajan, S. Further characterization of calcineurin B-like protein and its interacting partner CBL-interacting protein kinase from *Pisum sativum*. *Plant Signal. Behav.* **2007**, *2*, 358–361. [CrossRef] [PubMed]

37. Pandey, G.K. Emergence of a novel calcium signaling pathway in plants CBL-CIPK signaling network. *Physiol. Mol. Biol. Plants* **2008**, *14*, 51–68. [CrossRef] [PubMed]

38. Theocharis, A.; Cle'ment, C.; Barka, E.A. Physiological and molecular changes in plants grown at low temperatures. *Planta* **2012**, *235*, 1091–1105. [CrossRef] [PubMed]

39. Sangwan, V.; Örvar, B.L.; Beyerly, J.; Hirt, H.; Dhindsa, R.S. Opposite changes in membrane fluidity mimic cold andheat stress activation of distinct plant map kinase pathways. *Plant J.* **2002**, *31*, 629–638. [CrossRef] [PubMed]

40. Yang, T.; Ali, G.S.; Yang, L.; Du, L.; Reddy, A.S.N.; Poovaiah, B.W. Calciumcalmodulin-regulated receptor-like kinase CRLK1 interacts with MEKK1 in plants. *Plant Signal. Behav.* **2010**, *5*, 991–994. [CrossRef] [PubMed]

41. Rushton, P.J.; Somssich, I.E.; Ringler, P.; Shen, Q.J. Wrky transcription factors. *Trends Plant Sci.* **2010**, *15*, 247–258. [CrossRef] [PubMed]

42. Jakoby, M.; Weisshaara, B.; Dröge-Laserb, W.; Vicente-Carbajosac, J.; Tiedemannd, J.; Kroje, T.; Parcye, F. Bzip transcription factors in *Arabidopsis*. *Trends Plant Sci.* **2002**, *7*, 106–111. [CrossRef]

43. Van Buskirka, H.A.; Thomashow, M.F. *Arabidopsis* transcription factors regulating cold acclimation. *Physiol. Plant.* **2006**, *126*, 72–80. [CrossRef]

44. Zhang, L.; Zhao, G.; Jia, J.; Liu, X.; Kong, X. Molecular characterization of isolated wheat MYB genes and analysis of their expression during abiotic stress. *J. Exp. Bot.* **2012**, *63*, 203–214. [CrossRef] [PubMed]

45. Stockinger, E.J.; Gilmour, S.J.; Thomashow, M.F. Arabidopsis thaliana cbf1 encodes an ap2 domain-containing transcriptional activator that binds to the C-repeatydre, a cis-acting DNA regulatory element that stimulates transcription in response to low temperature and water deficit. *Proc. Natl. Acad. Sci. USA* **1997**, *94*, 1035–1040. [CrossRef] [PubMed]

46. Medina, J.N.; Bargues, M.N.; Terol, J.; Pe'rez-Alonso, M.; Salinas, J. The *Arabidopsis* CBF gene family is composed of three genes encoding AP2 domain-containing proteins whose expression is regulated by low temperature but not by abscisic acid or dehydration. *Plant Physiol.* **1999**, *119*, 463–469. [CrossRef] [PubMed]

47. Gilmour, S.J.; Zarka, D.G.; Stockinger, E.J.; Salazar, M.P.; Houghton, J.M.; Thomashow, M.F. Low temperature regulation of the arabidopsis cbf family of ap2 transcriptional activators as an early step in coldinduced cor gene expression. *Plant J.* **1998**, *16*, 433–442. [CrossRef] [PubMed]

48. Su, C.-F.; Wang, Y.-C.; Hsieh, T.-H.; Lu, C.-A.; Tseng, T.-H.; Yu, S.-M. A novel mybs3-dependent pathway confers cold tolerance in rice. *Plant Physiol.* **2010**, *153*, 145–158. [CrossRef] [PubMed]

49. Vannini, C.; Locatelli, F.; Bracale, M.; Magnani, E.; Marsoni, M.; Osnato, M.; Mattana, M.; Baldoni, E.; Coraggio, I. Overexpression of the rice osmyb4 gene increases chilling and freezing tolerance of arabidopsis thaliana plants. *Plant J.* **2004**, *37*, 115–127. [CrossRef] [PubMed]

50. Yang, A.; Dai, X.; Zhang, W.-H. A r2r3-type myb gene, osmyb2, is involved in salt, cold, and dehydration tolerance in rice. *J. Exp. Bot.* **2012**, *63*, 2541–2556. [CrossRef] [PubMed]

51. Gill, S.; Tuteja, N. Reactive oxygen species and antioxidant machinery in abiotic stress tolerance in crop plants. *Plant Physiol. Biochem.* **2010**, *48*, 909–930. [CrossRef] [PubMed]

52. Szabados, L.; Savour, A. Proline: A multifunctional amino acid. *Trends Plant Sci.* **2010**, *15*, 89–97. [CrossRef] [PubMed]

53. Griffith, M.; Yaish, M. Antifreeze proteins in overwintering plants: A tale of two activities. *Trends Plant Sci.* **2004**, *9*, 399–405. [CrossRef] [PubMed]

© 2017 by the authors. Licensee MDPI, Basel, Switzerland. This article is an open access article distributed under the terms and conditions of the Creative Commons Attribution (CC BY) license (http://creativecommons.org/licenses/by/4.0/).

forests

MDPI

Article

Transcriptome Sequencing and Comparative Analysis of *Piptoporus betulinus* in Response to Birch Sawdust Induction

Lixia Yang [1,2,†], Mu Peng [2,3,†], Syed Sadaqat Shah [4] and Qiuyu Wang [2,*]

[1]   Inner Mongolia Key Laboratory of Meadow Steppe Ecosystem and Global Change,
    College of Life and Environmental Science, Hulunbuir University, Inner Mongolia 021000, China;
    yanglixia1981@163.com
[2]   College of Life Science, Northeast Forestry University, Harbin 150040, China; pengmu1025@hotmail.com
[3]   Key Laboratory of Saline-alkali Vegetation Ecology Restoration in Oil Field (SAVER),
    Alkali Soil Natural Environmental Science Center (ASNESC), Northeast Forestry University,
    Harbin 150040, China
[4]   Key Laboratory of Vegetation Ecology, Ministry of Education, Institute of Grassland Science,
    Northeast Normal University, Changchun 130024, China; sadaqatafridi@yahoo.com
*   Correspondence: wqyll@sina.com; Tel.: +86-0451-8219-1755
†   These authors contributed equally to this work.

Academic Editor: Filippos A. (Phil) Aravanopoulos
Received: 18 August 2017; Accepted: 27 September 2017; Published: 7 October 2017

Abstract: *Piptoporus betulinus*, a brown-rot parasitic fungus of birch trees (*Betula* species), has been used as a common anti-parasitic and antibacterial agent. The lack of genetic resource data for *P. betulinus* has limited the exploration of this species. In this present study, we used Illumina Hiseq 2500 technology to examine the transcriptome assembly of *P. betulinus* in response to birch sawdust induction. By de novo assembly, 21,882 non-redundant unigenes were yielded, and 21,255 (97.1%) were annotated with known gene sequences. A total of 340 responsive unigenes were highly homologous with putative lignocellulose-degrading enzyme candidates. Additionally, 86 unigenes might be involved in the chemical reaction in xenobiotics biodegradation and metabolism, which suggests that this fungus could convert xenobiotic materials and has the potential ability to clean up environmental pollutants. To our knowledge, this was the first study on transcriptome sequencing and comparative analysis of *P. betulinus*, which provided a better understanding of molecular mechanisms underlying birch sawdust induction and identified lignocelluloses degrading enzymes.

Keywords: *Piptoporus betulinus*; transcriptome; comparative analysis; lignocellulose degradation

1. Introduction

Piptoporus betulinus, a brown-rot parasitic fungus of birch trees (*Betula* species), has been used as a common anti-parasitic and antibacterial agent for the treatment of wounds and various diseases, such as cancer, inflammation and so on [1,2]. Its extract has been demonstrated to be effective in preventing fatigue, strengthen immunity and relieving pain. Previous findings have suggested that this fungus has the ability to degrade of wood components, including cellulose and hemicellulose [3–5].

Lignocellulose represents an abundant carbon-neutral renewable resource that can be used for the production of bioenergy and biomaterials. It could be efficiently converted into ethanol using fungi as biological pre-treatment [6]. Furthermore, published findings have shown that brown-rot fungi have the potential as a new biocatalyst with unprecedented fermentability and a microbial starter [7]. Moreover, brown-rot fungi primarily break down cellulose and hemicellulose with carbohydrate active enzymes (CAZy), which catalyze cellulose into 6-carbon sugars [8,9]. In addition, it could greatly balance the conversion system of bioethanol production without genetic engineering support or external hydrolase. Therefore, it was considered as a promoter in lignocellulose biodegradation.

Although *P. betulinus* has various potential properties, the gene sequences in the publicly available databases for this fungus are rare. Only about 78 nucleotide sequences of *P. betulinus* to date have been deposited in National Center for Biotechnology Information's (NCBI) GenBank database, and most of them were isolated and cloned from ribosomal RNA genes [10,11]. In recent years, the lignocellulose-decaying transcriptomes of white-rot fungi have been thoroughly studied [12,13], while only few studies describing the transcriptomes of brown-rot fungi on wood have been reported [14,15]. This present study used Illumina Hiseq 2500 technology to examine the transcriptome of *P. betulinus* in response to birch sawdust induction. Transcriptome profiling of *P. betulinus* provides a better understanding of molecular mechanisms underlying birch sawdust induction and could identify lignocellulose-degrading enzymes in this species.

2. Materials and Methods

2.1. Fungal Strain

In August 2014, the fruiting body of the fungal strain was collected from *Betula platyphylla* at Liangshui Nature Reserve, Lesser Xing'an Mountains in Yichun city, Heilongjiang Province, China. The fruiting body was sterilized in sodium hypochlorite for 1 min, before being rinsed with sterile deionized water 3–5 times. This was cultured on potato dextrose agar plate and kept in the dark at 28 °C for 10 days. Following this, this strain was identified as *P. betulinus* according to Internal Transcribed Spacer (ITS) sequencing and alignment in NCBI (GenBank accession number MF967582).

2.2. RNA Extraction

The strain was inoculated into different liquid media (Potato dextrose broth (PDB) and PDB + birch sawdust (*Betula platyphlla*) (5 g/L)), which were cultured with continuous shaking (150 rpm) at 28 °C for 10 days. For each treatment (PDB and PDB + birch sawdust, respectively), the mycelia of *P. betulinus* were harvested by filtration, before being frozen in liquid nitrogen and prepared for RNA extraction. Total RNA was extracted using TRNzol reagent according to the manufacturer's protocol (TIANGEN, Beijing, China). Total RNA was extracted from each sample in triplicate, before synthesized cDNA were pooled together for sequencing. A quantitative real-time PCR (qRT-PCR) using RNA samples as templates was performed to detect genomic DNA. Samples with no amplified qRT-PCR products were used as a template for cDNA synthesis. The extracted RNA yield and purity were checked by NanoDrop 2000 (Thermo Scientific, Hudson, NH, USA). The qualified RNA samples were used for cDNA synthesis using PrimeScript™ RT reagent Kit (TakaRa). Subsequently, cDNA fragments were selected for PCR amplification and cDNA library were used for sequence analysis via Illumina HiSeq™ 2500(Illumina, San Diego, CA, USA).

2.3. Transcriptome Analysis

Raw reads were cleaned by removing adaptor sequences, short sequences and low-quality reads (reads containing Ns >5). The remaining clean reads were assembled into unigenes using short reads assembling program, which is also known as SOAPdenovo [16]. TIGR Gene Indices clustering tool (TGICL) was used to acquire a set of non-redundant unigenes [17]. After this, all the non-redundant unigenes were used for blast search (E-value < 10^{-5} or 10^{-3}) and annotation in various databases, including NCBI Nr database, SwissProt database, Kyoto Encyclopedia of Genes and Genomes (KEGG) database and Cluster of Orthologous Groups (COG) database. For the functional annotation, gene ontology (GO) terms were analyzed using the Blast2GO program [18]. Finally, Web Gene Ontology Annotation Plot (WEGO) was used to classify GO function for all unigenes [19].

A differential expression analysis between two treatments was performed using the DESeq2 R package (http://www.bioconductor.org/package/DESeq2/). The false discovery rate (FDR) control method was applied in the Benjamini and Hochberg method to correct the results for *p*-values. An FDR <0.01 and FC (fold change) ≥2 was set as the threshold to determine the significance of gene expression differences. The fragments per kilo bases per million reads (FPKM) were used to evaluate expressed values and quantify transcript levels, which normalizes and eliminates the influence of gene length and sequencing depth in calculating gene expression, allowing for direct comparison of gene expression between different treatments [20]. Genes with an adjusted *p*-value < 0.05 was assigned as differentially expressed according to the reports of Deng et al. [21]. For pathway enrichment analysis, all differentially expressed unigenes were submitted in the online KEGG Automatic Annotation Server (http://www.genome.jp/kegg/) and we searched for significantly enriched KEGG terms to obtain the pathway annotation (adjusted *p*-value < 0.05).

2.4. Quantitative Real-Time PCR Validation

A total of 12 representative birch sawdust induced-relevant unigenes (carbohydrate-active enzyme genes and lignocellulose-degrading enzyme genes) with significantly differential expression (adjusted *p*-value < 0.05) identified by RNA-seq were chosen for experimental validation using qRT-PCR with gene specific primers (Table S1). The qRT-PCR was performed using SYBR® Premix Ex Taq™ II Kit (Tli RNaseH Plus) (TakaRa, Tokyo, Japan) in a volume of 20 μL, which contained 10 μL of SYBR Premix Ex Taq (2×), 0.4 μL of ROX Reference Dye II (10×), 2 μL of cDNA template and 0.5 μM of each primer. The amplification was performed as follows: 95 °C for 30 s, 45 cycles of 95 °C for 5 s and 60 °C for 40 s. The qRT-PCR amplifications were conducted in an ABI 7500 detection system (Applied Biosystems, Carlsbad, CA, USA). All samples and reactions were performed in triplicate and the results were expressed relative to the expression levels of actin in each gene by using the $2^{-\triangle\triangle Ct}$ method [22].

2.5. Availability of Data

All transcriptome data files were submitted to the Sequence Read Archive database with the accession number SRP117136.

3. Results

3.1. Raw Reads Processing and de novo Assembly

For a better understanding of molecular mechanisms underlying birch sawdust induction and comparative transcriptome analysis in *P. betulinus*, two cDNA treatments prepared from PDB media and PDB + sawdust were sequenced with Illumina HiSeq™ 2500. An overview of the sequencing and assembly is given in Table 1 and Table S2. After filtering adapters and short sequences or low-quality bases, a total of 94,942,136 clean reads with 9,510,783,105 nucleotides were obtained in the two treatments (Table S2). Finally, de novo assembly yielded 21,882 non-redundant unigenes with an average length of 1949 bp and an N50 of 3007 bp (Table 1). Of these unigenes, 17,081 (88.1%) were >500 bp and 10,670 (48.8%) were >1500 bp. These obtained sequences provided abundant information on transcriptomes for further analysis of the birch sawdust-induced genes in *P. betulinus*.

Table 1. Overview of transcriptome sequencing and assembly for *P. betulinus*.

Length (bp)	Total Number	Percentage (%)
<200	0	0.0
200–500	4800	21.9
500–1000	3362	15.4
1000–1500	3049	23.9
1500–2000	2626	12.0
≥2000	8044	36.8
Total	21,882	
Total length of all unigenes		42,652,276
Median length of all unigenes (N50)		3007
Average length of all unigenes		1949.17

3.2. Functional Annotation

To identify the putative functions of unigenes in *P. betulinus*, we annotated all the assembled unigenes against the Nr, Swiss-port, KEGG and COG databases. With comparison against those four databases, a total of 21,255 (97.1%) unigenes were successfully annotated with known gene sequences. The number of unigenes with significant similarity to sequences in Nr, Swissport, KEGG, and COG databases were 21,255 (97.1%), 3351 (15.3%), 6256 (28.6%) and 7955 (36.4%), respectively (Table 2).

Table 2. Annotation of non-redundant unigenes.

Database	Number of Annotated Unigenes	Percentage of Annotated Unigenes
Nr	21,255	97.1%
Swissport	3351	15.3%
KEGG	6256	28.6%
COG	7955	36.4%

The GO analysis classified the functions of predicted unigenes into three main categories: biological processes, molecular function and cellular components (Figure 1). A total of 3351 sequences were assigned with 9111 GO terms, among which 2000 unigenes (22.0%) were assigned at least one GO term in the biological processes, 4253 (46.7%) in the cellular components and 4658 (51.1%) in molecular functions. Additionally, these unigenes were further classified into functional subcategories. In cellular components, the largest subcategory was cells (29.8%) and the second largest was cell parts (13.6%). Regarding molecular function, the largest number were found in catalytic activity (63.5%) and binding (46.8%). According to biological processes, genes involved in metabolic process and cellular process were highly represented, which accounted for 54.3% and 42.9% of the matched unigenes in the subcategory, respectively.

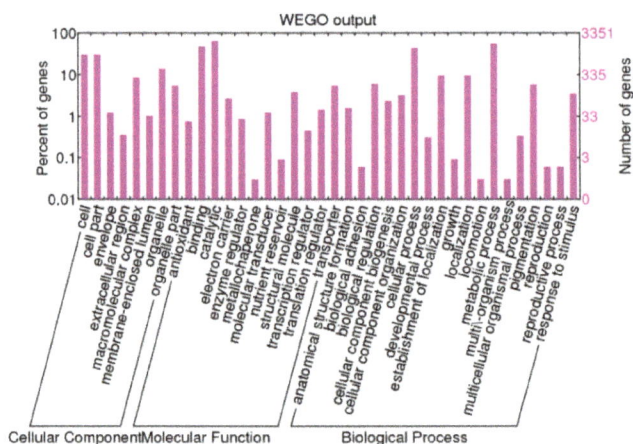

Figure 1. Functional annotation of non-redundant unigenes based on Gene Ontology (GO) classification. The results are summarized in three functional categories: cellular components, molecular function and biological process.

The genes belonging to the same biological pathway synergistically participated to accomplish the biological functions. The KEGG database was used to identify and predict the metabolism pathways in *P. betulinus*. Details of sequences involved in KEGG annotation are listed in Table S3. In KEGG metabolic pathways, a total of 6256 unigenes were matched in 309 different KEGG pathways (Table S3). Among these pathways, the most common was amino acid metabolic pathways with 240 members, followed by ribosome biogenesis (202) and energy metabolism (197). These might be involved in maintaining the basic metabolic process of *P. betulinus*.

In addition, in COG classification analysis, out of 21,882 Nr hits, 7955 unigenes were grouped into 25 COG categories. Among them, the prediction of the general function was the most populated group (18.8%), followed by amino acid transport and metabolism (8.16%), and carbohydrate transport and metabolism (6.9%). Two hundred and seventy-six unigenes were mapped into unknown functions, which might be involved in a specific gene in *P. betulinus*. The smallest groups were found in cell wall/membrane/envelope biogenesis and nuclear structures (Figure 2).

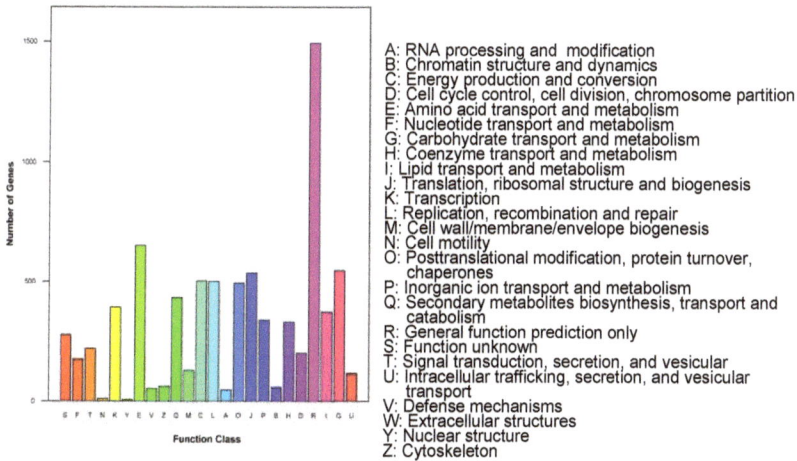

Figure 2. Histogram presentation of clusters of orthologous groups (COG) classification (7955 sequences have a COG classification among the 25 categories).

Based on GO classification, KEGG pathway and COG annotation, the catalytic function accounted for the highest percentage in the molecular functions, while amino acid transport and metabolism, carbohydrate transport and metabolism were the predominant categories. Therefore, further analysis was needed with specific focus on those pathways.

3.3. Changes in Gene Expression under Sawdust Induction

To characterize the differences of molecular response to the PDB (control) and PDB + birch sawdust medias, unigene expression levels were calculated by the FPKM method. Based on FPKM values, 776 unigenes were identified as differentially expressed genes (Table S4). Among them, 444 genes were up-regulated and 332 were down-regulated, which were identified by DESeq2 (http://www.bioconductor.org/package/DESeq2/) with adjusted p-value ≤ 0.05 and fold change value >2. Following this, all differentially expressed genes were mapped into 26 different metabolic pathway categories (Table S5). Of these annotated pathways, phenylpropanoid biosynthesis, starch and sucrose metabolism and lysosome were the main pathways that accounted for 7.8%, 7.3% and 6.4%, respectively. Notably, the pathway enrichment analysis revealed that phenylpropanoid biosynthesis as well as starch and sucrose metabolism were the important pathways in the degradation of wood components by *P. betulinus*.

3.4. Detection of Lignocellulose-Degrading Enzyme-Relevant Gene Sequences and qRT-PCR Analysis

For further insight into birch sawdust induction in *P. betulinus*, we detected and analyzed induction-related genes in this study. These matched enzymes were mainly involved in the degradation of cellulose, hemicellulose and starch. There were 340 highly homologous unigenes in putative wood-degrading enzymes (Table 3 and Table S6). Among them, 109 unigenes were identified as glycoside hydrolases (GHs), 37 belonged to glycosyl transferases (GTs), while 12 and 13 genes were assigned to carbohydrate esterases (CEs) and polysaccharide lyases (PLs), respectively. Fifty-three unigenes shared a similar identity to glucosidase, among which 52 homologs belonged to beta-glucosidase. Twenty-three unigenes were similar to 3-(3-hydroxy-phenyl) propionate hydroxylase, 20 unigenes to chitinase, 16 unigenes to mannosidase, 11 unigenes to polygalacturonase, 9 unigenes to mannosidase, 9 unigenes to pectinesterase.

Table 3. Putative wood-degrading genes in the *P. betulinus* transcriptome.

Enzymes	Number of Unigenes
glycoside hydrolases	109
glycosyl transferases	37
carbohydrate esterases	12
polysaccharide lyases	13
beta-glucosidase	52
alpha-glucosidase	1
mannosidase	16
glycosyltransferase	4
polygalacturonase	11
alpha-galactosidase	7
D-xylose 1-dehydrogenase (NADP)	4
N-glycosylase	2
Pectinesterase	9
3-(3-hydroxy-phenyl)propionate hydroxylase	22
alpha-amylase	5
xylose isomerase	7
beta-xylosidase	2
xyloglucan:xyloglucosyl transferase	2
glucosylceramidase	5
chitinase	20
Total	**340**

Several putative unigenes were involved in xenobiotic biodegradation and metabolism (Table S6). Enzymes involved in chloroalkane and chloroalkene degradation were *S*-(hydroxymethyl) glutathione dehydrogenase/alcohol dehydrogenase and aldehyde dehydrogenase with 7 and 41 unigenes, respectively [23]. One unigene was identified as carboxymethylenebutenolidase, which catalyzed the chemical reaction in toluene and chlorobenzene degradation [24]. Two unigenes, 1 and 23 unigenes, were found to exhibit a striking homology to amidase, nitrilase and 3-(3-hydroxy-phenyl) propionate hydroxylase, respectively. These above enzymes were related to styrene degradation. Alcohol dehydrogenase, which mainly degrade naphthalene compounds, were detected in *P. betulinus* transcriptome with 7 unigenes. Four unigenes belonged to aflatoxin B1 aldehyde reductase, which regulated the metabolism of xenobiotics by cytochrome P450.

To verify that the unigenes from sequencing were indeed differentially expressed genes and to analyze the difference of gene expression profile between control samples and induced samples, twelve unigenes related to lignocellulose-degrading enzyme-encoding genes were selected for qRT-PCR analysis (Figure 3). The results showed that 8 unigenes were up-regulated and 4 unigenes were down-regulated. Interestingly, the expression of c10553_g1_i5 and c10553_g1_i6 (mannosidase, EC: 3.2.1.21) under sawdust-induced samples increased by 6.3-fold and 1.5-fold compared to controls, suggesting that most of mannosidases were induced by sawdust in the lignocellulose-degrading process. Similarly, the expression of pectinesterase (c8512_g1_i14 and c8512_g1_i8) was also increased after being cultivated in birch sawdust. Several lignocellulose-degrading enzymes were detected to be down-regulated, including c7080_g1_i5 and c8929_g1_i1 (polygalacturonase); as well as c9335_g2_i1 and c5710_g1_i1 [3-(3-hydroxy-phenyl) propionate hydroxylase]. Therefore, those enzymes were jointly involved in the degradation of lignocellulose.

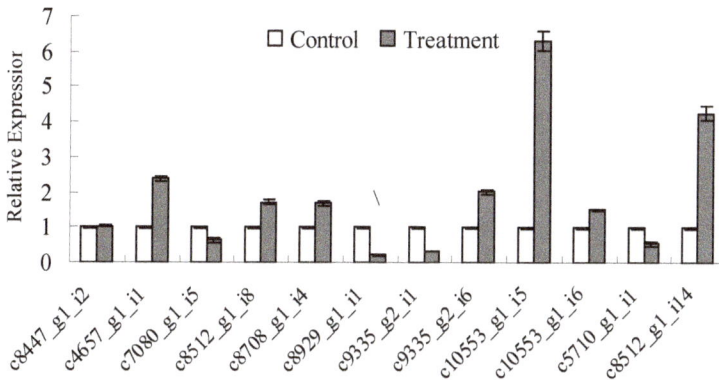

Figure 3. Quantitative real-time PCR validation of differentially expressed unigenes in the *P. betulinus* transcriptome. All reactions were performed in biological triplicate, and the error bars represent the standard deviations. This figure included c8447_g1_i2, c9355_g2_i6 β-D-Glucosidase; c10553_g1_i6, c10553_g1_i5, β-mannosidase; c8512_g1_i14, c8512_g1_i8, pectinesterase c4657_g1_i1, c8708_g1_i4, α-glycosidase; c7080_g1_i5, c8929_g1_i1, polygalacturonase; c9335_g2_i1 and c5710_g1_i1 [3-(3-hydroxy-phenyl) propionate hydroxylase].

4. Discussion

Transcriptomics, an important part of functional genomics, clarify the genetic transcription of cells at the overall level [25]. Data obtained by Illumina high-throughput sequencing has many advantages due to its large storage size, high efficiency and low cost, which is suitable for a species that has not had its whole genome sequenced. In this study, we firstly sequenced the *P. betulinus* transcriptome using Illumina Hiseq 2500 technology. After de novo assembly, 21,882 non-redundant unigenes were obtained, among which 21,255 were annotated into known functions involved in the lignocellulose-degrading process. These findings showed that a considerable number of genes were successfully identified through high-throughput sequencing and would be useful for the further functional analysis.

Based on KEGG pathway database, 6256 unigenes were mapped into 309 types of enzyme-encoding genes. Among them, many unigenes participated in degradation of wood components, which would provide a better platform for cloning of cellulose-degrading genes and related functional validation. The KEGG pathway results showed that amino acid transport and metabolism, carbohydrate transport and metabolism as the predominant pathways were closely associated with cellulose and hemicellulose degradation.

Annotation analysis of unigenes in the *P. betulinus* transcriptome showed that 340 transcripts were identified as putative wood-degrading genes, which is less than other brown-rot fungi [26,27]. Additionally, 53 unigenes were found to be similar to glucosidase, demonstrating that *P. betulinus* could degrade wood components. Interestingly, 2 unigenes were identified as pectinesterase in our transcriptome, which is comparable with other brown-rot fungi [28] and suggests that this enzyme was relatively conservative. Although we found that 20 unigenes had homology with chitinase, further confirmation of its ability to degrade chitin is required.

Lignin biodegradation was mainly detected in white-rot fungi, which produces multiple isoenzymes of lignin peroxidase, manganese peroxidase and laccase [29]. Although *P. betulinus* is unable to degrade lignin, it could secrete enzymes which may modify lignin or break lignin seal by the methylation pathway and sufficiently expose cellulose and hemicellulose for enzymatic action. Recently, most of the studies were focused on *Postia placenta*, such as genome, transcriptome, secretome analysis and nuclear magnetic resonance analysis [27,30]. Martinez et al. first reported transcriptional profile of brown-rot fungus *P. placenta*, and their results revealed that this fungus possessed unique extracellular enzyme systems, including an unusual repertoire of glycoside hydrolases, while exocellobiohydrolases and cellulose-binding domains were absent in this fungus [27]. When *P. placenta* was grown in medium containing cellulose as the sole carbon source, transcripts corresponding to many hemicellulases and endoglucanases were expressed at high levels compared to the glucose-grown culture [31]. It is well known that although brown-rot and white-rot fungi are both basidiomycetes, white-rot fungi can produce complex ligninolytic systems to degrade lignin [32]. The different degradation pathway mechanisms in these two fungi have attracted the attention of scientists. However, lignin peroxidase, manganese peroxidase, versatile peroxidase and laccase were not detected in the *P. placenta* genome [31], suggesting that this fungus could not induce lignin depolymerization, which was consistent with our finding.

In addition, several metabolic pathways might be involved in polycyclic aromatic hydrocarbon degradation, including phenylpropanoid biosynthesis, polycyclic aromatic hydrocarbon degradation, limonene and pinene degradation. However, in the past published papers, *P. betulinus* was thought to modify the structure of lignin instead of degrading lignin, which involve complex phenolic polymers composed of cinnamyl alcohol subunits, phenylpropanol and its derivatives [33]. Therefore, our paper speculated that those metabolic pathways in *P. betulinus* transcriptome were likely to participate in lignin modification. Importantly, we also found 8 potential wood-degrading related genes consisting of 3 laccases, 1 cellobiose dehydrogenase and 4 glucose oxidases. This indicated that *P. betulinus* could efficiently degrade wood components (Table S6), which is consistent with a previous report [34]. Although we detected wood-degrading enzymes encoding genes in transcriptome, enzymatic activity should be measured in a future experiment to confirm its enzyme activity level.

Eighty-six unigenes were matched with xenobiotics biodegradation and metabolism, including chloroalkane and chloroalkene degradation, toluene and chlorobenzene degradation, styrene degradation, naphthalene degradation and metabolism of xenobiotics by cytochrome P450. Prince and Roger [35] reported bioremediation using microbiological processes to clean-up organic and inorganic environmental pollutants by chloroalkane and chloroalkene degradation. Published findings have evaluated the effect of white-rot fungi on removal of gaseous chlorobenzene and naphthalene, while no reports have been published regarding this degradation using brown-rot fungi. Therefore, our next experiment would determine the ability of gaseous chlorobenzene or naphthalene in *P. betulinus* [36–38]. The findings of Krueger et al. [38] indicated that brown-rot fungus *Gloeophyllum trabeum* might be suitable for the biodegradation of styrene. Although white-rot fungi degrading xenobiotic compounds was originally attributed to its lignin-degrading enzyme system, Ide et al. [39] suggested that cytochrome P450 showed a novel catalytic property in xenobiotic metabolism. All annotated enzymes or degradation processes could be found in *P. betulinus* transcriptome, demonstrating that *P. betulinus* could not only harbor metabolic xenobiotic materials but also potentially degradation of environmental pollutants.

5. Conclusions

To our knowledge, this was the first study to describe the changes in transcriptome of brown-rot fungus *P. betulinus* when growing on birch saw dust. After an in-depth RNA-seq analysis, we obtained and annotated 3351 assembled unigenes to at least one database. A large number of potential lignicellulose-degrading enzyme-relevant genes were identified, suggesting that this species is suitable for wood degradation. In summary, our analysis results obtained from the transcriptome data provide an insight into the genetic background of brown-rot fungus *P. betulinus*.

Supplementary Materials: The following are available online at www.mdpi.com/1999-4907/8/10/374/s1, Table S1: The primers designed for qRT-PCR analysis, Table S2: Overview of output statistics on *P. betulinus* transcriptome sequencing, Table S3: The top 20 representative metabolic pathways of *P. betulunus*, Table S4: The list of significantly differently expressed genes, Table S5: The metabolic pathway of differently expressed genes in samples, Table S6: All the assembled unigenes functionally annotated in Nr database.

Acknowledgments: This study was supported by the National Natural Science Fund Project (31700410), Inner Mongolia Natural Science Fund Project (2016BS0303), Inner Mongolia Science and Technology Research Project of Higher Education (NJZY16298), the Fundamental Research Funds for the Central Universities (2572017AA23), the National Natural Science Foundation of China (31170568).

Author Contributions: Qiuyu Wang conceived and designed the experiments; Mu Peng and Syed Sadaqat Shah performed the experiments and analyzed the data; Lixia Yang wrote the paper.

Conflicts of Interest: The authors declare no conflict of interest.

References

1. Lemieszek, M.; Langner, E.; Kaczor, J.; Kandefer-Szerszen, M.; Sanecka, B.; Mazurkiewicz, W.; Rzeski, W. Anticancer effect of fraction isolated from medicinal birch polypore mushroom, *Piptoporus betulinus* (bull.: Fr.) p. Karst. (aphyllophoromycetideae): In Vitro studies. *Int. J. Med. Mushrooms* **2009**, *11*, 351–364. [CrossRef]
2. Wangun, H.V.K.; Berg, A.; Hertel, W.; Nkengfack, A.E.; Hertweck, C. Anti-inflammatory and anti-hyaluronate lyase activities of lanostanoids from *Piptoporus betulinus*. *J. Antibiot.* **2004**, *57*, 755–758. [CrossRef] [PubMed]
3. Meng, F.; Liu, X.; Wang, Q. Identification of wood decay related genes from *Piptoporus betulinus* (bull. Fr.) karsten using differential display reverse transcription pcr (ddrt-pcr). *Biotechnol. Biotechnol. Equip.* **2012**, *26*, 2961–2965. [CrossRef]
4. Shang, J.; Yan, S.; Wang, Q. Degradation mechanism and chemical component changes in *Betula platyphylla* wood by wood-rot fungi. *BioResources* **2013**, *8*, 6066–6077. [CrossRef]
5. Wymelenberg, A.V.; Gaskell, J.; Mozuch, M.; Bondurant, S.S.; Sabat, G.; Ralph, J.; Skyba, O.; Mansfield, S.D.; Blanchette, R.A.; Grigoriev, I.V.; Kersten, P.J.; Cullen, D. Significant alteration of gene expression in wood decay fungi *Postia placenta* and *Phanerochaete chrysosporium* by plant species. *Appl. Environ. Microbiol.* **2011**, *77*, 4499–4507. [CrossRef] [PubMed]
6. Klinke, H.B.; Thomsen, A.B.; Ahring, B.K. Inhibition of ethanol-producing yeast and bacteria by degradation products produced during pre-treatment of biomass. *Appl. Microbiol. Biotechnol.* **2004**, *66*, 10–26. [CrossRef] [PubMed]
7. Okamoto, K.; Kanawaku, R.; Masumoto, M.; Yanase, H. Efficient xylose fermentation by the brown rot fungus *Neolentinus lepideus*. *Enzym. Microb. Technol.* **2012**, *50*, 96–100. [CrossRef] [PubMed]
8. Dashtban, M.; Schraft, H.; Qin, W. Fungal bioconversion of lignocellulosic residues; opportunities & perspectives. *Int. J. Biol. Sci.* **2009**, *5*, 578–595. [PubMed]
9. Canam, T.; Town, J.; Iroba, K.; Tabil, L.; Dumonceaux, T. Pretreatment of lignocellulosic biomass using microorganisms: Approaches, advantages, and limitations. *Sustain. Degrad. Lignocellul. Biomass-Techn. Appl. Commer.* **2013**. [CrossRef]
10. Zeng, X.X.; Tang, J.X.; Yin, H.Q.; Liu, X.D.; Pei, J.; Liu, H.W. Isolation, identification and cadmium adsorption of a high cadmium-resistant *Paecilomyces lilacinus*. *Afr. J. Biotechnol.* **2010**, *9*, 6525–6533.
11. Kim, S.Y.; Park, S.Y.; Ko, K.S.; Jung, H.S. Phylogenetic analysis of antrodia and related taxa based on partial mitochondrial ssu rDNA sequences. *Antonie Leeuwenhoek* **2003**, *83*, 81–88. [CrossRef] [PubMed]
12. Korripally, P.; Hunt, C.G.; Houtman, C.J.; Jones, D.C.; Kitin, P.J.; Cullen, D.; Hammel, K.E. Regulation of gene expression during the onset of ligninolytic oxidation by *Phanerochaete chrysosporium* on spruce wood. *Appl. Environ. Microbiol.* **2015**, *81*, 7802–7812. [CrossRef] [PubMed]
13. Kuuskeri, J.; Häkkinen, M.; Laine, P.; Smolander, O.P.; Tamene, F.; Miettinen, S.; Nousiainen, P.; Kemell, M.; Auvinen, P.; Lundell, T. Time-scale dynamics of proteome and transcriptome of the white-rot fungus Phlebia radiata: Growth on spruce wood and decay effect on lignocellulose. *Biotechnol. Biofuels* **2016**, *9*, 192. [CrossRef] [PubMed]
14. Gaskell, J.; Blanchette, R.A.; Stewart, P.E.; Bondurant, S.; Adams, M.; Sabat, G.; Kersten, P.; Cullen, D. Transcriptome and secretome analyses of the wood decay fungus *Wolfiporia cocos* support alternative mechanisms of lignocellulose conversion. *Appl. Environ. Microbiol.* **2016**, *82*, 3979–3987. [CrossRef] [PubMed]

15. Skyba, O.; Cullen, D.; Douglas, C.J.; Mansfield, D. Gene expression patterns of wood decay fungi *Postia placenta* and *Phanerochaete chrysosporium* are influenced by wood substrate composition during degradation. *Appl. Environ. Microbiol.* **2016**, *82*, 4387–4400. [CrossRef] [PubMed]
16. Li, R.; Zhu, H.; Ruan, J.; Qian, W.; Fang, X.; Shi, Z.; Li, Y.; Li, S.; Shan, G.; Kristiansen, K. De novo assembly of human genomes with massively parallel short read sequencing. *Genome Res.* **2010**, *20*, 265–272. [CrossRef] [PubMed]
17. Pertea, G.; Huang, X.; Liang, F.; Antonescu, V.; Sultana, R.; Karamycheva, S.; Lee, Y.; White, J.; Cheung, F.; Parvizi, B. Tigr gene indices clustering tools (TGICL): A software system for fast clustering of large est datasets. *Bioinformatics* **2003**, *19*, 651–652. [CrossRef] [PubMed]
18. Conesa, A.; Götz, S.; García-Gómez, J.M.; Terol, J.; Talón, M.; Robles, M. Blast2Go: A universal tool for annotation, visualization and analysis in functional genomics research. *Bioinformatics* **2005**, *21*, 3674–3676. [CrossRef] [PubMed]
19. Ye, J.; Fang, L.; Zheng, H.; Zhang, Y.; Chen, J.; Zhang, Z.; Wang, J.; Li, S.; Li, R.; Bolund, L. Wego: A web tool for plotting go annotations. *Nucleic Acids Res.* **2006**, *34*, W293–W297. [CrossRef] [PubMed]
20. Trapnell, C.; Roberts, A.; Goff, L.; Pertea, G.; Kim, D.; Kelley, D.R.; Pimentel, H.; Salzberg, S.L.; Rinn, J.L.; Pachter, L. Differential gene and transcript expression analysis of RNA-seq experiments with tophat and cufflinks. *Nat. Protoc.* **2012**, *7*, 562–578. [CrossRef] [PubMed]
21. Deng, Y.; Yao, J.; Wang, X.; Guo, H.; Duan, D. Transcriptome sequencing and comparative analysis of *Saccharina japonica* (laminariales, phaeophyceae) under blue light induction. *PLoS ONE* **2012**, *7*, e39704. [CrossRef] [PubMed]
22. Livak, K.J.; Schmittgen, T.D. Analysis of relative gene expression data using real-time quantitative PCR and the $2^{-\Delta\delta Ct}$ method. *Methods* **2001**, *25*, 402–408. [CrossRef] [PubMed]
23. Cong, B.; Wang, N.; Liu, S.; Liu, F.; Yin, X.; Shen, J. Isolation, characterization and transcriptome analysis of a novel Antarctic *Aspergillus sydowii* strain MS-19 as a potential lignocellulosic enzyme source. *BMC Microbiol.* **2017**, *17*, 129. [CrossRef] [PubMed]
24. Schmidt, E.; Knackmuss, H.J. Chemical structure and biodegradability of halogenated aromatic compounds. Conversion of chlorinated muconic acids into maleoylacetic acid. *Biochem. J.* **1980**, *192*, 339. [CrossRef] [PubMed]
25. Carpentier, S.C.; Coemans, B.; Podevin, N.; Laukens, K.; Witters, E.; Matsumura, H. Functional genomics in a non-model crop: Transcriptomics or proteomics? *Physiol. Plant.* **2008**, *133*, 117. [CrossRef] [PubMed]
26. Hori, C.; Gaskell, J.; Igarashi, K.; Samejima, M.; Hibbett, D.; Henrissat, B.; Cullen, D. Genomewide analysis of polysaccharides degrading enzymes in 11 white-and brown-rot Polyporales provides insight into mechanisms of wood decay. *Mycologia* **2013**, *105*, 1412–1427. [CrossRef] [PubMed]
27. Martinez, D.; Challacombe, J.; Morgenstern, I.; Hibbett, D.; Schmoll, M.; Kubicek, C.P.; Ferreira, P.; Ruiz-Duenas, F.J.; Martinez, A.T.; Kersten, P. Genome, transcriptome, and secretome analysis of wood decay fungus *Postia placenta* supports unique mechanisms of lignocellulose conversion. *Proc. Natl. Acad. Sci. USA* **2009**, *106*, 1954–1959. [CrossRef] [PubMed]
28. Rytioja, J.; Hildén, K.; Yuzon, J.; Hatakka, A.; de Vries, R.P.; Mäkelä, M.R. Plant-polysaccharide-degrading enzymes from basidiomycetes. *Microbiol. Mol. Biol. Rev.* **2014**, *78*, 614–649. [CrossRef] [PubMed]
29. Yelle, D.J.; Wei, D.; John, R.; Hammel, K.E. Multidimensional nmr analysis reveals truncated lignin structures in wood decayed by the brown rot basidiomycete *Postia placenta*. *Environ. Microbiol.* **2011**, *13*, 1091–1100. [CrossRef] [PubMed]
30. Zhang, J.; Presley, G.N.; Hammel, K.E.; Ryu, J.S.; Menke, J.R.; Figueroa, M.; Hu, D.; Orr, G.; Schilling, J.S. Localizing gene regulation reveals a staggered wood decay mechanism for the brown rot fungus *Postia placenta*. *Proc. Natl. Acad. Sci. USA* **2016**, *113*, 10968. [CrossRef] [PubMed]
31. Tuor, U.; Winterhalter, K.; Fiechter, A. Enzymes of white-rot fungi involved in lignin degradation and ecological determinants for wood decay. *J. Biotechnol.* **1995**, *41*, 1–17. [CrossRef]
32. Lee, D.; Meyer, K.; Chapple, C.; Douglas, C.J. Antisense suppression of 4-coumarate:Coenzyme a ligase activity in Arabidopsis leads to altered lignin subunit composition. *Plant Cell* **1997**, *9*, 1985–1998. [CrossRef] [PubMed]
33. An, H.D.; Wei, D.S.; Xiao, T.T. Transcriptional profiles of laccase genes in the brown rot fungus *Postia placenta* mad-r-698. *J. Microbiol.* **2015**, *53*, 1–10. [CrossRef] [PubMed]

34. Prince, R.C. Bioremediation: An overview of how microbiological processes can be applied to the cleanup of organic and inorganic environmental pollutants. In *Encyclopedia of Environmental Microbiology*; John Wiley & Sons, Inc.: Hoboken, NJ, USA, 2003.

35. Wang, C.; Jin-Ying, X.I.; Hong-Ying, H.U.; Wen, X.H. Biodegradation of gaseous chlorobenzene by white-rot fungus *Phanerochaete chrysosporium*. *Biomed. Environ. Sci.* **2008**, *21*, 474–478. [CrossRef]

36. Hadibarata, T.; Teh, Z.C.; Rubiyatno; Zubir, M.M.; Khudhair, A.B.; Yusoff, A.R.; Salim, M.R.; Hidayat, T. Identification of naphthalene metabolism by white rot fungus *Pleurotus eryngii. Bioprocess Biosyst. Eng.* **2013**, *36*, 1455–1461. [CrossRef] [PubMed]

37. Hadibarata, T.; Yusoff, A.R.M.; Aris, A.; Kristanti, R.A. Identification of naphthalene metabolism by white rot fungus *Armillaria* sp.F022. *J. Environ. Sci.* **2012**, *24*, 728–732. [CrossRef]

38. Krueger, M.C.; Hofmann, U.; Moeder, M.; Schlosser, D. Potential of wood-rotting fungi to attack polystyrene sulfonate and its depolymerisation by *Gloeophyllum trabeum* via hydroquinone-driven fenton chemistry. *PLoS ONE* **2015**, *10*, e0131773. [CrossRef] [PubMed]

39. Ide, M.; Ichinose, H.; Wariishi, H. Molecular identification and functional characterization of cytochrome P450 monooxygenases from the brown-rot basidiomycete *Postia placenta. Arch. Microbiol.* **2012**, *194*, 243–253. [CrossRef] [PubMed]

© 2017 by the authors. Licensee MDPI, Basel, Switzerland. This article is an open access article distributed under the terms and conditions of the Creative Commons Attribution (CC BY) license (http://creativecommons.org/licenses/by/4.0/).

forests

MDPI

Article

Genetic Variation, Heritability and Genotype × Environment Interactions of Resin Yield, Growth Traits and Morphologic Traits for *Pinus elliottii* at Three Progeny Trials

Meng Lai [1,†], Leiming Dong [2,†], Min Yi [1], Shiwu Sun [3], Yingying Zhang [1], Li Fu [1], Zhenghua Xu [1], Lei Lei [1], Chunhui Leng [1] and Lu Zhang [1,*]

1   Key Laboratory of Silviculture, Co-Innovation Center of Jiangxi Typical Trees Cultivation and Utilization, College of Forestry, Jiangxi Agricultural University, Nanchang 330045, China; laimeng21@163.com (M.L.); yimin6104@163.com (M.Y.); 15797659575@163.com (Y.Z.); fuli1096@126.com (L.F.); fzxzh103@163.com (Z.X.); 13576002992@163.com (L.L.); lchunhui520@163.com (C.L.)
2   State Key Laboratory of Tree Genetics and Breeding, Research Institute of Forestry, Chinese Academy of Forestry, Beijing 100091, China; dongleiming2008@126.com
3   Baiyun Mountain State Owned Forest Farm, Jian 343011, China; awei0723@126.com
*   Correspondence: zhanglu856@mail.jxau.edu.cn; Tel.: +86-791-8381-3733
†   These authors contributed equally to this work.

Received: 26 July 2017; Accepted: 25 October 2017; Published: 30 October 2017

Abstract: To better understand the genetic control of resin yield, growth traits and morphologic traits for *Pinus elliottii* families, genetic relationships among these traits were examined in three 27-year-old progeny trials located in Jingdezhen, Jian and Ganzhou, Jiangxi Province, China. In total, 3695 trees from 112 families were assessed at the three sites. Significant site, family and family × site effects were found for resin yield, growth traits and morphologic traits. Resin yield and growth traits were found to be under moderate genetic control for the three sites combined, with family heritability and individual narrow-sense heritability ranging from 0.41 to 0.55 and 0.11 to 0.27, respectively. The coefficient of genotypic variation (CV_G) of stem volume (SV) and crown surface area (CSA) were higher than those of other traits at each site. Genetic correlation estimates indicated that selection for growth traits might lead to a large increment in resin yield (RY), and most morphologic traits had moderate to strong correlations with growth traits at each individual site. One possible strategy in tree breeding would be to maximize resin production through selection for growth traits.

Keywords: families; *Pinus elliottii*; resin yield; growth traits; morphologic traits; heritability; genotypic correlation

1. Introduction

Pine resin is an important non-timber secondary forest product. It produces turpentine (monoterpenes and sesquiterpenes) and rosin (diterpenes). These compounds are widely used in pharmaceutical, cosmetics, food, chemical and other such industries [1–3]. China is the leading producer of resin in the world. During the 1990s, China annually exported a total of 200,000 tons of resin to more than 40 countries, accounting for about 50% of the resin traded in the world [4]. Of the pine resin resources, *Pinus elliottii* is one of the main resin tree species in China.

P. elliottii (slash pine), which originated from the Southeastern United States, is one of the most important coniferous timber species in the *Pinus* genus. Slash pine was introduced to China in the late 1940s and began to be planted in Southern China on a large scale in the late 1970s [5]. In addition to its wide use for the wood, pulp and paper industry, this species has also long been employed as a main source of resin. In recent years, the resin price has risen annually. For example, the resin purchased from forest farmers has an average price of 10 yuan (about 1.5 dollars) per kilogram in 2016. Nowadays, it is more profitable for farmers than harvesting trees for timber. However, due to its high commercial value, slash pine has been subjected to overexploitation during past decades.

The quantity of constitutive resin produced by pine species is influenced by genetic and environmental factors [6]. Much of the attention devoted for identifying the variables that have a bearing on resin production has concentrated on environmental attributes and cultural practices aimed to stimulate tree growth. A prominent focus of this research includes potential determinants that might alter resin production over the course of a growing season, including those that can cause carbon resources to be shifted to secondary metabolic processes involved in resin synthesis away from the dominant primary processes that contribute to tree growth. Allocation of photosynthates to primary and secondary metabolic processes is believed to result in a trade-off between growth and resin formation following plant growth differentiation principles formulated first by Loomis [7] and later modified to explain the relationship between growth and resin flow in pines by Lorio [8]. Previous studies have suggested that growth traits and morphologic traits can influence resin yield, with trees with a larger diameter and crown size yielding more resin than their smaller pine species counterparts [9,10]. Thus, accurately exploring the genetic correlation among resin yield, growth traits and morphologic traits is needed, as they have major implications for the development of selection and breeding strategies.

Similarly, a large effort has also been committed to determine how strongly genetic effects influence variation of resin yield in pine trees. Researches have demonstrated that resin yield is a highly heritable trait, and important genetic gains can be obtained from the selection of high resin yielders. For example, Roberds and Strom [11] estimated that the repeatabilities of gum yield in loblolly pine (*Pinus teada* L) (0.64–0.67), longleaf pine (*Pinus palustris*) (0.46–0.77) and slash pine (0.55) are quite high. Squillace and Bengston [12] found that narrow-sense heritability for gum resin yield in slash pine is around 0.55, indicating that this trait is under fairly strong genetic control. Experimental observations in maritime pine (*Pinus pinaster* Ait.) and masson pine (*Pinus massoniana*) likewise indicate that resin-yielding capability is under substantial genetic regulation in these species [13,14]. General methods for increasing resin yield including genetically improve associated trees and develop resin pine plantations. In China, the genetic improvement to increase resin yield was initiated in the mid-to-late 1980s, and these programs primarily focus on the selection of trees that yield high amounts of resin based on the phenotypic performance of unimproved plantations [14]. Research has demonstrated that large differences occur among half-sib families in resin yield, and some superior families and clones for further studies on genetic improvement have been selected [15,16]. However, the majority of seedlings used for afforestation have not been subject to genetic improvement, and pine plantations have made limited contributions to the increment of resin yields. Slash pine improvement programs to promote resin yield are still nascent, and more breeding practices needed to be made for this species.

In China, tree breeding programs for slash pine have mainly emphasized improvements in tree growth and wood properties [2,17]. Several studies have focused on the resin-yielding capacity. However, most of these studies on resin yield and growth traits have either been performed at a single site or are based on a small number of samples or families. Consequently, limited information is available on the ecological variation, reaction norms and phenotypic plasticity of the resin yield, growth traits and morphologic traits of *P. elliottii* families. Moreover, the sample size in these previous studies was insufficient. For tree improvement, quantitative parameter estimates require an examination of more genotypes at different locations than those used in previous studies.

To understand the quantitative genetics for resin yield, growth traits and morphologic traits of *P. elliottii* in China, we conducted a large study using 3695 trees from 112 open-pollinated families established at three sites in Jiangxi Province with the following study aims: (1) to describe the resin yield, growth traits and morphologic traits in the studied area; (2) to quantify genetic variation and inheritance for resin yield, growth traits and wood traits; (3) to explore the genotype × environment interactions for these traits; (4) to examine the genetic relationship among resin yield, growth traits and wood traits. These findings will be helpful to decide the appropriate breeding strategies for improvement programs of slash pine in China.

2. Materials and Methods

2.1. Materials

The present study comprised of 112 open-pollinated families of *P. elliottii* plus trees randomly selected from a seed orchard in Georgia (12 plus trees), Mississippi (60 plus trees) and Florida (50 plus trees) in United States. Plus trees were selected for high resin yield, high growth rate, and straight stem form. Open-pollinated families from these trees were included in three trials established in the spring of 1990 at three locations in Jiangxi Province, Jingdezhen, Jian and Ganzhou, and the characteristics of the three trials are presented in Figure 1 and Table 1. The annual temperature and rainfall data in Table 1 were collected from the local meteorological bureau. The trials used a randomized complete block design with plots containing four trees per row and five replications, with a total of 2240 trees per site. The seedlings were planted with 3.0 m × 4.0 m spacing. Specific silvicultural treatments were not performed prior to the experiments. In 2016, all trees were measured; the survival rate for Jindezhen, Jian and Ganzhou were 55.8%, 56.1% and 53.1%, respectively; the survival rate was not high for the three trials, which reflected the snow disaster in 2008. The total number of trees sampled for the three sites were 1249, 1257 and 1189, respectively.

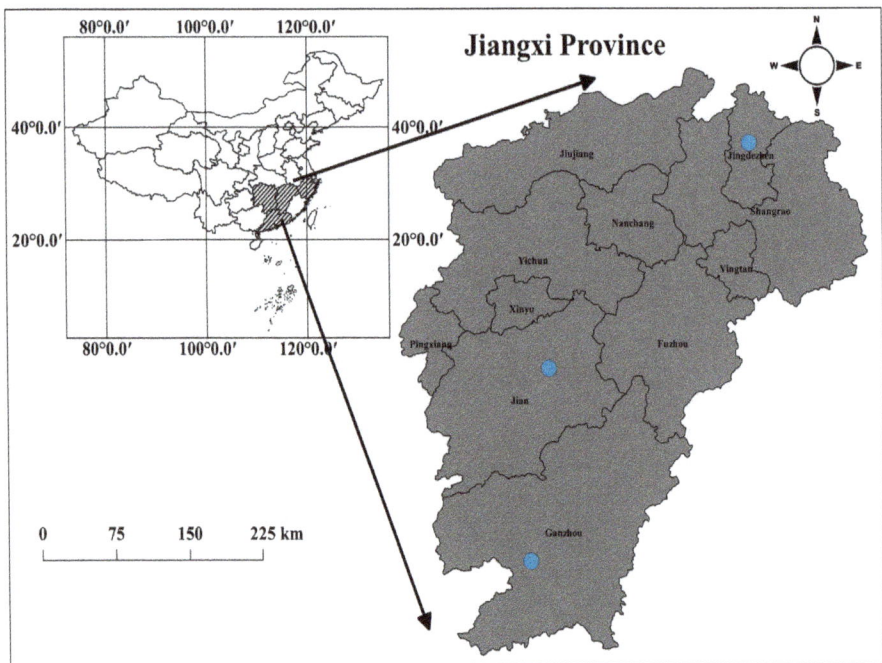

Figure 1. Location of the progeny trails (blue dots) in Jiangxi Province and main growing regions (shaded area) of *P. elliottii* in China.

Table 1. Location, climatic conditions and description of the three progeny trials.

Site	Latitude N (°)	Longitude W (°)	Mean Annual T (°C)	Rainfall mm/year	Survival (%)	Sample Trees
Jingdezhen	29.37	117.25	17.2	1805	55.8	1249
Jian	27.22	115.13	18.3	1487	56.1	1257
Ganzhou	25.38	114.93	20.2	1318	53.1	1189

2.2. Measurement of Resin Yield, Tree Growth and Evaluation of Morphologic Properties

Resin yield (RY) was measured using the bark streak wounding method for resin tapping [16]. After removing the outer bark at breast height using a sharp hatchet, a streak was incised in the bark-shaved face. The streak was 2 mm wide and 5 mm deep and had a side gutter angle of 45° and tapping load rate of less than 45% (Figure 2). Resin from each wound was funneled into open plastic bags attached to the base of the wounds. Rewounding was performed every 2 days throughout the experimental period in August, with 15 cuts performed on each tree. The sampled trees were measured for RY (in kilograms) at the end of August.

Figure 2. Bark streak wounding method for resin tapping.

Total height (HT) was measured in meters. Stem diameter at breast height (DBH) was measured in centimeter. The height (HGT) and DBH were used to calculate the stem volume (stem volume (SV), in m^3) of the outer bark of individual trees:

$$SV = \frac{0.527 \times \pi D^2 H}{4} \tag{1}$$

Morphologic traits including height under live crown (HLC), crown width (CW) and crown length (CL). The CW and CL were used to calculate the crown surface area (CSA) of an individual tree according to Chen et al. [18] and Huang et al. [19]:

$$CVA = \frac{\pi CW}{4} \sqrt{CL^2 + \frac{CW^2}{4}} \tag{2}$$

2.3. Statistical Models and Analysis

Univariate restricted maximum likelihood (REML) analyses were undertaken separately to obtain estimates of variance components for each trait using statistical software ASReml-R (v. 3.0, Queensland Department of Primary Industries and Fisheries, Brisbane, Australia) package in R [20,21]. The following linear mixed models (family models) were used for separate analyses (1) of individual site and for joint analyses (2) of the three sites together:

$$y = \mu + Xb + Z_1 f + Z_2 fb + e \tag{3}$$

$$y = \mu + X_1 s + X_2 b_{(s)} + Z_1 f + Z_2 fs + Z_3 fb_{(s)} + e \tag{4}$$

where y is the vector of observations, μ is the intercept, s, b, $b(s)$ are the fixed site, block and block within site effects, f, fb, fs, $fb(s)$, e refers to random family, family-site, family-block and family-block within site and residual effects, and X and Z are the known incidence matrices relating the observations in y to the fixed and random effects, respectively, assuming $f \sim N(0, \sigma_f^2 I)$, $fs \sim N(0, \sigma_{fs}^2 I)$, $fb_{(s)} \sim N(0, \sigma_{fb_{(s)}}^2 I)$, $e \sim N(0, \sigma_e^2 I)$, where σ_f^2, σ_{fs}^2, $\sigma_{fb_{(s)}}^2$, and σ_e^2 are family, family-site, family-block and family-block within site and residual error variances, respectively, I is an identity matrix.

The significance of the variance components for each trait was tested using the likelihood ratio test (LRT, [22]). Approximate standard errors for estimated variances were estimated using the Taylor series expansion method [23].

Open pollinated mating was assumed to have created true half-sibs and among which the genetic variation accounts for one quarter of the total additive genetic variance, thus additive genetic variance

$$\hat{\sigma}_a^2 = 4\hat{\sigma}_f^2 \tag{5}$$

phenotypic variance

$$\hat{\sigma}_p^2 = \hat{\sigma}_f^2 + \hat{\sigma}_{fb}^2 + \hat{\sigma}_e^2 \tag{6}$$

and

$$\hat{\sigma}_p^2 = \hat{\sigma}_f^2 + \hat{\sigma}_{fs}^2 + \hat{\sigma}_{fb_{(s)}}^2 + \hat{\sigma}_e^2 \tag{7}$$

for individual site and joint analyses, respectively. Variation at genetic, phenotypic and residual level (CV_G, CV_P and CV_E), individual narrow-sense heritability ($\hat{h}_i^2$), and family mean heritability ($\hat{h}_f^2$) were calculated, respectively, as follows [14]:

$$CV_G = \hat{\sigma}_f / \overline{X} \times 100\% \tag{8}$$

$$CV_P = \hat{\sigma}_p / \overline{X} \times 100\% \tag{9}$$

$$CV_E = \hat{\sigma}_e / \overline{X} \times 100\% \tag{10}$$

$$\hat{h}_i^2 = \frac{\hat{\sigma}_a^2}{\hat{\sigma}_p^2} \tag{11}$$

$$\hat{h}_f^2 = \frac{\hat{\sigma}_f^2}{\hat{\sigma}_f^2 + \hat{\sigma}_{fb}^2 / n_b + \hat{\sigma}_e^2 / (n_b \times n_k)} \tag{12}$$

$$\hat{h}_f^2 = \frac{\hat{\sigma}_f^2}{\hat{\sigma}_f^2 + \hat{\sigma}_{fs}^2/n_s + \hat{\sigma}_{fb_{(s)}}^2/(n_s \times n_b) + \hat{\sigma}_e^2/(n_s \times n_b \times n_k)} \qquad (13)$$

Formula (10) and (11) were used in individual site and joint analyses, respectively. Terms n_s, n_b and n_k are the number of sites, blocks within sites and the harmonic mean of the number of trees per family, respectively.

Genetic correlations (r_g) were estimated using bivariate analyses for all possible combinations of traits and calculated as [14]:

$$r_g = \frac{\hat{\sigma}_{g(xy)}}{\sqrt{\hat{\sigma}_{g(x)}^2 \times \hat{\sigma}_{g(y)}^2}} \qquad (14)$$

where the $\hat{\sigma}_{g(xy)}$ refers to the genotypic covariance between traits x and y, $\hat{\sigma}_{g(x)}^2$ denotes the genotypic variance component for trait x and $\hat{\sigma}_{g(y)}^2$ is the genotypic variance component for trait y.

To evaluate the extent of genotype × environment interaction (G × E) for each trait, between-site type-B genetic correlations obtained from bivariate models were used [24], and their approximate standard errors were estimated using the Taylor series expansion method [23]. The type-B genetic correlation was calculated as:

$$r_B = \frac{\hat{\sigma}_{g(a1a2)}}{\sqrt{\hat{\sigma}_{g(a1)}^2 \times \hat{\sigma}_{g(a2)}^2}} \qquad (15)$$

where $\hat{\sigma}_{g(a1a2)}$ is the covariance between genotypic effects from bivariate models of the same traits in different sites and $\hat{\sigma}_{g(a1)}^2$ and $\hat{\sigma}_{g(a2)}^2$ are estimated genotypic variances for the same traits in trial 1 and trial 2, respectively.

To evaluate the expected response in resin yield at each site by using different selection criteria, and developed implications for tree improvement. The expected gain in trait y was predicted from the correlated response to individual selection in trait x using the following formula [25]:

$$\Delta G = i\,\hat{h}_x\,\hat{h}_y\,r_g\,\hat{\sigma}_{py} \qquad (16)$$

where i is the selection intensity (according to the sample size in this study, the selection intensity is assumed to equal 1.76, which corresponds to the selection of ten parental trees out of 100); $\hat{h}_x$ and $\hat{h}_y$ are the square roots of individual tree heritability for direct-selection x trait and correlated y trait; r_g is the estimated genetic correlation between traits x and y; and $\hat{\sigma}_{py}$ is the phenotypic standard deviation for trait y.

3. Results

3.1. Mean Values and Site Effects

The mean values, range of variation and standard error of the mean for all traits in each of the three progeny trials are presented in Table 2. The trees at the Jian site had the lowest RY. The highest RY was observed at the Ganzhou site, with a 6.9 difference was observed between the lowest and highest average RY (Table 2). At the Jingdezhen site, the HGT of *P. elliottii* was the highest but the DBH was the lowest; thus, the SV of *P. elliottii* was not high at this site. Nevertheless, at the Jingdezhen site, the CSA was the highest, which was supported by the highest CW and moderate CL at this site, whereas at the Jian site, the CSA was the lowest. The HLC of trees at the Jingdezhen site was 33.66% higher than that of the tress at the Ganzhou site. An analysis that combined all three trials showed significant site effects ($p < 0.01$) for RY, growth traits and morphologic traits (Table 3).

3.2. Genetic Variation and Heritability

Significant differences among families in the RY, growth traits and morphologic traits were observed when analyzing the three sites together (Table 3). Growth traits showed higher estimated family heritability (h^2_f = 0.42–0.55) and estimated individual tree heritability (h^2_i = 0.14–0.27) than resin yield and most morphologic traits (Table 3).

The estimates of variances, heritability, coefficients of genotypic, residual, and phenotypic variations at each individual site are presented in Table 4. Significant differences were observed among the families for most of the traits examined at the three sites except for the DBH, HLC and CW at Jian and the RY at Ganzhou. In general, the coefficient of genetic variation (CV$_G$) values for the SV and CSA were higher than those for other traits at each site (Table 4). The family and individual tree heritabilities for all traits at the Jindezhen site ranged from 0.46 to 0.65 and 0.41 to 0.73, respectively, which were high compared with those estimated at the other two sites.

Table 2. Mean values, standard error of the mean (SE) and range of variation based on individual values of the resin yield, growth traits and morphologic traits of *P. elliottii* for the three progeny trials.

Trait (Site)	Jingdezhen			Jian			Ganzhou		
	Mean	SE	Range (Min–Max)	Mean	SE	Range (Min–Max)	Mean	SE	Range (Min–Max)
RY (kg)	1.17	0.006	0.3–2.2	1.16	0.011	0.3–2.8	1.24	0.009	0.3–3.25
HGT (m)	14.68	0.083	6.4–23.0	13.02	0.072	5.4–18.6	12.15	0.061	5.8–16.7
DBH (cm)	25.26	0.114	13.7–39.5	27.82	0.132	16.2–43.9	28.05	0.120	17.5–41.8
SV (m³)	0.418	0.005	0.06–1.37	0.441	0.005	0.09–1.12	0.412	0.004	0.11–0.96
HLC (m)	9.69	0.059	3.1–16.8	7.51	0.055	1.8–13.1	7.25	0.054	1.9–12.7
CW (m)	5.37	0.029	2.1–8.9	4.71	0.028	2.1–8.4	5.36	0.030	2.4–9.9
CL (m)	4.98	0.045	1.0–15.0	5.50	0.043	0.7–10.6	4.89	0.040	0.8–9.5
CSA (m²)	24.9	0.316	4.7–106.0	22.9	0.258	3.2–61.4	24.2	0.254	6.0–65.7

Resin yield (RY), diameter at breast height (DBH), height (HGT), stem volume (SV), height under live crown (HLC), crown width (CW), crown length (CL) and crown surface area (CSA).

Table 3. Estimates of variances ($\hat{\sigma}^2_f$, $\hat{\sigma}^2_{fs}$, $\hat{\sigma}^2_{fb_{(s)}}$, $\hat{\sigma}^2_e$) for family, family-site, family-block within site interactions and residual variances, respectively, family mean and individual narrow sense heritability ($\hat{h}^2_f$, $\hat{h}^2_i$) (with standard error in parenthesis), and coefficient of variation at genetic, phenotypic and residual level (CV$_G$, CV$_P$ and CV$_E$) for various traits from site combined analysis.

Trait	*p* Value of Site Effect	$\hat{\sigma}^2_f$ (SE)	$\hat{\sigma}^2_{fs}$ (SE)	$\hat{\sigma}^2_{fb_{(s)}}$ (SE)	$\hat{\sigma}^2_e$ (SE)	h^2_f (SE)	h^2_i (SE)	CV$_G$ (%)	CV$_E$ (%)	CV$_P$ (%)
RY	<0.01 **	0.003 (0.001)	0.003 (0.001)	0.006 (0.002)	0.1 (0.003)	0.41 (0.09)	0.11 (0.04)	4.58	26.45	28.00
HGT	<0.001 ***	0.45 (0.12)	0.33 (0.11)	2.07 (0.16)	3.86 (0.12)	0.55 (0.07)	0.27 (0.06)	5.06	14.77	19.46
DBH	<0.001 ***	0.65 (0.21)	0.56 (0.23)	1.06 (0.37)	16.3 (0.49)	0.45 (0.09)	0.14 (0.04)	2.99	14.94	15.94
SV	<0.001 ***	0.001 (0.001)	0.002 (0.001)	0.005 (0.001)	0.03 (0.001)	0.42 (0.09)	0.16 (0.05)	7.45	39.43	44.71
HLC	<0.001 ***	0.26 (0.06)	0.04 (0.06)	1.34 (0.1)	2.31 (0.07)	0.59 (0.06)	0.26 (0.06)	6.23	18.62	24.35
CW	<0.001 ***	0.03 (0.01)	0.04 (0.01)	0.09 (0.02)	0.88 (0.03)	0.41 (0.09)	0.13 (0.04)	3.53	18.20	19.83
CL	<0.001 ***	0.05 (0.03)	0.04 (0.17)	0.34 (0.05)	1.73 (0.05)	0.26 (0.12)	0.08 (0.04)	4.31	25.63	29.45
CSA	<0.001 ***	2.91 (1.24)	6.26 (1.59)	13.5 (2.08)	73.07 (2.22)	0.35 (0.10)	0.12 (0.05)	7.10	35.60	40.75

Degrees of freedom are 2 for site, 12 for block. Resin yield (RY), diameter at breast height (DBH), height (HGT), stem volume (SV), height under live crown (HLC), crown width (CW), crown length (CL) and crown surface area (CSA). *** p < 0.001; ** 0.001 < p < 0.01, level of significance of effects.

Table 4. Estimates of variances ($\hat{\sigma}_f^2$, $\hat{\sigma}_{fb}^2$, $\hat{\sigma}_e^2$) for family, family-block and residual variances, respectively, family mean and individual narrow sense heritability ($\hat{h}_f^2$, $\hat{h}_i^2$) (with standard error in parenthesis), and coefficient of variation at genetic, phenotypic and residual level (CV_G, CV_P and CV_E) for various traits at each site.

Site	$\hat{\sigma}_f^2$ (SE)	$\hat{\sigma}_{fb}^2$ (SE)	$\hat{\sigma}_e^2$ (SE)	h^2_f (SE)	h^2_i (SE)	CV_G (%)	CV_E (%)	CV_P (%)
Jian								
RY	0.009 (0.003)	0.003 (0.006)	0.157 (0.008)	0.35 (0.09)	0.21 (0.08)	8.16	34.16	35.37
HGT	0.38 (0.20)	3.17 (0.34)	3.25 (0.17)	0.28 (0.11)	0.22 (0.11)	4.71	13.84	20.01
DBH	0.49 (0.35) [ns]	0.05 (0.72)	21.54 (1.08)	0.20 (0.11)	0.10 (0.06)	2.51	16.68	16.89
SV	0.001 (0.001)	0.002 (0.001)	0.035 (0.002)	0.27 (0.10)	0.15 (0.07)	7.17	42.52	44.17
HLC	0.000 (0.000) [ns]	2.59 (0.21)	1.49 (0.08)	0.00 (0.00)	0.00 (0.00)	0.00	16.23	26.67
CW	0.014 (0.016) [ns]	0.06 (0.03)	0.86 (0.04)	0.13 (0.12)	0.06 (0.05)	2.52	19.65	20.52
CL	0.09 (0.05)	0.21 (0.09)	2.03 (0.11)	0.27 (0.10)	0.16 (0.08)	5.48	25.88	27.74
CSA	2.22 (1.46)	4.57 (2.93)	75.21 (3.84)	0.21 (0.11)	0.11 (0.07)	6.51	37.87	39.54
Ganzhou								
RY	0.003 (0.002) [ns]	0.001 (0.004)	0.110 (0.006)	0.20 (0.11)	0.10 (0.06)	4.38	26.53	27.02
HGT	0.28 (0.12)	1.19 (0.20)	3.01 (0.16)	0.34 (0.10)	0.25 (0.10)	4.30	14.27	17.41
DBH	0.64 (0.30)	0.000 (0.000)	16.67 (0.72)	0.28 (0.09)	0.15 (0.06)	2.85	14.55	14.83
SV	0.001 (0.001)	0.001 (0.001)	0.023 (0.001)	0.33 (0.10)	0.20 (0.08)	7.67	36.99	38.34
HLC	0.28 (0.10)	0.84 (0.15)	2.39 (0.13)	0.41 (0.09)	0.32 (0.11)	7.35	21.30	25.83
CW	0.03 (0.02)	0.000 (0.000)	1.03 (0.04)	0.24 (0.11)	0.12 (0.06)	3.28	18.91	19.19
CL	0.06 (0.04)	0.30 (0.08)	1.59 (0.08)	0.22 (0.12)	0.13 (0.08)	5.04	25.71	28.45
CSA	4.11 (1.701)	4.16 ((2.70)	68.92 (3.56)	0.35 (0.10)	0.22 (0.08)	8.36	34.25	36.25
Jingdezhen								
RY	0.005 (0.002)	0.016 (0.002)	0.031 (0.002)	0.46 (0.08)	0.41 (0.11)	6.31	14.92	19.37
HGT	1.59 (0.34)	1.85 (0.31)	5.30 (0.28)	0.64 (0.05)	0.72 (0.13)	8.60	15.69	20.14
DBH	2.41 (0.57)	3.98 (0.63)	9.88 (0.53)	0.57 (0.06)	0.59 (0.12)	6.14	12.44	15.96
SV	0.008 (0.002)	0.01 (0.002)	0.03 (0.001)	0.61 (0.05)	0.68 (0.13)	21.36	37.65	50.10
HLC	0.63 (0.14)	0.64 (0.15)	3.09 (0.16)	0.59 (0.06)	0.57 (0.12)	8.16	18.12	21.52
CW	0.16 (0.04)	0.25 (0.04)	0.72 (0.04)	0.56 (0.06)	0.55 (0.12)	7.38	15.77	19.77
CL	0.47 (0.10)	0.50 (0.09)	1.58 (0.08)	0.65 (0.05)	0.73 (0.13)	13.75	25.16	32.03
CSA	20.32 (4.65)	31.88 (4.86)	74.39 (3.96)	0.60 (0.06)	0.64 (0.13)	18.10	34.62	45.17

Degrees of freedom are 4 for block. Resin yield (RY), diameter at breast height (DBH), height (HGT), stem volume (SV), height under live crown (HLC), crown width (CW), crown length (CL) and crown surface area (CSA); [ns] no significance of effects.

3.3. Genotype × Environment Interaction

In the present study, the family × site interaction was significant for RY, growth traits and morphologic traits (Table 3). Inter-site genotypic correlations between HLC with pairs of Jindezhen-Jian and Ganzhou-Jian sites were not estimated because significant differences in HLC were not observed among the families at Jian (Table 5). Most of inter-site genotypic correlations between the same traits at Jingdezhen-Jian and Jingdezhen-Ganzhou sites were moderate or weak (Table 5). However, the type B genotypic correlations between Ganzhou and Jian for all traits were higher, indicating that these traits were stable across Ganzhou and Jian. Compared with the resin yield and morphologic properties, the growth traits generally showed higher inter-site genotypic correlations (0.18–0.68).

Table 5. Type B genetic correlations (with standard error in parenthesis) between sites for the resin yield, growth traits and morphologic traits of *P. elliottii*.

Traits Correlations	RY	HGT	DBH	SV	HLC	CW	CL	CSA
JDZ-JA	0.34 (0.14)	0.53 (0.18) **	0.18 (0.22)	0.30 (0.18)	0.00 (0.00)	0.17 (0.21)	0.19 (0.18)	0.22 (0.19)
JDZ-GZ	0.29 (0.20)	0.48 (0.16) *	0.29 (0.19)	0.35 (0.19) *	0.43 (0.17) *	0.16 (0.20)	0.13 (0.18)	0.15 (0.18)
GZ-JA	0.55 (0.18) **	0.68 (0.12) **	0.54 (0.15) **	0.54 (0.12) *	0.00 (0.00)	0.37 (0.15) *	0.22 (0.13)	0.23 (0.13)

Type B genetic correlations were used to measure the importance of G × E interactions for RY and measure the growth traits and morphologic traits of *P. elliottii* from the combined sites analysis. Resin yield (RY), diameter at breast height (DBH), height (HGT), stem volume (SV), height under live crown (HLC), crown width (CW), crown length (CL) and crown surface area (CSA). JDZ (Jingdezhen), GZ (Ganzhou) and JA (Jian). ** $0.001 < p < 0.01$; * $0.01 < p < 0.05$, level of significance of effects.

3.4. Genotypic Correlations between Traits, Genetic Gain, and Correlated Genetic Response

The estimated genetic correlations among the RY, growth traits and morphologic traits are presented in Table 6. Strong and positive genotypic correlations between RY and HGT, DBH and SV were observed across the three sites. This finding suggests that selection for growth traits might lead to a large increment in RY. Moderate to strong and positive correlations were observed between RY and the morphologic traits at specific localities except between HLC and RY at Ganzhou, which was negatively correlated. Most morphologic traits showed moderate to strong correlations with growth traits at each individual site.

Predicted genetic gains, assuming observed individual tree heritability and genotypic correlations and correlated genetic response in resin yield, with different selection criteria used, are presented for individual sites in Table 7. The predicted gains in RY were comparable at Jian and Jindezhen (11.08% and 9.53%, respectively), whereas the Ganzhou site had a lower gain (4.18%). Selection for growth traits resulted in a higher gains in RY at Jindezhen (10.97–11.53%). Overall, predicted gains from growth traits selection in RY were higher than the selection for morphologic traits.

Table 6. Estimated intertrait genetic correlations (with standard error in parenthesis) among resin yield, growth traits and morphologic traits across the three study sites.

Trait	HGT	DBH	SV	HLC	CW	CL	CSA
Jian							
RY	0.68 (0.26) **	0.82 (0.18) ***	0.73 (0.17) ***	0.00 (0.00)	0.34 (0.4)	0.61 (0.22) **	0.56 (0.24) **
HGT		0.95 (0.14) ***	0.98 (0.11) ***	0.00 (0.00)	0.20 (0.61)	0.99 (0.17) ***	0.95 (0.33) ***
DBH			0.97 (0.07) ***	0.00 (0.00)	−0.11 (0.85)	0.97 (0.24) ***	0.60 (0.31) **
SV				0.00 (0.00)	−0.01 (0.60)	0.98 (0.15) ***	0.73 (0.23) ***
HLC					0.00 (0.00)	0.00 (0.00)	0.00 (0.00)
CW						0.67 (0.46) *	0.98 (0.18) ***
CL							0.90 (0.09) ***
Ganzhou							
RY	0.58 (0.16) **	0.72 (0.17) ***	0.58 (0.20) **	−0.22 (0.27)	0.89 (0.20) ***	0.92 (0.19) ***	0.95 (0.13) ***
HGT		0.53 (0.19) **	0.85 (0.08) ***	0.88 (0.05) ***	−0.06 (0.27)	0.31 (0.18) *	0.12 (0.20)
DBH			0.89 (0.05) ***	0.10 (0.23)	0.66 (0.20) **	0.88 (0.16) ***	0.78 (0.12) ***
SV				0.52 (0.16) **	0.37 (0.25) *	0.72 (0.14) ***	0.54 (0.16) ***
HLC					−0.50 (0.25) *	−0.19 (0.19)	−0.36 (0.18) *
CW						0.91 (0.24) ***	0.98 (0.08) ***
CL							0.98 (0.05) ***

Table 6. *Cont.*

Trait	HGT	DBH	SV	HLC	CW	CL	CSA
Jindezhen							
RY	0.90 (0.04) ***	0.96 (0.02) ***	0.94 (0.03) ***	0.67 (0.09) ***	0.66 (0.08) ***	0.88 (0.04) ***	0.86 (0.04) ***
HGT		0.95 (0.02) ***	0.98 (0.01) ***	0.88 (0.03) ***	0.62 (0.08) ***	0.84 (0.04) ***	0.81 (0.05) ***
DBH			0.99 (0.01) ***	0.74 (0.07) ***	0.66 (0.08) ***	0.91 (0.03) ***	0.89 (0.03) ***
SV				0.79 (0.06) ***	0.63 (0.08) ***	0.91 (0.03) ***	0.87 (0.04) ***
HLC					0.45 (0.11) ***	0.48 (0.11) ***	0.49 (0.11) ***
CW						0.63 (0.08) ***	0.86 (0.04) ***
CL							0.94 (0.02) ***

Resin yield (RY), diameter at breast height (DBH), height (HGT), stem volume (SV), height under live crown (HLC), crown width (CW), crown length (CL) and crown surface area (CSA). *** $p < 0.001$; ** $0.001 < p < 0.01$; * $0.01 < p < 0.05$, level of significance of effects.

Table 7. Expected response ($\Delta G/\bar{x} \times 100$, ΔG is the genetic gains, and $\bar{x}$ is the mean values) in resin yield at each site when different selection criteria used.

Selection Criterion (Site)	Response (%)		
	Jian	Ganzhou	Jindezhen
RY	11.08	4.13	9.53
HGT	7.71	3.79	11.36
DBH	6.27	3.64	10.97
SV	6.83	3.39	11.53
HLC	0.00	−1.63	7.53
CW	2.01	4.03	7.28
CL	5.90	4.33	11.19
CSA	4.49	5.82	10.24

Resin yield (RY), diameter at breast height (DBH), height (HGT), stem volume (SV), height under live crown (HLC), crown width (CW), crown length (CL) and crown surface area (CSA).

4. Discussion

4.1. Mean Values and Site Effects

According to the three progeny trials in the present study, the mean RY values of *P. elliottii*, varied from 1.16 kg/tree to 1.24 kg/tree month (38.6–41.3 g/tree day) and were somewhat higher than some other pine species [14,16,26]. However, the present RY values for *P. elliottii* families were slightly lower than those reported for 15-year-old families of *P. massoniana* because these values were higher than 45 g/tree day [27]. These differences might reflect species variation within *Pinus*, differences in environment, the influence of climate variables, and differences in the sampling age. Moreover, the effects of tapping methods on RY should not be disregarded.

The mean annual growth rate for the HGT and DBH in the present study was slower than that of the 10-year-old and 13-year-old open-pollinated progeny tests for *P. elliottii* [28,29], and this finding might reflect the rapid growth of *P. elliottii* in the young stage (before age 12–13); however, after age 13, the growth was drastically reduced, which led to the slower mean annual growth rate for 27-year-old families of *P. elliottii* in the present study than that of the young plantation of *P. elliottii* [30]. Additionally, previous studies have suggested that *P. elliottii* exhibited different growth effects in different climatic provinces [31], and the slower annual growth rate for HGT and DBH in the present study might partially reflect the influence of climate variables on growth.

Site effects reflect the reaction of trees to the combined effects of edaphic and as local and regional climatic conditions [32]. Moreover, the physiological age, genotype and culture method may also impact the trees. Significant site effects were observed for RY, growth traits and morphologic traits (Table 3). The results indicated that edaphic and regional climatic conditions have significant RY, growth traits and morphologic property effects. These significant site effects on RY, growth traits and morphologic traits have previously been described and reported [33–35]. Comparisons of the growth rates between sites showed that the SV at Jindezhen and Ganzhou was lower and presented 5–7.5% less growth than that at Jian. This tendency corresponds to expectations that site conditions are more favorable for *P. elliottii* in Jian than the two other sites. As in Jiangxi Province, the most suitable place for the growth of *P. elliottii* was in the Jitai Basin, which lies in central Jiangxi Province and presents a climate and growth environment similar to that of their original regions.

The highest resin yield was not observed for the most productive Jian site, as expected, but at the least productive Ganzhou site. This disparity could result from of a variety of causes, but differences in within-site environmental heterogeneity are a likely contributing factor. The ability of an individual to produce resin is modulated by the environmental conditions because the environmental conditions have various effects on the properties of the resin duct, where resin is synthesized and stored. Sites with high radiation (favoring photosynthesis processes), high summer temperatures (delaying radial resin canal sealing), and low rainfall (inducing resin canal formation) will be highly productive or determine high resin production [26]. During the tapping season, warm springs and summers with low humidity percentages could correspond to higher resin yield values. Follow this pattern, Jindezhen site should have the highest resin yield, as the highest annual temperature and the least rainfall were observed at this site.

4.2. Genetic Variation and Heritability

The family effects in the joint analysis for RY, growth traits and morphologic traits were significant (Table 3). Previous studies [27,36,37] have also reported a significant family/clonal effect on the RY and growth traits of *Pinus* tree species, as well as for morphologic properties [38]. The present study shows the important genetic variation in resin yield among *P. elliottii* families and the heritabilities for resin yield were moderately high. Resin flow is extremely variable, depending on genetics, climate and environmental factors such as soil fertility and disturbances [39,40]. Therefore, the ability of a tree to produce resin and resin ducts is likely related to the degree of constraints imposed by genetic and environmental factors.

Many studies have shown that growth traits are under substantial genetic control for coniferous species. The heritabilities for growth traits of the three sites combined were moderate to high (Table 3). In general, the estimates of heritability were higher for HGT ($h^2_f = 0.55$, $h^2_i = 0.27$) than for DBH ($h^2_f = 0.45$, $h^2_i = 0.27$), indicating that HGT has higher genetic control than DBH and more genetic gain in terms of tree volume can be expected from the selection for HGT than for DBH. The heritability estimates of *P. elliottii* for morphologic traits in the present study for the three sites combined were not high except for HLC. Similar studies on other trees have drawn the same conclusions and indicated that most variation in morphological traits is caused by environmental factors not genetic factors, thereby limiting the achievable gains from selection [36,41].

For most observed traits, family heritabilities were higher than individual heritabilities for the three sites combined or at each individual site. This result indicates that higher genetic gains can be achieved with family selection in *P. elliottii* populations. In fact, the results of previous studies on other pines also support this view [15,42]. The heritability values of the observed traits in the present study varied from site to site. Overall, the heritabilities at Jindezhen were higher than those at Jian and Ganzhou, and this finding may have been related to the experiment site topography at Jindezhen, which was flat ground and may have had a relatively low environmental influence on these traits.

The CV_G, that is, the genetic variance standardized to the trait mean, is considered the most suitable parameter for comparisons of genetic variation and the ability to respond to natural or artificial selection [43]. In the present study, both the CV_G and heritability exhibited the same pattern, with higher values at Jindezhen and relatively lower values at Jian or Ganzhou. The CV_G values of the growth traits, which ranged from 2.99% to 7.45% in the present study, were of the same magnitude as the CV_G values of the morphologic traits, which ranged from 3.53% to 7.10%. These results indicate an equivalent genetic potential for the improvement of growth traits as that for morphologic traits. In the present study, the CV_G of SV was higher than the CV_G of HGT or DBH. The finding may reflect the fact that volume is a function of both HGT and DBH. The CV_G of HGT (5.06%) was higher than the

CV$_G$ of DBH (2.99%), indicating that the scope for selection of HGT among families is larger than that for DBH.

4.3. Genotype by Environment Interaction

Because *P. elliottii* breeding programs are focused on developing families suitable for different environments, G × E interactions may have a significant impact on the precision of breeding values estimates because they can reduce genetic gain. With incongruence between test locations and deployment zones, the G × E, if improperly accounted for, could result in bias estimates and decreased genetic gain [32]. The significant family × site interaction variation observed for the RY, growth traits and morphologic traits (Table 3) indicated that the families exhibited differences in ecological variation and reaction norms under different growth conditions. Significant G × E interactions have previously been reported for growth traits and resin yield [38]; however, few morphologic trait results have been reported for *Pinus* tree species.

The type B genetic correlation is a measure of the importance of environment interactions on genotype, and it provides a more precise estimate of G × E interactions [24]. Overall, the inter-site type B genetic correlations observed in the present study ranged from 0.13 to 0.68 (Table 5), suggesting a large amount of variation reflective of changes in rank among families across sites. The absence of significant positive type B genotypic correlations between genotypic values at the Jingdezhen-Jian and Jingdezhen-Ganzhou sites for most traits (Table 5) indicated a true G × E, and the differences between these sites might reflect the family × site interaction. However, the type B genotypic correlations between Ganzhou and Jian for all traits were higher than those at Jingdezhen-Jian and Jingdezhen-Ganzhou sites (Table 5), indicating that these traits were more stable across Ganzhou and Jian. The trails at Ganzhou and Jian are relatively close in distance so does the climatic condition, but no significant difference of mean annual temperature between these two trails and the one located at Jingdezhen was observed. However, the annual rainfall is around 1800 mm at Jingdezhen, which is higher than Ganzhou and Jian (~1400 mm) (Table 1). As far as we know, the optimum annual precipitation for slash pine is about 1400 mm at Jiangxi province. Therefore, we speculated that the rainfall was the most important factors driving G × E. But more environmental and climatic factors as soil type, soil fertility are required to investigate further to delimit the deployment regions.

The ranges of genetic type B correlation were low to moderate (0.29–0.55) for resin flow, indicating that the genetic performance measured in the present study was generally inconsistent across sites and the environmental conditions in the study locations were different. Similar results were observed among *P. sylvestris* and *P. elliottii* families planted at different sites [44,45]. Westbrook [46] performed an association analysis and observed that 81% of the SNPs (Single Nucleotide Polymorphism) significantly associated with resin flow in loblolly pine were specific to individual sites, implying that environmental context has a significant impact on the effects of the alleles underlying quantitative genetic variation in resin flow via complex molecular mechanisms. Generally, there are two approaches used to deal with G × E in tree improvement programs. The first one is to choose the best genotypes for each site-type based on well-characterized environments. As mentioned before, the lack of enough environmental information for the tested sites limited the utilize of independent selection. Alternatively, it was more suitable to identify and select those stable genotypes that perform well across all the environments. Tree HGT showed less G × E than resin flow and resin flow showed a strong positive correlation with tree HGT at Jingdezhen (Table 6), suggesting that selection based on tree HGT would increase the stability of the genetic entries around Jingdezhen without decreasing resin production.

4.4. Genotypic Correlations between Traits

Previous studies on *P. elliottii* have reported that RY and growth traits (DBH and HGT) present a strong positive correlation [15,28]. This finding indicates that genetic factors act concurrently to increase RY and growth. The explanations from previous studies on the relationship between RY and growth traits are often varied, and overall, the resource availability hypothesis (RAH) and growth-differentiation balance hypothesis (GDBH) are the main explanations for the relationship between RY and growth in pines. *Pinus* species have an extensive system of constitutive and induced defenses to resist attacks from bark beetles and associated fungi [47]. The primary defense is resin, which is synthesized and stored in an interconnected network of axial and radial resin ducts in the secondary phloem and xylem [48]. Both resin and resin ducts are costly to produce, and these costs could lead to trade-offs between growth and defense [49]. A tradeoff likely occurs between growth and defense based on the allocation of acquired resources because acquired resources are limiting; thus, the relationship between growth and defense is expected to be negative. In practice, however, studies on the relationship between growth and defense have reported mixed results, with certain studies finding a negative relationship [40,50] and other studies reporting a positive relationship [51,52]. These results highlight one of the difficulties of testing defense theories [53] because defenses could be either positively or negatively correlated with growth depending on the metric used. For example, Hood and Sala [54] reported that when only relative defense measures (i.e., duct density or relative duct area) were included, defense showed a negative correlation with growth, thus supporting the RAH. This finding predicts that constitutive defenses are favored in slower-growing species, which reflects the higher costs of replacing tissues. However, when absolute measures of defense (i.e., duct size, duct production and duct area) were included, defense showed a positive correlation with growth, which contradicts the RAH. Therefore, faster-growing trees have larger ducts and higher total duct areas, thereby leading to higher resin yield than slower-growing trees. Notably, slower-growing trees invest more in resin duct defenses per unit area of radial growth (i.e., duct density and relative duct area) and are likely to produce less resin than a larger tree with lower relative resin duct investment. The results of the present study support this pattern because the estimated genetic correlations between growth traits and resin yield were positive and strong.

In the present study, the estimated genetic correlations between most morphologic traits and RY were moderate to strong. Thus, a tree with a larger crown size typically produces more resin. This finding could be explained by the GDBH, which predicts that any resource limitation that reduces growth more than photosynthesis (e.g., water and nutrients) will increase the carbohydrates available for defense with little to no trade-off to growth. If the resource is limited, then the carbon demand for growth decreases but photosynthesis continues, presumably causing a shift of carbohydrate allocation from growth to defense (resin production and resin duct formation). Based on this logic, trees with larger crown sizes would produce more constitutive resin than slower-growing trees because trees with larger crown biomasses would have greater carbon surpluses later in the season to invest in resin and resin duct formation. Therefore, morphologic traits, particularly CW, CL and CSA, could be used as indirect assisted traits for selecting high-yielding trees.

4.5. Implications for Tree Improvement

The moderate heritability for resin yield and growth traits indicated that selective breeding according to the breeding merit ranks of parental trees (backward selection) would be effective. Forward selection was not recommended before the unknown paternal trees of the half-sibs were identified using pedigree reconstruction method. Unexpected G × E were observed for all the traits across the three trails suggesting determination of the environmental factors driving the important G × E was necessary to delimit deployment areas. The annual rainfall was probably the main factors but more climatic and soil factors should be investigated in future studies. Alternatively, the genotypes with stable performance over all the environments and higher breeding values were the proper candidates for further breeding and deployment programs.

On the whole, significant positive genetic correlations among resin yield, growth and morphologic traits indicate that selection for faster growth and larger crown size can increase resin yield. On the other hand, the inclusion of resin yield into tree breeding programs can lead to a partial enhancement of growth. Simultaneous improvement of resin yield, growth and morphologic traits of *P. elliottii* was concluded to be practical when strong positive correlations exist between these traits. Expected genetic gains and the correlated responses estimated in this study are based on genotypic correlations, genotypic variances, and individual tree heritability. The results obviously indicated that there would be important benefit in using the growth traits as selection traits together with resin yield if the objectives were to improve resin yield and growth increment.

5. Conclusions

We used 3695 samples from 112 families at three sites for a genetic study in resin yield, growth and morphologic properties of *P. elliottii* in Jiangxi Province, China to estimate the genetic parameters (heritability, type b genetic correlation, and among traits genetic correlation) and correlated response. These results are summarized as:

(1) Significant site, family and family × site effects were found for resin yield, growth and morphologic traits. The differences between the three sites might reflect the family × site interaction, and rainfall might be the main driver of the interaction.

(2) Heritabilities were moderate to high for resin yield and growth traits for the three sites combined, whereas with relatively lower heritabilities for morphologic traits.

(3) The genetic correlations of *P. elliottii* among resin yield, growth traits and morphologic traits were moderate to strong, these findings suggested that selection for growth and morphologic traits might lead to increment in resin yield, and genetic factors act concurrently to increase resin yield, growth and morphologic traits. Selection based on growth traits will have moderate to strong effects on resin yield.

Acknowledgments: This study was mainly supported by the Science and Technology Research Foundation of Education Department of Jiangxi Province. (No. GJJ150386) and Forestry Science and Technology Innovation Program of Forestry Department of Jiangxi Province (No. 201401). The authors would like to thank engineer Kai He from Baiyun Mountain state owned forest farm in Jiangxi province for his assistance with resin yield measurement. The authors would also like to thank Professor Xiaomei Sun from Chinese Academy of Forestry for her editing.

Author Contributions: M.L. and L.D. conducted the study and wrote the manuscript. L.Z. carried out the critical reading and grammatical correction of manuscript and was mainly responsible for who gained the fund providing the study need. M.Y., S.S., Y.Z., L.F., Z.X., L.L. and C.L. participated in discussions and helped to draft the manuscript. All authors read and approved the final manuscript.

Conflicts of Interest: The authors declare no conflict of interest.

References

1. Lei, L.; Pan, X.Q.; Zhang, L.; Ai, Q.; Li, B.; Yi, M.; Xu, Z.G. Genetic variation and comprehensive selection of turpentine composition in High-yielding Slash Pine (*Pinus elliottii*). *For. Res.* **2015**, *28*, 804–809.
2. Luo, X.Q.; Jiang, X.M.; Yin, Y.F.; Liu, Z.X. Genetic variation and comprehensive assessment in wood properties of 15 fmilies of *Pinus elliottii*. *For. Res.* **2003**, *16*, 694–699.
3. Rodrigues-Corrêa, K.C.S.; Lima, J.C.; Fett, A.G. Resins from pine: Production and industrial uses. In *Natural Products*, 1st ed.; Springer: Berlin/Heidelberg, Germany, 2013.
4. Liu, Y. Production, consumption and prediction of China's resin. *Chem. Ind. For. Prod.* **2001**, *35*, 31–33.
5. Wen, X.; Kuang, Y.; Shi, M.; Li, H. Biology of Hylobitelus xiaoi (Coleoptera: Curculionidae), a new pest of slash pine, *Pinus elliottii*. *J. Econ. Entomol.* **2004**, *97*, 1958–1964. [CrossRef] [PubMed]
6. Lombardero, M.J.; Ayres, M.P.; Lorio, P.L.; Ruel, J.J. Environmental effects on constitutive and inducible resin defences of *Pinus taeda*. *Ecol. Lett.* **2003**, *3*, 329–339. [CrossRef]
7. Loomis, W.E. Growth-differentiation balance vs. carbohydratenitrogen ratio. *Proc. Am. Soc. Hort. Sci.* **1932**, *29*, 240–245.
8. Lorio, P.L. Growth-differentiation balance: A basis for understanding southern pine beetle–tree interactions. *For. Ecol. Manag.* **1986**, *14*, 259–273. [CrossRef]
9. Coppen, J.J.; Hone, G.A. *Gum Naval Stores: Turpentine and Rosin from Pine Resin*; Non-Wood Forest Products 2; Natural Resources Institute, FAO: Rome, Italy, 1995.

10. Rodrigues-Corrêa, K.C.S.; Azenedo, P.C.N.; Sobreiro, L.E.; Pelissari, P. Resin yield of *Pinus elliottii* plantations in a subtropical climate: Effect of tree diameter, wound shape and concentration of active adjuvants in resin stimulating paste. *Ind. Crops Prod.* **2008**, *27*, 322–327. [CrossRef]

11. Roberds, J.H.; Strom, B.L. Repeatability estimates for resin yield measurements in three species of the southern pines. *For. Ecol. Manag.* **2006**, *14*, 259–273.

12. Squillace, A.E.; Bengston, G.W. Inheritance of gum yield and other characteristics of slash pine. In Proceedings of the Sixth Southern Conference on Forest Tree Improvement, Gainesville, FL, USA, 7–8 June 1961; School of Forestry, University of Florida: Gainesville, FL, USA, 1961; pp. 85–96.

13. Tadesse, W.; Nanos, N.; Aunon, F.J.; Alia, R.; Gil, L. Evaluation of high resin yielders of *Pinus pinaster* AIT. *For. Genet.* **2001**, *8*, 271–278.

14. Liu, Q.H.; Zhou, Z.C.; Fan, H.H.; Liu, Y.R. Genetic variation and correlation among resin yield, growth, and morphologic traits of *Pinus massoniana*. *Silv. Genet.* **2013**, *62*, 38–44. [CrossRef]

15. Li, Y.J.; Jiang, J.M.; Luan, Q.F. Determination and genetic analysis of resin productivity resin density and turpentine content in half-sib families of slash pine. *J. Beijing For. Univ.* **2012**, *34*, 48–51.

16. Zhuang, W.Y.; Zhang, Y.Y.; Zou, Y.X. Selection for high-resin yield of slash pine and analysis of factors concerned. *Acta. Agric. Univ. Jiangxiensis* **2007**, *29*, 55–60.

17. Xu, Y.M.; Bao, C.H.; Zhou, Z.X.; Wen, X.F.; Jiang, D.H.; Wang, G.M. Variation analyses of tree growth and wood properties among Slash Pine provenances and comprehensive selection of superior provenance. *J. Northeast For. Univ.* **2001**, *29*, 18–21.

18. Chen, D.S.; Sun, X.M.; Zhang, S.G. Evaluation on growth and stem form characteristics of species and hybrids of Larix spp. *For. Res.* **2016**, *29*, 10–16.

19. Huang, F.L.; Jiao, Y.J.; Liang, J.; Fan, J.F. Correlation and genetic difference of crown traits of Poplar clones. *J. Northwest For. Univ.* **2010**, *25*, 61–65.

20. Butler, D.G.; Cullis, B.R.; Gilmour, A.R.; Gogel, B.J. *ASReml-R Reference Manual*; Department of Primary Industries and Fisheries: Brisbane, Australia, 2009.

21. R Core Team. *R: A Language and Environment for Statistical Computing*; R Foundation for Statistical Computing: Vienna, Austria, 2014.

22. Self, S.G.; Liang, K.Y. Asymptotic properties of maximum likelihood estimators and likelihood ratio tests under nonstandard conditions. *J. Am. Stat. Assoc.* **1987**, *82*, 605–610. [CrossRef]

23. Gilmour, A.R.; Gogel, B.J.; Cullis, B.R.; Thompson, R.; Butler, D.; Cherry, M.; Collins, D.; Dutkowski, G.; Harding, S.A.; Haskard, K. *ASReml User Guide Release 3.0. VSN*; International Ltd.: Hemel Hempstead, UK, 2009.

24. Burdon, R.D. Genetic correlation as a concept for studying genotype-environment interaction in forest tree breeding. *Silv. Genet.* **1977**, *26*, 168–175.

25. Falconer, D.S. *Introduction to Quantitative Genetics*, 3rd ed.; Longman: London, UK, 1981.

26. Rodríguez-García, A.; Martin, J.A.; López, R.; Mutke, S.; Pinillos, F.; Gil, L. Influence of climate variables on resin yield and secretory structures in tapped *Pinus pinaster* Ait. in central Spain. *Agric. For. Meteorol.* **2015**, *202*, 83–93. [CrossRef]

27. Zeng, L.H.; Zhang, Q.; He, B.X.; Lian, H.M.; Cai, Y.L.; Wang, Y.S.; Luo, M. Age Trends in genetic parameters for growth and resin-yielding capacity in Masson Pine. *Silv. Genet.* **2013**, *62*, 7–17. [CrossRef]

28. Zhang, J.Z.; Shen, F.Q.; Jiang, J.M.; Luan, Q.F. Heritability estimates for real resin capacity and growth traits in high-gum-yielding slash pine. *J. Zhejiang A F Univ.* **2010**, *27*, 367–373.

29. Li, Y. The Research on Growth Traits and Resin Yield Rules of *Pinus Elliottii* Families. Master's Thesis, Central South University of Forestry and Technology, Changsha, China, 2016.

30. Chen, W.Y.; Zhang, X.G.; Chen, Z. Growth analysis for plantation of *Pinus elliottii*. *J. Sichuan For. Sci. Technol.* **1997**, *4*, 43–47.

31. Zheng, Y.G.; Xu, Y.B. Study on demarcation of the main regions for introduction of slash pine based on climatic factors in China. *J. South China Agric. Univ.* **1996**, *17*, 41–46.

32. Pliura, A.; Zhang, S.Y.; MacKay, J.; Bousquet, J. Genotypic variation in wood density and growth traits of poplar hybrids at four clonal trails. *For. Ecol. Manag.* **2007**, *238*, 92–106. [CrossRef]

33. Wang, J.H.; Gu, W.C.; Li, B.; Guo, W.Y. Study on selection of *Alnus Cremastogyne* provenance/family analysis of growth adaption and genetic stability. *Sci. Silv. Sin.* **2000**, *36*, 59–66.

34. Wu, S.J.; Xu, J.M.; Li, G.Y.; Risto, V. Genotypic variation in wood properties and growth traits of Eucalyptus hybrid clones in southern china. *New For.* **2011**, *42*, 35–50. [CrossRef]

35. Xin, N.N.; Zhang, R.; Fan, H.H.; Chen, K. Family variation and selection of growth and quality characteristics of 5-Year-Old *Schima superba* seedlings. *For. Res.* **2014**, *27*, 316–322.

36. Jin, G.Q.; Qin, G.F.; Liu, W.H.; Chu, D.Y. Genetic analysis of growth traits on tester strain progeny of *Pinus massoniana*. *Sci. Silv. Sin.* **2008**, *44*, 71–79.

37. Raymond, C.A. Genotype by environment interactions for *Pinus radiate* in New South Wales, Australia. *Tree Genet. Gemomes* **2011**, *7*, 819–833. [CrossRef]

38. Baltunis, B.S.; Brawner, J.T. Clonal stability in *Pinus radiata* across New Zealand and Australia. I. Growth and form traits. *New For.* **2010**, *40*, 305–322. [CrossRef]

39. Hood, S.; Sala, A.; Heyerdahl, E.K.; Boutin, M. Low-severity fire increases tree defense against bark beetle attacks. *Ecology* **2015**, *96*, 1846–1855. [CrossRef] [PubMed]

40. Moreira, X.; Zas, R.; Solla, A.; Sampedro, L. Differentiation of persistent anatomical defensive structures is costly and determined by nutrient availability and genetic growth-defence constraints. *Tree Physiol.* **2015**, *35*, 112–123. [CrossRef] [PubMed]

41. Isik, K.; Isik, F. Genetic variation in *Pinus brutia* Ten. in Turkey: II. Branching and Crown Traits. *Silv. Genet.* **1999**, *48*, 293–302.

42. Weng, Y.H.; Ford, R.; Tong, Z.K.; Krasowski, M. Genetic Parameters for Bole Straightness and Branch Angle in Jack Pine Estimated Using Linear and Generalized Linear Mixed Models. *For. Sci.* **2016**, *1*, 1–8. [CrossRef]

43. Houle, D. Comparing evolvability and variability of quantitative traits. *Genetics* **1992**, *130*, 195–204. [PubMed]

44. Bridgen, M.R. Genetic variation of Scotch pine resin physiology. *Diss. Abstr. Int. B* **1980**, *40*, 75–78.

45. Romanelli, R.C.; Sebbenn, A.M. Genetic parameters and selection gains for resin production in *Pinus elliottii* var. elliottii, in south of Sao Paulo state. *Rev. Inst. Florest.* **2004**, *16*, 11–23.

46. Westbrook, J.W.; Resende, M.F.R.; Munoz, P.; Walker, A.R.; Wegrzyn, J.L. Association genetics of resin flow in loblolly pine: Discovering genes and predicting phenotype for improved resistance to bark beetles and bioenergy potential. *New Phytol.* **2013**, *199*, 89–100. [CrossRef] [PubMed]

47. Franceschi, V.R.; Krokene, P.; Christiansen, E.; Krekling, T. Anatomical and chemical defenses of conifer bark against bark beetles and other pests. *New Phytol.* **2005**, *167*, 353–376. [CrossRef] [PubMed]

48. Bannan, M.W. Vertical resin ducts in the secondary wood of the Abietineae. *New Phytol.* **1936**, *35*, 11–46. [CrossRef]

49. Strauss, S.Y.; Rudgers, J.A.; Lau, J.A.; Irwin, R.E. Direct and ecological costs of resistance to herbivory. *Trends Ecol. Evol.* **2002**, *17*, 278–285. [CrossRef]

50. Kempel, A.; Schädler, M.; Chrobock, T.; Fischer, M.; Kleunen, M. Tradeoffs associated with constitutive and induced plant resistance against herbivory. *Proc. Natl. Acad. Sci. USA* **2011**, *108*, 5685–5689. [CrossRef] [PubMed]

51. Ferrenberg, S.; Kane, J.M.; Mitton, J.B. Resin duct characteristics asso-ciated with tree resistance to bark beetles across lodgepole and limber pines. *Oecologia* **2014**, *174*, 1283–1292. [CrossRef] [PubMed]

52. Rodríguez-García, A.; López, R.; Martín, J.A.; Pinillos, F.; Gil, L. Resin yield in *Pinus pinaster* is related to tree dendrometry, stand density and tapping-induced systemic changes in xylem anatomy. *For. Ecol. Manag.* **2014**, *313*, 47–54. [CrossRef]

53. Stamp, N. Out of the quagmire of plant defense hypotheses. *Q. Rev. Biol.* **2003**, *78*, 23–55. [CrossRef] [PubMed]

54. Hood, S.; Sala, A. Ponderosa pine resin defenses and growth: Metrics matter. *Tree Physiol.* **2015**, *35*, 1223–1235. [CrossRef] [PubMed]

© 2017 by the authors. Licensee MDPI, Basel, Switzerland. This article is an open access article distributed under the terms and conditions of the Creative Commons Attribution (CC BY) license (http://creativecommons.org/licenses/by/4.0/).

forests

MDPI

Article

Genetic Parameters of Growth Traits and Stem Quality of Silver Birch in a Low-Density Clonal Plantation

Pauls Zeltiņš *, Roberts Matisons, Arnis Gailis, Jānis Jansons, Juris Katrevičs and Āris Jansons

Latvian State Forest Research Institute "Silava", 111 Rigas Street, LV-2169 Salaspils, Latvia; robism@inbox.lv (R.M.); arnis.gailis@silava.lv (A.G.); janis.jansons@mnkc.lv (J.J.); juris.katrevics@silava.lv (J.K.); aris.jansons@silava.lv (Ā.J.)
* Correspondence: pauls.zeltins@silava.lv; Tel.: +371-22-315-010

Received: 4 December 2017; Accepted: 22 January 2018; Published: 23 January 2018

Abstract: Silver birch (*Betula pendula* Roth) is productive on abandoned agriculture land, and thus might be considered as an option for profitable plantation forestry. Application of the most productive genotypes is essential. However, information about genetic gains in low-density plantations is still lacking. A 40-year-old low-density (400 trees ha^{-1}) plantation of 22 grafted silver birch plus-tree clones growing on former agricultural land in the central Latvia was studied. Although grafted plantations are not common in commercial forestry, the trial provided an opportunity to assess genetic parameters of middle-aged birch. The plantation that had reached the target diameter for final harvest (DBH (diameter at breast height) = 27.7 ± 5.5 cm) had an 85% survival rate, and stemwood productivity was 5.25 m^3 ha^{-1}year^{-1}. Still, rootstock × scion interaction and cyclophysis might have caused some biases. Broad-sense heritability (H^2) ranged from 0.02 for probability of spike knots to 0.40 for branch angle. Estimated H^2 for monetary value of stemwood was 0.16. In general, the correlations between growth and stem quality traits were weak, implying independent genetic control, though branchiness strongly correlated with diameter at breast height. The monetary value of stemwood strongly correlated with productivity traits. The observed correlations suggested that productivity and stem quality of birch might be improved simultaneously by genetic selection.

Keywords: mature *Betula pendula*; clonal forestry; tree breeding; target diameter

1. Introduction

The economic importance of plantation forestry on abandoned agricultural land is increasing [1]. Application of the most productive genotypes is essential for profitability of such plantations [2]. In the Baltics, hybrids of *Populus* L. are highly productive, yet they are strongly damaged by wildlife and require continuous protection [3]. Silver birch (*Betula pendula* Roth) has substantially lower environmental risks, yet is productive on agricultural land [4], and might be considered as an alternative. When appropriately cultivated (e.g., in a low-density plantation), birch can rapidly reach target diameter, reducing rotation time and increasing profitability of a plantation [4]. However, information about very low-density plantations is lacking.

Many traits including productivity and branchiness are highly heritable, emphasizing the potential to improve growth and stem quality [5,6]. Nevertheless, some traits can have common genetic control [7], which might differ regionally [4,6]. Furthermore, genetic parameters, such as heritability or genotypic coefficients of variation at final-harvest age, are unknown for silver birch. Genetic gains can be estimated theoretically from young trials, but the information about actual realization of these gains at mature age is available for tropical tree species [8,9], although is still lacking for silver birch.

The aim of this study was to estimate genetic parameters at the final-harvest age for stem quality and growth traits of silver birch clones planted in a low-density (400 trees ha^{-1}) plantation on former agricultural land. We hypothesized that the gain of productivity and stem quality of silver birch in a low-density plantation can be substantially improved by tree breeding.

2. Materials and Methods

The study site was located in the central part of Latvia (57°32′ N, 24°44′ E). The topography was flat (elevation < 100 m above sea level). The mean annual temperature was 6.2 °C; the mean monthly temperature ranged from 4.6 °C to 17.5 °C in February and July, respectively. The mean annual precipitation was ca. 690 mm.

The trial was established in 1972 on agricultural land, equivalent to *Oxalidosa* stand type with mesotrophic loamy soil. One year after grafting, clones of 22 birch plus-trees from the central part of Latvia (56°37′–57°28′ N; 24°50′–26°24′ E) were planted in a 5 × 5 m grid (400 trees ha^{-1}) as single-tree plots in 13–56 randomly distributed replications. Clones were randomized spatially all over the planting site. Initially, the plantation was intended as a seed orchard, but abandoned soon thereafter; hence no management, except some initial cleaning, was performed. The area of the plantation was 1.8 ha (720 planting spots).

At the age of 40 years in 2012, for each tree (1) diameter at breast height (DBH; cm); (2) height (m); (3) height of the lowest living branch (m); (4) mean branch angle (°); (5) mean projection of crown (MPC; m); (6) occurrence of spike knots; (7) double tops; and (8) stem cracks (present/absent), and arbitrary scores using 6-point-scales of (9) stem straightness and (10) branchiness were measured.

Data analysis was conducted in program R, v. 3.3. [10]. For each tree, the volume of stemwood assortments was calculated according to the model by Ozolins [11]. Wood defects and stem quality traits were considered when determining the structure of stemwood assortment (according to the practices of commercial forestry in Latvia). According to the estimated volume of assortments, the monetary value of stemwood (MV) of each sampled tree was calculated as an integrative parameter. Prices of different assortments according to top diameter, as used in the calculation of MV, were 20, 26, 45, 60, and 70 euro m^{-3} for firewood (<13 cm), pulpwood (<13 cm), logs 14–18 cm, logs 19–25 cm, and logs >26 cm, respectively.

Heritability coefficients H^2 (broad-sense individual-tree heritability) for the studied variables were calculated [7]:

$$H^2 = \sigma_G^2/\sigma_P^2, \tag{1}$$

where σ_G^2 is genotypic variance and σ_P^2 is phenotypic variance constituted of genotypic and environmental variance.

Genetic gain was estimated according to formula [7]:

$$R = S \cdot H^2, \tag{2}$$

where S is selection differential, which is the mean phenotypic value of the selected clones expressed as a deviation from the trial mean. For each variable, superiority of the top three clones against trial mean was assessed.

Genotypic and phenotypic clone mean Pearson correlations were estimated for the studied variables [7]. Genotypic correlations between the traits were calculated using the formula:

$$r_G = \frac{\sigma_{G(x,y)}}{\sqrt{\sigma_{G(x)}^2\sigma_{G(y)}^2}}, \tag{3}$$

where $\sigma_{G(x,y)}$ is the genetic covariance between traits x and y; $\sigma_{G(x)}^2$ and $\sigma_{G(y)}^2$ are the genotypic components of variance estimated for the traits. Standard errors for the genotypic correlation estimates were obtained with the delta method [12].

Genotypic coefficients of variation (*CVg*), describing the extent of genetic variability of a variable in relation to the mean of trial, were calculated as:

$$CVg = \sqrt{\sigma_G^2} \cdot 100 / \bar{x}, \tag{4}$$

where $\bar{x}$ is the phenotypic mean.

The corresponding components of genotypic and environmental variance were extracted using a random model:

$$y_{ij} = \mu + c_i + \varepsilon_{ij}, \tag{5}$$

where y_{ij} is observation of each trait of the *ij*th tree, μ is the overall mean, and c_i is the random clone effect. For the quantitative variables (e.g., DBH, tree height), a linear mixed model was used. For the binomial variables (e.g., survival, probability of cracks, etc.), a generalized linear mixed model applying binomial residual distribution and "logit" link function was fitted. For both models, R package lme4 was used [13]. For stem straightness and branchiness, ordinal logistic regression was applied [14] using R package ordinal [15]. The environmental variance of the link functions was determined as $\pi^2/3$, or 3.29. Genetic covariance $\sigma_{G(x,y)}$ between any two traits x and y was estimated using function varcomp in package lme4.

3. Results

The studied planation had 84.4% survival at the age of 40 years. The mean (±standard deviation) height and DBH of trees was 26.2 ± 2.2 m and 27.7 ± 5.6 cm, respectively. The total standing stemwood volume of the plantation was 210 m³ ha^{-1}, and the mean annual stemwood increment was 5.25 m³ ha^{-1} year^{-1}. Accordingly, MV was estimated ca. 9600 euro ha^{-1}, mainly contributed by the logs of smaller, medium, and large dimensions (44%, 25%, and 21%, respectively).

The estimated H^2 and CVg differed among the variables (Table 1). The highest heritability was estimated for branch angle, mean projection of crown (MPC), branchiness, and stem straightness ($0.40 \leq H^2 \leq 0.29$, respectively), while the lowest heritability was estimated for survival, probability of spike knots and cracks (<0.08). Intermediate $H^2 = 0.16$ for MV was similar to commonly reported tree height, height of the lowest living branch, and DBH (0.14, 0.14, and 0.21. respectively). The CVg of the quantitative variables ranged from 3.2% to 21.8% for tree height and MV, respectively (Table 1). For DBH and height of the lowest living branch, intermediate genotypic variation (ca. 9%) around the phenotypic mean was estimated, while it was higher for branch angle and MPC at −14.8% and 19.2%, respectively. For each variable, selection of top three clones resulted in 3.8%, 0.6%, and 2.7% genetic gain for DBH, tree height, and MV, respectively (Table 2).

The estimated genotypic correlations among the studied variables were similar to phenotypic clone mean Pearson correlations (Table 3); the latter are described. Correlations among tree height, DBH, and MV were high ($r > 0.63$); nevertheless, DBH and MV ($r > 0.66$) correlated with MPC. Branchiness correlated with DBH ($r = 0.79$), yet not with tree height (*p*-value = 0.41). Moderate to strong ($0.30 < |r| < 0.78$) negative correlations were observed between height of the lowest living branch and DBH, double tops, stem straightness, branchiness, and MPC. Occurrence of double tops showed moderate to strong correlations with stem straightness, branchiness, and MPC ($r = 0.70, 0.67,$ and 0.56, respectively), but a negative correlation ($r = -0.68$) with occurrence of spike knots. Mostly, weak and non-significant correlations were observed between the occurrence of stem cracks as well as branch angle and other variables.

Table 1. Statistics, coefficients of heritability (H^2), and genotypic variation (CV_g, %) of the morphometric variables (traits), and monetary value of 40-year-old grafted birch plus-trees from the low-density plantation. The monetary value of stemwood was calculated considering stem quality.

	Mean	Min	Max	Standard Deviation	Heritability Coefficient H^2 ± Standard Error	Genotypic Coefficient of Variation CV_g ± Standard Error (%)
Quantitative variables						
Stem diameter at breast height, cm	27.7	14.2	45.8	5.6	0.21 ± 0.06	9.5 ± 1.5
Tree height, m	26.2	15.3	31.6	2.2	0.14 ± 0.05	3.2 ± 0.5
Height of the lowest living branch, m	11.2	1.8	18.0	2.7	0.14 ± 0.05	9.3 ± 1.4
Branch angle, °	43.2	15.0	80.0	10.4	0.40 ± 0.08	14.8 ± 2.3
Mean projection of crown, m	2.9	1.1	6.3	0.8	0.39 ± 0.08	19.2 ± 3.0
Monetary value of stemwood, euro	28.2	3.7	95.4	14.6	0.16 ± 0.05	21.8 ± 3.4
Qualitative variables						
Survival, % of trees *	84.4	59.6	100.0	–	0.08 ± 0.03	–
Spike knot, % of trees *	23.2	5.2	42.8	–	0.02 ± 0.02	–
Double tops, % of trees *	34.9	6.0	75.1	–	0.14 ± 0.05	–
Stem straightness, score *	3.2	2.5	4.7	–	0.29 ± 0.07	–
Branchiness, score *	3.3	2.5	5.3	–	0.33 ± 0.08	–
Stem cracks, % of trees *	24.9	0.0	50.3	–	0.08 ± 0.03	–

* Mean values for clones.

Table 2. Clone means with standard errors (SEs) for studied traits.

Clone	Number of Trees	Survival, %	Diameter at Breast Height, cm Mean	SE	Height, m Mean	SE	Height of the Lowest Living Branch, m Mean	SE	Branch Angle, ° Mean	SE	Mean Projection of Crown, m Mean	SE	Monetary Value of Stemwood, Euro Mean	SE	Stem Straightness, Score Mean	SE	Branchiness, Score Mean	SE	Double Tops, % of Trees	Spike Knots, % of Trees	Stem Cracks, % of Trees
1	36	79.4	26.6	0.7	25.7	0.3	11.5	0.4	38.8	0.6	2.7	0.1	22.6	1.4	3.1	0.1	3.3	0.1	16.7	27.8	41.7
2	36	86.9	29.9	0.7	27.1	0.3	11.6	0.3	40.4	0.9	2.9	0.1	34.5	2.2	3.0	0.1	3.2	0.1	11.1	36.1	33.3
3	21	78.2	31.6	1.2	26.0	0.5	8.8	0.6	38.6	1.5	3.8	0.2	36.2	2.6	4.3	0.2	4.5	0.3	66.7	19.0	4.8
4	20	83.3	33.2	0.9	25.7	0.4	8.3	0.3	41.8	1.3	4.5	0.2	36.4	2.1	4.7	0.2	5.3	0.2	75.0	5.0	5.0
5	16	92.3	30.9	1.2	27.1	0.5	11.5	0.8	49.7	1.9	3.5	0.2	35.3	3.7	3.4	0.2	3.9	0.3	37.5	25.0	18.8
6	28	91.9	29.1	1.0	26.5	0.5	10.6	0.5	42.9	1.7	3.3	0.2	31.8	3.1	3.8	0.2	3.7	0.2	53.6	21.4	21.4
7	41	90.8	27.5	0.9	27.4	0.3	11.8	0.4	43.3	1.2	3.1	0.1	28.4	2.3	3.3	0.1	3.3	0.1	56.1	24.4	0.0
8	24	81.8	22.2	1.0	24.3	0.4	11.9	0.6	44.8	1.8	2.3	0.1	16.5	2.5	3.2	0.2	2.5	0.1	20.8	37.5	8.3
9	16	91.4	28.6	2.0	26.1	0.5	9.5	0.6	38.1	2.1	3.3	0.3	27.4	3.5	3.8	0.2	3.8	0.4	62.5	12.5	12.5
10	16	68.1	23.9	1.1	23.7	0.9	10.7	0.7	38.8	1.6	2.3	0.1	18.7	2.5	3.3	0.2	2.8	0.2	25.0	18.8	43.8

Table 2. Cont.

Clone	Number of Trees	Survival, %	Diameter at Breast Height, cm		Height, m		Height of the Lowest Living Branch, m		Branch Angle, °		Mean Projection of Crown, m		Monetary Value of Stemwood, Euro		Stem Straightness, Score		Branchiness, Score		Double Tops, % of Trees	Spike Knots, % of Trees	Stem Cracks, % of Trees
			Mean	SE	Mean	SE	Mean	SE	Mean	SE	Mean	SE	Mean	SE	Mean	SE	Mean	SE			
11	35	88.7	25.7	0.6	25.7	0.4	11.2	0.4	38.0	1.8	2.6	0.1	22.0	1.3	3.3	0.2	3.4	0.1	57.1	14.3	14.3
12	39	88.3	24.2	0.7	24.9	0.3	11.7	0.4	38.5	1.0	2.6	0.1	19.6	1.7	3.4	0.2	3.1	0.1	41.0	15.4	20.5
13	18	90.2	23.9	1.3	26.3	0.8	12.2	0.6	37.8	1.0	2.2	0.1	20.7	3.4	2.8	0.3	2.7	0.2	33.3	11.1	16.7
14	12	89.2	27.8	1.4	25.5	0.7	10.3	0.6	48.8	3.1	3.0	0.2	27.3	4.4	3.8	0.3	3.4	0.2	33.3	41.7	41.7
15	29	100.0	30.4	0.6	26.7	0.3	9.3	0.5	37.6	1.0	3.3	0.1	34.5	2.1	2.7	0.1	3.5	0.1	55.2	6.9	48.3
16	23	59.6	28.8	1.4	27.0	0.5	11.9	0.5	41.1	1.9	2.6	0.1	32.8	3.5	2.9	0.2	3.0	0.2	43.5	13.0	30.4
17	33	69.0	28.6	0.6	26.8	0.2	11.7	0.4	61.5	2.0	3.4	0.1	30.8	1.7	2.6	0.1	3.1	0.1	6.1	33.3	30.3
18	46	82.7	28.4	0.8	27.2	0.2	12.3	0.5	57.8	1.5	2.8	0.1	31.7	2.4	3.0	0.1	2.8	0.1	26.1	23.9	28.3
19	36	87.3	28.2	1.0	26.0	0.3	11.1	0.4	41.4	1.0	2.6	0.1	30.8	2.8	2.5	0.1	2.9	0.1	16.7	22.2	41.7
20	28	81.1	24.8	0.7	26.0	0.4	10.9	0.6	41.1	0.9	2.5	0.1	21.2	2.0	3.3	0.2	3.0	0.1	28.6	28.6	25.0
21	31	94.4	26.6	0.8	26.1	0.3	12.8	0.4	36.5	1.3	2.2	0.1	26.2	2.2	2.5	0.2	2.7	0.2	9.7	32.3	19.4
Ka1	14	82.9	33.3	2.0	27.4	0.5	11.0	0.6	49.3	2.5	3.8	0.3	42.0	6.8	3.6	0.2	4.1	0.3	28.6	42.9	50.0
Total	598	84.4	27.7	0.2	26.2	0.1	11.2	0.1	43.2	0.4	2.9	0.0	28.2	0.6	3.2	0.0	3.3	0.0	34.9	23.2	24.9

Table 3. Genotypic correlations (standard errors by delta method in brackets) in the upper diagonal part and phenotypic clone mean Pearson correlations (significant correlations with $p \leq 0.05$ in bold) in the lower diagonal part (*—calculation stopped due to infinite likelihood).

	Tree Height	Stem Diameter at Breast Height	Stem Cracks	Height of the Lowest Living Branch	Branch Angle	Double Tops	Spike Knots	Stem Straightness	Branchiness	Mean Projection of Crown	Monetary Value of Stemwood
Tree height	1	0.65 (0.16)	0.02 (0.30)	0.14 (0.27)	0.43 (0.21)	0.03 (0.27)	0.17 (0.45)	-0.16 (0.25)	0.16 (0.25)	0.35 (0.22)	0.79 (0.11)
Stem diameter at breast height	0.63	1	0.11 (0.29)	-0.56 (0.19)	0.23 (0.23)	0.35 (0.23)	-0.23 (0.43)	0.44 (0.20)	0.79 (0.10)	0.86 (0.07)	0.93 (0.03)
Stem cracks	0.03	0.10	1	-0.11 (*)	0.08 (0.02)	-0.68 (0.20)	0.38 (0.44)	-0.60 (0.21)	-0.30 (0.27)	-0.20 (0.27)	0.47 (*)
Height of the lowest living branch	0.17	-0.51	0.07	1	0.29 (0.23)	-0.75 (0.13)	0.90 (0.37)	-0.76 (0.12)	-0.85 (0.24)	-0.77 (0.11)	-0.32 (0.25)
Branch angle	0.36	0.22	0.15	0.25	1	-0.37 (0.22)	0.67 (0.31)	-0.09 (0.23)	-0.25 (0.07)	0.27 (0.22)	0.29 (0.23)
Double tops	0.05	0.35	-0.51	-0.69	-0.34	1	-1.19 (0.35)	0.78 (0.12)	0.74 (0.13)	0.60 (0.17)	-0.10 (*)
Spike knot	0.06	-0.07	0.29	0.40	0.46	-0.68	1	-0.47 (0.43)	-0.64 (0.40)	-0.35 (0.22)	0.06 (0.48)
Stem straightness	-0.15	0.42	-0.41	-0.71	-0.08	0.70	-0.15	1	0.87 (0.07)	0.60 (0.14)	0.12 (0.25)
Branchiness	0.19	0.79	-0.22	-0.78	-0.05	0.67	-0.26	0.82	1	0.93 (0.03)	0.28 (0.28)
Mean projection of crown	0.36	0.86	-0.15	-0.71	0.26	0.56	-0.16	0.70	0.93	1	0.65 (0.14)
Monetary value of stemwood	0.74	0.93	0.32	-0.30	0.28	0.14	0.03	0.14	0.54	0.66	1

4. Discussion

The calculated H^2 (Table 1) implied potential for substantial improvement of productivity and stem quality, hence yields of birch plantations by tree breeding [5]. Nevertheless, H^2 of the variables differed (Table 1), implying unequal potential for the improvement of the traits [7]. Branch angle, branchiness, projection of crown, and stem straightness, which largely influence timber quality [2], were highly heritable and had intermediate CVg (Table 1), implying potential for considerable improvement [7]. High CVg was also observed for MV (21.8), indicating potential financial benefits from breeding. Nevertheless, strong correlation between branchiness and DBH, MPC, and stem straightness indicated possible negative effects on stem quality when selecting fast growing trees with straight stems (Table 3). Additionally, height of the lowest living branch had significant negative correlations with the same variables, supporting the abovementioned consideration. Earlier studies reported a significant moderate correlation between DBH and number of branches [5,16]. Significant negative genotypic correlation between productivity traits and stem straightness (r_G ranging from -0.45 to -0.72) was noticed in Sweden [16]. However, other stem quality traits such as spike knots, stem cracks, and double tops did not show significant relation to productivity traits and MV, suggesting the possibility for simultaneous improvement [16,17].

The heritability of survival was low (Table 1), suggesting the prevailing effect of the micro-site conditions, as shown by Stener and Jansson [16] for birch in Sweden. Environmental factors can strongly affect performance of the species, masking the genetic effect and resulting in low heritability parameters [6]. The estimated genetic parameters (Table 1) might have been already affected by the pre-selection of planting material (plus-trees) with improved branching and stem properties, as a seed orchard was initially intended. Although the utilization of grafted silver birch is not a common practice in commercial forestry, the trial provided information about genetic parameters at middle age that has not been previously published. This might have caused some imprecisions in genetic parameters due to uncontrolled rootstock × scion effect. Although the issue has been scarcely studied for forest trees [18], for loblolly pine, the rootstock × scion effect has been negligible compared to the effects of clone and site factors [19]. This was also supported by good survival of grafts indicating compatibility between rootstock and scions. The negative effect of cyclophysis due to different biological ages of rootstock and scion [20–23] appeared insubstantial, as indicated by the productivity of the plantation. Similarly, a weak effect of cyclophysis on growth and survival of vegetatively propagated silver birch has been shown in boreal conditions [24,25]. Still, grafts might have lower branchiness and branch thickness [26].

The single-tree-plot design of the plantation might have also affected genetic parameters of the traits, as the measurements from such plots are influenced by competition among different genotypes [27]. However, low planting density likely had postponed the onset of inter-tree competition, therefore reducing exaggeration of the genotypic variance of growth traits [5,16,28]. Hence, the estimated H^2 and CVg were somewhat lower than reported in earlier studies, in which H^2 ranged 0.07–0.56 for tree height, and 0.11–0.59 for DBH, while CVg for the respective traits has been reported to range between 5 and 14, and between 9 and 21, respectively [5,16,28]. Still, heritability of height and DBH varies widely among different trials [16]. Considering varying genetic control of the studied traits, H^2 and CVg of MV were intermediate (0.16 and 21.8), as similarly observed in Sweden [5].

For silver birch, genetic gains of around 10% for height and 20% for DBH of the top 10% clones at the age of 7–11 years are reported [5,16], while corresponding realized gains in our study site at the moment of possible final-harvest was around 17 and 5 times lower, respectively. This may imply weak age-age correlations, as well as reflect lower heritability and high variability due to strong environmental effects (Table 2). However, earlier measurements from the studied trial were not available for comparison.

The studied plantation appeared ready for the final harvest already at the age of 40 years. Higher productivity (up to 8.90 m^3 ha^{-1} year^{-1} [29] vs. 5.25 m^3 ha^{-1} year^{-1} in studied trial) and good stem quality might be achieved in conventional plantations with higher planting densities [30,31], although increasing planting distance does not influence the height growth [31]. Nevertheless, decreased competition and application of the pre-selected planting material apparently improved the assortment structure of the studied birch, shifting its distribution towards the higher value, thus suggesting efficiency of the low-density clonal plantation for the production of solid wood and possible further economic improvement in a low-density short-rotation plantation. Together with selected planting material, reduced establishment costs with wider spacing might be a strong driving factor for choosing lower planting densities. Increased value does not only result from increases in volume production, but also from improved stem quality leading to more valuable logs [9]. Besides, breeding effect on productivity might not fully express in dense stands, since birch maintain vigorous growth when presented with low within-stand competition [4].

5. Conclusions

Although the utilization of grafted silver birch is not a common practice in commercial forestry, the studied forty-year-old, low-density grafted clonal plantation appeared efficient for the production of solid wood. Considering heritability and genetic gains of the studied traits, the gain of birch plantations might be substantially improved by breeding. The non-significant correlations between stem quality and dimensions of trees suggested that the traits could be improved simultaneously. However, the strong correlation between branchiness and DBH implied that stem quality would be reduced when selecting for productivity. Still, rootstock × scion interaction and cyclophysis effects are uncertain and might be potentially significant. Considering the potential for strong environmental effects on the performance of birch, verification of the results in diverse growing conditions is required.

Acknowledgments: The study was carried out in accordance to contract No. 1.2.1.1/16/A/009 between "Forest Sector Competence Centre" Ltd. and the Central Finance and Contracting Agency, funded by the European Regional Development Fund (ERDF) within the framework of the project "Forest Sector Competence Centre". We acknowledge JSC "Latvijas valsts meži" for information about standing stock of the conventional stands. Jānis Donis and Virgilijus Baliuckas helped with data analysis.

Author Contributions: Ā.J. conceived the original research idea. All authors contributed to the experimental design. J.K. and J.J. were responsible for data collection. Ā.J., R.M., and P.Z. analyzed the data. A.G., Ā.J., R.M., and P.Z. wrote the paper.

Conflicts of Interest: The authors declare no conflict of interest.

References

1. Sedjo, R.A. The potential of high-yield plantation forestry for meeting timber needs. *New For.* **1999**, *17*, 339–360. [CrossRef]
2. Savill, P.; Evans, J.; Auclair, D.; Falck, J. *Plantation Silviculture in Europe*; Oxford University Press: Oxford, UK, 1997.
3. Tullus, A.; Rytter, L.; Tullus, T.; Weih, M.; Tullus, H. Short-rotation forestry with hybrid aspen (*Populus tremula* L. × *P. tremuloides* Michx.) in Northern Europe. *Scand. J. For. Res.* **2012**, *27*, 10–29. [CrossRef]
4. Hynynen, J.; Niemisto, P.; Vihera-Aarnio, A.; Brunner, A.; Hein, S.; Velling, P. Silviculture of birch (*Betula pendula* Roth and *Betula pubescens* Ehrh.) in northern Europe. *Forestry* **2010**, *83*, 103–119. [CrossRef]
5. Stener, L.; Hedenberg, Ö. Genetic Parameters of Wood, Fibre, Stem Quality and Growth Traits in a Clone Test with *Betula pendula*. *Scand. J. For. Res.* **2003**, *18*, 103–110. [CrossRef]
6. Koski, V.; Rousi, M. A review of the promises and constraints of breeding silver birch (*Betula pendula* Roth) in Finland. *For. Int. J. For. Res.* **2005**, *78*, 187–198. [CrossRef]
7. Falconer, D.S.; Mackay, T.F.C. *Introduction to Quantitative Genetics*, 4th ed.; Longman Group Ltd.: London, UK, 1996.
8. Kimberley, M.O.; Moore, J.R.; Dungey, H.S. Quantification of realised genetic gain in radiata pine and its incorporation into growth and yield modelling systems. *Can. J. For. Res.* **2015**, *45*, 1676–1687. [CrossRef]
9. Moore, J.R.; Dash, J.P.; Lee, J.R.; McKinley, R.B.; Dungey, H.S. Quantifying the influence of seedlot and stand density on growth, wood properties and the economics of growing radiata pine. *For. Int. J. For. Res.* **2017**, 1–14. [CrossRef]
10. R Development Core Team. *R: A Language and Environment for Statistical Computing*; R Foundation for Statistical Computing: Vienna, Austria, 2016.

11. Ozolins, R. Forest stand assortment structure analysis using mathematical modelling. *For. Stud.* **2002**, *7*, 33–42.
12. Lynch, M.; Walsh, B. *Genetics and Analysis of Quantitative Traits*; Sinauer Associates: London, UK, 1998.
13. Bates, D.; Mächler, M.; Bolker, B.; Walker, S. Fitting Linear Mixed-Effects Models using lme4. *J. Stat. Softw.* **2014**, *67*, 1–48.
14. Long, J.S. *Regression Models for Categorical and Limited Dependent Variables*; Sage Publications: London, UK, 1997; ISBN 0803973748.
15. Christensen, R.H.B. Ordinal—Regression Models for Ordinal Data. R Package Version 2015. Available online: https://cran.r-project.org/web/packages/ordinal/ordinal.pdf (accessed on 15 October 2017).
16. Stener, L.-G.; Jansson, G. Improvement of *Betula pendula* by clonal and progeny testing of phenotypically selected trees. *Scand. J. For. Res.* **2005**, *20*, 292–303. [CrossRef]
17. Viherä-Aarnio, A.; Velling, P. Growth and stem quality of mature birches in a combined species and progeny trial. *Silva Fenn.* **1999**, *33*, 225–234. [CrossRef]
18. Jayawickrama, K.J.S.; Jett, J.B.; McKeand, S.E. Rootstock effects in grafted conifers: A review. *New For.* **1991**, *5*, 157–173. [CrossRef]
19. Jayawickrama, K.J.; McKeand, S.E.; Jett, J.B. Rootstock effects on scion growth and reproduction in 8-year-old grafted loblolly pine. *Can. J. For. Res.* **1997**, *27*, 1781–1787. [CrossRef]
20. Olesen, P.O. On cyclophysis and topophysis. *Silvae Genet.* **1978**, *27*, 173–178.
21. Greenwood, M.S.; Hutchison, K.W. Maturation as a Developmental Process. In *Clonal Forestry I*; Springer: Berlin/Heidelberg, Germany, 1993; pp. 14–33. ISBN 978-3-642-84177-4.
22. Viherä-Aarnio, A.; Ryynänen, L. Seed production of micropropagated plants, grafts and seedlings of birch in a seed orchard. *Silva Fenn.* **1994**, *28*, 257–263. [CrossRef]
23. Wendling, I.; Trueman, S.J.; Xavier, A. Maturation and related aspects in clonal forestry—Part I: Concepts, regulation and consequences of phase change. *New For.* **2014**, *45*, 449–471. [CrossRef]
24. Jones, O.P.; Welander, M.; Waller, B.J.; Ridout, M.S. Micropropagation of adult birch trees: Production and field performance. *Tree Physiol.* **1996**, *16*, 521–525. [CrossRef] [PubMed]
25. Viherä-Aarnio, A.; Velling, P. Micropropagated silver birches (*Betula pendula*) in the field—Performance and clonal differences. *Silva Fenn.* **2001**, *35*, 385–401. [CrossRef]
26. Viherä-Aarnio, A.; Ryynänen, L. Growth, crown structure and seed production of birch seedlings, grafts and micropropagated plants. *Silva Fenn.* **1995**, *29*, 3–12. [CrossRef]
27. Vergara, R.; White, T.L.; Huber, D.A.; Shiver, B.D.; Rockwood, D.L. Estimated realized gains for first-generation slash pine (*Pinus elliottii* var. *elliottii*) tree improvement in the southeastern United States. *Can. J. For. Res.* **2004**, *34*, 2587–2600. [CrossRef]
28. Malcolm, D.C.; Worrell, R. Potential for the improvement of silver birch (*Betula pendula* Roth.) in Scotland. *Forestry* **2001**, *74*, 439–453. [CrossRef]
29. Oikarinen, M. Growth and yield models for silver birch (*Betula pendula*) plantations in southern Finland. *Commun. Inst. For. Fenn.* **1983**, *113*, 1–75.
30. Niemistö, P. Influence of initial spacing and row-to-row distance on the crown and branch properties and taper of silver birch (*Betula pendula*). *Scand. J. For. Res.* **1995**, *10*, 235–244. [CrossRef]
31. Niemistö, P. Influence of initial spacing and row-to-row distance on the growth and yield of silver birch (*Betula pendula*). *Scand. J. For. Res.* **1995**, *10*, 245–255. [CrossRef]

© 2018 by the authors. Licensee MDPI, Basel, Switzerland. This article is an open access article distributed under the terms and conditions of the Creative Commons Attribution (CC BY) license (http://creativecommons.org/licenses/by/4.0/).

forests

MDPI

Article

In Vitro Tetraploid Induction from Leaf and Petiole Explants of Hybrid Sweetgum (*Liquidambar styraciflua* × *Liquidambar formosana*)

Yan Zhang [1], Zewei Wang [1], Shuaizheng Qi [1], Xiaoqi Wang [1], Jian Zhao [1], Jinfeng Zhang [1,*], Bailian Li [1,2], Yadong Zhang [3], Xuezeng Liu [3] and Wei Yuan [4]

[1] Beijing Advanced Innovation Center for Tree Breeding by Molecular Design, National Engineering Laboratory for Tree Breeding, Key Laboratory of Genetics and Breeding in Forest Trees and Ornamental Plants of Ministry of Education, Key Laboratory of Forest Trees and Ornamental Plants Biological Engineering of State Forestry Administration, College of Biological Sciences and Technology, Beijing Forestry University, Beijing 100083, China; zhangyan19890802@126.com (Y.Z.); linxuewzw@163.com (Z.W.); 13051857171@163.com (S.Q.); liqiu8826623@163.com (X.W.); lbyyuanwei@163.com (J.Z.); Bailian_Li@ncsu.edu (B.L.)

[2] Department of Forestry and Environmental Resources, North Carolina State University, Raleigh, NC 27695, USA

[3] Henan Longyuan Flowers & Trees Co., Ltd., Yanling 461200, China; lyhmkjb@163.com (Y.Z.); lxuezeng@126.com (X.L.)

[4] Zhengzhou China Green Expo Garden, Zhengzhou 451460, China; lbyyuanwei@163.com

* Correspondence: zjf@bjfu.edu.cn

Received: 4 June 2017; Accepted: 20 July 2017; Published: 28 July 2017

Abstract: *Liquidambar* is an important forestry species used to generate many commercial wood products, such as plywood. Inducing artificial polyploidy is an effective method to encourage genetic enhancements in forestry breeding. This report presents the first in vitro protocol for the induction of genus *Liquidambar* tetraploids based on the established in vitro regeneration system of hybrid sweetgum (*Liquidambar styraciflua* × *Liquidambar formosana*). The leaves and petioles from three genotypes were pre-cultured in woody plant medium (WPM) supplemented with 0.1 mg/L thidiazuron (TDZ), 0.8 mg/L benzyladenine (BA), and 0.1 mg/L α-naphthalene acetic acid (NAA) for a variable number of days (4, 6 or 8 days), and exposed to varying concentrations of colchicine (120, 160, 200 mg/L) for 3, 4 or 5 days; the four factors were investigated using an orthogonal experimental design. Adventitious shoots were rooted in 1/2 WPM medium supplemented with 2.0 mg/L indole butyric acid (IBA) and 0.1 mg/L NAA. The ploidy level was assessed using flow cytometry and chromosome counting. Four tetraploids and nine mixoploids were obtained from the leaves. Pre-treatment of the leaves for 8 days and exposure to 200 mg/L colchicine for 3 days led to the most efficient tetraploid induction. Producing 11 tetraploids and five mixoploids from petioles, the best tetraploid induction treatment for petioles was almost the same as that with the leaves, except that pre-culturing was required for only 6 days. In total, 15 tetraploids were obtained with these treatments. This study described a technique for the induction of tetraploid sweetgum from the leaves or petioles of parental material. Based on the success of polyploid breeding in other tree species, the production of hybrid sweetgum allotetraploids constitutes a promising strategy for the promotion of future forestry breeding.

Keywords: chromosome doubling; sweetgum; allotetraploid

1. Introduction

Liquidambar styraciflua, belonging to the genus *Liquidambar*, is found widely in the southern regions of the United States. *L. styraciflua* has become one of the most important commercial hardwoods in the United States [1,2], because it has a fast growing rate and provides many useful materials such as wood, plywood, pulp and paper production [2]. *Liquidambar formosana*, mainly distributed in East Asia, is an important tree in China owing to its fast-growing properties and use in timber and medicinal production and landscaping [3]. Interestingly, *L. styraciflua* can be interfertile with *L. formosana* [4,5]. Due to the potential of heterosis from interspecies hybridisation, hybrid sweetgum can show robust growth [5].

Allopolyploids are the products of merging two or more genomes by intraspecific or interspecific hybridisation [6,7]. Allopolyploid breeding has been applied successfully in forestry due to its advantages of high biomass and fitness. In vitro regeneration using the reproductive organs of woody plants has been successful in many species. Induction of tetraploidy has been successful in a number of species, including *Populus* [8,9], *Paulownia tomentosa* [10], and Citrus [11]. Unlike meiotic (sexual) chromosome doubling, in vitro asexual polyploidy breeding is not limited by season, and it benefits from a low mixoploids-inducing rate; it has been applied widely in forestry breeding.

Successful induction of tetraploids has been achieved using various chemical reagents, such as colchicine [12], oryzalin [13], and trifluralin [14]. Colchicine is the most widely applied chemical for in vitro tetraploid induction. Moreover, many explant types have been used as materials for in vitro polyploidy induction, such as leaves [8], petioles [15], calluses [12], shoots [16], hypocotyl segments, and cotyledonary nodes [17]. Although leaves and petioles are not used commonly as explants for tetraploid induction, they exhibit a high tetraploid-inducing rate and low number of mixoploids, indicating that they can be used for efficient polyploidy breeding [9,18]. We have improved the media that were applied to the in vitro regeneration of *L. styraciflua* [19,20] and *L. formosana* [21], and the modified medium was suitable for establishing an efficient regeneration system from leaves and petioles of hybrid sweetgum (*L. styraciflua* × *L. formosana*). Therefore, it is possible to acquire in vitro colchicine-induced tetraploid sweetgum.

This study describes an efficient method for the in vitro induction of tetraploids. First, we established multiple hybrid sweetgum genotypes. Then, we investigated the effects of explant type, explant genotype, time of pre-incubation, time of colchicine treatment and concentration of colchicine using an orthogonal experimental design. The tetraploids were determined by flow cytometry and chromosome counting. We hope that the biomass, resistance and ornamental value of *Liquidambar* will be improved by chromosome doubling after in vitro colchicine treatment.

2. Materials and Methods

2.1. Plant Materials

Floral branches of the male parent (*L. formosana*) were collected from five genotypes at Shanghai Chen Shan Botanical Garden (Songjiang District, Shanghai, China); pollen was collected; the same volume of pollen was measured and they were mixed together. In April, the male inflorescences of *L. styraciflua* were removed, and controlled pollination was applied to acquire hybrid seeds by pollination with the pollen of *L. formosana*. *L. styraciflua* was grown at the Shanghai Chen Shan Botanical Garden (Songjiang District, Shanghai, China). The fruits were collected in mid-September 2015, and the seeds were stored at 4 °C.

2.2. Establishment of Aseptic Seedlings and In Vitro Multiplication

Plump seeds were selected and sterilised in 75% ethanol (*v/v*) for 45 s, rinsed once with sterile distilled water followed by 2% sodium hypochlorite for 8 min, and then washed three times with sterile distilled water. The seeds were added to basal woody plant medium (WPM) supplemented with 4 g/L agar and 2 g/L polygel, and 30 g/L sucrose (pH 5.8–5.9) in a 9-mm culture dish; no phytohormone was added; in this experiment, all media were semi-solidified. After 20 days, the germinated seedlings were placed in magenta boxes with 50 mL basal WPM [22]. A total of 150 plant genotypes were sub-cultured in rooting medium containing half-strength WPM medium supplemented with 2.0 mg/L indole butyric acid (IBA) and 0.1 mg/L naphthalene acetic acid (NAA). Regeneration medium for leaves and petioles consisted of WPM medium with 0.1 mg/L TDZ, 0.8 mg/L benzyladenine (BA) and 0.1 mg/L NAA supplemented with 2 g/L agar and 4 g/L double coagulation, and 30 g/L sucrose. After 40 days, the explants were transferred into elongation medium supplemented with WPM basal salts, 0.4 mg/L BA and 0.1 mg/L NAA; the concentrations of agar and double coagulation were changed to 2 g/L agar and 2 g/L, respectively.

Three hybrid plant genotypes (named Z1, Z2, and Z3) were selected for tetraploid induction (Table 1). The shoot-inducing rate of all leaves and petioles reached 85%, and they displayed similar morphological characteristics during development.

Shoot-inducing rate: Number of adventitious shoots ($\geq$1 cm)/number of explants $\times$ 100%

Table 1. Influencing factors and level values.

Levels	Factors			
	A	B	C	D
	Genotype	Colchicine Concentration (mg/L)	Pre-Culture Duration (day)	Exposure Time (day)
1	Z1	120	4	3
2	Z2	160	6	4
3	Z3	200	8	5

2.3. Colchicine Application

After sub-culturing for 60 days, the second and third leaves and petioles of Z1, Z2, and Z3 were selected as the materials for colchicine treatment. The leaf samples were proximal halves cut twice through the main vein, and the petioles were cut into 0.8–1.0 cm pieces; both sample types were pre-cultured in WPM supplemented with 0.1 mg/L TDZ, 0.8 mg/L BA and 0.1 mg/L NAA for 4, 6, or 8 days, and then treated for 3, 4 or 5 days with various concentrations of colchicine (120, 160 or 200 mg/L). The genotype, number of days of pre-incubation, colchicine exposure time, and concentration of colchicine were included as variables in the $L(9_3)^4$ orthogonal experimental design (Table 1); two explant types, leaf and petiole were applied to the same design, respectively. Cultures with colchicine treatment were performed in 100 mL flasks and incubated at 25 $\pm$ 2 °C, under dark conditions. Adventitious shoots longer than 1.5 cm were harvested and placed into the rooting medium.

Explants survival rate: Number of survival explants (leaf or petiole)/all explants $\times$ 100% (n = 3)

2.4. Flow Cytometric Analysis of Ploidy Level

We collected leaves that were growing vigorously in vitro, and the ploidy level was analysed by Cyflow Ploidy Analyser (Partec, Görlitz, Germany). The leaves were chopped with a sharp razor blade (Gillette, Boston, MA, USA) in a plastic dish containing 1.25 mL of modified Galbraith's buffer (9.15 g/L $MgCl_2 \cdot 6H_2O$, 4.19 g/L 3-(N-morpholino) propansulfonic acid (MOPS), 8.82 g/L sodium citrate, 0.1% polyethylene glycol p-(1,1,3,3-tetramethylbutyl)-phenyl ether (Triton X-100), pH 7.0). The crude nuclei solution was filtered through a 50-μm nylon filter. Subsequently, the leachate was stained with 100 μL 4′,6-diamidino-2-phenylindole (DAPI, 10 μg/mL) for 10 s to detect the ploidy level. Leaves from the diploid full-sib family were used as a control to adjust the number of DAPI channels to 50. Sub-culturing was performed twice over a 5-month period, and the ploidy level was examined twice in this period. Finally, tetraploids, mixoploids, and diploids were placed vertically into Jiffy Mix (Shippagan, NB, Canada). After 2 months, the plants were transferred into plastic pots (height: 12 cm, top width: 10 cm, bottom width: 9.25 cm) containing a 2:1:1 sterilised mixture of peat, vermiculite, and perlite (autoclaved at 121 °C for 30 min). After 2 months, the final ploidy level was determined.

2.5. Chromosome Counting

Shoot growth occurred in rooting medium for nearly 2 weeks, and 5–10 mm of the root tips were examined. The root tips were rinsed and fixed in Carnoy's solution [23] for 24 h at 4 °C. After washing three times, the root tips were hydrolysed in 1 N HCl for 15 min at 60 °C, then the root tips were washed three times for 5 min. The treated root tips were cut into ~1.5 mm sections, stained with 1 drop of carbol fuchsin [24] solution for 15 min, and observed with a microscope using a 100× oil immersion lens (Olympus, Tokyo, Japan).

2.6. Statistical Analysis

Range analysis was used to evaluate the importance of each factor. The value of Range ($R = k \cdot (\text{max}) - k\,(\text{mix});$) and k_x were positively related to the importance of the factors and levels, respectively. The significance of differences among the treatments were evaluated using the homogeneity test and analysis of variance (ANOVA) using SPSS statistical software (ver. 18.0; SPSS Inc., Chicago, IL, USA). $P < 0.05$ was considered to indicate statistical significance. Arcsine transformation ($\theta = \sin^{-1}\sqrt{P}$ θ: angle, P: percentage) to determine percentages before ANOVA was performed using Microsoft Excel 2010 software (Microsoft Corp., Washington, DC, USA).

3. Results

3.1. Morphologic Observations in the Initial Stage of Regeneration

Less than 10% of the cultures were contaminated, and 150 genotypes of hybrid sweetgums were established in vitro and sub-cultured. Three genotypes were selected for colchicine treatment. The leaves of the hybrid sweetgum were trisected during the early stage. After incubation for 4 days, no nodules were observed on the wound of the main vein of the leaves and petioles. From 4 to 6 days, nodules appeared and expanded rapidly, and growth continued until 8 days. The nodules in the leaves near the petioles were larger than those found on other wounds (Figure 1). The first bud appeared nearly 20 days after the shoot explants were placed in medium.

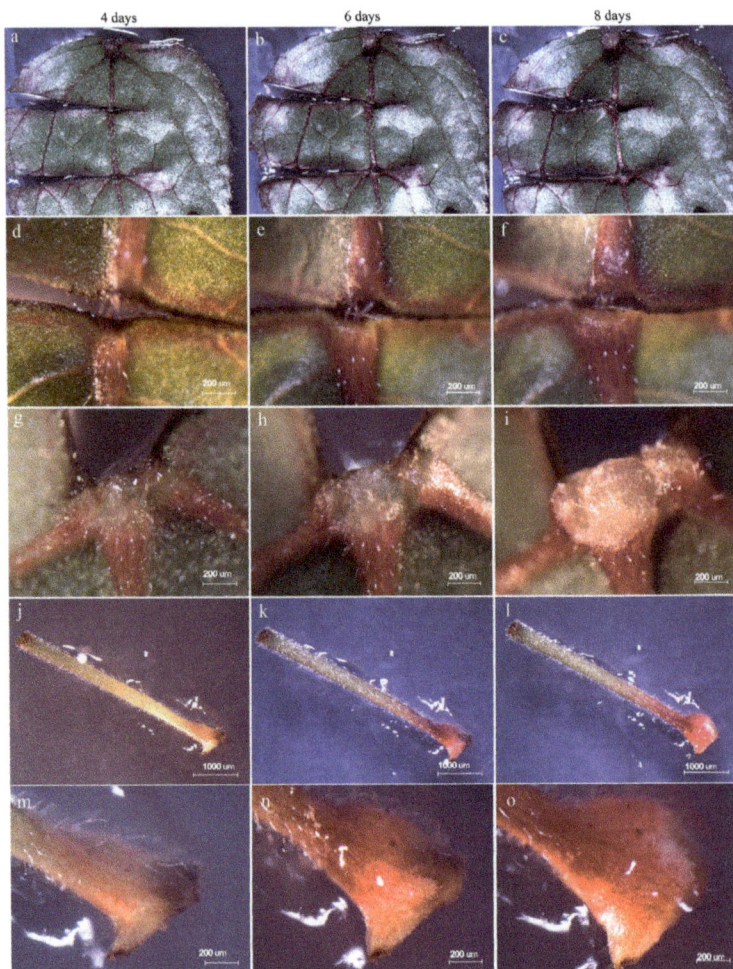

Figure 1. Morphology of leaves and petioles cultured in regeneration medium after 4, 6, and 8 days. (**a–c**) morphology of leaves cultured for 4, 6, and 8 days, respectively; (**d-f**) morphology of the second wound in the leaf main vein at 4, 6, and 8 days respectively; (**g-I**) morphology of the first wound in the leaf main vein at 4, 6, and 8 days, respectively; (**j-l**) morphology of the petiole cultured for 4, 6, and 8 days, respectively; (**m-o**) morphology of one end of the petiole cultured for 4, 6, and 8 days, respectively.

3.2. Survival Rate and Regeneration of Colchicine-Treated Explants

Under treatment 4, the survival rates were <10% (Tables 2 and 3). Table 4 shows that the concentration of colchicine (120–200 mg/L) had no significant effect on the survival rate of leaf explants. In contrast, genotype, pre-culture time and exposure time had significant effects on the survival rate in leaves (Table 4) and in petioles (Table 5). Exposure time exerted the most significant effect on survival rate for both leaves and petioles (Table 6).

Table 2. Design of orthogonal table $L_9(3)^4$ for leaves.

Treatment	Factors				Number of Shoots Examined	Survival Rate %	No. of Tetraploid	No. of Mixoploid	Tetraploid Induction %
	A	B	C	D					
c1	1	1	1	1	50	95.00	1	1	2.00
c2	1	2	2	2	50	75.00	0	1	0.00
c3	1	3	3	3	30	41.67	1	2	3.33
c4	2	1	2	3	15	8.33	0	0	0.00
c5	2	2	3	1	50	81.67	1	1	2.00
c6	2	3	1	2	50	90.00	0	1	0.00
c7	3	1	3	2	50	75.00	0	1	0.00
c8	3	2	1	3	50	70.00	0	0	0.00
c9	3	3	2	1	50	85.00	1	2	2.00

Table 3. Design of orthogonal table $L_9(3)^4$ for petioles.

Treatment	Factors				Number of Shoots Examined	Survival Rate %	No. of Tetraploid	No. of Mixoploid	Tetraploid Induction %
	A	B	C	D					
c1	1	1	1	1	50	85.00	1	0	2.00
c2	1	2	2	2	50	56.67	2	1	4.00
c3	1	3	3	3	30	28.33	1	0	3.33
c4	2	1	2	3	15	10.00	0	0	0.00
c5	2	2	3	1	50	56.67	3	2	6.00
c6	2	3	1	2	50	75.00	0	0	0.00
c7	3	1	3	2	50	48.33	0	0	0.00
c8	3	2	1	3	50	51.67	0	0	0.00
c9	3	3	2	1	50	70.00	4	2	8.00

Table 4. The variation analyses of survival rates for different genotype leaves (*Liquidambar styraciflua* × *L. formosana*), concentration, pre-culture duration and exposure time.

Variation Source	*df*	MS	*F*	Sig.
Genotype	2	294.147	4.876	0.020 *
Concentration	2	193.362	3.205	0.064
Pre-culture duration	2	1087.422	18.027	0.000 *
Exposure time	2	2769.584	45.913	0.000 *
Error	18	60.322		
Total	27			

df: degrees of freedom; MS: mean square; Sig.: significance; * Represents a significant difference at $p < 0.05$.

Table 5. The variation analyses of survival rates for different genotype petioles (*Liquidambar styraciflua* × *L. formosana*), concentration, pre-culture duration and exposure time.

Variation Source	*df*	MS	*F*	Sig.
Genotype	2	138.765	6.238	0.009 *
Concentration	2	97.631	4.389	0.028 *
Pre-culture duration	2	811.981	36.502	0.000 *
Exposure time	2	1612.270	72.478	0.000 *
Error	18	22.245		
Total	27			

df: degrees of freedom; MS: mean square; Sig.: significance;* Represents a significant difference at $p < 0.05$.

Table 6. The range analysis of the hybrid sweetgum survival rate by orthogonal test.

Explant Type		A	B	C	D
	K_1	211.67	178.33	255.00	261.67
	K_2	180.00	226.67	168.33	240.00
	K_3	230.00	216.67	198.33	120.00
Leaf	k_1	70.56	59.44	85.00	87.22
	k_2	60.00	75.56	56.11	80.00
	k_3	76.67	72.22	66.11	40.00
	R	16.67	16.11	28.89	47.22
	K_1	170.00	143.33	211.67	211.67
	K_2	141.67	165.00	136.67	180.00
	K_3	170.00	173.33	133.33	90.00
Petiole	k_1	56.67	47.78	70.56	70.56
	k_2	47.22	55.00	45.56	60.00
	k_3	56.67	57.78	44.44	30.00
	R	9.44	10.00	26.11	40.56

R: Range. Range = $k \cdot (\max) - k (\text{mix})$; $K_{1A} = X_{A1} + X_{A2} + X_{A3}$, $K_{2A} = X_{A4} + X_{A5} + vX_{A6}$, $K_{3A} = X_{A7} + X_{A8} + vX_{A9} \dots$; $k_x = K_x / \text{number of level}$.

3.3. Analysis by Flow Cytometry and Polyploid Determination

The ploidy levels were determined by three flow cytometry tests (Figure 2). The first test was administered in shoots that were rooted in medium for 1 month, the second test was administered after sub-culturing for 2 months, and the third test was administered after transplantation in the soil for 2 months. Therefore, the chromosomes were counted over a total of 6 months. The results showed that tetraploids and mixoploids were induced in the three genotypes of hybrid sweetgum (Tables 2 and 3). The most effective treatment for inducing polyploidy in the petioles was treatment 9, which consisted of pre-culturing genotype Z3 for 6 days, followed by 200 mg/L colchicine treatment for 3 days. This treatment resulted in 8% tetraploid and 4% mixoploid induction rates (Table 3). For the leaves, one tetraploid was acquired in c1, c3, c5, and c9 (Table 2). The chromosome number of tetraploid hybrid sweetgum was $2n = 4x = 52$ (Figure 3a), and the chromosome number of diploids was $2n = 2x = 26$ (Figure 3b).

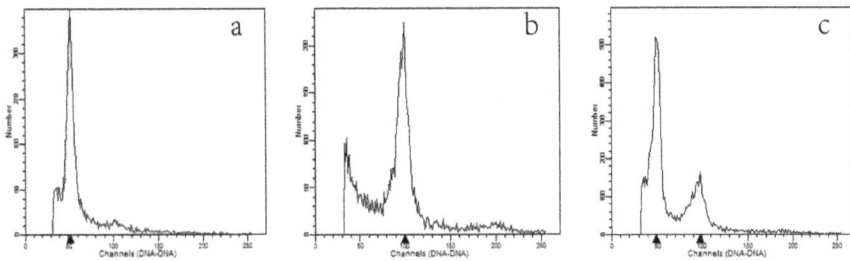

Figure 2. Histograms of flow cytometric analysis of *Liquidambar styraciflua* × *Liquidambar formosana* (a) diploid plant (control); (b) tetraploid plant; (c) diploid + tetraploid plant.

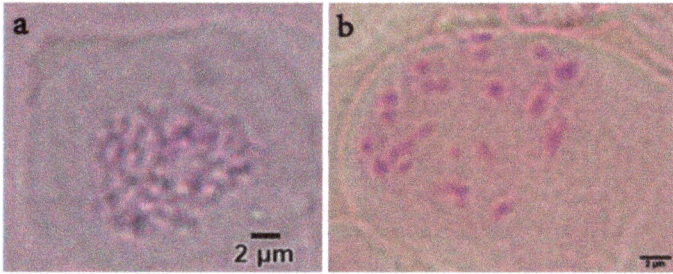

Figure 3. Chromosome counting of regenerated *Liquidambar styraciflua* × *L. formosana* plant
(**a**) Chromosomes of a tetraploid plant. (**b**) Chromosomes of a diploid plant.

Range analysis of the results showed that the best concentration of colchicine and exposure
time was 200 mg/L and 3 days, respectively, for both leaf and petiole explants. The number of
tetraploids was lower than for those exposed to colchicine for 5 days (Table 7). Range analysis also
showed that the exposure time was the most important influencing factor in leaf and petiole tetraploid
numbers. Moreover, the morphology of the adventitious shoots of leaves were similar to that of
petioles (Figure 4a,b), and tetraploid plantlets was significantly different from that found in the diploid
plantlets, showing deeper green leaf colour and shorter root length and internodal distance (Figure 4c).

Table 7. The range analysis of the hybrid sweetgum tetraploid-inducing rate by orthogonal test.

Explant Type		A	B	C	D
	K_1	5.33	2.00	2.00	6.00
	K_2	2.00	2.00	2.00	0.00
	K_3	2.00	5.33	5.33	3.33
Leaf	k_1	1.78	0.67	0.67	2.00
	k_2	0.67	0.67	0.67	0.00
	k_3	0.67	1.78	1.78	1.11
	R	1.11	1.11	1.11	2.00
	K_1	9.33	2.00	2.00	16.00
	K_2	6.00	10.00	12.00	4.00
	K_3	8.00	11.33	9.33	3.33
Petiole	k_1	3.11	0.67	0.67	5.33
	k_2	2.00	3.33	4.00	1.33
	k_3	2.67	3.78	3.11	1.11
	R	1.11	3.11	3.33	4.22

R: Range. Range = k (max) − k (mix); $K_{1A} = X_{A1} + X_{A2} + X_{A3}, K_{2A} = X_{A4} + X_{A5} + X_{A6}, K_{3A} = X_{A7} + X_{A8} + X_{A9} \ldots ; k_x$
= K_x /number of level.

Figure 4. Morphology of shoot regeneration in Z3 leaves and petioles after 60 days. (**a**) leaf; (**b**) petiole;
(**c**) tetraploid; (left) and diploid (right).

4. Discussion

This study describes a method of in vitro tetraploid induction from leaf explants of hybrid sweetgum (*L. styraciflua* × *L. formosana*). Morphological mutation has been found in tetraploid plantlets. Interestingly, this phenomenon has appeared in herbs, fruit and timber trees [9,11,18]. In this paper, we investigated the effects of explant type, genotype, pre-incubation time, exposure time in colchicine and concentration of colchicine. The exposure time had the greatest effect on the tetraploid-inducing rate, this may be due to the lower survival rate of explants after over exposure to colchicine.

All three genotypes acquired tetraploids, with similar results to those obtained using *Actinidia chinensis* and *Populus* [8,15]. Few studies have considered both leaves and petioles together, perhaps due to limited efficiency of regeneration, such as in *Actinidia chinensis* [15]. Therefore, previous studies have provided no evidence as a basis for comparing the induction rates of leaves and petioles. In this study, the ability to induce tetraploidy was higher in petioles than in leaves. It is possible that the nine treatments were more suited to petioles, or the chromosomes and petiole cells were more responsive to colchicine treatment. This method has potential for successful acquisition of polyploids using a high-efficiency regeneration system for leaves and petioles of some plant species. In this study, we found lower induction rates for leaves and petioles pre-cultured for 4 days versus those pre-cultured for 6 or 8 days. Interestingly, we observed that the wounds in the main vein expanded slowly (Figure 1). The relationship between wound expansion and induction rate requires further confirmation in future studies.

The survival rate of leaves was higher than that of petioles subjected to nine treatments. For petioles, the pre-culture duration, colchicine concentration and exposure time significantly affected the survival rate. Similar conclusions were reported by [25]. However, colchicine concentration was not a significant factor for leaf survival. Interestingly, increased pre-treatment time correlated roughly with decreased survival rates, perhaps because inactivation of in vitro explant cells occurred slowly. The duration of colchicine treatment was the most important factor influencing the survival rate, and leaves exposed to colchicine for 5 days acquired only two tetraploids.

Unsynchronised cell divisions may lead to the induction of mixoploids. Mixoploid induction has been reported in *Ranunculus asiaticus* [14], *Echinacea purpurea* [18], and *Pyrus pyrifolia* N. cv. Hosui [26]. Mixoploid induction always accompanies tetraploid induction, and the in vitro regeneration rate of mixoploids was markedly lower than that obtained directly from treated seeds and shoot tips. Although a higher rate of mixoploid induction occurs during direct regeneration from organs versus from somatic embryos, somatic embryogenesis is often limited by its high threshold of regeneration. We hypothesise that increasing the concentration of cytokinin, and selecting younger and more robust colchicine-treated explants, might improve the efficiency of tetraploid induction. Sivolapov and Blagodarova reported that stable mixoploids in poplar showed growth advantages in field observations [27].

In this study, we explored conditions that were conducive for sweetgum polyploidy breeding. Future studies will be aimed at optimising the treatment conditions and regeneration system for other genotypes. In addition, as various regeneration systems have been successfully established in *Liquidambar* [5,20,28], our future studies will incorporate other regeneration systems besides the leaf and petiole to induce polyploidy sweetgum.

5. Conclusions

To the best of our knowledge, this is the first report of in vitro tetraploid induction from leaf and petiole explants of hybrid sweetgum (*L. styraciflua* × *L. formosana*). This method is effective, not limited by flowering period and easy to operate. In the future, this method could be used for tetraploid induction in multiple genotypes, and these tetraploid plants will be observed and measured continuously with the aim of selecting fast growing, high biomass, strong resistance, superior sweetgum with peculiar ornamental value. Furthermore, hybrid tetraploid sweetgum could be a potential source for the promotion of sweetgum breeding and producing triploids by crossing with diploids.

Acknowledgments: This work was supported by National Key R&D Program of China (2017YFD0600404), Medium and Long Scientific Research Project for Young Teachers in Beijing Forestry University (2015ZCQ-SW-02), the Project of National Natural Science Foundation of China (31370658), "948" Project of China (2014-4-59), Program for Changjiang Scholars and Innovative Research Team in University (IRT13047), and the Project of Beijing Gardening and Greening Bureau (CEG-2016-01).

Author Contributions: Jinfeng Zhang, Bailian Li and Jian Zhao contributed to the experiment design; Yadong Zhang, Xuezheng Liu and Wei Yuan contributed to the collection of seeds; Shuaizheng Qi and Xiaoqi Wang contributed to hybridization; Zewei Wang contributed to the analysis of the ploidy level; Yan Zhang contributed to the induction of tetraploids, paper writing and other aforementioned works.

Conflicts of Interest: The authors declare no conflict of interest.

References

1. Harlow, W.M.; Harrar, E.S.; Hardin, J.W.; White, F.M. *Textbook of Dendrology*, 8th ed.; McGraw-Hill: New York, NY, USA, 1996; p. 534.

2. Merkle, S.A.; Neu, K.A.; Battle, P.J.; Bailey, R.L. Somatic embryogenesis and plantlet regeneration from immature and mature tissues of sweetgum (*Liquidambar styraciflua*). *Plant Sci.* **1998**, *132*, 169–178. [CrossRef]

3. Zheng, Y.Q.; Pan, B.; Itohl, T. Chemical induction of traumatic gum ducts in Chinese sweetgum, *Liquidambar formosana*. *IAWA J.* **2015**, *36*, 58–68.

4. Santamour, F.S. Interspecific hybridization in *Liquidambar*. *For. Sci.* **1972**, *18*, 23–26.

5. Vendrame, W.A.; Holliday, C.P.; Merkle, S.A. Clonal propagation of hybrid sweetgum (*Liquidambar styraciflua* × *L. formosana*) by somatic embryogenesis. *Plant Cell Rep.* **2001**, *20*, 691–695.

6. Song, Q.; Chen, Z.J. Epigenetic and developmental regulation in plant polyploids. *Curr. Opin. Plant Biol.* **2015**, *24*, 101–109. [CrossRef] [PubMed]

7. Tamayo-Ordóñez, M.C.; Espinosa-Barrera, L.A.; Tamayo-Ordóñez, Y.J.; Ayil-Gutiérrez, B.; Sánchez-Teyer, L.F. Advances and perspectives in the generation of polyploidy plant species. *Euphytica* **2016**, *209*, 1–22. [CrossRef]

8. Xu, C.P.; Huang, Z.; Liao, T.; Li, Y.; Kang, X.Y. In vitro tetraploid plants regeneration from leaf explants of multiple genotypes in *Populus*. *Plant Cell Tissue Organ Cult.* **2016**, *125*, 1–9. [CrossRef]

9. Cai, X.; Kang, X.Y. In vitro tetraploid induction from leaf explants of *Populus pseudo-simonii* Kitag. *Plant Cell Rep.* **2011**, *30*, 1771–1778. [CrossRef] [PubMed]

10. Tang, Z.Q.; Chen, D.L.; Song, Z.J.; He, Y.C.; Cai, D.T. In vitro induction and identification of tetraploid plants of *Paulownia tomentosa*. *Plant Cell Tissue Organ Cult.* **2010**, *102*, 213–220. [CrossRef]

11. Wu, J.H.; Mooney, P. Autotetraploid tangor plant regeneration from in vitro *Citrus* somatic embryogenic callus treated with colchicine. *Plant Cell Tissue Organ Cult.* **2002**, *70*, 99–104. [CrossRef]

12. Yang, X.M.; Cao, Z.Y.; An, L.Z.; Wang, Y.M.; Fang, X.W. In vitro tetraploid induction via colchicine treatment from diploid somatic embryos in grapevine (*Vitis vinifera* L.). *Euphytica* **2006**, *152*, 217–224. [CrossRef]

13. Stanys, V.; Weckman, A.; Staniene, G.; Duchovskis, P. In vitro induction of polyploidy in japanese quince (*Chaenomeles japonica*). *Plant Cell Tissue Organ Cult.* **2006**, *84*, 263–268. [CrossRef]

14. Dhooghe, E.; Denis, S.; Eeckhaut, T.; Reheul, D.; Van Labeke, M. In vitro induction of tetraploids in ornamental *Ranunculus*. *Euphytica* **2009**, *168*, 33–40. [CrossRef]

15. Wu, J.H.; Ferguson, A.R.; Murray, B.G. Manipulation of ploidy for kiwifruit breeding: In vitro chromosome doubling in diploid *Actinidia chinensis*. Planch. *Plant Cell Tissue Organ Cult.* **2011**, *106*, 503–511. [CrossRef]

16. Zhang, Q.Y.; Luo, F.X.; Liu, L.; Guo, F.C. In vitro induction of tetraploids in crape myrtle (*Lagerstroemia indica* L.). *Plant Cell Tissue Organ Cult.* **2010**, *101*, 41–47.

17. De Carvalho, J.F.R.P.; de Carvalho, C.R.; Otoni, W.C. In vitro induction of polyploidy in annatto (*Bixa orellana*). *Plant Cell Tissue Organ Cult.* **2005**, *80*, 69–75. [CrossRef]

18. Nilanthi, D.; Chen, X.L.; Zhao, F.C.; Yang, Y.S.; Wu, H. Induction of tetraploids from petiole explants through colchicine treatments in *Echinacea purpurea* L. *J. Biomed. Biotechnol.* **2009**. [CrossRef] [PubMed]

19. Brand, M.H.; Lineberger, R.D. In vitro adventitious shoot formation on mature-phase leaves and petioles of *Liquidambar styraciflua* L. *Plant Sci.* **1988**, *57*, 173–179. [CrossRef]

20. Kim, M.K.; Sommer, H.E.; Bongarten, B.C.; Merkle, S.A. High-frequency induction of adventitious shoots from hypocotyl segments of *Liquidambar styraciflua L.* by thidiazuron. *Plant Cell Rep.* **1997**, *16*, 536–540. [CrossRef]

21. Xu, L.; Liu, G.F.; Bao, M.Z. Adventitious shoot regeneration from in vitro leaves of formosan sweetgum (*Liquidambar formosana* L.). *Hortscience* **2007**, *42*, 721723.

22. Lloyd, G.; McCown, B. Commercially feasible micropropagation of mountain laurel; Kalmia latifolia; by use of shoot-tip culture. *Proc. Int. Plant Propag. Soc.* **1980**, *30*, 421–427.

23. Erlanson, E.W. Cytological Conditions and Evidences for Hybridity in North American Wild Roses. *Bot. Gaz.* **1929**, *87*, 443–506. [CrossRef]

24. Carr, D.H.; Walker, J.E. Carbol fuchsin as a stain for human chromosomes. *Stain Technol.* **1961**, *36*, 233–236. [CrossRef] [PubMed]

25. Głowacka, K.; Jeżowski, S.; Kaczmarek, Z. In vitro induction of polyploidy by colchicine treatment of shoots and preliminary characterisation of induced polyploids in two *Miscanthus* species. *Ind. Crops Prod.* **2010**, *32*, 88–96. [CrossRef]

26. Kadota, M.; Niimi, Y. In vitro induction of tetraploid plants from a diploid Japanese pear cultivar (*Pyrus pyrifolia* N. cv. Hosui). *Plant Cell Rep.* **2002**, *21*, 282–286.

27. Sivolapov, A.I.; Blagodarova, T.A. Different levels of mixoploiy in hybrid poplars. In *Cytogenetic Studies of Forest Trees and Shrub Species*; Borzan, Z., Schlarbaum, S.E., Eds.; Faculty of Foresty Inc.: Zagreb, Croatia, 1997; pp. 311–316.

28. Merkle, S.A.; Battle, P.J.; Ware, G.O. Factors influencing production of inflorescence-derived somatic seedlings of sweetgum. *Plant Cell Tissue Organ Cult.* **2003**, *73*, 95–99. [CrossRef]

© 2017 by the authors. Licensee MDPI, Basel, Switzerland. This article is an open access article distributed under the terms and conditions of the Creative Commons Attribution (CC BY) license (http://creativecommons.org/licenses/by/4.0/).

MDPI

St. Alban-Anlage 66

4052 Basel

Switzerland

Tel. +41 61 683 77 34

Fax +41 61 302 89 18

www.mdpi.com

Forests Editorial Office

E-mail: forests@mdpi.com

www.mdpi.com/journal/forests

www.ingramcontent.com/pod-product-compliance
Lightning Source LLC
Chambersburg PA
CBHW051713210326
41597CB00032B/5469